AF355993

Antioxidant Activity of Fermented Foods and Food Microorganisms

Antioxidant Activity of Fermented Foods and Food Microorganisms

Guest Editor

Myung-Ji Seo

Basel • Beijing • Wuhan • Barcelona • Belgrade • Novi Sad • Cluj • Manchester

Guest Editor
Myung-Ji Seo
Division of Bioengineering
Incheon National University
Incheon
Korea, South

Editorial Office
MDPI AG
Grosspeteranlage 5
4052 Basel, Switzerland

This is a reprint of the Special Issue, published open access by the journal *Antioxidants* (ISSN 2076-3921), freely accessible at: www.mdpi.com/journal/antioxidants/special_issues/Fermented_Foods_and_Food_Microorganisms.

For citation purposes, cite each article independently as indicated on the article page online and using the guide below:

Lastname, A.A.; Lastname, B.B. Article Title. *Journal Name* **Year**, *Volume Number*, Page Range.

ISBN 978-3-7258-2926-2 (Hbk)
ISBN 978-3-7258-2925-5 (PDF)
https://doi.org/10.3390/books978-3-7258-2925-5

Contents

 *antioxidants*

Editorial

Fermented Foods and Food Microorganisms: Antioxidant Benefits and Biotechnological Advancements

Myung-Ji Seo [1,2,3]

1   Division of Bioengineering, Incheon National University, Incheon 22012, Republic of Korea; mjseo@inu.ac.kr
2   Department of Bioengineering and Nano-Bioengineering, Incheon National University, Incheon 22012, Republic of Korea
3   Research Center for Bio Materials & Process Development, Incheon National University, Incheon 22012, Republic of Korea

Citation: Seo, M.-J. Fermented Foods and Food Microorganisms: Antioxidant Benefits and Biotechnological Advancements. *Antioxidants* **2024**, *13*, 1120. https://doi.org/10.3390/antiox13091120

Received: 9 September 2024
Accepted: 13 September 2024
Published: 16 September 2024

Fermented foods have been a part of human civilization since ancient times, offering enhanced flavors, extended shelf-life, and improved nutritional value through the action of microorganisms [1]. The fermentation process has been pivotal in transforming various food substrates, ranging from dairy and vegetables to cereals and legumes, into staples of diets around the globe [2–4]. One of the most compelling health benefits of fermentation lies in its ability to significantly boost the antioxidant capacity of foods [5,6].

The rising demand for plant-based protein and non-dairy milk alternatives has highlighted soybeans and soymilk for their nutritional benefits. Fermentation using lactic acid bacteria (LAB) like *Lacticaseibacillus paracasei* modifies the nutritional profile of soy-derived products and provides a rich source of antioxidant peptides, which are small enough to be absorbed and utilized by the body [7]. Peptides from *L. paracasei* MK1-fermented soymilk exhibited antioxidant activity through mechanisms involving the inhibition of reactive oxygen species (ROS), which are known to exacerbate skin aging by inducing oxidative damage to skin cells. Lee et al. [8] demonstrated that certain synthesized peptides contributed to anti-aging effects, enhancing procollagen production and reducing pro-inflammatory cytokine levels in UVB-damaged skin cells, thus mitigating the negative effects of photoaging by reducing oxidative stress, promoting skin elasticity, and reducing wrinkle formation. Letizia et al. [9] fermented soymilk using *Lactiplantibacillus plantarum* LP95 and demonstrated a significant enhancement of the antioxidant properties of the soymilk. As a strain with notable β-glucosidase activity, *L. plantarum* LP95 efficiently converted isoflavone glucosides such as genistein, glycitein, and daidzein into their more bioavailable aglycones during fermentation. The process improved the organoleptic properties and shelf-life stability, and increased isoflavone concentrations and antioxidant capacity, indicating the ability of the fermented product to mitigate oxidative damage in numerous health conditions.

Fermentation offers a biotechnological approach to enhance the bioaccessibility of antioxidant compounds in plants, freeing them from glycosides and plant matrix complexes. In almonds, microbiota modulation varied between cultivars, and the Guara cultivar supported a distinct gut microbial community structure, promoting beneficial genera like *Dorea*, which is associated with dietary fiber degradation. Guara exhibited the highest total antioxidant capacity and the highest concentration of short-chain fatty acids (SCFAs), promoting gut health by serving as energy sources for intestinal cells and modulating the immune response [10]. In herbal infusions, lacto-fermentation by *L. plantarum* enhanced the phenolic content of infusions with thyme, rosemary, echinacea, and pomegranate peel, and led to modulated anti-inflammatory effects and an increased antioxidant capacity [11].

The use of fermentation has also been explored for the transformation of by-products and waste materials to obtain new products with improved quality [12]. The use of microbial inoculants in fermentation has further improved the taste, texture, and nutritional profile of foods. Functional fruit wines combine the flavors of fruit with added health benefits, and upcycled by-products are valuable alternatives for nutraceutical development and reducing environmental waste. The health benefit potential of grape pomace, a by-product of the wine industry, was confirmed through fermentation with LAB which enhanced its functional properties for use in sourdough bread [13]. The fiber- and polyphenol-rich grape pomace, typically discarded or used for animal feed, was able to be used as a functional food ingredient. Fermented pomace exhibited significantly higher antioxidant activity and down-regulation of pro-inflammatory cytokines, in addition to the enhanced flavor profiles of the produced bread. Likewise, Li et al. [14] used various parts of green bananas, including the peel, pulp, and whole fruit, to develop functional fruit wines. Overall, the wines showed higher antioxidant activities than their raw fruit counterparts, and the peels, typically considered waste, maintained the highest levels of specific phenolics, tannins, and flavonoids. Peel wine had the greatest potency against oxidative stress induced by hydrogen peroxide in cell-cultured based assays, along with a better capacity to scavenge superoxide anions, highlighting their potential as a rich source of antioxidants with anti-inflammatory and cardioprotective effects.

The aromatic profile of Catarratto wines, a non-aromatic white grape variety from Sicily, was enhanced by using *Metschnikowia pulcherrima* in combination with the sequential inoculation of *Saccharomyces cerevisiae* and the addition of a glutathione-rich inactivated yeast [15]. The inactivated yeast helped limit oxidation, thus preserving the color and stability of the aromatic compounds, particularly by protecting esters and volatile thiols from oxidative degradation. The resulting wines had an enhanced aromatic complexity, fruity and floral aromas, and improved oxidative stability and preservation. The addition of GIY protected the wine from oxidative damage and allowed for the preservation of key aromatic compounds, while *M. pulcherrima* significantly contributed to the production of desirable esters. As an alternative to wine, kombucha is a popular drink with various health benefits and a growing variety of substrates used in its production. Xiong et al. [16] developed kombucha beverages using bamboo leaf and mulberry leaf as alternative substrates for fermentation. These caffeine-free alternatives to traditional kombucha possess strong antioxidant activities and various bioactive compounds.

Fermentation not only enhances the concentration of antioxidants but also modifies their molecular structure, increasing their solubility and bioavailability, making them more effective in human bodies [17]. In individuals with metabolic conditions such as hyperglycemia and hyperlipidemia, fermentation plays a key role in elevating the bioavailability of antioxidants for overall metabolic health and the management of non-communicable diseases like diabetes and cardiovascular disease [18]. Black goji berry juice fermented with *Lactobacillus rhamnosus* significantly enhanced its sensory qualities and functional properties, particularly in inhibiting the key steps involved in lipid digestion and absorption [19]. Fermentation converted complex polyphenols into simpler, more bioactive compounds and increased lactic acid production and probiotic viability. The fermented product showed enhanced abilities to bind bile acids, inhibit cholesterol micelle formation, and increase pancreatic lipase inhibition, all crucial activities in reducing cholesterol absorption and lipid digestion. The fermented berries also demonstrated improved inhibition of an enzyme that degrades incretin hormones crucial for glucose regulation. This inhibition extends the activity of incretins like GLP-1 and GIP, enhancing insulin secretion, reducing blood sugar levels, and allowing for the potential dietary intervention of managing hyperglycemia in diabetic individuals. Wang et al. [20] presented the role of quorum-sensing transcription factor SdiA in regulating NADPH metabolism and (S)-equol production in the probiotic *Escherichia coli* Nissle 1917. (S)-Equol, a metabolite derived from dietary isoflavones like daidzein, has significant antioxidative and estrogenic activities. The researchers identified a bottleneck in (S)-equol production linked to NADPH availability and provided insights into optimizing engineered probiotics for (S)-equol production by manipulating quorum-sensing pathways and improving NADPH regeneration to enhance the health benefits of dietary isoflavones, especially in individuals who cannot naturally produce (S)-equol.

Fermentation has also been exploited for the benefit of livestock, aiding in their general health and also in reducing the need for antibiotics. The study on *Limosilactobacillus reuteri* PSC102 highlights its significant probiotic, antioxidant, and antibacterial properties, suggesting its potential as a functional feed additive, particularly for swine [21]. A standout feature of this strain is its strong antioxidant activity, as evidenced by its ability to scavenge free radicals, and through its reducing power. These antioxidant activities are vital for mitigating oxidative stress-related disorders which are a common challenge in farm animals, especially during periods of rapid growth or infection. The antioxidant properties of the microorganism suggest that it could enhance the health and growth performance of animals by counteracting the damaging effects of ROS. Qiu et al. [22] demonstrated that yeast culture supplementation in laying hens had significant benefits on antioxidant capacity, egg production performance, egg quality, intestinal morphology, and microbial composition. These improvements were attributed to enhanced nutrient absorption and metabolism, facilitated by the beneficial components of the cultures, such as amino acids, peptides, and vitamins, which optimize digestive and metabolic functions, and the oligosaccharides and other bioactive compounds that enhance antioxidant enzyme activity and neutralize free radicals. Yeast culture supplementation increased serum total antioxidant capacity and glutathione peroxidase activity, while reducing malondialdehyde levels, which are markers of oxidative stress. The improvement in antioxidant activity highlights the potential of the cultures to mitigate oxidative damage in laying hens, thereby supporting overall health and longevity. Zhao et al. [23] delved into the role of *Lactobacillus* in reducing oxidative stress, with a focus on its potential in preventing and treating diseases such as inflammatory bowel disease (IBD), cancer, and liver-related diseases. The antioxidant properties of *Lactobacillus* have been widely studied in cell lines, animal models, and some clinical reports, showing promising results in mitigating oxidative damage and inflammation. The *Lactobacillus* genus commonly found in the gastrointestinal tract can produce antimicrobial metabolites and modulate the host immune system, as some strains possess oxidative stress resistance genes, while others produce exopolysaccharides (EPS) that protect cells from oxidative damage. These microorganisms have shown potential in increasing the host's antioxidant capacities by enhancing enzyme activity or directly neutralizing ROS, and studies have shown that *Lactobacillus* strains can positively influence the redox balance and mitigate oxidative stress. Siziya et al. [24] explored the antioxidant potential of C_{30} carotenoids produced by microorganisms, highlighting their significance and advantages in various industries such as food, feed, cosmetics, and pharmaceuticals. While carotenoids are known for their antioxidant, anti-inflammatory, and anticancer properties, which protect cells from oxidative damage, C_{40} carotenoids dominate both research foci and commercial markets compared to their less abundant C_{30} counterparts [25,26]. The paper discusses the ROS quenching capabilities and oxidative stress reduction potential of the C_{30} carotenoids, which may aid in preventing cellular damage. The study discusses the metabolic pathways of microbial carotenoids and the potential for large-scale, eco-friendly production through biotechnological advancements. Nemes et al. [27] focused on integrated technologies for cereal bran valorization, aiming to improve the bioaccessibility and bioavailability of bioactive components such as phenolic acids, dietary fibers, and biopeptides that are naturally bound within the bran matrix. These bioactive compounds have demonstrated potential health benefits, including anti-inflammatory, antioxidant, anticancer, and cardiometabolic protective effects, which tend to be limited due to their insoluble and bound forms in the bran. An integrated approach was proposed, involving pretreatment methods such as physical, chemical, and enzymatic techniques, followed by microbial fermentation to release these bound components for enhanced utilization in functional foods and supplements. Solid-state and submerged fermentation improve the release of bioactive compounds from bran with microorganisms such as *Aspergillus*, *Lactobacillus*, and *Saccharomyces*, which can break down cell wall components, releasing phenolics, peptides, and other bioactive compounds. Biopeptides were noted as having antihypertensive, antioxidant, and antimicrobial properties, with oats and rice bran being particularly rich sources of bioactive peptides. Phenolic compounds in bran scavenge free radicals, reducing oxidative stress and potentially lowering the risk of chronic diseases while bran components like ferulic acid and biopeptides have been linked to reduced inflammation in clinical studies.

The papers of this Special Issue provide a glimpse into the current state of research into fermented foods and the roles of food microorganisms in the specific aspect of antioxidant activity. As the demand for natural, health-promoting foods continues to rise, fermented products represent a fusion of ancient culinary practices and cutting-edge science, offering sustainable solutions to improve public health and combat diseases driven by oxidative stress. The growing body of research in this field underscores the transformative potential of fermentation biotechnology, not only for improving food preservation and safety but also for delivering potent therapeutic benefits.

Acknowledgments: I gratefully thank Inonge Noni Siziya in Division of Food and Nutrition, Chonnam National University, Republic of Korea for critical reading and improving the English of the manuscript.

Conflicts of Interest: The author declares no conflicts of interest.

References

1. Marco, M.L.; Heeney, D.; Binda, S.; Cifelli, C.J.; Cotter, P.D.; Foligné, B.F.; Gänzle, M.; Kort, R.; Pasin, G.; Pihlanto, A.; et al. Health benefits of fermented foods: Microbiota and beyond. *Curr. Opin. Biotechnol.* **2017**, *44*, 94–102. [CrossRef] [PubMed]
2. Nkhata, S.G.; Ayua, E.; Kamau, E.H.; Shingiro, J.-B. Fermentation and germination improve nutritional value of cereals and legumes through activation of endogenous enzymes. *Food Sci. Nutr.* **2018**, *6*, 2446–2458. [CrossRef] [PubMed]
3. Sharma, R.; Garg, P.; Kumar, P.; Bhatia, S.K.; Kulshrestha, S. Microbial fermentation and its role in quality improvement of fermented foods. *Fermentation* **2020**, *6*, 106. [CrossRef]
4. Adebo, J.A.; Njobeh, P.B.; Gbashi, S.; Oyedeji, A.B.; Ogundele, O.M.; Oyeyinka, S.A.; Adebo, O.A. Fermentation of cereals and legumes: Impact on nutritional constituents and nutrient bioavailability. *Fermentation* **2022**, *8*, 63. [CrossRef]
5. Hur, S.J.; Lee, S.Y.; Kim, Y.-C.; Choi, I.; Kim, G.B. Effect of fermentation on the antioxidant activity in plant-based foods. *Food Chem.* **2014**, *160*, 346–356. [CrossRef]
6. Saud, S.; Xiaojuan, T.; Fahad, S. The consequences of fermentation metabolism on the qualitative qualities and biological activity of fermented fruit and vegetable juices. *Food Chem. X* **2024**, *21*, 101209. [CrossRef]
7. Coda, R.; Rizzello, C.G.; Pinto, D.; Gobbetti, M. Selected lactic acid bacteria synthesize antioxidant peptides during sourdough fermentation of cereal flours. *Appl. Environ. Microbiol.* **2012**, *78*, 1087–1096. [CrossRef]
8. Lee, S.; Choi, S.-P.; Jeong, H.; Yu, W.K.; Kim, S.W.; Park, Y.-S. The radical scavenging activities and anti-wrinkle effects of soymilk fractions fermented with *Lacticaseibacillus paracasei* MK1 and their derived peptides. *Antioxidants* **2023**, *12*, 1392. [CrossRef]
9. Letizia, F.; Fratianni, A.; Cofelice, M.; Testa, B.; Albanese, G.; Di Martino, C.; Panfili, G.; Lopez, F.; Iorizzo, M. Antioxidative properties of fermented soymilk using *Lactiplantibacillus plantarum* LP95. *Antioxidants* **2023**, *12*, 1442. [CrossRef]
10. Delgado-Osorio, A.; Navajas-Porras, B.; Pérez-Burillo, S.; Hinojosa-Nogueira, D.; Toledano-Marín, Á.; Pastoriza de la Cueva, S.; Paliy, O.; Rufián-Henares, J.Á. Cultivar and harvest time of almonds affect their antioxidant and nutritional profile through gut microbiota modifications. *Antioxidants* **2024**, *13*, 84. [CrossRef]
11. Ozturk, T.; Ávila-Gálvez, M.Á.; Mercier, S.; Vallejo, F.; Bred, A.; Fraisse, D.; Morand, C.; Pelvan, E.; Monfoulet, L.-E.; González-Sarrías, A. Impact of lactic acid bacteria fermentation on (poly) phenolic profile and in vitro antioxidant and anti-inflammatory properties of herbal infusions. *Antioxidants* **2024**, *13*, 562. [CrossRef] [PubMed]
12. Faria, D.J.; Carvalho, A.P.A.; Conte-Junior, C.A. Valorization of fermented food wastes and byproducts: Bioactive and valuable compounds, bioproduct synthesis, and applications. *Fermentation* **2023**, *9*, 920. [CrossRef]
13. Torreggiani, A.; Demarinis, C.; Pinto, D.; Papale, A.; Difonzo, G.; Caponio, F.; Pontonio, E.; Verni, M.; Rizzello, C.G. Upcycling grape pomace through sourdough fermentation: Characterization of phenolic compounds, antioxidant activity, and anti-inflammatory potential. *Antioxidants* **2023**, *12*, 1521. [CrossRef] [PubMed]
14. Li, Z.; Qin, C.; He, X.; Chen, B.; Tang, J.; Liu, G.; Li, L.; Yang, Y.; Ye, D.; Li, J. Development of green banana fruit wines: Chemical compositions and in vitro antioxidative activities. *Antioxidants* **2022**, *12*, 93. [CrossRef]
15. Naselli, V.; Prestianni, R.; Badalamenti, N.; Matraxia, M.; Maggio, A.; Alfonzo, A.; Gaglio, R.; Vagnoli, P.; Settanni, L.; Bruno, M. Improving the aromatic profiles of Catarratto wines: Impact of Metschnikowia pulcherrima and glutathione-rich inactivated yeasts. *Antioxidants* **2023**, *12*, 439. [CrossRef]
16. Xiong, R.-G.; Wu, S.-X.; Cheng, J.; Saimaiti, A.; Liu, Q.; Shang, A.; Zhou, D.-D.; Huang, S.-Y.; Gan, R.-Y.; Li, H.-B. Antioxidant activities, phenolic compounds, and sensory acceptability of Kombucha-Fermented beverages from bamboo leaf and mulberry leaf. *Antioxidants* **2023**, *12*, 1573. [CrossRef]

17. Bryant, K.L.; Hansen, C.; Hecht, E.E. Fermentation technology as a driver of human brain expansion. *Commun. Biol.* **2023**, *6*, 1190. [CrossRef]

18. Hu, X.; Hu, J.; Leng, T.; Liu, S.; Xie, M. *Rosa roxburghii*-deible fungi fermentation broth attenuates hyperglycemia, hyperlipidemia and affects gut microbiota in mice with type 2 diabetes. *Food Biosci.* **2023**, *52*, 102432. [CrossRef]

19. Kamonsuwan, K.; Balmori, V.; Marnpae, M.; Chusak, C.; Thilavech, T.; Charoensiddhi, S.; Smid, S.; Adisakwattana, S. Black goji berry (*Lycium ruthenicum*) juice fermented with *Lactobacillus rhamnosus* GG enhances inhibitory activity against dipeptidyl peptidase-IV and key steps of lipid digestion and absorption. *Antioxidants* **2024**, *13*, 740. [CrossRef]

20. Wang, Z.; Dai, Y.; Azi, F.; Dong, M.; Xia, X. Deciphering the crucial roles of the quorum-sensing transcription factor SdiA in NADPH metabolism and (*S*)-equol production in *Escherichia coli* Nissle 1917. *Antioxidants* **2024**, *13*, 259. [CrossRef]

21. Ali, M.S.; Lee, E.-B.; Lim, S.-K.; Suk, K.; Park, S.-C. Isolation and identification of *Limosilactobacillus reuteri* PSC102 and evaluation of its potential probiotic, antioxidant, and antibacterial properties. *Antioxidants* **2023**, *12*, 238. [CrossRef] [PubMed]

22. Qiu, Q.; Zhan, Z.; Zhou, Y.; Zhang, W.; Gu, L.; Wang, Q.; He, J.; Liang, Y.; Zhou, W.; Li, Y. Effects of yeast culture on laying performance, antioxidant properties, intestinal morphology, and intestinal flora of laying hens. *Antioxidants* **2024**, *13*, 779. [CrossRef] [PubMed]

23. Zhao, T.; Wang, H.; Liu, Z.; Liu, Y.; De, J.; Li, B.; Huang, X. Recent perspective of *Lactobacillus* in reducing oxidative stress to prevent disease. *Antioxidants* **2023**, *12*, 769. [CrossRef] [PubMed]

24. Siziya, I.N.; Hwang, C.Y.; Seo, M.-J. Antioxidant potential and capacity of microorganism-sourced C_{30} Carotenoids-a review. *Antioxidants* **2022**, *11*, 1963. [CrossRef] [PubMed]

25. Bhatt, T.; Patel, K. Carotenoids: Potent to prevent diseases review. *Nat. Prod. Bioprospect.* **2020**, *10*, 109–117. [CrossRef]

26. Umeno, D.; Tobias, A.V.; Arnold, F.H. Evolution of the C_{30} carotenoid synthase CrtM for function in a C_{40} pathway. *J. Bacteriol.* **2002**, *184*, 6690–6699. [CrossRef]

27. Nemes, S.A.; Călinoiu, L.F.; Dulf, F.V.; Fărcas, A.C.; Vodnar, D.C. Integrated technology for cereal bran valorization: Perspectives for a sustainable industrial approach. *Antioxidants* **2022**, *11*, 2159. [CrossRef]

 *antioxidants*

Article

Effects of Yeast Culture on Laying Performance, Antioxidant Properties, Intestinal Morphology, and Intestinal Flora of Laying Hens

Quan Qiu [1,2], Zhichun Zhan [2], Ying Zhou [2], Wei Zhang [2], Lingfang Gu [2], Qijun Wang [2], Jing He [1], Yunxiang Liang [1], Wen Zhou [3,*] and Yingjun Li [1,*]

[1] National Key Laboratory of Agricultural Microbiology, College of Life Science and Technology, Huazhong Agricultural University, Wuhan 430070, China; quanqiu@webmail.hzau.edu.cn (Q.Q.); hejingjj@mail.hzau.edu.cn (J.H.); fa-lyx@163.com (Y.L.)

[2] Wuhan Sunhy Biology Co., Ltd., Wuhan 430070, China; zzc@sunhy.cn (Z.Z.); tech@sunhy.cn (Y.Z.); zhangw8804@163.com (W.Z.); glf8254@163.com (L.G.); wqj@sunhy.cn (Q.W.)

[3] Green Chemical Reaction Engineering, Engineering and Technology Institute Groningen (ENTEG), University of Groningen, Nijenborgh 4, 9747 AG Groningen, The Netherlands

* Correspondence: wen.zhou@rug.nl (W.Z.); yingjun@mail.hzau.edu.cn (Y.L.)

Abstract: Yeast culture (YC) plays a significant role in enhancing the performance and health of poultry breeding. This study investigated the impact of different YC supplementation concentrations (basal diet with 1.0 g/kg and 2.0 g/kg of YC, YC1.0, and YC2.0) on egg production performance, egg quality, antioxidant properties, intestinal mucosal structure, and intestinal flora of laying hens. Both YC1.0 and YC2.0 groups significantly enhanced the egg protein height, Haugh unit, and crude protein content of egg yolks compared to the control group ($p < 0.05$). The supplementation with YC2.0 notably increased the egg production rate, reduced feed-to-egg ratio, and decreased the broken egg rate compared to the control group ($p < 0.05$). Additionally, YC supplementation enhanced serum total antioxidant capacity (T-AOC) and glutathione peroxidase (GSH-PX) activity while reducing malondialdehyde (MDA) content ($p < 0.05$). Moreover, YC supplementation promoted duodenal villus height and villus ratio in the duodenum and jejunum ($p < 0.05$). Analysis of cecal microorganisms indicated a decrease in Simpson and Shannon indices with YC supplementation ($p < 0.05$). YC1.0 reduced the abundance of Proteobacteria, while YC2.0 increased the abundance of Bacteroidales ($p < 0.05$). Overall, supplementation with YC improved egg production, quality, antioxidant capacity, intestinal morphology, and cecal microbial composition in laying hens, with significant benefits observed at the 2.0 g/kg supplementation level.

Keywords: yeast culture; laying hens; egg production; egg quality; intestinal mucosal structure; intestinal flora

Citation: Qiu, Q.; Zhan, Z.; Zhou, Y.; Zhang, W.; Gu, L.; Wang, Q.; He, J.; Liang, Y.; Zhou, W.; Li, Y. Effects of Yeast Culture on Laying Performance, Antioxidant Properties, Intestinal Morphology, and Intestinal Flora of Laying Hens. *Antioxidants* **2024**, *13*, 779. https://doi.org/10.3390/antiox13070779

Academic Editor: Claus Jacob

Received: 28 May 2024
Revised: 23 June 2024
Accepted: 26 June 2024
Published: 27 June 2024

1. Introduction

Antibiotics have historically played a crucial role in improving livestock and poultry production by reducing disease incidence and enhancing performance [1]; however, their overuse has led to many issues like antibiotic residues, the emergence of drug-resistant strains, and environmental contamination, posing significant risks to food safety and human health [2]. Consequently, many countries are phasing out or restricting antibiotic use in animal farming [3]. While this policy has enhanced food safety, it has also negatively impacted production efficiency, including decreased feed conversion rates and increased disease treatment costs [4]. Thus, finding effective antibiotic alternatives is imperative for sustaining the health and productivity of the livestock industry.

Yeast culture (YC) is a microecological preparation comprising various metabolites derived from yeast cells grown on a specific culture medium and fully fermented under controlled conditions [5]. YC plays a role in maintaining a healthy balance of gut microbiota and supporting the growth of beneficial bacteria in the gut to enhance digestive health [6]. It serves as a promising antibiotic alternative in its prebiotic effects and immune modulation. Comprising yeast cell contents, metabolites, fermentation substrates, and minor inactive cells, it is abundant in functional compounds such as amino acids, peptides, oligosaccharides, and vitamins, enhancing the immune response and making animals less susceptible to diseases that would otherwise require antibiotic treatment [6]. Widely utilized in ruminant, swine, and poultry farming, its supplementation enhances heat stress resistance in dairy cows [7], early lactation milk yield in dairy goats [8], reproductive performance in sows, the growth of weaned piglets [9], and immune function in broiler chickens [10].

Numerous studies have demonstrated the benefits of YC supplementation in reducing the need for antibiotics, like improved growth performance, enhanced disease resistance, and better production outcomes, indirectly contributing to improved parameters of egg quality [6,11]. Moreover, YC has many economic and environmental benefits in cost savings and sustainable farming [12]. Overall, YC offers a natural alternative to antibiotics by enhancing gut health, supporting the immune system, and improving overall animal performance, thus translating to consistent egg production and quality [13,14]. This approach not only addresses the growing concern of antibiotic resistance but also promotes better health and productivity in livestock, contributing to more sustainable and economically viable farming practices.

Current research predominantly focuses on utilizing YC in broiler diets, leaving a gap in understanding its impact on laying hens. This study investigates the effects of dietary YC supplementation on egg production, quality, antioxidant properties, intestinal mucosal structure, and the microbiota of laying hens. Findings aim to furnish valuable insights into the application of YC in laying hen diets, enhancing theoretical foundations for its widespread adoption.

2. Materials and Methods

2.1. Diets, Experimental Design, and Hens

Yeast culture (YC), sourced from Wuhan Sunhy Biological Co., Ltd. (Wuhan, China), is derived from the *Kluyveromyces marxianus* strain and various cereal substrates. Through controlled liquid–solid two-phase deep fermentation and low-temperature drying, the process generated bio-products abundant in small peptides, amino acids, organic acids, and mannans. The key nutritional indicators include the following: crude protein $\geq$ 16.0%; crude fiber $\leq$ 11.0%; crude ash $\leq$ 9.0%; moisture $\leq$ 11.0%; small peptides $\geq$ 30.0%; mannan $\geq$ 0.8%; lactic acid $\geq$ 3.0%.

A total of 300 healthy 40-week-old Hy-Line brown laying hens, with the same egg production rates and body weights, were selected using a single-factor completely randomized design. They were then randomly divided into three treatment groups, each with five replicates of twenty hens. The control group (control) was fed a corn–soybean meal basal diet formulated according to the National Research Council (NRC, 1994) nutritional requirements (refer to Table 1). Detailed information of premix is provided in Table S1. The experimental groups, YC1.0 and YC2.0, were supplemented with 1.0 g/kg and 2.0 g/kg of YC, respectively, in the basal diet. The trial period lasted for 35 days. The final weight of each laying hen was approximately 1.8 kg, measured under standardized conditions at the end of the testing period.

The laying hen housing system comprises a three-tiered cage structure (length, width, and height are 100 cm, 50 cm, and 60 cm, respectively) with artificial and natural lighting, maintaining a consistent 15-h photoperiod (05:00 a.m.–20:00 p.m.) at an intensity of 10–15 Lux. Temperature (15–20 °C) and humidity (40–60%) were regulated by mechanical and natural ventilation systems. Feeding occurred twice daily (07:30 a.m. and 15:30 p.m.), with amounts adjusted to minimize leftover feed. Temperature and humidity levels were monitored at these times. Fecal management and routine immunizations were performed as standard protocols.

Egg production rates were monitored to ensure statistical parity among groups. Continuous access to food and water was provided throughout the experiment. All experimental procedures were approved by the University's Animal Care and Use Committee (SCXK2020-0084), and performed in accordance with internationally accepted guidelines and ethical principles.

Table 1. The composition and nutrient levels of basal diets (air-dried basis).

Items	Content (%)
Ingredients	
Corn	60.7
Soybean	23.5
Soybean oil	1
$CaCO_3$	9
$CaHPO_4$	0.8
Premix	5
Total	100
Nutrient content	
Metabolizable energy (MJ/kg)	10.93
Crude protein	16.09
Lysine	0.78
Methionine and Cystine	0.52
Calcium	3.55
Available Phosphorus	0.52

2.2. Measurements of Laying Eggs

2.2.1. The Laying Rate of Eggs

To assess the performance of egg production, total egg weight, egg count, feed consumption, remaining feed inventory, flock size, average egg weight, egg production rate (including broken eggs), average intake, and feed-to-egg ratio were daily measured.

Egg production rate (%) = Total number of eggs $\times$ 100/(number of days $\times$ number of laying hens);

Average egg weight (g) = Total egg weight/number of eggs;

Average daily feed intake (g) = Total feed intake $\times$ 100/(number of days $\times$ number of of laying hens);

Feed-to-egg ratio = Total feed consumption/total egg weight;

Broken egg rate (%) = Number of broken eggs $\times$ 100/total number of eggs.

2.2.2. The Quality of Eggs

On the 35th day of the experiment, two eggs per group were randomly selected for comprehensive egg quality assessment. Measurements included egg weight, shape index, Haugh units (HU), eggshell weight, moisture, protein, and fat content of the egg white and yolk. The egg shape index was determined using vernier calipers with 0.01 mm precision, measuring the longitudinal and maximum transverse diameters. Eggshells were cleaned, dried overnight at 50–53 °C, and weighed. Protein content in the egg yolk and white was analyzed by Kjeldahl™ 8100 (FOSS, Hillerød, Denmark). The fat content was determined with a Soxhlet apparatus (Soxtherm SOX 406, Gerhardt, Königswinter, Germany). The parameters were measured as follows:

Egg shape index (%) = Transverse diameter $\times$ 100/longitudinal diameter;

Haugh units (HU) = 100 $\times$ LOG [H + 7.57 $-$ 1.7 $\times$ w$^{0.37}$] (H is the average height of concentrated albumen, w is the weight of the egg);

Eggshell gravity (%) = Eggshell dry weight $\times$ 100/egg weight;

Egg white/egg yolk moisture content (%) = (egg white/egg yolk wet weight $-$ egg white/egg yolk dry weight) $\times$ 100/(egg white/egg yolk dry weight);

Egg white/egg yolk crude protein content (%) = c $\times$ v $\times$ 14 $\times$ 6.25 $\times$ 100/m (c is the concentration of the standard hydrochloric acid solution (mol/L); v is the volume of the standard hydrochloric acid solution used (mL); 14 is the molar mass of nitrogen (g/mol);

6.25 is the conversion factor from nitrogen to crude protein; mm is the mass of egg white or egg yolk (g));

Fat content (%) = $(m_1 - m_0) \times M_1 \times 100/(M_2 \times M_0)$ (m_1 represents the weight of the dried ether cup after fat extraction; m_0 is the weight of the dried ether cup before extraction; M_1 denotes the weight of the egg yolk; M_2 indicates the weight of the egg yolk sample; and M_0 signifies the initial weight of the egg. All the units are in grams);

Egg yolk gravity (%) = weight of egg yolk $\times$ 100/weight of whole egg.

2.3. Determination of Antioxidant Capacity

To analyze the serum antioxidant capacity, one hen from each replicate group was randomly selected at 07:00 am on day 36 of the experiment. Blood samples (4 mL) were drawn from the wing vein after 2-h water deprivation and centrifuged at 3000 rpm/min for 10 min to obtain serum. The supernatant was aliquoted into 1.5 mL centrifuge tubes and stored at $-20\ ^\circ$C for analysis of serum antioxidant indicators. Serum total antioxidant capacity (T-AOC), malondialdehyde (MDA) content, superoxide dismutase (SOD) activity, and glutathione peroxidase (GSH-PX) activity were measured using kits from Nanjing Jiancheng Bioengineering Institute (Nanjing, China), following the instructions of the manufacturer.

2.4. Analysis of Intestinal Mucosal Structure

To investigate the intestinal mucosal structure, the duodenum, jejunum, and ileum were collected from one of the laying hens of each replicate group. A 1–2 cm section from the middle of each intestinal segment was placed in 4% paraformaldehyde solution. These intestinal segments were embedded in paraffin, sectioned (5 μm thick), and stained with hematoxylin–eosin (HE). Under a light microscope, 15 villi per section were randomly selected to measure villus height and crypt depth. Villus height was measured from the villus tip to the crypt entrance, and crypt depth from the crypt base to the submucosa. The villus height/crypt depth (V/C) ratio was calculated to assess intestinal morphology.

2.5. Analysis of Gut Microbiota

To investigate the influence of YC on the gut microbiota composition in laying hens, 16S rDNA high-throughput sequencing analysis was used. Cecal contents were collected from one of the laying hens per replicate group, placed in sterile 2 mL centrifuge tubes, rapidly frozen in liquid nitrogen, and stored at $-80\ ^\circ$C for further analysis. Total DNA was extracted from the frozen cecal contents using the QIAamp Fast DNA Stool Kit (Qiagen, Hamburg, Germany). DNA quality was assessed by 0.8% agarose gel electrophoresis and quantified with a UV spectrophotometer. The V3 region of the 16S rRNA gene of the cecal microflora was sequenced using 454 high-throughput sequencing technology. Low-quality sequences were filtered out using Mothur software (Version 1.35.1), and sequences with >97% similarity were clustered into operational taxonomic units (OTUs). Bacterial community diversity was analyzed and compared between the control and treatment groups using the Silva and RDP databases. Sequencing and analysis were performed by Beijing Meiji Biotechnology company (Beijing, China).

2.6. Statistical Analysis

Experimental data were organized in Excel 2016, and a one-way analysis of variance (ANOVA) was performed using SPSS 16.0. Tukey's HSD test was used for multiple comparisons. Results are presented as mean $\pm$ standard deviation, with significance set at $p < 0.05$.

3. Results

3.1. Effects of Yeast Culture (YC) Supplementation to Diet on Egg Production Performance of Laying Hens

Table 2 illustrates the effects of the yeast culture (YC) supplementation on laying hens. Adding 1.0 g/kg YC (YC1.0) significantly reduced the broken egg rate by 63.48% compared to the control ($p < 0.05$), without affecting egg production rate, egg weight, or daily feed intake ($p > 0.05$). Supplementation with 2.0 g/kg YC (YC2.0) significantly improved the egg production rate, feed-to-egg ratio, and broken egg rate ($p < 0.05$), with no significant changes in egg weight and daily feed intake ($p > 0.05$).

Table 2. Effect of yeast culture (YC) supplementation on laying performance of laying hens.

Items	Control	YC1.0	YC2.0	SEM	*p* Value
Egg laying rate (%)	90.25 ± 1.36 [b]	91.78 ± 1.25 [ab]	92.62 ± 1.33 [a]	0.408	0.042
Average daily feed intake (g)	127.49 ± 3.87	124.56 ± 4.46	123.61 ± 3.13	1.023	0.290
Average egg weight (g)	61.54 ± 0.64	62.43 ± 0.30	62.05 ± 0.96	0.190	0.162
Ratio of feed to egg	2.29 ± 0.03 [a]	2.19 ± 0.06 [a]	2.15 ± 0.09 [b]	0.022	0.019
Broken egg rate (%)	1.15 ± 0.23 [a]	0.42 ± 0.23 [b]	0.26 ± 0.18 [b]	0.116	<0.001

Data are means of 5 replicates of 20 samples each replicate. SEM, standard error of mean. [a,b] Labels in a row, the differences are significant ($p < 0.05$); labels containing the same lowercase letters or no letters, the differences are not significant ($p > 0.05$).

3.2. Effects of YC Supplementation to Diet on Egg Quality of Laying Hens

Table 3 demonstrates the effect of YC supplementation on egg quality parameters. Adding 1.0 g/kg of YC significantly improved albumen height by 9.73% and Haugh units by 4.69%, with no significant influence on the specific gravity of egg yolk, eggshell specific gravity, fat content, and egg yolk crude protein content compared to the control group. Supplementation with 2.0 g/kg of YC significantly increased albumen height by 10.32%, Haugh units by 4.98%, and egg yolk crude protein content by 4.27%, without affecting egg yolk specific gravity, eggshell specific gravity, fat content, and egg white protein content.

Table 3. Effect of yeast culture (YC) supplementation on egg quality of laying hens [1].

Items	Control	YC1.0	YC2.0	SEM	*p* Value
Shape index	1.32 ± 0.06	1.32 ± 0.04	1.32 ± 0.06	0.003	0.842
Yolk ratio (%)	27.84 ± 1.15	28.11 ± 1.58	28.97 ± 1.26	0.346	0.412
Eggshell ratio (%)	10.31 ± 0.21	10.48 ± 0.23	10.58 ± 0.83	0.126	0.693
Albumen height (mm)	6.78 ± 0.30 [b]	7.44 ± 0.34 [a]	7.48 ± 0.30 [a]	0.114	0.007
Haugh unit	83.17 ± 1.96 [b]	87.07 ± 1.51 [a]	87.31 ± 1.98 [a]	0.669	0.006
Egg white moisture (%)	87.46 ± 0.40	87.82 ± 0.53	87.37 ± 0.34	0.115	0.250
Egg yolk moisture (%)	45.81 ± 0.35	45.37 ± 0.35	44.99 ± 1.04	0.183	0.186
Fat content (%)	8.08 ± 0.37	8.34 ± 0.49	8.49 ± 0.52	0.121	0.402
Protein content in egg white (%)	9.54 ± 0.32	9.54 ± 0.37	9.99 ± 0.24	0.094	0.067
Protein content in egg yolk (%)	16.85 ± 0.18 [b]	17.28 ± 0.55 [ab]	17.57 ± 0.34 [a]	0.121	0.039

[1] Data are means of 5 replicates of 20 samples each replicate. SEM, standard error of mean. [a,b] Labels in a row, the differences are significant ($p < 0.05$); labels containing the same lowercase letters or no letters, the differences are not significant ($p > 0.05$).

3.3. Effects of YC Supplementation to Diet on Serum Antioxidant Capacity of Laying Hens

Figure 1 shows the effects of YC supplementation on serum antioxidant parameters in laying hens, with detailed information provided in Table S2. Supplementing the diet with 1.0 g/kg and 2.0 g/kg of YC significantly increased total antioxidant capacity (T-AOC) by 35.91% and 48.39%; glutathione peroxidase (GSH-PX) activity by 15.95% and 19.13%; and reduced malondialdehyde (MDA) content by 39.74% and 44.55% ($p < 0.05$). There was no significant effect on superoxide dismutase (SOD) activity compared to the control group ($p > 0.05$).

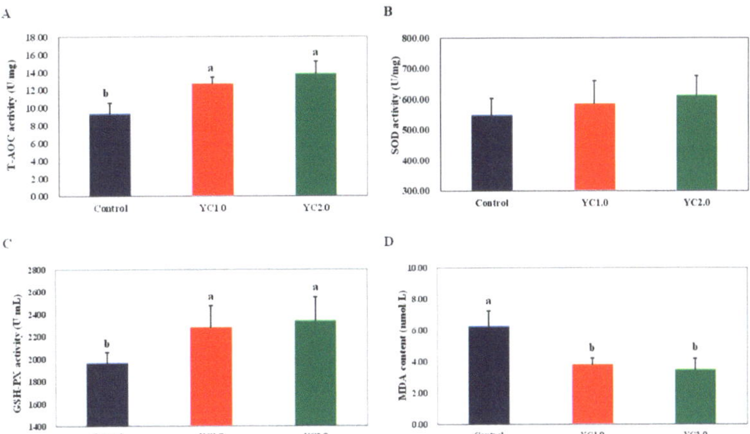

Figure 1. The effect of YC supplementation on serum antioxidant capacity of laying hens. (**A**) is total antioxidant capacity (TOC) activity, (**B**) is superoxide dismutase (SOD) activity, (**C**) is glutathione peroxidase (GSH-PX) activity, and (**D**) is malondialdehyde (MDA) content. [a,b] Labels in a row, the differences are significant ($p < 0.05$); labels containing the same lowercase letters or no letters, the differences are not significant ($p > 0.05$).

3.4. Effects of YC Supplementation to Diet on Intestinal Mucosal Structure of Laying Hens

Figure 2 illustrates the effects of YC supplementation on intestinal morphology in laying hens, with detailed data presented in Table 4. Supplementing with 1.0 g/kg and 2.0 g/kg kg/ton of YC significantly increased duodenal villus height by 17.91% and 22.48%; duodenal villus height to crypt depth ratio (V/C) by 21.93% and 30.08%; and jejunal V/C by 22.69% and 26.89% ($p < 0.05$). Meanwhile, there were no significant changes in jejunal villus height, ileal villus height, ileal V/C, and crypt depth in the duodenum, jejunum, and ileum ($p > 0.05$).

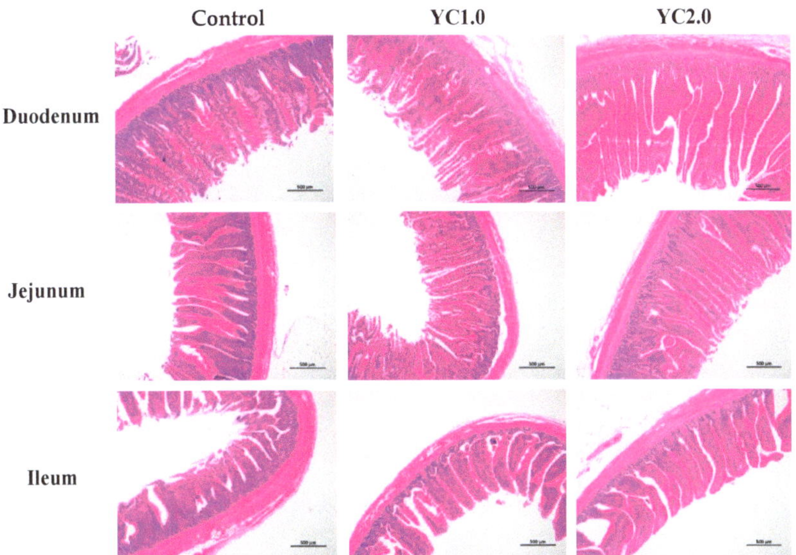

Figure 2. The effect of YC supplementation on intestinal mucosal structure of laying hens.

Table 4. Effect of YC supplementation on intestinal mucosal structure of laying hens.

	Items	Control	YC1.0	YC2.0	SEM	*p* Value
	Villus height (μm)	1040.39 ± 93.32 [b]	1226.71 ± 140.77 [a]	1274.25 ± 108.21 [a]	38.651	0.018
Duodenum	Crypt depth (μm)	139.77 ± 9.07	134.41 ± 10.59	131.04 ± 8.34	2.439	0.363
	Villus/Crypt	7.48 ± 0.99 [b]	9.12 ± 0.73 [a]	9.73 ± 0.70 [a]	0.321	0.003
	Villus height (μm)	833.49 ± 120.03	932.63 ± 121.38	982.61 ± 90.40	31.386	0.141
Jejunum	Crypt depth (μm)	140.95 ± 13.69	129.24 ± 21.76	130.20 ± 8.72	4.006	0.447
	Villus/Crypt	5.95 ± 0.94 [b]	7.30 ± 1.08 [a]	7.55 ± 0.57 [a]	0.285	0.031
	Villus height (μm)	573.08 ± 106.65	659.34 ± 107.01	648.33 ± 102.28	27.185	0.399
Ileum	Crypt depth (μm)	109.67 ± 19.07	103.45 ± 8.61	101.48 ± 9.94	3.330	0.612
	Villus/Crypt	5.33 ± 1.20	6.39 ± 0.99	6.40 ± 0.91	0.282	0.217

Data are means of 5 replicates of 20 samples each replicate. SEM, standard error of mean. [a,b] Labels in a row, the differences are significant ($p < 0.05$); labels containing the same lowercase letters or no letters, the differences are not significant ($p > 0.05$).

3.5. Effects of YC Supplementation to Diet on Cecal Microbiota Composition of Laying Hens

Figure 3 illustrates the influence of YC supplementation to diet on the cecal microbiota of laying hens. As shown in Figure 3A, YC supplementation significantly reduced the Simpson and Shannon indices ($p < 0.05$), indicating decreased microbial diversity. However, there was no significant effect on the Chao1 and ACE indices ($p > 0.05$). Figure 3B presents Principal Coordinate Axis Analysis (PCoA), which revealed no significant clustering differences between the experimental groups, indicating that YC supplementation did not significantly affect the β-diversity of the cecal microbial communities in laying hens.

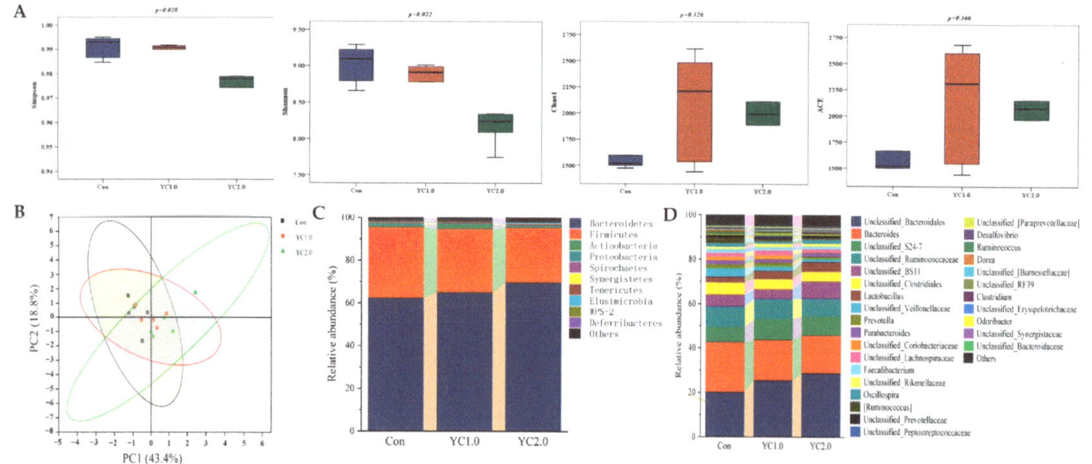

Figure 3. The effect of YC supplementation to diet on cecal microbiota of laying hens. (**A**) is the box plots of α diversity index. (**B**) is Principal Coordinate Axis Analysis (PCoA) two-dimensional sorting diagram. (**C**) is the relative abundance of cecal microflora at phylum level. (**D**) is the relative abundance of cecal microflora at genus level.

Figure 3C,D, along with Table S3, display the relative abundance of cecal microbiota at the phylum and genus levels. Figure 3C indicates that the predominant phyla are Bacteroidetes, Firmicutes, Actinobacteria, and Proteobacteria. Supplementation with 1.0 g/kg of YC significantly increased the abundance of Actinobacteria ($p < 0.05$) and reduced Proteobacteria ($p < 0.05$), while changes in Bacteroidetes and Firmicutes were not significant ($p > 0.05$). Supplementation with 2.0 g/kg of YC resulted in non-significant changes in these phyla ($p > 0.05$).

At the genus level, bacterial genera with relative abundance higher than 1% included *Bacteroidales*, *Bacteroides*, *Unclassified_S24-7*, *Ruminococcaceae*, *Unclassified_BS11*, *Clostridiales*, *Lactobacillus*, *Veillonellaceae*, *Prevotella*, and *Parabacteroides* (Figure 3D). Supplementation with 1.0 g/kg of YC increased the relative abundance of *Bacteroidales*, Unclassified_S24-7, *Lactobacillus*, and *Rikenellaceae* ($p > 0.05$), while reducing *Bacteroides*, *Clostridiales*, *Faecalibacterium* ($p > 0.05$), *Veillonellaceae*, *Oscillospira*, and *Ruminococcus* ($p < 0.05$). Supplementation with 2.0 g/kg of YC showed similar trends, with significant increases in *Bacteroidales* ($p < 0.05$) and reductions in *Veillonellaceae* and *Ruminococcus* ($p < 0.05$).

4. Discussion

4.1. Effects of YC Supplementation to Diet on Egg Production Performance of Laying Hens

Yeast culture (YC) is rich in peptides, amino acids, vitamins, oligosaccharides, organic acids, minerals, and other beneficial factors that enhance livestock and poultry growth by improving feed palatability and promoting nutrient digestion and absorption, thereby boosting production performance [6,15]. Studies have shown that dietary YC increases egg production rate and egg weight while reducing the feed-to-egg ratio in laying hens [16]. Zhang et al. mentioned that 0.3% YC supplementation for 4 weeks significantly enhanced egg production rate, reduced feed-to-egg ratio due to YC supplementation, and upregulated intestinal digestive enzyme activities and intestinal health-related gene expression [17]. Similarly, Liu et al. reported that 1% YC significantly increased the egg production rate and decreased the feed-to-egg ratio, thereby improving overall egg production performance [18].

In this study, supplementation with 2.0 g/kg of YC significantly increased egg production rate, reduced feed-to-egg ratio, and decreased broken egg rate, demonstrating that YC improved egg production performance under these conditions. Zhang et al. also demonstrated that YC supplementation significantly increased amylase and chymotrypsin activity in the duodenum, enhanced the expression of duodenal tight junction proteins, and improved intestinal structure [17]. Liu et al. reported that YC significantly improved the apparent digestibility of dietary crude protein, confirming that YC enhances feed nutrient digestibility, and, consequently, production performance in laying hens [18].

4.2. Effects of YC Supplementation to Diet on Egg Quality of Laying Hens

Egg quality encompasses the shape, size, cleanliness, eggshell strength, and content quality, including the consistency of egg white, yolk size and color, and fat content. Egg yolk concentration is a key nutritional aspect. Higher yolk-to-egg ratios indicate better quality, and Haugh units measure protein quality and freshness, serving as a primary international standard for egg quality assessment. In this experiment, YC supplementation significantly increased egg protein height, Haugh units, and the crude protein content of egg yolk. With higher YC levels, the specific gravity of egg yolk, eggshell, and fat content also increased, while the egg shape index as well as the moisture content of egg white and egg yolk remained unchanged. Studies by Zhong et al. [19], Özsoy et al. [16], and Liu et al. [6] similarly reported that YC enhances egg protein height and Haugh units. Liu et al. observed a dose-dependent increase in these parameters with an addition of YC. Zhong et al. [19] also noted significant increases in yolk and eggshell weight.

These improvements are attributed to enhanced digestion and nutrient deposition facilitated by YC, which is rich in small peptides, amino acids, vitamins, and other nutrients that promote metabolic processes in laying hens [20,21]. However, some studies, such as Yalçin et al. [22] and Zhang et al. [17], found no significant effects on egg protein height, Haugh units, or eggshell quality, likely due to variations in yeast species, supplementation levels, hen age, and management conditions.

4.3. Effects of YC Supplementation to Diet on Serum Antioxidant Capacity of Laying Hens

Glutathione peroxidase (GSH-Px) and superoxide dismutase (SOD) are essential antioxidant enzymes that remove free radicals, protect cells from oxidative damage, and aid in cell repair [23]. Malondialdehyde (MDA) is a marker of lipid peroxidation, indicating oxidative stress levels [24]. The total antioxidant capacity (T-AOC) reflects the overall antioxidant defense against oxidative damage. Normally, free radical production and elimination are balanced but stress increases free radical production, leading to oxidative damage and higher MDA levels. To counter this, animals produce endogenous antioxidants, including enzymes like GSH-Px and SOD. Measuring these enzymes and MDA levels in serum assesses the antioxidant capacity of livestock and poultry.

Research indicates that YC enhances antioxidant enzyme activity and reduces MDA levels, improving overall antioxidant capacity [15]. Liu et al. found that dietary YC significantly increased serum T-AOC activity in laying hens but did not affect GSH-Px activity and MDA levels [6]. In contrast, Liu et al. reported that YC significantly increased serum GSH-Px and SOD activities while reducing MDA levels [18].

In this experiment, dietary YC significantly increased serum T-AOC and GSH-Px activities and reduced MDA levels in laying hens. SOD activity also showed an upward trend with higher YC levels. These findings align with previous studies, indicating that YC enhances antioxidant capacity and reduces oxidative damage. This effect is attributed to the rich oligosaccharide content in YC, particularly mannan oligosaccharides, which possess strong antioxidant and free-radical scavenging abilities [25,26]. Additionally, mannan oligosaccharides and glucans improve antioxidant properties by enhancing intestinal mucosa structure and promoting the absorption of nutrients like zinc, selenium, and copper [27].

4.4. Effects of YC Supplementation to Diet on Intestinal Mucosal Structure of Laying Hens

The small intestine is the primary site for nutrient digestion and absorption in animals. Key indicators of intestinal function include villus height, crypt depth, and the villus height-to-crypt depth ratio. Higher villi indicate more absorptive epithelial cells, enhancing nutrient absorption; shallower crypts suggest more mature cells with stronger secretion capabilities. A higher villus height-to-crypt depth ratio indicates a better intestinal mucosal structure and stronger digestion and absorption capacity [28]. Research demonstrates that YC improves intestinal structure, protects mucosa, and promotes development. Gao et al. found that 2.5 g/kg of YC significantly increased duodenal villus height and the villus-to-crypt ratio in broiler chickens [29]. He et al. reported that 2.0% of YC significantly enhanced jejunal and ileal villus height and the villus-to-crypt ratio in pigs [26]. Liu et al. showed similar effects in weaned piglets [30], and Zhang et al. noted increased jejunal villus height and villus-to-crypt ratio in geese with 2% YC supplementation [31].

In this experiment, dietary YC significantly increased duodenal villus height and the villus-to-crypt ratio in the duodenum and jejunum, consistent with previous studies. Mannan oligosaccharides in YC inhibited pathogenic bacterial colonization, protecting intestinal mucosa, and maintaining its normal function [32]. They also promoted beneficial bacteria growth, which ferment oligosaccharides to produce organic acids like acetic, propionic, and butyric acids. These acids support probiotic bacteria growth and provide energy for intestinal epithelial cells, promoting mucosal development [11]. Additionally, small peptides of YC have enhanced intestinal epithelial cell proliferation and inhibit apoptosis, aiding intestinal development according to the research from Wang et al. [33]. YC also supplies nucleotides necessary for rapid growth, promoting intestinal cell division, differentiation, and maturation [34]. Moreover, Sindaye et al. found that lysozyme could enhance intestinal morphology in laying hens [35].

4.5. Effects of YC Supplementation to Diet on Cecal Microbiota Composition of Laying Hens

The balance of intestinal microflora is crucial for animal health. Under normal conditions, the intestinal flora maintains a dynamic equilibrium, with bacterial abundances remaining stable. External stimuli can disrupt this balance, leading to an imbalance in intestinal flora. In the cecum of laying hens, the dominant bacterial phyla were Firmicutes, Bacteroidetes, Proteobacteria, and Actinobacteria, as previously reported [36,37]. During peak egg production, healthy hens showed 47.5–62.0% of Bacteroidetes and 30.8–60.4% of Firmicutes, based on reporting by others [38–40].

The previous studies indicate that dietary YC increases Bacteroidetes and decreases Firmicutes and Proteobacteria in the cecum of laying hens [6,18]. Bacteroidetes include probiotic bacteria involved in protein and polysaccharide degradation, enhancing nutrient metabolism and egg production as reported by Panasevich et al. [41] and Xu et al. [42]. Proteobacteria, which includes many pathogens like *Escherichia coli* and *Salmonella*, decrease in abundance with YC, promoting intestinal health according to the reports from Bi et al. [43]. Meanwhile, Tian et al. mentioned that the regulation of cecal microbiota might result in the production performance and egg quality [44].

The previous studies indicate that *Bacteroides* (11.1–32.5%), *Riken* (10.1–15.0%), *Lachnospira* (5.0–8.0%), and *Lactobacillus* (2.0–10.0%) dominate the cecum of laying hens at the genus level [38–40]. YC oligosaccharides promote probiotic growth and inhibit harmful bacteria, modulating intestinal flora composition. Specifically, YC increases the abundance of *Lactobacillus*, although not always significantly, suggesting a positive impact on intestinal health and hen growth [6,19,45].

5. Conclusions

This study revealed that adding yeast culture (YC) to the diet can improve the composition of intestinal flora and intestinal mucosal structure of laying hens and improve antioxidant capacity, thereby improving the production performance and egg quality of laying hens. Adding 2.0 g/kg of YC had an impact on production performance and egg quality, antioxidant capacity, intestinal mucosal structure, and bacterial flora structure better than at the addition level of 1.0 g/kg.

Supplementary Materials: The following supporting information can be downloaded at: https://www.mdpi.com/article/10.3390/antiox13070779/s1, Table S1. The composition of premix. Table S2. Effect of yeast culture (YC) supplementation on serum antioxidant capacity of laying hens. Table S3. The Effect of yeast culture (YC) supplementation on relative abundance of cecal microflora at phylum and genus levels of laying hens.

Author Contributions: Conceptualization, methodology, software, validation, formal analysis, investigation, resources, data curation, writing—original draft preparation, Q.Q.; writing—review and editing, Z.Z.; writing—review and editing, Y.Z.; writing—review and editing, visualization, W.Z. (Wei Zhang); writing—review and editing, L.G.; writing—review and editing, Q.W.; supervision, writing—review and editing, J.H.; writing—review and editing, Y.L. (Yunxiang Liang); writing—review and editing, visualization, W.Z. (Wen Zhou); supervision, project administration, funding acquisition, writing—review and editing, Y.L. (Yingjun Li). All authors have read and agreed to the published version of the manuscript.

Funding: This research was funded by Hubei Province Science and Technology Talent Service Enterprise Project (2023DJC127) and Fundamental Research Funds for the Central Universities (2662022SKPY001).

Institutional Review Board Statement: The animal study protocol was approved by the Ethics Committee of NAME OF Huazhong Agricultural University (protocol code HZAUCH-2020-0021, date 10 May 2020).

Informed Consent Statement: Not applicable.

Data Availability Statement: The original contributions presented in the study are included in the article/Supplementary Material, further inquiries can be directed to the corresponding author.

Conflicts of Interest: Authors Q.Q., Z.Z., Y.Z., W.Z., L.G. and Q.W. were employed by the company Wuhan Sunhy Biological Co., Ltd. (Wuhan, China). The remaining authors declare that the research was conducted in the absence of any commercial or financial relationships that could be construed as a potential conflict of interest.

References

1. Elahi, E.; Li, G.; Han, X.; Zhu, W.; Liu, Y.; Cheng, A.; Yang, Y. Decoupling Livestock and Poultry Pollution Emissions from Industrial Development: A Step Towards Reducing Environmental Emissions. *J. Environ. Manag.* **2024**, *350*, 119654. [CrossRef]
2. Pandey, S.; Doo, H.; Keum, G.B.; Kim, E.S.; Kwak, J.; Ryu, S.; Choi, Y.; Kang, J.; Kim, S.; Lee, N.R.; et al. Antibiotic Resistance in Livestock, Environment and Humans: One Health Perspective. *J. Anim. Sci. Technol.* **2024**, *66*, 266–278. [CrossRef]
3. Da Silva, R.A.; Arenas, N.E.; Luiza, V.L.; Bermudez, J.A.Z.; Clarke, S.E. Regulations on the Use of Antibiotics in Livestock Production in South America: A Comparative Literature Analysis. *Antibiotics* **2023**, *12*, 1303. [CrossRef]
4. Cardinal, K.M.; Andretta, I.; da Silva, M.K.; Stefanello, T.B.; Schroeder, B.; Ribeiro, A.M.L. Estimation of Productive Losses Caused by Withdrawal of Antibiotic Growth Promoter from Pig Diets—Meta-Analysis. *Sci. Agricola* **2021**, *78*, e20200266. [CrossRef]
5. Wang, C.; Fan, J.; Ma, K.; Wang, H.; Li, D.; Li, T.; Ma, Y. Effects of Adding Allium Mongolicum Regel Powder and Yeast Cultures to Diet on Rumen Microbial Flora of Tibetan Sheep (*Ovis aries*). *Front. Veter- Sci.* **2024**, *11*, 1283437. [CrossRef]
6. Liu, Y.; Cheng, X.; Zhen, W.; Zeng, D.; Qu, L.; Wang, Z.; Ning, Z. Yeast Culture Improves Egg Quality and Reproductive Performance of Aged Breeder Layers by Regulating Gut Microbes. *Front. Microbiol.* **2021**, *12*, 633276. [CrossRef]
7. Dias, J.D.; Silva, R.B.; Fernandes, T.; Barbosa, E.F.; Graças, L.E.; Araujo, R.C.; Pereira, R.A.; Pereira, M.N. Yeast Culture Increased Plasma Niacin Concentration, Evaporative Heat Loss, and Feed Efficiency of Dairy Cows in a Hot Environment. *J. Dairy Sci.* **2018**, *101*, 5924–5936. [CrossRef]
8. Stella, A.; Paratte, R.; Valnegri, L.; Cigalino, G.; Soncini, G.; Chevaux, E.; Dell'orto, V.; Savoini, G. Effect of Administration of Live Saccharomyces Cerevisiae on Milk Production, Milk Composition, Blood Metabolites, and Faecal Flora in Early Lactating Dairy Goats. *Small Rumin. Res.* **2007**, *67*, 7–13. [CrossRef]
9. Waititu, S.; Yin, F.; Patterson, R.; Yitbarek, A.; Rodriguez-Lecompte, J.; Nyachoti, C. Dietary Supplementation with a Nucleotide-Rich Yeast Extract Modulates Gut Immune Response and Microflora in Weaned Pigs in Response to a Sanitary Challenge. *Animal* **2017**, *11*, 2156–2164. [CrossRef]
10. Attia, Y.A.; Al-Khalaifah, H.; Ibrahim, M.S.; Al-Hamid, A.E.A.; Al-Harthi, M.A.; El-Naggar, A. Blood Hematological and Biochemical Constituents, Antioxidant Enzymes, Immunity and Lymphoid Organs of Broiler Chicks Supplemented with Propolis, Bee Pollen and Mannan Oligosaccharides Continuously or Intermittently. *Poult. Sci.* **2017**, *96*, 4182–4192. [CrossRef]
11. Shen, Y.B.; Piao, X.S.; Kim, S.W.; Wang, L.; Liu, P.; Yoon, I.; Zhen, Y.G. Effects of Yeast Culture Supplementation on Growth Performance, Intestinal Health, and Immune Response of Nursery Pigs. *J. Anim. Sci.* **2009**, *87*, 2614–2624. [CrossRef] [PubMed]
12. Mukherjee, A.; Verma, J.P.; Gaurav, A.K.; Chouhan, G.K.; Patel, J.S.; Hesham, A.E.-L. Yeast a Potential Bio-Agent: Future for Plant Growth and Postharvest Disease Management for Sustainable Agriculture. *Appl. Microbiol. Biotechnol.* **2020**, *104*, 1497–1510. [CrossRef] [PubMed]
13. Nandy, S.K.; Srivastava, R. A Review on Sustainable Yeast Biotechnological Processes and Applications. *Microbiol. Res.* **2018**, *207*, 83–90. [CrossRef] [PubMed]
14. Alagawany, M.; Bilal, R.M.; Elnesr, S.S.; Elwan, H.A.M.; Farag, M.R.; Dhama, K.; Naiel, M.A.E. Yeast in Layer Diets: Its Effect on Production, Health, Egg Composition and Economics. *World's Poult. Sci. J.* **2023**, *79*, 135–153. [CrossRef]
15. Zhang, J.; Wan, K.; Xiong, Z.; Luo, H.; Zhou, Q.; Liu, A.; Cao, T.; He, H. Effects of Dietary Yeast Culture Supplementation on the Meat Quality and Antioxidant Capacity of Geese. *J. Appl. Poult. Res.* **2021**, *30*, 100116. [CrossRef]
16. Özsoy, B.; Karadağoğlu, Ö.; Yakan, A.; Önk, K.; Çelik, E.; Şahin, T. The Role of Yeast Culture (*Saccharomyces cerevisiae*) on Performance, Egg Yolk Fatty Acid Composition, and Fecal Microflora of Laying Hens. *Rev. Bras. Zootec.* **2018**, *47*, e20170159. [CrossRef]
17. Zhang, J.-C.; Chen, P.; Zhang, C.; Khalil, M.M.; Zhang, N.-Y.; Qi, D.-S.; Wang, Y.-W.; Sun, L.-H. Yeast Culture Promotes the Production of Aged Laying Hens by Improving Intestinal Digestive Enzyme Activities and the Intestinal Health Status. *Poult. Sci.* **2020**, *99*, 2026–2032. [CrossRef] [PubMed]
18. Liu, J.; Jin, Y.; Yang, J. Influence of Spent Ginger Yeast Cultures on the Production Performance, Egg Quality, Serum Composition, and Intestinal Microbiota of Laying Hens. *Anim. Biosci.* **2022**, *35*, 1205–1214. [CrossRef] [PubMed]
19. Zhong, S.; Liu, H.; Zhang, H.; Han, T.; Jia, H.; Xie, Y. Effects of Kluyveromyces Marxianus Isolated from Tibetan Mushrooms on the Plasma Lipids, Egg Cholesterol Level, Egg Quality and Intestinal Health of Laying Hens. *Rev. Bras. Cienc. Avic. Braz. J. Poult. Sci.* **2016**, *18*, 261–268. [CrossRef]
20. Jensen, G.; Patterson, K.; Yoon, I. Yeast Culture Has Anti-Inflammatory Effects and Specifically Activates NK Cells. *Comp. Immunol. Microbiol. Infect. Dis.* **2008**, *31*, 487–500. [CrossRef]
21. Bilal, R.M.; Elwan, H.A.M.; Elnesr, S.S.; Farag, M.R.; El-Shall, N.A.; Ismail, T.A.; Alagawany, M. Use of Yeast and Its Derived Products in Laying Hens: An Updated Review. *World's Poult. Sci. J.* **2022**, *78*, 1087–1104. [CrossRef]
22. Yalçin, S.; Özsoy, B.; Erol, H. Yeast Culture Supplementation to Laying Hen Diets Containing Soybean Meal or Sunflower Seed Meal and Its Effect on Performance, Egg Quality Traits, and Blood Chemistry. *J. Appl. Poult. Res.* **2008**, *17*, 229–236. [CrossRef]
23. Ibtisham, F.; Nawab, A.; Niu, Y.; Wang, Z.; Wu, J.; Xiao, M.; An, L. The Effect of Ginger Powder and Chinese Herbal Medicine on Production Performance, Serum Metabolites and Antioxidant Status of Laying Hens under Heat-Stress Condition. *J. Therm. Biol.* **2019**, *81*, 20–24. [CrossRef]
24. Tsikas, D. Assessment of Lipid Peroxidation by Measuring Malondialdehyde (MDA) and Relatives in Biological Samples: Analytical and Biological Challenges. *Anal. Biochem.* **2017**, *524*, 13–30. [CrossRef] [PubMed]

25. Qin, L.; Ji, W.; Wang, J.; Li, B.; Hu, J.; Wu, X. Effects of Dietary Supplementation with Yeast Glycoprotein on Growth Performance, Intestinal Mucosal Morphology, Immune Response and Colonic Microbiota in Weaned Piglets. *Food Funct.* **2019**, *10*, 2359–2371. [CrossRef] [PubMed]

26. He, W.; Gao, Y.; Guo, Z.; Yang, Z.; Wang, X.; Liu, H.; Sun, H.; Shi, B. Effects of Fermented Wheat Bran and Yeast Culture on Growth Performance, Immunity, and Intestinal Microflora in Growing-Finishing Pigs. *J. Anim. Sci.* **2021**, *99*, skab308. [CrossRef] [PubMed]

27. Ogbuewu, I.P.; Okoro, V.M.; Mbajiorgu, E.F.; Mbajiorgu, C.A. Yeast (*Saccharomyces cerevisiae*) and Its Effect on Production Indices of Livestock and Poultry—A Review. *Comp. Clin. Pathol.* **2019**, *28*, 669–677. [CrossRef]

28. Caspary, W.F. Physiology and Pathophysiology of Intestinal Absorption. *Am. J. Clin. Nutr.* **1992**, *55* (Suppl. S1), 299S–308S. [CrossRef] [PubMed]

29. Gao, J.; Zhang, H.J.; Yu, S.H.; Wu, S.G.; Yoon, I.; Quigley, J.; Gao, Y.P.; Qi, G.H. Effects of Yeast Culture in Broiler Diets on Performance and Immunomodulatory Functions. *Poult. Sci.* **2008**, *87*, 1377–1384. [CrossRef]

30. Liu, G.; Yu, L.; Martínez, Y.; Ren, W.; Ni, H.; Al-Dhabi, N.A.; Duraipandiyan, V.; Yin, Y. Dietary *Saccharomyces cerevisiae* Cell Wall Extract Supplementation Alleviates Oxidative Stress and Modulates Serum Amino Acids Profiles in Weaned Piglets. *Oxidative Med. Cell. Longev.* **2017**, *2017*, 3967439. [CrossRef]

31. Zhang, J.; Yuan, Y.; Wang, F.; He, H.; Wan, K.; Liu, A. Effect of Yeast Culture Supplementation on Blood Characteristics, Body Development, Intestinal Morphology, and Enzyme Activities in Geese. *J. Anim. Physiol. Anim. Nutr.* **2022**, *107*, 598–606. [CrossRef]

32. Castillo, M.; Martín-Orúe, S.M.; Taylor-Pickard, J.A.; Pérez, J.F.; Gasa, J. Use of Mannanoligosaccharides and Zinc Chelate as Growth Promoters and Diarrhea Preventative in Weaning Pigs: Effects on Microbiota and Gut Function. *J. Anim. Sci.* **2008**, *86*, 94–101. [CrossRef]

33. Wang, H.; Jia, G.; Chen, Z.-L.; Huang, L.; Wu, C.-M.; Wang, K.-N. The Effect of Glycyl-Glutamine Dipeptide Concentration on Enzyme Activity, Cell Proliferation and Apoptosis of Jejunal Tissues from Weaned Piglets. *Agric. Sci. China* **2011**, *10*, 1088–1095. [CrossRef]

34. Che, L.; Hu, L.; Liu, Y.; Yan, C.; Peng, X.; Xu, Q.; Wang, R.; Cheng, Y.; Chen, H.; Fang, Z.; et al. Dietary Nucleotides Supplementation Improves the Intestinal Development and Immune Function of Neonates with Intra-Uterine Growth Restriction in a Pig Model. *PLoS ONE* **2016**, *11*, e0157314. [CrossRef]

35. Sindaye, D.; Xiao, Z.; Wen, C.; Yang, K.; Zhang, L.; Liao, P.; Zhang, F.; Xin, Z.; He, S.; Ye, S.; et al. Exploring the Effects of Lysozyme Dietary Supplementation on Laying Hens: Performance, Egg Quality, and Immune Response. *Front. Vet. Sci.* **2023**, *10*, 1273372. [CrossRef]

36. Videnska, P.; Sedlar, K.; Lukac, M.; Faldynova, M.; Gerzova, L.; Cejkova, D.; Sisak, F.; Rychlik, I. Succession and Replacement of Bacterial Populations in the Caecum of Egg Laying Hens over Their Whole Life. *PLoS ONE* **2014**, *9*, e115142. [CrossRef]

37. Dai, D.; Qi, G.H.; Wang, J.; Zhang, H.J.; Qiu, K.; Wu, S.G. Intestinal Microbiota of Layer Hens and Its Association with Egg Quality and Safety. *Poult. Sci.* **2022**, *101*, 102008. [CrossRef]

38. Hamid, H.; Zhang, J.; Li, W.; Liu, C.; Li, M.; Zhao, L.; Ji, C.; Ma, Q. Interactions between the Cecal Microbiota and Non-Alcoholic Steatohepatitis Using Laying Hens as the Model. *Poult. Sci.* **2019**, *98*, 2509–2521. [CrossRef]

39. Joat, N.; Van, T.T.H.; Stanley, D.; Moore, R.J.; Chousalkar, K. Temporal dynamics of gut microbiota in caged laying hens: A field observation from hatching to end of lay. *Appl. Microbiol. Biotechnol.* **2021**, *105*, 4719–4730. [CrossRef]

40. Miao, L.; Gong, Y.; Li, H.; Xie, C.; Xu, Q.; Dong, X.; Elwan, H.A.; Zou, X. Alterations in Cecal Microbiota and Intestinal Barrier Function of Laying Hens Fed on Fluoride Supplemented Diets. *Ecotoxicol. Environ. Saf.* **2020**, *193*, 110372. [CrossRef]

41. Panasevich, M.R.; Meers, G.M.; Linden, M.A.; Booth, F.W.; Perfield, J.W., II; Fritsche, K.L.; Wankhade, U.D.; Chintapalli, S.V.; Shankar, K.; Ibdah, J.A.; et al. High-fat, high-fructose, high-cholesterol feeding causes severe Nash and cecal microbiota dysbiosis in juvenile Ossabaw swine. *Am. J. Physiol. Endocrinol. Metab.* **2018**, *314*, E78–E92. [CrossRef] [PubMed]

42. Xu, X.; Dai, M.; Lao, F.; Chen, F.; Hu, X.; Liu, Y.; Wu, J. Effect of Glucoraphanin from Broccoli Seeds on Lipid Levels and Gut Microbiota in High-Fat Diet-Fed Mice. *J. Funct. Foods* **2020**, *68*, 103858. [CrossRef]

43. Bi, Y.; Tu, Y.; Zhang, N.; Wang, S.; Zhang, F.; Suen, G.; Shao, D.; Li, S.; Diao, Q. Multiomics Analysis Reveals the Presence of a Microbiome in the Gut of Fetal Lambs. *Gut* **2021**, *70*, 853–864. [CrossRef]

44. Tian, Y.; Li, G.; Zhang, S.; Zeng, T.; Chen, L.; Tao, Z.; Lu, L. Dietary Supplementation with Fermented Plant Product Modulates Production Performance, Egg Quality, Intestinal Mucosal Barrier, and Cecal Microbiota in Laying Hens. *Front. Microbiol.* **2022**, *13*, 955115. [CrossRef]

45. Wu, C.-C.; Weng, W.-L.; Lai, W.-L.; Tsai, H.-P.; Liu, W.-H.; Lee, M.-H.; Tsai, Y.-C. Effect of *Lactobacillus plantarum* Strain K21 on High-Fat Diet-Fed Obese Mice. *Evid.-Based Complement. Altern. Med.* **2015**, *2015*, 391767. [CrossRef]

 *antioxidants*

Article

Black Goji Berry (*Lycium ruthenicum*) Juice Fermented with *Lactobacillus rhamnosus* GG Enhances Inhibitory Activity against Dipeptidyl Peptidase-IV and Key Steps of Lipid Digestion and Absorption

Kritmongkhon Kamonsuwan [1], Vernabelle Balmori [2], Marisa Marnpae [3], Charoonsri Chusak [1], Thavaree Thilavech [4], Suvimol Charoensiddhi [5], Scott Smid [6] and Sirichai Adisakwattana [1,*]

[1] Center of Excellence in Phytochemical and Functional Food for Clinical Nutrition, Department of Nutrition and Dietetics, Faculty of Allied Health Sciences, Chulalongkorn University, Bangkok 10330, Thailand; 6278303837@student.chula.ac.th (K.K.); charoonsri.c@chula.ac.th (C.C.)

[2] Department of Food Science and Technology, Southern Leyte State University, Sogod 6606, Philippines; vbalmori@southernleytestateu.edu.ph

[3] The Halal Science Center, Chulalongkorn University, Bangkok 10330, Thailand; marisa.m@chula.ac.th

[4] Department of Food Chemistry, Faculty of Pharmacy, Mahidol University, Bangkok 10400, Thailand; thavaree.thi@mahidol.edu

[5] Department of Food Science and Technology, Faculty of Agro-Industry, Kasetsart University, Bangkok 10900, Thailand; suvimol.ch@ku.th

[6] Discipline of Pharmacology, School of Biomedicine, Faculty of Health and Medical Sciences, The University of Adelaide, Adelaide 5000, SA, Australia; scott.smid@adelaide.edu.au

* Correspondence: sirichai.a@chula.ac.th; Tel.: +66-2-218-1099 (ext. 111)

 check for updates

Citation: Kamonsuwan, K.; Balmori, V.; Marnpae, M.; Chusak, C.; Thilavech, T.; Charoensiddhi, S.; Smid, S.; Adisakwattana, S. Black Goji Berry (*Lycium ruthenicum*) Juice Fermented with *Lactobacillus rhamnosus* GG Enhances Inhibitory Activity against Dipeptidyl Peptidase-IV and Key Steps of Lipid Digestion and Absorption. *Antioxidants* **2024**, *13*, 740. https://doi.org/10.3390/antiox13060740

Academic Editor: Myung-Ji Seo

Received: 20 May 2024
Revised: 14 June 2024
Accepted: 15 June 2024
Published: 19 June 2024

Abstract: With the global increase in hyperglycemia and hyperlipidemia, there is an urgent need to explore dietary interventions targeting the inhibition of dipeptidyl peptidase-IV (DPP-IV) and lipid digestion and absorption. This study investigated how *Lactobacillus rhamnosus* GG (LGG) affects various aspects of black goji berry (BGB) (*Lycium ruthenicum* Murr.) juice, including changes in physicochemical and functional properties, as well as microbiological and sensory attributes. Throughout the fermentation process with 2.5–10% (w/v) BGB, significantly improved probiotic viability, lactic acid production, and decreased sugar content. While total flavonoids increase, anthocyanins decrease, with no discernible change in antioxidant activities. Metabolite profiling reveals elevated phenolic compounds post-fermentation. Regarding the inhibition of lipid digestion and absorption, fermented BGB exhibits improved bile acid binding, and disrupted cholesterol micellization by approximately threefold compared to non-fermented BGB, while also increasing pancreatic lipase inhibitory activity. Furthermore, a decrease in cholesterol uptake was observed in Caco-2 cells treated with fermented BGB (0.5 mg/mL), with a maximum reduction of 16.94%. Fermented BGB also shows more potent DPP-IV inhibition. Sensory attributes are significantly improved in fermented BGB samples. These findings highlight the potential of BGB as a bioactive resource and a promising non-dairy carrier for LGG, enhancing its anti-hyperglycemic and anti-hyperlipidemic properties.

Keywords: black goji berry; *Lactobacillus rhamnosus*; anthocyanins; flavonoids; non-volatile metabolite profiling

1. Introduction

One of the most concerning health challenges globally is the increasing incidence of non-communicable diseases (NCDs). These persistent conditions, which are not contagious, present a substantial global health concern requiring urgent attention and collaborative effort. Hyperlipidemia and hyperglycemia are crucial contributors to the development and progression of NCDs such as cardiovascular diseases, stroke, and diabetes [1]. The high prevalence of hyperlipidemia is a concerning global health issue, contributing significantly to morbidity and mortality rates worldwide [2]. There is a current trend towards investigating novel methods to address hyperlipidemia beyond traditional approaches. One promising approach to managing hyperlipidemia involves targeting key steps in lipid digestion and absorption. Recent research suggests that inhibiting pancreatic lipase, cholesterol micellization, and bile acid reabsorption could be effective. These strategies aim to reduce dietary fat absorption, potentially lowering circulating lipid levels and decreasing the risk of cardiovascular disease [3]. Elevated blood glucose levels, characteristic of diabetes, are associated with severe complications including cardiovascular disease, neuropathy, nephropathy, and retinopathy. Effective management of hyperglycemia is imperative in diabetes care to mitigate these adverse outcomes. One treatment strategy involves targeting the enzyme dipeptidyl peptidase-IV (DPP-IV), which is pivotal in regulating incretin hormones like glucagon-like peptide-1 (GLP-1) and glucose-dependent insulinotropic polypeptide (GIP). Inhibiting DPP-IV decreases the breakdown of these incretin hormones, extending their effects. This leads to increased insulin secretion, decreased glucose production, and consequently, better blood sugar management. Current research has demonstrated that DPP-IV inhibitors, such as sitagliptin and vildagliptin, are effective in improving glycemic control in diabetic patients [4]. Aligned with these strategies, significant focus has been placed on utilizing probiotics and functional foods, including phytochemicals, to regulate lipid and carbohydrate metabolisms and support cardiovascular health. Moreover, probiotics and phytochemicals have been found to exhibit both lipid-lowering and glucose-lowering effects, making them promising adjuncts in the management of metabolic disorders [5,6].

Probiotic-fermented fruit juice has become a popular and commercially available food product, drawing attention from a broad spectrum of consumers, including those with dietary restrictions [7]. The probiotic fermentation process not only extends shelf life but also enhances biological properties and improves sensory attributes [8]. Probiotic enzymes are integral in breaking down diverse compounds within fruit matrices, resulting in the generation of active substances such as polysaccharides, organic acids, short-chain fatty acids, and phenolic compounds. Concurrently, the fermentation process decreases sugar content and reduces antinutritional factors like alkaloids, tannins, and oxalates [9]. These bioconversions could transform the functional components of fruit juice together with enhancing their bioaccessibility and bioavailability. Furthermore, the probiotic-mediated bioconversion contributes to increased antimicrobial, anticancer, and antioxidant activities in food products, generating esters, alcohols, and terpenes that improve sensory quality [10,11].

The genus *Lactobacillus*, a subset of lactic acid bacteria (LAB), stands out as a widely utilized category of probiotics, recognized for its "generally recognized as safe" (GRAS) status and potential health advantages [12]. Specifically, *Lactobacillus rhamnosus* GG (LGG), the initial patented strain within the genus, emerges as a potential probiotic with the capacity to endure and thrive in both acidic and alkaline environments, exhibiting robust growth characteristics, and an ability to adhere to the intestinal epithelial layer [13]. These attributes position LGG favorably for application in the industry-level production of fermented foods. Extensive research has demonstrated the health-promoting effects of LGG, in both in vitro and human studies [14]. Notably, LGG has shown efficacy in safeguarding the mucosa, bolstering intestinal crypt survival, and fostering immune responses. Moreover, it has been reported to alleviate conditions such as diarrhea, gastrointestinal infections, irritable bowel syndrome, and inflammatory bowel disease [15]. Consequently, LGG emerges as a promising strain for the development of food products with potential health benefits.

Various fruits, including pomegranate, cherry, mulberry, blackberry, blueberry, and goji berry, serve as carriers for probiotics [16]. These non-dairy carriers are renowned for their abundance of phytochemical compounds, which possess potent antioxidant activities. Within the goji berry family (*Lycium* spp.), black goji berry (*Lycium ruthenicum* Murr.) (BGB) emerges as a particularly promising raw material for fermented fruit juice production.

Specifically, it has been identified to exhibit higher phytochemical and antioxidant activities compared to other goji berry varieties, such as red goji berry [17]. When directly compared to other berry fruits like blackberry or blueberry, BGB demonstrates elevated anthocyanin contents, along with other phytochemical compounds such as flavonoids, phenolic acids, and carotenoids [18]. These phytochemical components contribute to various biological properties and health benefits in BGB, similar to those found in fermented food products, including anti-diabetic, anti-hyperlipidemic, and prebiotic-like properties [19].

Despite these promising attributes, there is limited evidence supporting BGB juice as a carrier for probiotics, especially *Lactobacillus* spp. Thus, this study aimed to investigate the impact of LGG on the fermentation process of BGB juice. Furthermore, it explored changes in physicochemical properties, non-volatile compounds, antioxidant capacity, probiotic viability, and sensory attributes. Additionally, the study assessed the effects on key steps of lipid digestion and absorption, including pancreatic lipase inhibition, cholesterol micelle formation, cholesterol uptake in enterocytes, and bile acid binding. Moreover, it examined the inhibitory effects of BGB and its fermentation on DPP-IV activity. The findings hold promise for the development of novel fermented juice with potential health benefits, particularly in modulating DPP-IV activity and lipid digestion and absorption.

2. Materials and Methods

2.1. Materials

The BGB was commercially purchased from a reputable Chinese dispensary (Chin Heng Huat, Bangkok, Thailand). The starter culture was a commercial probiotic *Lactobacillus rhamnosus* GG, obtained from Chr. Hansen A/S, Horsholm, Denmark, in freeze-dried direct vat set (DVS) format. Folin–Ciocalteu reagent, 1,1-diphenyl 2-picrylhydrazyl (DPPH), ascorbic acid, 6-hydroxyl- 2,5,7,8-tetramethylchromane-2-carboxylic acid (Trolox), 2-2′-Azino-bis (3-ethylbenzothiazoline-6-sulfonic acid) (ABTS), 2,4,6-tripyridyl-s-triazine (TPTZ), phosphatidylcholine, and oleic acid were purchased from Sigma-Aldrich Chemical (St. Louis, MO, USA). Cholesterol test kits (Cholesterol liquicolor®) were obtained from Human Diagnostics (Wiesbaden, Germany), and the total bile acid kit was purchased from BIOBASE (Jinan, China). The DPP-IV Inhibitor screening assay kit was procured from Cayman Chemical (Ann Arbor, MI, USA), while the *Lactobacillus* MRS agar was obtained from Himedia (Thane, India). Caco-2 cells were obtained from the Sigma-Aldrich (Sydney, NSW, Australia). Ezetimibe was purchased from Merck, Sharp, & Dohme (Kenilworth, NJ, USA). 22-(N–N7-nitrobenz-2-oxa-1,3-diazol-4-yl) amino)-23,24-bisnor-5-cholen-3-ol (NBD-cholesterol) was obtained from Invitrogen (Eugene, OR, USA).

2.2. Preparation of BGB Powder

The extraction method was modified from the procedure outlined in a previous study [17]. Dried fruit was boiled with distilled water at a concentration of 5% (*w/v*) within a temperature range of 95 to 100 °C for 60 min. The resulting aqueous solution of BGB was filtered through Whatman 125 mm filter paper. Subsequently, the sample underwent lyophilization using a freeze-drying technique (Hong Ta Enterprise, Samutprakarn, Thailand). The BGB powder was then stored at −20 °C for further investigation.

2.3. Fermentation

For the fermentation process, prepared BGB powder was employed to create infusions with concentrations of 2.5%, 5%, and 10% *w/v*, supplemented with 5% *w/v* refined sugar. Following pasteurization conditions outlined in a prior study at 65 °C for 30 min [20], LGG freeze-dried granules were directly added to the pasteurized BGB infusion at a concentration of 0.005% *w/v* inoculation to reach the bacterial concentration at 7 log CFU/mL. Then, slow agitation for 30 min ensured even distribution of the culture. BGB samples were transferred in the air-tight glass jar under anaerobic conditions. The incubation temperature was maintained at 37 °C for 24 h in a digital incubator (DAIHAN Scientific, Wonju, Republic of Korea). Samples were taken at 0 and 24 h fermentation for further analysis. The process of lyophilization, using a freeze-drying method, was conducted on fermented samples in preparation for the subsequent cholesterol uptake assay in the Caco-2 cell line.

2.4. Determination of Physicochemical Properties
2.4.1. pH, Total Soluble Solid (TSS), and Color

pH measurements were performed using a pH meter (S40 Seven MultiTM, Metter-Toledo, Switzerland). Total Soluble Solids (TSS) were determined with a hand-held refractometer (HSR-500, Atago Co. Ltd., Tokyo, Japan), and the results were expressed as °Brix. Color measurements were conducted using a colorimeter (CM-3500D, Konica Minolta Co. Ltd., Tokyo, Japan), with results expressed as L^* (Lightness/Brightness), a^* (Greenness to Redness), and b^* (Blueness to Yellowness). To ensure accuracy, all samples underwent triplicate measurements, and the calibration process adhered to the protocols specified for each method.

2.4.2. Total Carbohydrate and Reducing Sugar

Total carbohydrates were determined using the phenol-sulfuric acid method while reducing sugars were analyzed using the DNS method [21]. For total carbohydrate analysis, 400 µL of the sample was mixed with 10 µL of 80% phenol (%w/w) and 1 mL of sulfuric acid. After a 10 min stand and cooling at room temperature, absorbance was read at 490 nm. Total carbohydrate content was calculated using a glucose standard curve and expressed as mg/mL sample.

To analyze reducing sugars, 250 µL of the sample was mixed with 250 µL of DNS reagent (1% 3,5-dinitrosalicylic acid, 0.2% phenol, 0.05% Na_2SO_3, and 1% NaOH in an aqueous solution). The mixtures were heated at 100 °C for 10 min to stop the reaction. Then, 250 µL of 40% potassium sodium tartrate solution was added to stabilize the color, and absorbance was recorded at 540 nm. Glucose solution served as a standard, and the reducing sugar content was expressed as mg glucose equivalents/mL sample.

2.4.3. Analysis of Lactic Acid

The lactic acid concentration was determined using a previously published method [22]. The BGB sample was centrifugated at $6500 \times g$ for 20 min at 4 °C and filtered through a 0.2 µm PVDF syringe filter. Subsequently, 5 µL of the sample was injected into a Nexera UHPLC System (Shimadzu, Kyoto, Japan) equipped with an Inertsil® ODS-3 C_{18} reverse-phase column (250 mm × 4.6 mm, 5 µm) at 35 °C. The mobile phase, consisting of 10 mM KH_2PO_4 (pH 2.4) with phosphoric acid (mobile phase A) and acetonitrile (mobile phase B), flowed at a rate of 0.7 mL/min. The gradient program was set as follows: 0.01–7 min, 25% B; 8–11 min, 50% B; 12 min, 25% B, held until 15 min, with UV–visible detection at 210 nm. Identification of results was achieved by comparing peak retention times with standards, and quantification employed the external standard method.

2.4.4. Total Phenolic Content (TPC)

Briefly, 50 µL of the sample was mixed with Folin–Ciocalteu reagent, which had been diluted $10\times$ in distilled water, and allowed to stand for 10 min. Then, 50 µL of Na_2CO_3 was added, and the absorbance was measured at 760 nm after a 30 min incubation in the dark at room temperature. The results were determined using a standard curve of gallic acid and expressed as g gallic acid equivalent (GAE)/100 mL [23].

2.4.5. Total Anthocyanin Content (TAC)

TAC was determined using the pH differential method [24]. Two buffer systems, 0.025 M potassium chloride at pH 1.0 and 0.4 M sodium acetate at pH 4.5, were prepared. Subsequently, 500 µL of the sample was mixed with each buffer (500 µL), and the absorbance at 520 nm and 700 nm was measured against a distilled water-filled blank cell. The results were expressed as mg cyanidin-3-glucoside equivalent (C3GE)/100 mL. Anthocyanin content (*c*) was calculated following the formula (Equations (1) and (2)):

$$\Delta A = (A_{520nm} - A_{700nm})\text{pH1.0} - (A_{520nm} - A_{700nm})\text{pH4.5} \tag{1}$$

$$c(\text{mg}/100 \text{ mL}) = \left(\Delta A \times MW \times DF \times 10^3\right)/\varepsilon \times 1 \tag{2}$$

where *MW* represents the molecular weight of cyanidin-3-glucoside (449.2 g/mol), *DF* represents the dilution factor, 10^3 represents the factor for conversion from g to mg, ε represents extinction coefficients of cyanidin-3-glucoside (26,900 L/mol·cm), and l represents pathlength in cm.

2.4.6. Total Flavonoid Content (TFC)

TFC was assessed following a published method [25]. A 50 μL sample aliquot was mixed with 10 μL 10% *w/v* AlCl$_3$ solution, 10 μL 1 M sodium acetate, and 150 μL absolute ethanol. After a 30 min incubation, the absorbance was measured at 430 nm, and the results were expressed as mg quercetin equivalent (QE)/100 mL.

2.4.7. Antioxidant Activities
Ferric-Reducing Antioxidant Power (FRAP)

The antioxidant potential of the sample was measured using the FRAP assay [26]. The FRAP reagent, composed of 0.3 M sodium acetate buffer solution (pH 3.6), 10 mM TPTZ solution in 40 mM HCl, and 20 mM FeCl$_3$ solution in a 10:1:1 ratio, was mixed with 10 μL of samples. After a 30 min incubation in the darkness at room temperature, the absorbance was recorded at 595 nm. The results were calculated using a FeSO$_4$ standard curve and expressed as mmol FeSO$_4$/100 mL.

DPPH Radical Scavenging Activity

The DPPH radical scavenging activity was determined using the stable radical DPPH (2,2-diphenyl-1-picrylhydrazyl) as described previously [26]. In summary, 10 μL of the sample was mixed with 90 μL of 0.2 mM DPPH reagent (diluted with ethanol). Ascorbic acid (1 mg/mL) served as the standard. After a 30 min incubation at room temperature, the absorbance was measured at 515 nm. The results were expressed as mg ascorbic acid equivalent (AAE)/100 mL.

Trolox Equivalent Antioxidant Activity (TEAC)

For TEAC, the method adapted from a previous publication that induced the radical anion (ABTS$^{\circ+}$) by adding 2.4 mM potassium persulfate (K$_2$S$_2$O$_8$) and 7 mM ABTS [26]. After 16 h of dark incubation, 10 μL of the sample was mixed with 90 μL of ABTS$^{\circ+}$ solution. The decrease in solution absorbance was measured at 734 nm. The results were calculated using the Trolox standard curve and expressed as mg Trolox equivalent (TE)/100 mL.

2.5. Non-Volatile Metabolite Profiling Using Liquid Chromatography Coupled with High-Resolution Fourier Transform Mass Spectrometry (LC-HRFTMS)

Untargeted metabolomics is a highly utilized approach for studying variations in the metabolite composition of plants. This method involves analyzing all metabolites present in a biological sample without specific targeting, allowing for a comprehensive understanding of the diverse metabolic activities within plants [27]. Using the tools of metabolomics through the employment of LC-HRFTMS is an efficient approach. Goji berry samples underwent Solid Phase Extraction (SPE) to enhance metabolite detection, followed by separation on a Thermo Vanquish Horizon LC system coupled to a Thermo Orbitrap IDX HR-FTMS (Thermo Fisher Scientific, Waltham, MA, USA). MS and MS/MS analyses were conducted in both negative and positive electrospray ionization (ESI) modes. The Kinetex F5 column (2.6 μm, 150 mm × 2.1 mm ID) provided steric selectivity for separating structural isomers. Key settings included a 2 μL injection for MS analysis and 4 μL for MS/MS analysis, a flow rate of 0.4 mL/min, and an oven temperature of 30 °C, with a total LC runtime of 42 min. The mobile phase, composed of 0.1% formic acid, 0.5% Methanol in Milli-Q water (A), and 0.1% formic acid, 2% Milli-Q water, and 40% Acetonitrile in Methanol (B), followed a specific gradient: 0.00–6.25 min, 0–1% B; 6.25–20.00 min, 1.0–7.5% B; 20.00–30.00 min, 7.5–60% B; 30.00–33.00 min, 60–90% B; 33.00–38.00 min, 90% B; 38.00–38.20 min, 90–0% B, and held until 42 min. The ESI interface featured a +3500 V capillary, an ion transfer tube temperature of 275 °C, and a vaporizer temperature of 300 °C. The auxiliary gas flow was set to 7 units with a sheath gas flow of 25 units. This method, paired with the specified MS conditions, enables the capture of a diverse range of slightly polar to non-polar non-volatile molecules. Putative identification of the compounds was assigned based on internal and external spectral databases.

2.6. *Determination of Biological Activities*
2.6.1. Bile Acid Binding Activity

The bile acid binding ability assay was conducted with minor adjustments, following a previous publication [25]. Bile extract (1 mg/mL) in 0.1 M phosphate-buffered saline (PBS), pH 7.4) served as the bile acid in this experiment. Cholestyramine (3 mg/mL) was employed as a positive control. Incubation of the samples with bile extract occurred at 37 °C for 90 min. The resulting mixture underwent filtration through a 0.22 μm syringe filter. The collected filtrate was then analyzed spectrophotometrically at 540 nm using a bile acid analysis kit.

2.6.2. Pancreatic Lipase Inhibitory Activity

The inhibitory effect on pancreatic lipase was investigated following a previously established protocol [25]. The 5 μL of samples were mixed with 45 μL of pancreatic lipase solution (7.5 mg/mL). Subsequently, 50 μL of 4-methylumbelliferone oleate (4-MUO) solution (0.2 mM in 0.1 M PBS, pH 7.0) was added to initiate the enzyme reaction. The incubation was carried out at 37 °C for 20 min. To halt the enzymatic reaction, 100 μL of 0.1 M sodium citrate (pH 4.2) was introduced. Orlistat served as the positive control. The amount of 4-methylumbelliferone by the lipase was quantified using fluorescence spectroscopy with an excitation wavelength of 320 nm and an emission wavelength of 450 nm.

2.6.3. Cholesterol Micellization Inhibitory Activity

To establish a cholesterol solubilization model, artificial micelles were generated based on a previously conducted study [25]. In brief, a mixture of 2 mM cholesterol, 1 mM oleic acid, and 2.4 mM phosphatidylcholine in methanol was prepared. The mixture was dried under nitrogen gas. Then, 15 mM PBS (pH 7.4) with 6.6 mM taurocholate salt was redissolved. The mixture underwent sonication before incubating at 37 °C overnight. Following this, BGB samples were introduced to the artificial micelle solution and incubated for 120 min at 37 °C. Cholesterol content was assessed using cholesterol test kits, with gallic acid employed as the positive control. The results were expressed as the percentage inhibition of cholesterol micellization.

2.6.4. Dipeptidyl Peptidase-IV (DPP-IV) Inhibitory Activity

DPP-IV inhibitory activity was assessed utilizing a DPP-IV inhibitor screening assay kit, employing a fluorescence-based method for screening inhibitors. In brief, 10 μL of the samples were added to a 96-well plate containing 30 μL of diluted assay buffer and 10 μL of diluted DPP(IV). The reaction was initiated by adding 50 μL of diluted substrate solution, followed by incubation at 37 °C for 30 min. The fluorescence intensity, with an excitation of 355 nm and an emission wavelength of 460 nm, was monitored using a spectrofluorometer. Sitagliptin was used as the positive control.

2.7. Microbiological Analysis

The microbiological analysis for this study was based on the U.S. Food and Drug Administration's (U.S. FDA) Laboratory Method [28]. The number of microorganisms in the products was evaluated using the aerobic plate count (APC) technique. Viable cells of lactic acid bacteria were quantified using MRS (De Man, Rogosa, and Sharpe) agar. Samples were diluted with peptone solution (0.1% w/v). All equipment and materials were sterilized using an autoclave at 121 °C for 15 min. The dilutions were shaken and subjected to the pour plate technique using warm MRS agar. Plates were incubated at 37 °C for 48 h and colonies were counted using a colony counter.

2.8. Caco-2 Cell Culture

2.8.1. Cell Viability

The cytotoxic effects of BGB were evaluated using the 3-(4,5-dimethylthiazol-2-yl)-2,5-diphenyltetrazolium bromide (MTT) colorimetric assay. Caco-2 cells, a human epithelial colorectal adenocarcinoma cell line, were employed for this study. Cells were seeded in 96-well plates at a density of 2×10^4 cells/mL and incubated overnight in Dulbecco's Modified Eagle's Medium (DMEM) supplemented with 10% fetal bovine serum (FBS), 1% non-essential amino acid (NEAA), and 1% penicillin–streptomycin solution (Complete medium). Following adherence, the cells were exposed to varying concentrations of BGB (ranging from 0.025 to 2 mg/mL) or vehicle control for 24 h. Subsequently, the culture medium was removed, and the cells were incubated with 0.25 mg/mL MTT solution (prepared in serum-free DMEM) for 3 h at 37 °C in a humidified atmosphere containing 5% CO_2. The formazan crystals formed by the reduction of MTT by metabolically active cells were solubilized by adding 100 µL of dimethyl sulfoxide (DMSO) to each well. The absorbance was measured at 570 nm using a microplate reader. The results were presented as the percentage of cell viability compared to the untreated cells (% of control).

2.8.2. Cholesterol Uptake in Caco-2 Cells

Caco-2 cells were maintained in complete DMEM high glucose. Cells were passaged every 3–4 days upon reaching approximately 80% confluence. Cells were seeded onto 24-well plates at a density of 25,000 cells per well and then incubated at 37 °C for 7 days to allow for differentiation. During this period, the medium in the well was replaced every 2 days. Experiments were carried out following the previous publication with slight modifications [29]. After the differentiation period, cells were starved with serum-free DMEM low glucose for 24 h. Subsequently, cells underwent two washes with Hanks' balanced salt solution (HBSS; composed of 140 mM NaCl, 5 mM KCl, 1.2 mM Na_2HPO_4, 2 mM $CaCl_2$, 1.2 mM $MgSO_4$, 20 mM HEPES, and 0.2% bovine serum albumin, pH 7.4) and were then incubated in HBSS at 37 °C 1 h before the experiments. After incubation, cells were treated with free cholesterol (containing 0.5 mM taurocholate salt and 25 µM NBD-cholesterol) supplemented with varying concentrations of BGB (0.05–0.5 mg/mL). Ezetimibe (50 µM) served as a positive control. Cells underwent five washes with cold HBSS, and the fluorescence was measured at an excitation wavelength of 485 nm and an emission wavelength of 535 nm. Cell lysis was performed using a buffer solution (composed of 10 mM tris-HCl pH 7.4, 150 mM NaCl, 1% Triton-X-100, 1 mM EDTA, and 0.1% SDS), employing two freeze–thaw cycles. Lysates were then centrifuged at 12,000 rpm at 4 °C for 10 min, and supernatants were collected for protein concentration determination using a BCA kit (Thermo Fisher Scientific, Waltham, MA, USA), with bovine serum albumin as the standard. Total protein quantification enabled normalization based on the total number of cells used. Results were expressed as the percentage of cholesterol uptake relative to control values.

2.9. Sensory Evaluation

The sensory quality of the samples was assessed by a panel of 50 untrained individuals, both male and female, aged between 18 and 50. The research protocol received approval from the Research Ethics Review Committee for Research Involving Human Research Participants, Group 1, Chulalongkorn University (COA No. 234/65), and all participants provided written informed consent before the evaluation. Utilizing a 9-point hedonic scale, ranging from 1 (disliked extremely) to 9 (liked extremely), the sensory analysis evaluated BGB juice and its fermentation at various concentrations. Samples were assigned random 3-digit numbers and presented to panelists in a sensory evaluation laboratory. The panelists were guided on the use of the hedonic scale, instructed to cleanse their palate with water and salted crackers between samples, and assessed for color, appearance, odor, taste, sweetness, sourness, off-flavor, and overall acceptability. The sensory test occurred in individual booths with ample ventilation, under white light, and at room temperature.

2.10. Statistical Analyses

The results were expressed as mean values $\pm$ SEM, $n = 3$. Data were analyzed using the SPSS 21 statistical program (SPSS, Inc., Chicago, IL, USA). To identify any significant differences among treatments, a one-way analysis of variance (ANOVA) was performed at a 95% confidence level. Duncan's multiple range test was employed to discriminate among the means of the various factors. The paired *t*-test was used to compare the differences in parameters before and after fermentation. Visualizations were created using Sigma Plot (version 12.0) and GraphPad Prism (version 9.5.1). The principal component analysis and volcano plot were analyzed using R (version 4.2.3).

3. Results and Discussion

3.1. Alterations in Physicochemical Properties Following Fermentation of BGB

The results indicate a substantial decrease in the pH value of the fermented juice, as detailed in Table 1. Prior to fermentation, the initial pH of the BGB juice ranged from 5.24 to 5.37. After 24 h fermentation at all treatment concentrations, there was a noticeable decrease in pH, falling within the range of 3.87 to 3.93. Before the fermentation process, an inverse relationship was observed between the concentration of BGB and the pH value of the mixture. As the concentration of BGB increased, the pH exhibited a decreasing trend. This can be attributed to the inherent acidity of the BGB, primarily due to its content of organic acids [19]. The higher concentration of these acidic compounds contributed to a lower pH, rendering the mixture more acidic. Furthermore, the concentration of BGB had a directly proportional effect on the TSS, reducing sugars, and total carbohydrate content. With an increase in BGB concentration, a significant elevation in these parameters was observed. This can be ascribed to the natural sugars present in BGB, such as glucose and fructose [19], becoming more concentrated as the mixture thickened. The higher concentration of these saccharides contributed to an increase in TSS, reducing sugars, and total carbohydrate levels.

Table 1. Physiochemical parameters of different concentrations of BGB juice at 0 and 24 h fermentation.

Samples	pH	TSS (°Brix)	Total Sugar (mg/mL)	Reducing Sugar (mg/mL)	Lactic Acid (mg/mL)	L*	a*	b*	Bacterial Enumeration (LogCFU/mL)
BGB 2.5%, 0 h	5.37 ± 0.01 Aa	7.88 ± 0.02 Ca	65.84 ± 1.21 Ca	21.46 ± 0.17 Ca	N.D.	0.77 ± 0.08 Ab	0.45 ± 0.06 Ab	0.43 ± 0.05 Bb	7.43 ± 0.01 Bb
BGB 2.5%, 24 h	3.88 ± 0.01 Bb	7.56 ± 0.01 Cb	57.31 ± 1.41 Cb	19.60 ± 0.26 Cb	2.66 ± 0.03 C	1.95 ± 0.10 Aa	3.30 ± 0.06 Aa	1.07 ± 0.06 Aa	8.52 ± 0.07 Aa
BGB 5%, 0 h	5.32 ± 0.01 Ba	10.22 ± 0.03 Ba	73.64 ± 0.92 Ba	46.48 ± 0.51 Ba	N.D.	0.33 ± 0.03 Bb	0.16 ± 0.03 Bb	0.24 ± 0.05 Ab	7.46 ± 0.01 Bb
BGB 5%, 24 h	3.87 ± 0.00 Bb	9.86 ± 0.03 Bb	65.39 ± 1.43 Bb	40.08 ± 0.05 Bb	5.17 ± 0.35 B	0.78 ± 0.01 Ba	2.10 ± 0.03 Ba	0.55 ± 0.03 Ba	8.75 ± 0.02 Aa
BGB 10%, 0 h	5.24 ± 0.01 Ca	14.36 ± 0.05 Aa	92.36 ± 1.31 Aa	95.37 ± 1.38 Aa	N.D.	0.21 ± 0.02 Bb	0.07 ± 0.09 Bb	0.25 ± 0.05 Ab	7.45 ± 0.00 Bb
BGB 10%, 24 h	3.93 ± 0.01 Ab	13.90 ± 0.06 Ab	80.27 ± 1.12 Ab	82.88 ± 0.46 Ab	8.11 ± 0.67 A	0.54 ± 0.01 Ca	1.35 ± 0.11 Ca	0.27 ± 0.01 Ca	8.74 ± 0.08 Aa

The results are expressed as mean $\pm$ S.E.M. ($n = 3$). Means with different uppercase letters at the same time point (A–C: treatment effects) and lowercase letters at the same treatment (a,b: time effects) are significantly different ($p < 0.05$). BGB: Black Goji Berry extract; TSS: Total Soluble Solids; N.D.: Not detected.

Concurrently, the concentration of lactic acid after 24 h fermentation exhibited a significant increase for all BGB concentrations, with values ranging from 2.66 $\pm$ 0.03 to

8.11 $\pm$ 0.67 mg/mL. This decline in pH and increase in lactic acid concentration can be attributed to the heightened production of lactic acid, a metabolic by-product of LGG via the homofermentative pathway [30]. Lactic acid is generated through the metabolic utilization of saccharides present in the BGB. Consequently, the fermentation process with LGG led to a noteworthy reduction in total soluble solids (TSS), as well as total and reducing sugar content across all BGB concentrations. These findings align with a previous study where the fermentation of goji berry juice (*Lycium barbarum* L.) occurred in the presence of various lactic acid bacteria mixtures. The results suggest that diverse bacterial strains directly metabolized reducing sugars, resulting in the production of lactic acid and other chemical compounds through a sequence of biochemical reactions [31]. However, the concentration of BGB did not exhibit a significant impact on the physicochemical properties of the mixture following the fermentation process, suggesting that the changes observed were predominantly influenced by the fermentation conditions rather than the initial BGB concentration.

The analysis of color parameters revealed a significant increase in L^*, a^*, and b^* values across all concentrations of BGB before and after 24 h of fermentation (Table 1). The elevation in the L^* value may be attributed to the breakdown of juice components, particularly anthocyanin pigments, during fermentation, resulting in a lighter color. Similar observations were reported in a previous study where pomelo fermentation by lactobacilli led to an increase in the lightness value, potentially enhancing consumer perception and acceptance of the product [32]. The increase in color intensity within the red hue (a^*) can be linked to the acidic conditions produced during fermentation, conducive to the alteration in the formation of anthocyanin pigments. Theoretically, the chemical structure of anthocyanin is pH-dependent. Under acidic conditions, particularly at pH = 1, anthocyanin exists in the form of a flavylium cation, contributing to the production of red and purple colors. This elucidates the significant increase in a^* values observed after fermentation, concurrently with the reduction in pH value [33]. As the pH increases within the range of 2–4, the quinoidal blue species becomes prevalent. Additionally, at a pH between 5 and 6, colorless carbinol pseudobase and a chalcone appear. Given that anthocyanin pigments exhibit a reddish color in acidic conditions.

3.2. Alterations in Viable Lactobacilli Following Fermentation of BGB

This study presents the changes in the viable count of lactobacilli before and after 24 h of fermentation, as detailed in Table 1. Before fermentation, the initial viable cell counts ranged from 7.43 to 7.46 $\log_{10}$ CFU/mL, with similar counts observed in all concentrations of BGB. Following 24 h fermentation, a significant increase in viable counts was noted, reaching 8.52, 8.75, and 8.74 $\log_{10}$ CFU/mL in BGB concentrations of 2.5%, 5%, and 10%, respectively. No significant variation was observed among the different concentrations.

Several factors influence the viability of probiotics including pH, titratable acidity, oxygen level, water activity, salt, and sugar content. Therefore, the presence of sugar in BGB may exert a substantial impact on the growth and viability of probiotic bacteria. As evidenced by the previous study, a relationship was observed between the survival of LGG in the presence of glucose and its ability to utilize sugars. Thus, LGG demonstrated an ability to utilize carbohydrate components, thereby contributing to its overall survivability [34].

Consequently, the findings suggest that BGB serves as a viable probiotic carrier for *L. rhamnosus* GG bacteria, exhibiting substantial proliferation within the juice matrix. The achieved viable cell counts surpassed the recommended threshold for conferring health benefits, falling within the range of log 6–log 7 of probiotic bacteria per mL or gram of food [35].

3.3. Alterations in Phytochemical Composition Following Fermentation of BGB

The impact of LGG fermentation on phytochemical compounds and antioxidant activities in BGB is outlined in Table 2. TPC, TAC, and TFC levels exhibited a concentration-dependent trend, peaking at a 10% concentration of BGB. While TPC levels remained constant after fermentation across all treatments, TFC levels showed a significant increase, and TAC experienced a significant decrease after fermentation. The consistent TPC trend aligns with findings from a study on camu-camu fruit, where TPC levels remained stable following a 72 h fermentation period [36]. However, a decline at 24 h was noted, potentially due to initial phenolic utilization and rearrangement into polymeric forms, followed by the release of soluble free phenolic compounds at 48 h.

Table 2. Biochemical and Antioxidant parameters of different concentrations of BGB beverage at 0 and 24 h fermentation.

Samples	TPC (g GAE/100 mL)	TAC (mg C3GE/100 mL)	TFC (mg QE/100 mL)	FRAP (mmol FeSO$_4$/100 mL)	DPPH (mg AAE/100 mL)	TEAC (mg TE/100 mL)
BGB 2.5%, 0 h	0.13 ± 0.00 [Ca]	6.84 ± 0.12 [Ca]	2.21 ± 0.16 [Cb]	1.18 ± 0.02 [Ca]	67.55 ± 1.92 [Ca]	637.70 ± 27.94 [Ca]
BGB 2.5%, 24 h	0.13 ± 0.00 [Ca]	6.17 ± 0.07 [Cb]	4.43 ± 0.11 [Ca]	1.22 ± 0.02 [Ca]	68.37 ± 1.92 [Ca]	637.22 ± 18.94 [Ca]
BGB 5%, 0 h	0.25 ± 0.00 [Ba]	12.77 ± 0.10 [Ba]	5.76 ± 0.25 [Bb]	2.36 ± 0.04 [Ba]	119.88 ± 0.88 [Ba]	810.99 ± 30.59 [Ba]
BGB 5%, 24 h	0.25 ± 0.00 [Ba]	11.40 ± 0.11 [Bb]	7.75 ± 0.21 [Ba]	2.38 ± 0.02 [Ba]	121.42 ± 2.43 [Ba]	809.17 ± 18.92 [Ba]
BGB 10%, 0 h	0.41 ± 0.00 [Aa]	22.79 ± 0.28 [Aa]	9.42 ± 0.05 [Ab]	4.47 ± 0.01 [Aa]	230.77 ± 5.30 [Ab]	1158.33 ± 22.69 [Aa]
BGB 10%, 24 h	0.42 ± 0.00 [Aa]	18.92 ± 0.09 [Ab]	10.36 ± 0.18 [Aa]	4.59 ± 0.05 [Aa]	240.19 ± 3.31 [Aa]	1160.86 ± 15.55 [Aa]

The results are expressed as mean ± S.E.M. (n = 3). Means with different uppercase letters at the same time point (A–C: treatment effects) and lowercase letters at the same treatment (a,b: time effects) are significantly different ($p < 0.05$). TPC: Total phenolic content; TAC: Total Anthocyanin content; TFC: Total Flavonoid content; FRAP: Ferric-Reducing Antioxidant Power; DPPH: DPPH radical scavenging activity; TEAC: Trolox Equivalent Antioxidant Activity; GAE: Gallic acid equivalent; C3GE: Cyanidin-3-glucoside equivalent; QE: Quercetin equivalent; AAE: Ascorbic Acid Equivalent; TE: Trolox Equivalent.

The decline in anthocyanin observed may be attributed to the enzymatic fermentation process, particularly involving β-glucosidase. This enzymatic activity hydrolyses the β-1,4-glycosidic bond and leads to a decrease in detectable anthocyanins, facilitating their conversion into the main phenolic acid product [37]. This aligns with studies on red cabbage sprouts and blueberry/blackberry juices [38], where probiotic fermentation resulted in reduced anthocyanin content and an upgraded trend in certain phenolic acids such as syringic acid, ferulic acid, and gallic acid [39]. An increase in TFC across all treatments suggests flavonoid glycosides may undergo degradation during fermentation, or the production of flavonoids may arise from the degradation of complex polyphenols [40]. This is consistent with a study on jujube juice fermented with LGG and *L. plantarum-1*, where total flavonoid content significantly increased [41]. Further investigations into the kinetic changes in *L. rhamnosus* GG activity within BGB are warranted.

3.4. Alterations in Antioxidant Activities Following Fermentation of BGB

The antioxidant activities of BGB before and after fermentation are presented in Table 2, indicating a concentration-dependent increase, with the highest observed at 10% BGB. This variation is influenced by factors such as environmental conditions, fruit varieties, and extraction methods [19]. Flavonoids, polysaccharides, carotenoids, and AA-2βG contribute to BGB's antioxidant activities, functioning through mechanisms like radical scavenging, metal chelation, and interactions with other antioxidants [42,43]. Despite these variations, no significant differences were observed in antioxidant activity parameters after fermentation. This consistent pattern aligns with the stability trend of TPC. Similar trends in TPC and the antioxidant activity during the fermentation of fruit matrices with potential probiotics have been reported. For instance, the fermentation of camu-camu fruit by lactic acid bacteria showed no changes in total soluble phenolic content and antioxidant activity [36]. In contrast, the fermentation of goji berry by multiple strains of probiotics resulted in a significant increase in TPC and antioxidant activity. Various antioxidant indexes correlated with the concentrations of both free and bound forms of phenolic compounds [31]. Therefore, the sustained high antioxidant capacity, potentially associated with health benefits, was maintained after fermentation with LGG.

3.5. Alterations in Biological Properties Following Fermentation of BGB

The biological activity of fermented BGB, including bile acid binding, inhibition of cholesterol micellization, pancreatic lipase activity, and DPP-IV, is depicted in Figure 1. All BGB concentrations showed the ability to bind primary bile acid, with a significant increase of 38.24%, 28.03%, and 20.81% in BGB 2.5%, 5%, and 10% after fermentation (Figure 1A). The binding values of fermented samples were comparable to cholestyramine at a concentration of 3 mg/mL. This suggests that fermented BGB may disrupt the endogenous bile acid pool, potentially stimulating bile acid synthesis from cholesterol, and contributing to reduced blood cholesterol levels [44]. The phytochemical components of BGB, especially the increasing of flavonoids, might interact with bile acids through ionic, hydrogen, and hydrophobic interactions, forming insoluble polyphenol–bile acid complexes and increasing fecal bile excretion, disrupting micelle formation [44]. Additionally, a previous study observed a high bile acid adsorption capacity in flavonoid-rich lupin cotyledons (*Lupinus angustifolius* L.) and suggested the formation of hydrophobic interactions between polyphenols and bile acid [45].

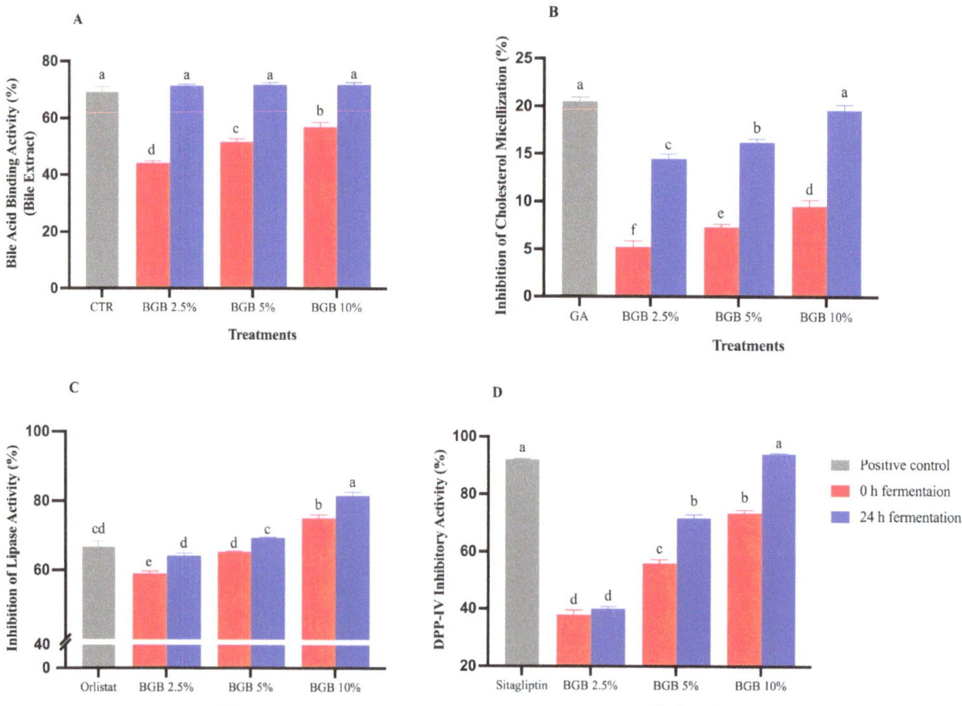

Figure 1. Potential biological activities of BGB beverage at 0 and 24 h of fermentation: The percentage of bile acid binding (**A**), the inhibition of cholesterol micellization (**B**), the inhibition of lipase activity (**C**), and DPP-IV inhibitory activity (**D**). The results were expressed as mean ± S.E.M. ($n = 3$). Significant differences are presented with different superscripted letters ($p < 0.05$). CTR: Cholestyramine; GA: Gallic acid.

In cholesterol absorption, micelles are formed in the intestine by bile salts, cholesterol, and phospholipids. Our study reveals a significant increase in the inhibition of cholesterol micellization formation after fermentation across all concentrations (Figure 1B). Similar findings were observed in the fermentation of gac fruit beverage with lactobacilli, demonstrating an increased capacity to disrupt cholesterol micellization formation [22]. In general, probiotics employ enzymes like hydroxysteroid dehydrogenase and conjugated bile acid hydrolase to break down bile acids and hydrolyze bile salts, disrupting the enterohepatic circulation of bile acids. Probiotic bacteria also reduce cholesterol absorption by binding it and incorporating it into the cell membrane, playing a preventive role in micelle production [46].

After fermentation, BGB significantly enhanced the concentration-dependent inhibition of pancreatic lipase across all treatments after fermentation (Figure 1C). This inhibitory effect may be attributed to polyphenols in BGB, forming complexes with pancreatic lipase and impairing its enzymatic activity [47]. Particularly, flavonoids have the potential to disrupt the structural integrity of lipase enzymes and diminish the substrate (olein) affinity for the enzyme. Consequently, the activity of the lipase was reduced [48]. Fermented BGB, through binding bile acid and inhibiting cholesterol micellization and pancreatic lipase activity, contributes to lipid-lowering effects. Fermented BGB exhibited concentration-dependent inhibition of DPP-IV, with notable increases observed at 5% and 10% BGB concentrations (Figure 1D). The inhibitory capability of fermented 10% BGB was comparable to sitagliptin (100 μM), an antidiabetic drug. This may result from hydrogen bonds formed between amino acid residues in DPP-IV and polyphenols present in fermented BGB [49]. Additionally, *Lactobacillus* spp. strains, especially *L. rhamnosus*, are reported to have DPP-IV inhibitory activity [50]. The fermentation of BGB contributes to enhanced DPP-IV inhibitory activity, a phenomenon modulated by the presence of bioactive components and the viability of probiotics. Theoretically, DPP-IV is an enzyme involved in the degradation of endogenous GLP-1. The inhibition effect on this enzyme may lead to an increase in active GLP-1, while concurrently reducing GLP-1 clearance. This results in a lowering of both fasting and postprandial glucose concentration, thereby offering potential benefits for individuals experiencing hyperglycemic conditions [51].

3.6. Effect of BGB on Cell Viability and Cholesterol Uptake in Caco-2 Cells

The MTT assay evaluated the cytotoxicity of non-fermented and fermented BGB on Caco-2 cells. As shown in Figure 2A,B, both samples exhibited no significant cytotoxicity at concentrations ranging from 0.025 to 2 mg/mL. The non-cytotoxic range identified was considered appropriate for subsequent cholesterol uptake assays, ensuring that any observed effects were not influenced by compromised cell viability.

The impact of BGB on the uptake of cholesterol into Caco-2 cells is illustrated in Figure 2C. Fermented BGB exhibited a significant reduction in cholesterol uptake compared to the baseline samples at 0 h. This reduction was observed in a concentration-dependent manner, with the highest reduction recorded at 16.94% with 0.5 mg/mL of fermented BGB. Interestingly, this reduction was comparable to the effect observed with 50 μM of Ezetimibe, a known cholesterol absorption inhibitor, which resulted in a reduction of 18.86%. The process of cholesterol uptake in cells is mediated by the Niemann–Pick C1-Like 1 (NPC1L1) protein transporter situated on the apical membrane of enterocytes [52]. Based on our findings, the observed increase in flavonoid content after fermentation may contribute to the reduction in cholesterol uptake in intestinal cells. Previous studies have demonstrated that pre-incubation of Caco-2 cells with flavonoids can lead to a concentration-dependent reduction in cholesterol uptake, thereby diminishing cholesterol absorption by influencing the intestinal epithelial cells [53]. LGG itself exhibits a capacity to reduce cholesterol uptake [54]. Furthermore, the bioconversion process during fermentation with LGG can alter the phytochemical composition of the samples. Notably, certain polyphenolic compounds have been reported to reduce cholesterol uptake by inhibiting the NPC1L1. These polyphenols include luteolin, quercetin, catechin, epigallocatechin gallate, and chlorogenic acid [55]. Interestingly, our analysis revealed an increase in the levels of these polyphenols after 24 h fermentation (Table S1).

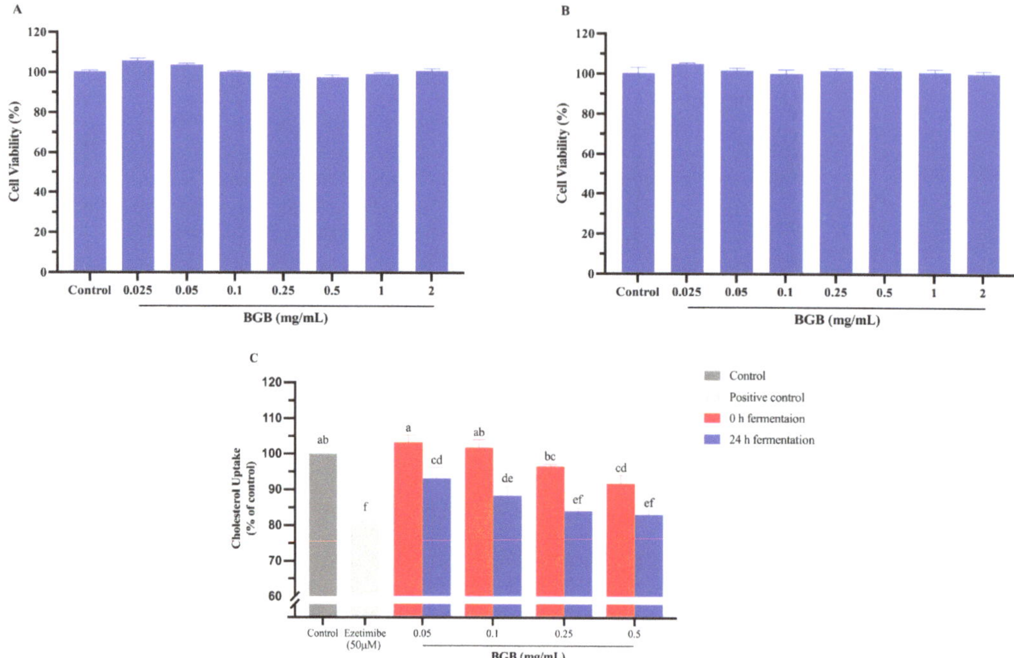

Figure 2. Effect of BGB on Caco-2 cells line: Cell viability after the treatment of cell with BGB at 0 h fermentation (**A**) and the treatment of cell with BGB at 24 h fermentation (**B**), and the impact of BGB on cholesterol uptake in Caco-2 cells (**C**). The results are presented as mean $\pm$ SEM (n = 3). Significant differences are presented with different superscripted letters ($p < 0.05$).

3.7. Non-Volatile Compound Profiling

A total of 718 compounds were detected in the samples by Reverse-Phase High-Performance Liquid Chromatography–High-Resolution Fourier Transform Mass Spectrometry (HPLC-HRFTMS). Fold changes (expressed as Log_2 fold changes) and ratios between the samples before and after fermentation were calculated to allow comparisons among samples. Adjusted p-values were also calculated to allow for statistical comparisons between the two samples. Based on our analysis using LCMS/MS, we detected a total of 339 and 379 metabolites in negative- and positive-ion modes, respectively. These metabolites belong to various classes of chemical compounds, as classified based on chemical taxonomy. The phytochemical profile of BGB comprises an array of bioactive compounds (Table S1), including flavonoids such as catechin, naringenin, and prunin, as well as anthocyanins, predominantly cyanidin. Additionally, BGB contains various phenolic acids, including gallic acid, gentisic acid, sinapinic acid, and caffeic acid. Figure 3A illustrates the quantitative distribution of the metabolites, with the top three chemical classes being organic heterocyclic compounds, amino acids, peptides, proteins, and organic acids. However, our primary focus was on investigating the correlation of phenolic compounds throughout the fermentation process.

Metabolites exhibiting significant differences (adjusted $p < 0.05$) are detailed in Table S1. The negative log_2 fold change indicates an upregulation of compounds after 24 h of fermentation. Notably, a majority of the phenolics of interest (24 out of 44 compounds) demonstrate a significant increase after fermentation, including gallic acid, gentisic acid, chlorogenic acid, caffeic acid, sinapinic acid, protocatechuic acid, and hesperitin (Table S1). As depicted in Figure 3B, p-coumaric acid experienced the most pronounced decrease after BGB fermentation. This alteration is attributed to the activity of hydroxycinnamic reductase from *Lactobacillus* spp. These findings are consistent with previous studies on *L. plantarum* fermentation, which observed the conversion of phenolic acids (such as caffeic acid, *p*-coumaric acid, and ferulic acid) into caffeic acid, epicatechin, catechin, and rosmarinic acid [56].

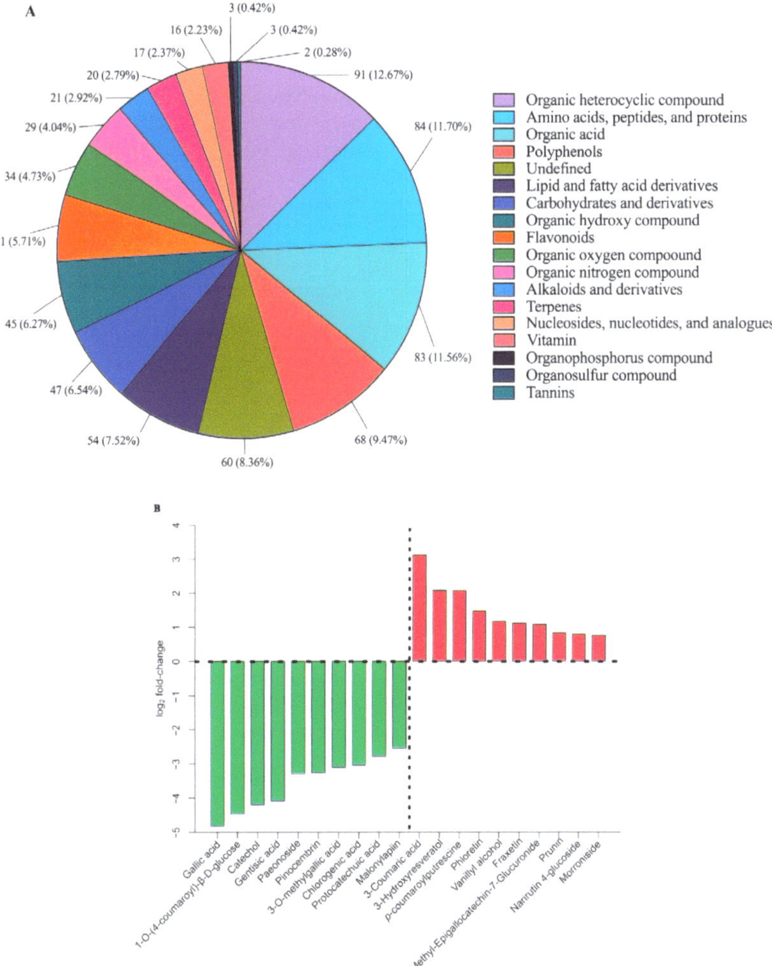

Figure 3. The quantitative distribution of identified metabolites across distinct chemical classes. Different color blocks represent different chemical classifications, the percentage denotes the proportion of metabolites within that classification relative to the overall metabolites. Unassigned metabolites were categorized as undefined (**A**) and Top 10 phytochemical compounds with the largest fold change. Green and red bars represent upregulated and downregulated metabolites after fermentation, respectively (**B**).

Figure 4 highlights the interesting phenolic compounds with yellow circles. The increase in the fold change of phenolic acids and flavonoids can be attributed to the bio-transformation properties exhibited by probiotics. These properties enable the utilization of phenolic compounds, which are subsequently converted into compounds often displaying greater bioactivity than their parent compounds [57]. While phenolic acids showed increasing trends after 24 h fermentation, other chemical compounds in BGB may be responsible for their antioxidant activities. Specifically, certain amino acids (tyrosine, tryptophan, methionine, lysine, cysteine, and histidine) and dipeptides containing these amino acid moieties could contribute significantly to the observed antioxidant effects. Antioxidant-active peptides typically consist of 5–16 amino acid residues and are known to inhibit lipid peroxidation, scavenge free radicals, and chelate transition metal ions [58]. Our findings indicated a reduction in certain antioxidant-active amino acids and peptides, such as tyrosine, tryptophan, and valine-tryptophan (Table S1). This alteration may be attributed to the utilization of the LGG where amino acids are potentially utilized as an energy source [59]. Consequently, the consistent antioxidant activities observed after bacteria fermentation may be attributed to the decrease in specific amino acids and the concurrent increase in phenolic contents, including phenolic acids and flavonoids. It is plausible that the balanced increase and decrease in these compounds contribute to the maintenance of antioxidant activity levels despite individual fluctuations.

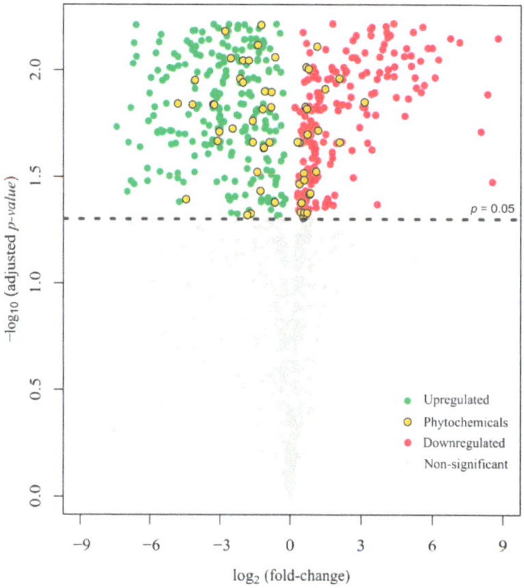

Figure 4. Volcano plot showing log$_2$ fold change and adjusted *p*-values of metabolite compounds quantified using LC-HRFTMS. Negative log$_2$ fold change indicates the increment of compounds. Green circles represent upregulated metabolites and red circles represent downregulated metabolites after fermentation. Yellow circles represent interested phenolic compounds.

Additionally, a decrease in anthocyanin was observed after BGB fermentation. This could be attributed to the degradation of anthocyanin glucoside into its aglycone forms and free phenolic acids. Similar findings were reported in a previous study, where fermentation of jussara pulp by lactobacilli led to the conversion of cyanidin-3-glucoside and cyanidin-3-rutinoside into the main phenolic acid product, protocatechuic acid [60]. Furthermore, phenolic metabolites might influence the growth and metabolism of probiotics. A previous study showed that the flavanol catechin stimulated the growth of *L. plantarum* by facilitating expedited sugar consumption, increasing sugar utilization, and triggering malic acid decarboxylation [61]. Therefore, alterations in phenolic components could potentially stimulate the growth of probiotics in BGB.

3.8. Sensory Evaluation

The sensory evaluation of BGB samples across various attributes is depicted in Table S2. BGB presents a unique blend of sweet, tangy, and pungent flavors, making it highly attractive to consumers. However, there is an inverse correlation between acceptability scores and BGB concentration. Higher concentrations of BGB resulted in a significant reduction in almost all evaluated attributes, including appearance, color, taste, sweetness, sourness, off-flavor, and overall acceptability. This trend may be attributed to the concentration-dependent nature of phenolic compounds (TPC, TFC, and TAC) in BGB samples. Polyphenols also contribute to sensory characteristics such as color, flavor, odor, astringency, and bitterness [62]. Consequently, the increased presence of phenolic components in BGB may be associated with perceptions of bitterness and astringency, leading to reduced consumer acceptance.

Conversely, sensory acceptability significantly improved after fermentation. The 10% BGB initially obtained the lowest acceptability score before fermentation. However, after fermentation, its attribute scores improved and became comparable to those of the 2.5% and 5% BGB across all parameters. Due to lactic acid production by lactobacilli, the sourness score exhibited a significant increase. In essence, the fermentation of BGB has the potential to enhance sensory acceptability scores across various attributes, including appearance, color, odor, taste, sweetness, sourness, off-flavor, and overall acceptability.

3.9. Principal Component Analysis (PCA)

The principal component analysis (PCA) depicted in Figure 5 reveals that two principal components (PCs) accounted for over 95% of the variance. PC1 explained 76.6% and PC2 explained 19.6% of the total sample variance. Along PC1, variance was primarily associated with sample concentration, while along PC2, variance reflected changes in parameters between 0 and 24 h fermentation. Variable loadings and the biplot indicated that DPP-IV inhibition and FRAP were the key parameters contributing to PC1. The concentration of BGB positively correlated with DPP-IV inhibition and FRAP. Conversely, bile extract binding ability and sensory acceptability were the main contributors to PC2, albeit in a partially opposite direction to FRAP. Fermented samples exhibited higher bile acid binding and sensory acceptability, as depicted in Figure 1A and Table S2. This analysis reinforces the observed trends in parameter concentrations and the impact of fermentation.

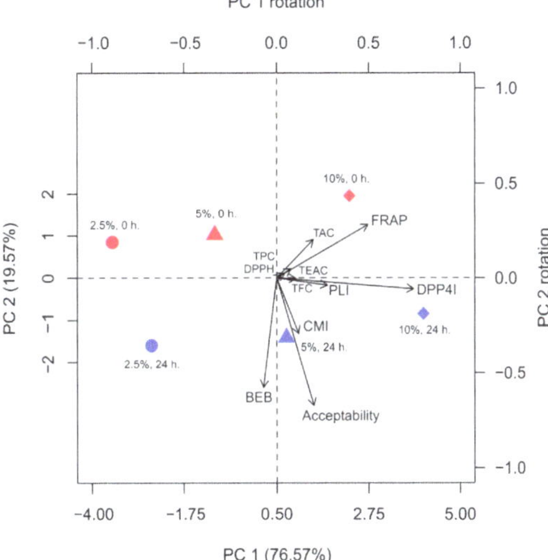

Figure 5. Principal component analysis (PCA) and biplot of BGB samples at 0 and 24 h of fermentation. Principal axes were calculated from 3 groups of parameters: phytochemical, biological, and sensory acceptability. TPC: Total phenolic content; TFC: Total flavonoid content; TAC: Total anthocyanin content; DPPH: DPPH radical scavenging activity; FRAP: Ferric-reducing antioxidant power; TEAC: Trolox equivalent antioxidant capacity; CMI: Cholesterol micellization inhibition; PLI: Pancreatic lipase inhibition; BEB: Bile extract binding; DPP4I: Dipeptidyl peptidase-IV inhibition.

4. Conclusions

BGB is an effective carrier for LGG, enhancing lactic acid production and probiotic viability over 24 h of fermentation. Fermentation reduced pH, total soluble solids (TSS), total carbohydrates, and reducing sugars, turning BGB purple due to increased acidity. Biochemically, TPC remained stable, maintaining consistent antioxidant activity. However, TFC increased, while TAC decreased post-fermentation. The fermentation process alters BGB's chemical composition, resulting in heightened levels of phenolic acids like gallic acid and chlorogenic acid, as well as flavonoids, amino acids, and dipeptides. Notably, fermented BGB exhibits enhanced functional properties, including improved bile extract binding, inhibition of cholesterol micellization and pancreatic lipase, and decreased cholesterol uptake in intestinal cells without any cytotoxic effects. These findings suggest potential benefits for managing hyperlipidemia by targeting lipid digestion and absorption. Additionally, fermented BGB shows promise in inhibiting dipeptidyl peptidase-IV (DPP-IV), indicating its potential role in managing hyperglycemia. Additionally, fermented BGB showed an improvement in sensory quality. Overall, these results highlight the potential of LGG fermentation to enhance the functional properties of BGB, paving the way for the development of innovative functional foods aimed at addressing hyperlipidemia and hyperglycemia.

Supplementary Materials: The following supporting information can be downloaded at: https://www.mdpi.com/article/10.3390/antiox13060740/s1, Table S1: Non-volatile compounds identified in BGB beverage at 0 and 24 h of fermentation; Table S2: Sensory acceptability of BGB beverage at 0 and 24 h of fermentation.

Author Contributions: Conceptualization, Methodology, and Formal analysis, K.K. and S.A.; Validation, Investigation, Visualization, and Writing—original draft, K.K.; Investigation, V.B., M.M. and C.C.; Supervision, S.C., S.S. and S.A.; Writing—review and editing, T.T. and S.A. All authors have read and agreed to the published version of the manuscript.

Funding: This research project was supported by the 90th Anniversary of Chulalongkorn University Scholarship [Ratchadaphiseksomphot Endowment Fund; Grant Number GCUGR1125661052D] and National Research Council of Thailand (NRCT); Grant Number N42A640325, Thailand.

Institutional Review Board Statement: This study was conducted in accordance with the Declaration of Helsinki and approved by the Research Ethics Review Committee for Research Involving Human Subjects at Chulalongkorn University (COA No. 234/2565; 7 December 2022).

Informed Consent Statement: Informed consent was obtained from all subjects involved in the study.

Data Availability Statement: Data are contained within the article and Supplementary Materials.

Acknowledgments: We extend our gratitude to the Second Century Fund (C2F) (Chulalongkorn University). We are also grateful to Metabolomics South Australia, which is funded through Bioplatforms Australia Pty Ltd. (BPA), a National Collaborative Research Infrastructure Strategy (NCRIS), and investment from the South Australian State Government and The Australian Wine Research Institute. We would like to thank the Halal Science Center, Chulalongkorn University, for analytical instruments support.

Conflicts of Interest: The authors declare no conflicts of interest.

References

1. O'Keefe, J.H.; Bell, D.S. Postprandial hyperglycemia/hyperlipidemia (postprandial dysmetabolism) is a cardiovascular risk factor. *Am. J. Card.* **2007**, *100*, 899–904. [CrossRef]
2. Zheng, J.; Wang, J.; Zhang, Y.; Xia, J.; Guo, H.; Hu, H.; Shan, P.; Li, T. The Global Burden of Diseases attributed to high low-density lipoprotein cholesterol from 1990 to 2019. *Front. Public Health* **2022**, *10*, 891929. [CrossRef]
3. Ros, E. Intestinal absorption of triglyceride and cholesterol. Dietary and pharmacological inhibition to reduce cardiovascular risk. *Atheroscler* **2000**, *151*, 357–379. [CrossRef]
4. Singh, A.K. Dipeptidyl peptidase-4 inhibitors: Novel mechanism of actions. *Indian J. Endocrinol. Metab.* **2014**, *18*, 753–759. [CrossRef]

5. Bahmani, M.; Mirhoseini, M.; Shirzad, H.; Sedighi, M.; Shahinfard, N.; Rafieian-Kopaei, M. A review on promising natural agents effective on hyperlipidemia. *J. Evid. Based Complement. Altern. Med.* **2015**, *20*, 228–238. [CrossRef]
6. Panwar, H.; Calderwood, D.; Grant, I.R.; Grover, S.; Green, B.D. Lactobacilli possess inhibitory activity against dipeptidyl peptidase-4 (DPP-4). *Ann. Microbiol.* **2016**, *66*, 505–509. [CrossRef]
7. Szutowska, J. Functional properties of lactic acid bacteria in fermented fruit and vegetable juices: A systematic literature review. *Eur. Food Res. Technol.* **2020**, *246*, 357–372. [CrossRef]
8. Mannaa, M.; Han, G.; Seo, Y.S.; Park, I. Evolution of food fermentation processes and the use of multi-omics in deciphering the roles of the microbiota. *Foods* **2021**, *10*, 2861. [CrossRef]
9. Emmanuel, E.; Deborah, S. Phytochemical and anti-nutritional studies on some commonly consumed fruits in lokoja, kogi state of Nigeria. *Gen. Med. Open* **2018**, *2*, 2–5. [CrossRef]
10. Idrees, M.; Imran, M.; Atiq, N.; Zahra, R.; Abid, R.; Alreshidi, M.; Roberts, T.; Abdelgadir, A.; Tipu, M.K.; Farid, A.; et al. Probiotics, their action modality and the use of multi-omics in metamorphosis of commensal microbiota into target-based probiotics. *Front. Nutr.* **2022**, *9*, 959941. [CrossRef]
11. Lan, T.; Lv, X.; Zhao, Q.; Lei, Y.; Gao, C.; Yuan, Q.; Sun, X.; Liu, X.; Ma, T. Optimization of strains for fermentation of kiwifruit juice and effects of mono- and mixed culture fermentation on its sensory and aroma profiles. *Food Chem. X* **2023**, *17*, 100595. [CrossRef]
12. Shehata, M.G.; El Sohaimy, S.A.; El-Sahn, M.A.; Youssef, M.M. Screening of isolated potential probiotic lactic acid bacteria for cholesterol lowering property and bile salt hydrolase activity. *Ann. Agric. Sci.* **2016**, *61*, 65–75. [CrossRef]
13. Doron, S.; Snydman, D.R.; Gorbach, S.L. *Lactobacillus* GG: Bacteriology and clinical applications. *Gastroenterol. Clin. N. Am.* **2005**, *34*, 483–498. [CrossRef]
14. Segers, M.E.; Lebeer, S. Towards a better understanding of *Lactobacillus rhamnosus* GG -host interactions. *Microb Cell Fact.* **2014**, *13* (Suppl. S1), S7. [CrossRef]
15. Capurso, L. Thirty years of *Lactobacillus rhamnosus* GG: A review. *J. Clin. Gastroenterol.* **2019**, *53* (Suppl. S1), S1–S41. [CrossRef]
16. Doriya, K.; Kumar, D.S.; Thorat, B.N. A systematic review on fruit-based fermented foods as an approach to improve dietary diversity. *J. Food Process. Preserv.* **2022**, *46*, e16994. [CrossRef]
17. Liu, B.; Xu, Q.; Sun, Y. Black goji berry (*Lycium ruthenicum*) tea has higher phytochemical contents and in vitro antioxidant properties than red goji berry (*Lycium barbarum*) tea. *Food Qual. Saf.* **2020**, *4*, 193–201. [CrossRef]
18. Tang, P.; Giusti, M. Black goji (*Lycium ruthenicum* Murr.) polyphenols: Potent antioxidants and natural colorants. In *Phytochemicals in Goji berries*, 1st ed.; Ye, X., Jiang, Y., Eds.; CRC Press: Boca Raton, FL, USA, 2020; pp. 109–134.
19. Vidovic, B.B.; Milincic, D.D.; Marcetic, M.D.; Djuris, J.D.; Ilic, T.D.; Kostic, A.Z.; Pesic, M.B. Health benefits and applications of goji berries in functional food products development: A review. *Antioxidants* **2022**, *11*, 248. [CrossRef]
20. Liu, Y.; Fang, H.; Liu, H.; Cheng, H.; Pan, L.; Hu, M.; Li, X. Goji berry juice fermented by probiotics attenuates dextran sodium sulfate-induced ulcerative colitis in mice. *J. Funct. Foods* **2021**, *83*, 104491. [CrossRef]
21. Jain, A.; Jain, R.; Jain, S. Quantitative analysis of reducing sugars by 3, 5-dinitrosalicylic acid (DNSA method). In *Basic Techniques in Biochemistry, Microbiology, and Molecular Biology*; Springer: New York, NY, USA, 2020; pp. 181–183.
22. Marnpae, M.; Chusak, C.; Balmori, V.; Kamonsuwan, K.; Dahlan, W.; Nhujak, T.; Hamid, N.; Adisakwattana, S. Probiotic gac fruit beverage fermented with *Lactobacillus paracasei*: Physiochemical properties, phytochemicals, antioxidant activities, functional properties, and volatile flavor compounds. *LWT* **2022**, *169*, 113986. [CrossRef]
23. Chusak, C.; Chanbunyawat, P.; Chumnumduang, P.; Chantarasinlapin, P.; Suantawee, T.; Adisakwattana, S. Effect of gac fruit (*Momordica cochinchinensis*) powder on in vitro starch digestibility, nutritional quality, textural and sensory characteristics of pasta. *LWT* **2020**, *118*, 108856. [CrossRef]
24. Lee, J.; Durst, R.; Wrolstad, R. AOAC 2005.02: Total monomeric anthocyanin pigment content of fruit juices, beverages, natural colorants, and wines- ph differential method. *J. AOAC Int.* **2005**, *88*, 37–39.
25. Adisakwattana, S.; Intrawangso, J.; Hemrid, A.; Chanathong, B.; Mäkynen, K. Extracts of edible plants inhibit pancreatic lipase, cholesterol esterase and cholesterol micellization, and bind bile acids. *Food Technol Biotech.* **2012**, *50*, 11–16.
26. Chayaratanasin, P.; Barbieri, M.A.; Suanpairintr, N.; Adisakwattana, S. Inhibitory effect of *Clitoria ternatea* flower petal extract on fructose-induced protein glycation and oxidation-dependent damages to albumin in vitro. *BMC Complement. Altern. Med.* **2015**, *15*, 27. [CrossRef]
27. Beniddir, M.A.; Kang, K.B.; Genta-Jouve, G.; Huber, F.; Rogers, S.; van der Hooft, J.J.J. Advances in decomposing complex metabolite mixtures using substructure- and network-based computational metabolomics approaches. *Nat. Prod. Rep.* **2021**, *38*, 1967–1993. [CrossRef]
28. Peeler, J.T.; Maturin, L. BAM Chapter 3: Aerobic Plate Count: U.S. Food and Drug Administration. Available online: https://www.fda.gov/food/laboratory-methods-food/bam-chapter-3-aerobic-plate-count (accessed on 3 October 2023).
29. Thilavech, T.; Adisakwattana, S. Cyanidin-3-rutinoside acts as a natural inhibitor of intestinal lipid digestion and absorption. *BMC Complement. Altern. Med.* **2019**, *19*, 242.
30. Coudeyras, S.; Marchandin, H.; Fajon, C.; Forestier, C. Taxonomic and strain-specific identification of the probiotic strain *Lactobacillus rhamnosus* 35 within the *Lactobacillus casei* group. *Appl. Environ. Microbiol.* **2008**, *74*, 2679–2689. [CrossRef]
31. Liu, Y.; Cheng, H.; Liu, H.; Ma, R.; Ma, J.; Fang, H. Fermentation by multiple bacterial strains improves the production of bioactive compounds and antioxidant activity of goji juice. *Molecules* **2019**, *24*, 3519. [CrossRef]

32. Balmori, V.; Marnpae, M.; Chusak, C.; Kamonsuwan, K.; Katelakha, K.; Charoensiddhi, S.; Adisakwattana, S. Enhancing phytochemical compounds, functional properties, and volatile flavor profiles of pomelo (*Citrus grandis* (L.) Osbeck) juices from different cultivars through fermentation with *Lacticaseibacillus paracasei*. *Foods* **2023**, *12*, 4278. [CrossRef]

33. Enaru, B.; Dretcanu, G.; Pop, T.D.; Stanila, A.; Diaconeasa, Z. Anthocyanins: Factors affecting their stability and degradation. *Antioxidants* **2021**, *10*, 1967. [CrossRef]

34. Corcoran, B.M.; Stanton, C.; Fitzgerald, G.F.; Ross, R.P. Survival of probiotic *lactobacilli* in acidic environments is enhanced in the presence of metabolizable sugars. *Appl. Environ. Microbiol.* **2005**, *71*, 3060–3067.

35. Lahteinen, T.; Malinen, E.; Koort, J.M.; Mertaniemi-Hannus, U.; Hankimo, T.; Karikoski, N.; Pakkanen, S.; Laine, H.; Sillanpaa, H.; Soderholm, H.; et al. Probiotic properties of Lactobacillus isolates originating from porcine intestine and feces. *Anaerobe* **2010**, *16*, 293–300. [CrossRef]

36. Fujita, A.; Sarkar, D.; Genovese, M.I.; Shetty, K. Improving anti-hyperglycemic and anti-hypertensive properties of camu-camu (*Myriciaria dubia* Mc. Vaugh) using lactic acid bacterial fermentation. *Process Biochem.* **2017**, *59*, 133–140. [CrossRef]

37. Ávila, M.; Hidalgo, M.; Sánchez-Moreno, C.; Pelaez, C.; Requena, T.; de Pascual-Teresa, S. Bioconversion of anthocyanin glycosides by *Bifidobacteria* and *Lactobacillus*. *Food Res. Int.* **2009**, *42*, 1453–1461. [CrossRef]

38. Hunaefi, D.; Gruda, N.; Riedel, H.; Akumo, D.N.; Saw, N.M.M.T.; Smetanska, I. Improvement of antioxidant activities in red cabbage sprouts by lactic acid bacterial fermentation. *Food Biotech.* **2013**, *27*, 279–302. [CrossRef]

39. Wu, Y.; Li, S.; Tao, Y.; Li, D.; Han, Y.; Show, P.L.; Wen, G.; Zhou, J. Fermentation of blueberry and blackberry juices using *Lactobacillus plantarum*, *Streptococcus thermophilus* and *Bifidobacterium bifidum*: Growth of probiotics, metabolism of phenolics, antioxidant capacity in vitro and sensory evaluation. *Food Chem.* **2021**, *348*, 129083. [CrossRef]

40. Rodriguez, H.; Curiel, J.A.; Landete, J.M.; de las Rivas, B.; Lopez de Felipe, F.; Gomez-Cordoves, C.; Mancheno, J.M.; Munoz, R. Food phenolics and lactic acid bacteria. *Int. J. Food Microbiol.* **2009**, *132*, 79–90. [CrossRef]

41. Zhao, M.N.; Zhang, F.; Zhang, L.; Liu, B.J.; Meng, X.H. Mixed fermentation of jujube juice (*Ziziphus jujuba* Mill.) with *L. rhamnosus* GG and *L. plantarum*-1: Effects on the quality and stability. *Int. J. Food Sci. Technol.* **2019**, *54*, 2624–2631. [CrossRef]

42. Kulczyński, B.; Gramza-Michałowska, A. Goji berry (*Lycium barbarum*): Composition and health effects—A review. *Pol. J. Food Nutr. Sci.* **2016**, *66*, 67–75. [CrossRef]

43. Niki, E.; Noguchi, N. Evaluation of antioxidant capacity. What capacity is being measured by which method? *IUBMB Life* **2000**, *50*, 323–329. [CrossRef]

44. Insull, W., Jr. Clinical utility of bile acid sequestrants in the treatment of dyslipidemia: A scientific review. *South. Med. J.* **2006**, *99*, 257–273. [CrossRef]

45. Naumann, S.; Schweiggert-Weisz, U.; Eisner, P. Characterisation of the molecular interactions between primary bile acids and fractionated lupin cotyledons (*Lupinus angustifolius* L.). *Food Chem.* **2020**, *323*, 126780. [CrossRef]

46. De Boever, P.; Verstraete, W. Bile salt deconjugation by *Lactobacillus plantarum* 80 and its implication for bacterial toxicity. *J. Appl. Microbiol.* **1999**, *87*, 345–352.

47. Martinez-Gonzalez, A.I.; Alvarez-Parrilla, E.; Diaz-Sanchez, A.G.; de la Rosa, L.A.; Nunez-Gastelum, J.A.; Vazquez-Flores, A.A.; Gonzalez-Aguilar, G. In vitro inhibition of pancreatic lipase by polyphenols: A kinetic, fluorescence spectroscopy and molecular docking study. *Food Technol. Biotech.* **2017**, *55*, 519–530. [CrossRef]

48. Buchholz, T.; Melzig, M.F. Polyphenolic compounds as pancreatic lipase inhibitors. *Planta Med.* **2015**, *81*, 771–783. [CrossRef]

49. Fan, J.; Johnson, M.H.; Lila, M.A.; Yousef, G.; de Mejia, E.G. Berry and citrus phenolic compounds inhibit dipeptidyl peptidase IV: Implications in diabetes management. *J. Evid. Based Complement. Altern. Med.* **2013**, *2013*, 479505. [CrossRef]

50. Yan, F.; Li, N.; Yue, Y.; Wang, C.; Zhao, L.; Evivie, S.E.; Li, B.; Huo, G. Screening for potential novel probiotics with dipeptidyl peptidase IV-Inhibiting activity for type 2 diabetes attenuation *in vitro* and *in vivo*. *Front. Microbiol.* **2019**, *10*, 2855. [CrossRef]

51. Vella, A. Mechanism of action of DPP-4 inhibitors—New insights. *J. Clin. Endocrinol. Metab.* **2012**, *97*, 2626–2628. [CrossRef]

52. Jia, L.; Betters, J.L.; Yu, L. Niemann-pick C1-like 1 (NPC1L1) protein in intestinal and hepatic cholesterol transport. *Ann. Rev. Physiol.* **2011**, *73*, 239–259. [CrossRef]

53. Nekohashi, M.; Ogawa, M.; Ogihara, T.; Nakazawa, K.; Kato, H.; Misaka, T.; Abe, K.; Kobayashi, S. Luteolin and quercetin affect the cholesterol absorption mediated by epithelial cholesterol transporter Niemann-pick C1-like 1 in Caco-2 cells and rats. *PLoS ONE* **2014**, *9*, e97901. [CrossRef]

54. Le, B.; Yang, S.H. Identification of a novel potential probiotic *Lactobacillus plantarum* FB003 isolated from salted-fermented shrimp and its effect on cholesterol absorption by regulation of NPC1L1 and PPARα. *Probiotics Antimicrob. Proteins* **2019**, *11*, 785–793. [CrossRef]

55. Kobayashi, S. The Effect of polyphenols on hypercholesterolemia through inhibiting the transport and expression of Niemann-Pick C1-Like 1. *Int. J. Mol. Sci.* **2019**, *20*, 4939. [CrossRef]

56. Filannino, P.; Tlais, A.Z.A.; Morozova, K.; Cavoski, I.; Scampicchio, M.; Gobbetti, M.; Cagno, R.D. Lactic acid fermentation enriches the profile of biogenic fatty acid derivatives of avocado fruit (*Persea americana* Mill.). *Food Chem.* **2020**, *317*, 126384. [CrossRef]

57. Leonard, W.; Zhang, P.; Ying, D.; Adhikari, B.; Fang, Z. Fermentation transforms the phenolic profiles and bioactivities of plant-based foods. *Biotechnol. Adv.* **2021**, *49*, 107763. [CrossRef]

58. Chen, H.M.; Muramoto, K.; Yamauchi, F.; Fujimoto, K.; Nokihara, K. Antioxidative properties of histidine-containing peptides designed from peptide fragments found in the digests of a soybean protein. *J. Agric. Food Chem.* **1998**, *46*, 49–53. [CrossRef]

59. Sun, J.; Chen, H.; Qiao, Y.; Liu, G.; Leng, C.; Zhang, Y.; Lv, Z.; Feng, Z. The nutrient requirements of *Lactobacillus rhamnosus* GG and their application to fermented milk. *J. Dairy Sci.* **2019**, *102*, 5971–5978. [CrossRef]
60. Braga, A.R.C.; de Souza Mesquita, L.M.; Martins, P.L.G.; Habu, S.; de Rosso, V.V. *Lactobacillus* fermentation of jussara pulp leads to the enzymatic conversion of anthocyanins increasing antioxidant activity. *J. Food Compos. Anal.* **2018**, *69*, 162–170.
61. Lopez de Felipe, F.; Curiel, J.A.; Munoz, R. Improvement of the fermentation performance of *Lactobacillus plantarum* by the flavanol catechin is uncoupled from its degradation. *J. Appl. Microbiol.* **2010**, *109*, 687–697. [CrossRef]
62. Issaoui, M.; Delgado, A.M.; Caruso, G.; Micali, M.; Barbera, M.; Atrous, H.; Ouslati, A.; Chammem, N. Phenols, flavors, and the mediterranean diet. *J. AOAC Int.* **2020**, *103*, 915–924. [CrossRef]

 *antioxidants*

MDPI

Article

Impact of Lactic Acid Bacteria Fermentation on (Poly)Phenolic Profile and In Vitro Antioxidant and Anti-Inflammatory Properties of Herbal Infusions

Tarik Ozturk [1,†], María Ángeles Ávila-Gálvez [2,†], Sylvie Mercier [3,†], Fernando Vallejo [2], Alexis Bred [3], Didier Fraisse [3], Christine Morand [3], Ebru Pelvan [1], Laurent-Emmanuel Monfoulet [3,*] and Antonio González-Sarrías [2,*]

1 Life Sciences, TÜBİTAK Marmara Research Center, P.O. Box 21, 41470 Gebze-Kocaeli, Türkiye; tarik.ozturk@tubitak.gov.tr (T.O.); ebru.pelvan@tubitak.gov.tr (E.P.)
2 Laboratory of Food and Health, Research Group on Quality, Safety and Bioactivity of Plant Foods, Department of Food Science and Technology, CEBAS-CSIC, Campus de Espinardo, P.O. Box 164, 30100 Murcia, Spain; mavila@cebas.csic.es (M.Á.Á.-G.); fvallejo@cebas.csic.es (F.V.)
3 Université Clermont Auvergne, INRAE, UNH, F-63000 Clermont-Ferrand, France; sylvie.mercier@inrae.fr (S.M.); alexis.bred@uca.fr (A.B.); didier.fraisse@uca.fr (D.F.); christine.morand@inrae.fr (C.M.)
* Correspondence: laurent-emmanuel.monfoulet@inrae.fr (L.-E.M.); agsarrias@cebas.csic.es (A.G.-S.)
† These authors contributed equally to this study.

Abstract: Recently, the development of functional beverages has been enhanced to promote health and nutritional well-being. Thus, the fermentation of plant foods with lactic acid bacteria can enhance their antioxidant capacity and others like anti-inflammatory activity, which may depend on the variations in the total content and profile of (poly)phenols. The present study aimed to investigate the impact of fermentation with two strains of *Lactiplantibacillus plantarum* of several herbal infusions from thyme, rosemary, echinacea, and pomegranate peel on the (poly)phenolic composition and whether lacto-fermentation can contribute to enhance their in vitro antioxidant and anti-inflammatory effects on human colon myofibroblast CCD18-Co cells. HPLC-MS/MS analyses revealed that fermentation increased the content of the phenolics present in all herbal infusions. In vitro analyses indicated that pomegranate infusion showed higher antioxidant and anti-inflammatory effects, followed by thyme, echinacea, and rosemary, based on the total phenolic content. After fermentation, despite increasing the content of phenolics, the antioxidant and anti-inflammatory effects via reduction pro-inflammatory markers (IL-6, IL-8 and PGE$_2$) were similar to those of their corresponding non-fermented infusions, with the exception of a greater reduction in lacto-fermented thyme. Overall, the findings suggest that the consumption of lacto-fermented herbal infusions could be beneficial in alleviating intestinal inflammatory disorders.

Keywords: polyphenols; thyme; rosemary; pomegranate; echinacea; CCD18-Co; inflammation; antioxidant

Citation: Ozturk, T.; Ávila-Gálvez, M.Á.; Mercier, S.; Vallejo, F.; Bred, A.; Fraisse, D.; Morand, C.; Pelvan, E.; Monfoulet, L.-E.; González-Sarrías, A. Impact of Lactic Acid Bacteria Fermentation on (Poly)Phenolic Profile and In Vitro Antioxidant and Anti-Inflammatory Properties of Herbal Infusions. *Antioxidants* **2024**, *13*, 562. https://doi.org/10.3390/antiox13050562

Academic Editor: Myung-Ji Seo

Received: 16 April 2024
Revised: 29 April 2024
Accepted: 30 April 2024
Published: 2 May 2024

1. Introduction

In recent decades, consumers' increasing choice and consumption of nutritionally enriched and health-promoting foods has sparked global interest in the development of fermented functional foods [1]. Thus, although since ancient times, humans have consumed and produced foods and beverages that have been subjected to fermentation, such as dairy products, beer, and bread, the use of fermentative starters is becoming an aspect of growing interest in the field of food biotechnology to develop new functional foods and beverages [2]. Briefly, the fermentation method in foods and beverages is based on the action of microorganisms, which induce the conversion of food components by microbial enzymes, causing desirable biochemical changes that can provide many benefits to foods extending their shelf-life, nutritional value, and sensory properties and providing many beneficial components for health [2,3]. In this sense, fermented foods may enhance health benefits, including antioxidant, anti-inflammatory, anti-allergenic, anti-microbial, and anti-diabetic effects, among others. This is boosted by the potential probiotic effects of their constituent microorganisms, if they are still present, as well as by the enzymatic bioconversion products to biologically active metabolites (i.e., exopolysaccharides, vitamins, minerals, phenolic compounds, bioactive peptides, organic acids, free amino acids, etc.) [2–5]. Most of the functional foodstuffs are produced by species and (or) strains belonging to lactic acid bacteria (LAB), which are generally recognized as safe (GRAS) and confer a multitude of functional and sensory properties [6,7]. In this regard, for outside plant-based fermented foods or drinks which are currently widely studied, the LAB starter cultures are directed towards the fermentation of novel substrates from plant foods, herbs, or spices in order to improve their health-promoting effects by stimulating the release or production of bioactive metabolites [8–10].

Herbs or plants parts such as roots, leaves, or flowers have been traditionally used to prevent illness, maintain health, and cure some diseases, and many of these have been used ubiquitously to prepare herbal infusions or teas. Herbal teas or infusions commonly refer to "non-*Camellia sinensis* derived infusions", prepared from boiling fresh or dried parts of edible plants. These beverages are becoming increasingly popular worldwide due to their diverse taste, caffeine-free nature, and potential beneficial effects attributed mainly to being rich in multiple bioactive compounds [11,12]. In general, herbal teas have attracted the interest of researchers due to their beneficial properties, such as antioxidant, anti-inflammatory, anti-microbial, and anti-cardiovascular-disease effects, among others. Most of these health-related effects have been attributed to their high content in phytochemicals, such as (poly)phenols [11–16]. Among various herbal teas, thyme, rosemary, echinacea, and pomegranate peel are often chosen for their appealing flavors and potential health benefits, including antioxidant and anti-inflammatory properties, which may be attributed to their phenolic compounds. Thus, rosmarinic acid and luteolin derivatives are the main (poly)phenols in rosemary and thyme [17,18], while ellagitannins and ellagic acid are the primary phenolics in pomegranate peel [19]. Additionally, chicoric and caftaric acids are the most abundant in echinacea extracts [20,21]. However, most of these phenolic compounds are poorly absorbed and usually occur as glycosides, considered biologically inactive, and their bioavailability requires the initial hydrolysis of the sugar moiety by intestinal β-glucosidases producing their aglycones. In addition, they can undergo bioconversion mediated by gut microbiota, in which they are broken down into smaller molecules via enzymatic reactions to facilitate their bioavailability and biological activity in both the gastrointestinal tract and systemic tissues [22]. Therefore, considering these limitations, the fermentative biotransformation approach using LAB may be an effective strategy to improve the bioaccessibility, bioavailability, and bioactivity of phenolic compounds present in herbal infusions.

The objective of this study was to evaluate, for the first time, the effect of the lacto-fermentation on the phenolic profile of four herbal infusions produced from two aromatic herbs (thyme and rosemary) and two plant extracts (*Echinacea pupurea* flower extract and pomegranate peel extract) rich in phenolic compounds in order to elucidate whether lacto-fermentation of these infusions can contribute to enhance their antioxidant and anti-inflammatory activities. To obtain lacto-fermented herbal teas, different strains of *Lactiplantibacillus plantarum* were used. Subsequently, the phenolic profile of each herbal infusion was evaluated by analyzing it before and after lactic fermentation. The in vitro antioxidant effects and their anti-inflammatory effects on human colon fibroblasts (CCD18-Co), which induced an inflammatory cytokine, were also explored. Our results provide a scientific basis for highlighting the effective bioactivity of lacto-fermented herbal infusions.

2. Materials and Methods

2.1. Reagents

Analytical HPLC-grade chemicals, including acetonitrile (ACN), formic acid, dimethyl sulfoxide (DMSO), and methanol (MeOH), were obtained from J.T. Baker (Deventer, The Netherlands). BMS-345541 was purchased from Selleck (Houston, TX, USA). Ethanol ($\geq$99.8%), and Man, Rogosa, and Sharpe (MRS) broth were purchased from Merck (Darmstadt, Germany). Gallic acid was purchased from Extrasynthese (Genay, France). All other reagents were purchased from Sigma-Aldrich (St. Louis, MO, USA), unless stated otherwise. Ultrapure Millipore water was used throughout the study, generated by a Milli-Q water (18.2 MmΩ) device (Merck, Darmstadt, Germany).

2.2. Plant Materials and Preparation of Extracts

Pomegranate peel (*Punica granatum* L.) was purchased from a local supplier (Doğan Baharatçılık Kimyevi Maddeler Tic. ve San. A.Ş., İstanbul, Türkiye). Whole aerial parts of thyme (*Thymus vulgaris*) were purchased from a local market (produced by Sanita Tarım Ürünleri Baharat, Kozmetik San.Tic.A.Ş., İzmir, Türkiye). Rosemary (*Rosmarinus officinalis* L.) leaves were obtained from BATEM (Republic of Türkiye Ministry of Agriculture and Forestry West Mediterranean Agricultural Research Institute). Echinacea (*Echinacea purpurea* L.) flower extract was obtained from a local producer (HMC Naturel Tarım ve Hayvancılık San. Tic. Ltd. Şti., Dodurga, Çorum, Türkiye).

Thyme plant materials were directly used without grinding before fermentation. Pomegranate peel was ground before fermentation, as described below. Rosemary leaves were extracted with 80% ethanol, which was subsequently evaporated. The remaining aqueous phase was lyophilized (Christ Epsilon 2–4 Lyo-Screen-Control (LSC), Osterode am Harz, Germany) to obtain a dry powder used for fermentation. Echinacea flowers were extracted with MilliQ water using a pilot-scale continuous counter current extractor (Niro Atomizer, AC-27, Soeborg, Denmark) and spray-dried (Minor Spray Dryer, Niro Atomizer, Soeborg, Denmark) to obtain a powder for further fermentation.

2.3. Strains Descriptions

Two different strains of lactic acid bacteria were isolated from local plant sources in Türkiye by TUBİTAK MAM and were identified by 16S rDNA sequencing. Strain A, identified as the *Lactiplantibacillus plantarum* 129 J1 strain, was isolated from olive and developed through an evolutionary engineering strategy, where superior strains were selected among mutant populations by the gradual application of selective pressure to mimic the natural evolutionary process. The resulting strain, J1, was able to almost completely survive passage through the upper gastrointestinal tract. Strain B, identified as the *Lactiplantibacillus plantarum* P1 strain, was isolated from fermented traditional black carrot juice, and was selected due to its ability to decrease pH rapidly and suppress competitive flora.

2.4. Fermented-Infusions Preparation

Firstly, 50 g of thyme or pomegranate peel, as detailed above, was dissolved in water at a ratio of 1:20 (w/v) and incubated at 90 °C for 15 min. The rosemary and echinacea were dissolved in water at a ratio of 1:50 (w/v) and incubated at 90 °C for 1 min. Next, for all beverages, 500 mL of MilliQ water at room temperature was added to the samples, and the bottles were cooled at 30 °C. The prepared beverages were inoculated a rate of 1% of each strain grown for 48 h at 30 °C. The pH was measured before and after fermentation to confirm its decrease (from pH values above 5 to values below 4.4, depending on the beverage) as a result of the fermentation process (Supplementary Table S1).

Selected samples were transferred to 50 mL screw-cap polyethylene centrifuge tubes and centrifuged (Hettich Rotina 420R, Sérézin du Rhône, France) at 3500× *g* for 2 min to clarify the extracts. The supernatants were transferred to new tubes and pasteurized in a water bath at 85 °C for 10 min. Next, the samples were aseptically stored at −80 °C for further analyses.

2.5. Determination of Total Phenolic Content

Analysis of Polyphenolic Content (Total) and by HPLC-MS/MS

The total phenolic content (TPC) of herbal infusions was determined using a previously reported method with minor modifications [23]. Briefly, infusion samples were diluted twice in purified water, and 2 mL of these diluted solutions were mixed with 1 mL of undiluted Folin–Ciocalteu reagent. The volume was finally adjusted to 25 mL with a sodium carbonate solution (150 g/L). After incubation for 30 min at room temperature, absorbance was recorded at 740 nm using a Jasco V-630 spectrophotometer (Lisses, France), and the result was expressed in mg of gallic acid equivalents (GAE) per g of dry material using a standard curve of gallic acid (0.005–0.1 mg/mL).

2.6. HPLC-MS/MS Analysis

The phenolic content of plant material (raw extracts) from echinacea, thyme, rosemary, and pomegranate peel was analyzed as described elsewhere [24]. Briefly, 50 mg of each sample was dissolved in a solution containing 10 mL of MeOH/DMSO/H$_2$O (40:40:20, $v/v/v$) supplemented with 0.1% HCl (v/v). The samples were then vortexed for 2 min, subjected to ultrasonic bath sonication for 5 min, and centrifuged at $4000\times g$ for 5 min at room temperature. This extraction process was repeated using a 5 mL extraction solution to maximize the phenolic compound yield. Finally, the resulting supernatant of all extracts was filtered through a 0.45 μm polyvinylidene difluoride (PVDF) filter prior to HPLC-MS/MS analysis. Each sample underwent extraction and analysis in triplicate to ensure consistency.

On the other hand, for the analysis of the phenolic profile of non-fermented and lacto-fermented beverages, 1 mL was extracted with MeOH in a 1:1 (v/v) proportion to remove contaminants from the fermentation process. The samples were homogenized using a vortex for 2 min and centrifuged at $10000\times g$ for 15 min, and the supernatant was recovered. Next, each sample was evaporated in a speed vacuum and re-suspended in 150 μL of MeOH. The final samples in MeOH were filtered using 0.45 μm PVDF filters and transferred to vials before HPLC analysis.

HPLC analyses were performed on an Agilent 1200 HPLC system with a photodiode array detector (DAD) (Agilent Technologies, Waldbronn, Germany) and an ion trap (IT) mass spectrometer detector in series (Bruker Daltonik, Bremen, Germany). A reverse-phase Poroshell 120 C18 column (100 × 3.0 mm, 2.7 μm) was utilized. The mobile phases consisted of water:formic acid (99:1) as A and acetonitrile (ACN) as B, with a flow rate of 0.6 mL/min. The gradient was as follows: 0–1% B at 0 min, 1–40% B at 0–20 min, 40–90% B at 20–30 min, and 90% B at 30–33 min, followed by a return to initial conditions (1% B) in 1 min with column re-equilibration for 5 min. The injection volume was 12 μL. UV-Vis spectra were acquired in the range of 200 to 600 nm. In the mass spectrometer, nitrogen served as the drying and nebulizing gas, with a pressure of 65 psi, flow of 11 L/mL, and temperature of 350 °C.

The identification of compounds was performed by taking information about their elution order; UV spectra; molecular weight; fragmentation by MS/MS; and, whenever possible, chromatographic comparison with authentic standards. External calibration curves with appropriate standards belonging to the different families of phenolic compounds were used for the quantification. Rosmarinic acid from thyme and rosemary extracts was quantified with its own standard at 320 nm. Punicalin and punicalagins from pomegranate were quantified with their own standards, and ellagic acid (and derivatives) with ellagic acid, at 360 nm. Flavanones were quantified at 340 nm with hesperidin and eriocitrin, flavones at 340 nm with apigenin, and flavonols at 360 nm with quercetin. The hydroxycinnamic acids in echinacea extract were quantified at 320 nm with chicoric acid or caffeic acid.

2.7. Antioxidant Activity

The DPPH (2,2-diphenyl-1-picrylhydrazyl) scavenging activity was evaluated according to a previously published protocol [25]. Briefly, 50 μL of reference and digestive solutions were mixed with 2.5 mL of the fresh radical mixture (25 μg/mL in MeOH). After incubation at room temperature for 30 min, the absorbance was measured at 515 nm. The DPPH scavenging capacity was expressed as micromoles of trolox (6-hydroxy-2,5,7,8-tetramethyl-3,4-dihydrochromene-2-carboxylic acid) equivalent (μmol TE) per gram of dry material using a standard curve of Trolox (100–3000 μmol/L).

The ferric reducing antioxidant power (FRAP) assay was performed according to Katalinić et al. [26] with minor modifications. Stock solutions included a 300 mM acetate buffer (3.1 g of $C_2H_3NaO_2 \cdot 3H_2O$ and 16 mL of $C_2H_4O_2$) pH 3.6, a 10 mM TPTZ (2,4,6-tripyridyl-s-triazine) in 40 mM HCl solution, and a 20 mM $FeCl_3 \cdot 6H_2O$ solution. A fresh FRAP working solution was then prepared by mixing 25 mL of acetate buffer, 2.5 mL of TPTZ solution, and 2.5 mL of $FeCl_3 \cdot 6H_2O$ solution and kept at 37 °C before use. Next, 10 μL of each sample was added to react with 200 μL of the FRAP solution. After 30 min at room temperature in the dark, the absorbance of the ferrous colored product was recorded at 593 nm. Results were expressed in μmol TE/g of sample using a standard trolox curve (50–750 μmol/L).

2.8. Xanthine Oxidase (XO) Inhibitory Activity

Inhibition of XO activity was determined according to the method described by Sowndhararajan et al. [27], with slight modifications. Reference and digestion samples (70 μL) were incubated in the dark at 25 °C with 120 μL of PBS (phosphate buffer solution), 120 mM pH 7.5, and 100 μL of a 1.5 mM xanthine solution. After 5 min, 10 μL of XO 0.30 UI/mL was added. The progress of the reaction (uric acid production) was measured at 293 nm, and the percentage of inhibition was calculated. Six sample concentrations (0.58–2.33 mg/mL) were used for analysis, and activities were expressed as IC_{50} in mg of dry material/mL.

2.9. Cell Culture

Myofibroblasts of the colon CCD18-Co cell line were obtained from the American Type Culture Collection (ATCC, Rockville, MD, USA). CCD18-Co cells were maintained in Eagle's minimum essential medium (EMEM) supplemented with 10% fetal bovine serum (FBS), antibiotics (streptomycin and penicillin at 100 mg/mL and 100 U/mL, respectively), 1.5 g/L sodium bicarbonate, 1 mM sodium pyruvate, 0.1 mM nonessential amino acids, and 2 mM L-glutamine. Cells were maintained at 37 °C in the presence of 5% CO_2. The range of population doubling levels (PDL) used in all experiments was from 26 to 32.

2.10. Cell Viability and Inflammatory Assay

CCD18-Co cells were subcultured at 2000 cells per well on 96-well plates and incubated with media as described above for 1 day. To select the highest nontoxic concentrations of non-fermented and fermented from each herbal infusion, the osmolarity and pH of a range of percentages (5, 2.5, 1 and 0.5%) were evaluated using a vapor pressure osmometer 5520 (VAPRO Wescor, Logan Utah, UT, USA) and a pH indicator paper (Neutralit, pH 5.5-9.0, Merck), respectively. Additionally, to confirm that the treatments did not exert an antiproliferative and/or cytotoxic effect, the CCD18-Co cell viability and proliferation were measured using the MTT reduction assay at 24 h, as described by González-Sarrías et al. [28]. Once these parameters were optimized, the attached cells in 96-well plates were incubated in EMEM supplemented with 0.1% FBS (*v/v*) for 24 h. Then, cells were treated with the sterilized (filtered by 0.22 μm) non-fermented and fermented samples at non-cytotoxic concentrations (2.5%) and co-treated with 1 ng/mL IL-1β for 16 h. Cells in the absence of IL-1β were used as negative controls (CT). BMS-345541 (BMS) at 5 μM was assayed as a positive control of the anti-inflammatory effect. After the inflammatory assay, the culture medium was collected and frozen at −80 °C for further analysis. Although doses of 2.5% obtained non-statistical differences to control cells of cell viability and proliferation values, the cell proliferation data obtained using an MTT reduction assay after each treatment were used to normalize the values of inflammatory markers. Assays were repeated three times (n = 3), with 6 measurements within each replicate.

2.11. Effect of the Lacto-Fermented Beverages on Cytokine Production and PGE2 Biosynthesis in IL-1β-Stimulated Cells

Pro-inflammatory cytokines, including IL-8 and IL-6, were measured in culture medium using their corresponding ELISA kits from PeproTech (Rocky Hill, NJ, USA), and the absorbance at 405 and 650 nm (reference wavelength) was detected using a microplate reader (Infinite M1000 Pro, Tecan, Grodig, Austria). The analysis of PGE2 in the culture medium was measured using an ELISA kit from Cayman (San Diego, CA, USA) and the same absorbance-detecting microplate reader. The data, expressed as average ± SD, were the results of three independent biological replicates (n = 3). The culture medium of the different treatments (carried out in each replicate) were pooled from six to eight different wells.

2.12. Statistical Analysis

All experimental data were expressed as the mean ± standard deviation (S.D.). The two-tailed unpaired Student's *t* test was used for statistical analysis of the data using SPSS Software, version 27.0 (SPSS Inc., Chicago, IL, USA) or Prism 6, version 6.01 (GraphPad). Graphs were constructed using GraphPad Prism 9.1.1 software (GraphPad Software, San Diego, CA, USA). To examine the correlation between phenolic profiles and antioxidant and anti-inflammatory effects, Pearson correlation analysis using MetaboAnalyst 6.0. was performed. The difference was considered to be statistically significant at a *p* value < 0.05.

3. Results

3.1. Screening of Lacto-Fermented Beverages Based on pH

From a regulatory point of view, for measuring the impact of food safety criteria on public health, foods are classified into low- and high-acid foods according to their pH. Foods lower than pH = 4.6 are considered as high-acid ones, since the spores of an extremely dangerous microorganism called *Clostridium botulinum* cannot germinate or produce toxins below this pH range, even if the food is pasteurized [29]. For this reason, a first screening measuring the pH value as an indicator for preservation and safety was carried out after the fermentation process (with each strain A or B) for each plant material. Thus, all lacto-fermented beverages showed pH values low enough to be considered as high-acid foods, and, therefore, safe without the need to add additives (Supplementary Table S1). Next, they were pasteurized and collected for further analyses.

3.2. Phenolic Characterization of Plant Material

A comprehensive HPLC-MS/MS analysis of the different plant material used for the lacto-fermentation of thyme, pomegranate, rosemary, and echinacea beverages is detailed in Supplementary Table S2. All phenolic compounds were identified using a comprehensive analytical approach, which involved assessing their retention time, mass, mass fragmentation patterns, and UV-Vis characteristics. To validate these identifications, comparisons were made with authentic standards whenever feasible. In thyme, the predominant compounds identified were chrysoeriol glucoside, followed by eriodictyol and rosmarinic acid. In pomegranate peel extract, punicalagin and ellagic acid emerged as the most abundant phenolic compounds detected. In the case of rosemary, the principal compounds observed were the flavonol isorhamnetin-3-O-glucoside and the flavone luteolin glucoside. Lastly, the echinacea extract exhibited chicoric acid and caftaric acid as the most prevalent compounds (Supplementary Table S2).

3.3. Phenolic Profile Comparison of Non-Fermented and Lacto-Fermented Beverages

The total (poly)phenol content (TPC) and the phenolic profile of both fermented and non-fermented infusions were determined (Tables 1 and 2, respectively). Both the TPC and the phenolic profile differed for the four infused beverages. For TPC assessed using the Folin–Ciocalteu assay, lactic acid fermentation with strain A and B resulted in a significant higher TPC (from 4% to 36% increase) compared to the non-fermented infusions (Table 1). The highest effect was observed after fermentation for thyme infusion. On the other hand, both fermentations slightly modulated the TPC of pomegranate and echinacea infusion, but not rosemary infusions. Furthermore, no significant difference was observed for TPC after fermentation between strain A and strain B.

Table 1. Total phenolic content of fermented and non-fermented beverages.

	Non-Fermented	Fermented A *	Fermented B *
Thyme	35.15	44.92 (+27.8%)	48.07 (+36.8%)
Rosemary	36.11	37.62 (+4.2%)	34.16 (−5.4%)
Echinacea	70.09	78.76 (+12.4%)	79.29 (+13.1%)
Pomegranate peel	90.01	98.96 (+9.9%)	99.63 (+10.7%)

Values are expressed as mg eq GAE/g of dry matter. * Effect of the lactic acid fermentations on the TPC (two-way ANOVA): fermented (A or B) vs. non-fermented: $p = 0.0295$; fermented A vs. fermented B: $p > 0.05$.

Table 2. Comparison of major phenolics detected in fermented and non-fermented beverages.

Herbal Infusion	Phenolic	Non-Fermented	Fermented A	Fermented B
Thyme	Luteolin glucoside	6.19 ± 0.82	19.22 ± 0.97 [a]	17.45 ± 2.36 [a]
	Chrysoeriol glucoside	8.72 ± 5.62	42.39 ± 3.69 [a]	59.15 ± 2.91 [a,b]
	Eriodictyol	22.30 ± 2.00	27.99 ± 0.85 [a]	18.89 ± 1.50 [b]
	Rosmarinic acid [a]	37.08 ± 3.21	51.40 ± 1.09 [a]	58.84 ± 2.61 [a]
	Quercetin [a]	0.17 ± 0.03	0.63 ± 0.01 [a]	0.47 ± 0.03 [a,b]
	Salvianolic acid A	6.37 ± 0.65	12.84 ± 0.39 [a]	14.80 ± 1.03 [a,b]
Rosemary	Isorhamnetin-3-glucoside	42.12 ± 4.57	68.66 ± 10.30	43.37 ± 4.64 [b]
	Hispidulin-7-*O*-glucoside	43.58 ± 1.11	88.48 ± 7.25 [a]	78.69 ± 3.99 [a]
	Rosmarinic acid *	19.99 ± 2.60	50.80 ± 1.40 [a]	97.41 ± 7.42 [a,b]
	Hesperidin *	2.72 ± 0.47	39.44 ± 1.72 [a]	25.74 ± 3.61 [a,b]
	Luteolin-3-acetyl-*O*-glucuronide	4.45 ± 0.60	43.26 ± 0.85 [a]	48.21 ± 10.58 [a]
	Luteolin glucoside	1.93 ± 0.21	32.49 ± 0.82 [a]	21.52 ± 2.93 [a,b]
	Isorhamnetin-3-*O*-rutinoside	1.71 ± 0.28	29.38 ± 1.48 [a]	22.31 ± 3.15 [a]
	Rosmanol peak1	0.36 ± 0.02	3.72 ± 0.10 [a]	2.55 ± 0.38 [a]
	Rosmanol peak 2	0.25 ± 0.03	3.61 ± 0.29 [a]	2.62 ± 0.41 [a]
Echinacea	Caftaric acid	0.15 ± 0.006	0.58 ± 0.02 [a]	0.27 ± 0.03 [a,b]
	Chlorogenic acid	0.03 ± 0.003	0.04 ± 0.01	0.009 ± 0.004
	Neochlorogenic acid	0.02 ± 0.002	0.22 ± 0.03 [a]	0.05 ± 0.004 [a,b]
	Caffeic acid *	0.05 ± 0.007	0.30 ± 0.10	0.07 ± 0.01
	Chicoric acid *	1.52 ± 0.08	2.43 ± 0.15 [a]	1.49 ± 0.09 [b]
	Feruloylcaffeoyltartaric acid 1	0.11 ± 0.01	0.34 ± 0.05 [a]	0.11 ± 0.01 [b]
	Feruloylcaffeoyltartaric acid 2	0.08 ± 0.005	0.24 ± 0.01 [a]	0.08 ± 0.02 [b]
Pomegranate peel	Punicalin [a]	38.05 ± 7.83	14.69 ± 0.50 [a]	11.66 ± 1.25 [a,b]
	Punicalagin isomers [a]	67.29 ± 1.59	68.63 ± 4.86	63.66 ± 3.64
	Punigluconin	4.13 ± 1.66	3.04 ± 1.36	2.74 ± 0.96
	Pedunculagin II	nd	3.36 ± 0.18	2.96 ± 0.39
	Galloyl-HHDP-hexose	2.06 ± 0.67	3.18 ± 0.33	3.74 ± 0.80
	Ellagic acid-hex	2.76 ± 0.32	5.88 ± 0.40 [a]	5.83 ± 0.19 [a]
	Gallagic acid	nd	1.05 ± 0.08	1.09 ± 0.12
	Granatin-B	0.51 ± 0.12	1.36 ± 0.09 [a]	1.57 ± 0.25 [a]
	Ellagic acid-pentose	nd	2.96 ± 0.46	3.41 ± 0.12
	Ellagic acid [a]	9.58 ± 0.65	14.98 ± 0.47 [a]	15.75 ± 0.38 [a]

Values are expressed as mg/g extract. [a] Significant difference ($p < 0.05$) compared with non-fermented beverages. [b] Significant difference ($p < 0.05$) between beverages fermented with Strain A and fermented with Strain B. * Identified and quantified with their authentical standard.

Regarding the (poly)phenol profiles obtained by HPLC-MS/MS, as depicted in Table 2, both strain A and B exhibited increases in most of the quantified phenolics, albeit showing some differences between strains. Specifically, in thyme, differences were observed in certain phenolics, such as eriodictyol, quercetin, and salvianolic acid A. Regarding rosemary, differences were noted in isorhamnetin-3-glucoside, rosmarinic acid, and luteolin glucoside; in echinacea, all phenolics showed significant differences between strain A and B, except for caffeic acid and chlorogenic acid; and in pomegranate herbal tea, only punicalin showed a notable difference.

In both thyme-based fermented infusions, infusion phenolics such as luteolin, chryso-eriol, erodictyol, rosmarinic acid, quercetin, and salvianolic acid were quantified in greater amounts compared to the non-fermented one. Chrysoeriol glucoside exhibited the most pronounced increase with strain B, while eriodictyol was the only compound that did not show any significant increase compared to the non-fermented thyme herbal infusion. Interestingly, rosmarinic acid, the main compound found in non-fermented thyme, exhibited 1.4 and 1.6 times higher concentrations after fermentation with strains A and B, respectively, compared to the concentrations detected in the non-fermented beverage.

In regard to the phenolics detected in the rosemary infusion, rosmarinic acid exhibited a 2.5-fold increase with strain A and a 4.9-fold increase when it was fermented with strain B compared to the non-fermented beverage. Additionally, the difference between strains A and B was statistically significant (Table 2). Another pronounced increase was observed with hesperidin, showing a 14.5-fold increase with strain A and a 4.5-fold increase with strain B compared to the non-fermented rosemary beverage.

Surprisingly, in the case of echinacea infusions, a higher pronounced effect was observed mainly after fermentation with strain A. Thus, although caftaric acid, chicoric acid, feruloylcaffeoyltartaric acids, and neochlorogenic acid concentrations increased after lacto-fermentation with both strains compared to the non-femented sample, the use of strain A resulted in the most notable increase in all phenolic compounds.

Finally, for the herbal infusion from pomegranate peel, it is noteworthy that only ellagic acid hexose, ellagic acid, and granitin B exhibited increased concentrations following lacto-fermentation with both strain A and B in association with new compounds such as gallagic acid and ellagic acid-pentose. Conversely, punicalin demonstrated a significant decrease in its concentration with lacto-fermentation using both strains.

3.4. Antioxidant Capacity

The antioxidative effect of the four lacto-fermented infusions and their corresponding non-fermented ones was tested by measuring the in vitro DPPH activity, FRAP activity, and xanthine oxidase inhibition (XO) (Table 3). Thyme-fermented, rosemary-fermented, and echinacea-fermented infusions exhibited significant higher DPPH activity than their respective non-fermented infusions. The FRAP activity of thyme infusions was slightly improved by the fermentation. Regarding rosemary-fermented and echinacea-fermented infusions, they showed significantly increased FRAP activity compared to the non-fermented rosemary infusion and non-fermented echinacea infusion, respectively. Fermentation did not modulate the inhibition of XO of pomegranate and echinacea infusions, but significantly improved it in thyme and rosemary beverages after both fermentations.

3.5. Effects on Cell Viability in CCD18-Co Myofibroblasts

After the evaluation of cell viability, as well as the osmolarity and pH, using the MTT assay, the highest non-cytotoxic dose of both fermented and non-fermented infusions was selected to run the in vitro colonic inflammatory model. The selected concentration for each treatment of 2.5% in culture medium afforded pH values of 7 and osmolality values of 285–305 mmol/Kg, which are within the tolerance limits of this human cell model and showed no statistically significant differences on cell viability (over 95%) compared to untreated CCD18-Co cells.

Table 3. Antioxidant activities of lacto-fermented and non-fermented infusions.

	DPPH Activity (µmol eq Trolox/g of Extract)	FRAP Activity (µmol eq Trolox/g of Extract)	Xanthine Oxidase (XO) IC50 (mg/mL)
Thyme Non-fermented	92.87	439.16	1.45
Fermented A	323.93 [a]	519.90	0.98 [c]
Fermented B	306.46 [a]	507.33	0.97 [c]
Rosemary Non-fermented	100.26	203.63	2.11
Fermented A	347.92 [a]	401.86 [b]	1.35 [c]
Fermented B	355.41 [a]	367.54 [b]	1.36 [c]
Echinacea Non-fermented	262.95	383.30	0.32
Fermented A	439.38 [a]	642.82 [b]	0.35
Fermented B	445.41 [a]	644.37 [b]	0.37
Pomegranate peel Non-fermented	1104.25	1373.50	0.44
Fermented A	1233.28	1142.97	0.42
Fermented B	1060.01	1199.00	0.34

Effect of the lactic acid fermentations on the antioxidant activity (two-way ANOVA). [a] DPPH fermented (A or B) vs. non-fermented: $p = 0.0159$; fermented A vs. fermented B: $p > 0.05$. [b] FRAP fermented (A or B) vs. unfermented: $p = 0.0357$; fermented A vs. fermented B: $p > 0.05$. [c] XO fermented (A or B) vs. non-fermented: $p = 0.0494$; fermented A vs. fermented B: $p > 0.05$.

3.6. Effect on IL-1β-Induced IL-6, IL-8 and PGE2 Production in CCD18-Co Myofibroblasts

The anti-inflammatory effect of the four lacto-fermented beverages and their corresponding non-fermented beverage samples at the subtoxic dose of 2.5% on IL-1β-induced CCD18-Co myofibroblasts was tested by measuring the IL-6, IL-8, and PGE2 production for 18 h. The exposure of the cells to IL-1β led to an increase ($p < 0.05$) in the release of both pro-inflammatory cytokines and PGE2 compared to both untreated samples (CT) (Figure 1). The inflamed cells co-treated with non-fermented beverages samples of each plant material extract showed a reduction ($p < 0.05$) in the concentration of inflammatory markers, with the exception of rosemary for the IL-8 levels. Among the treatments, pomegranate peel beverage showed higher reductions for IL-6 and IL-8, followed by thyme, echinacea and rosemary beverages, respectively (Figure 1A,B). On the contrary, regarding the PGE2 production, the treatment with non-fermented beverages of rosemary and thyme showed the highest reduction (Figure 1C). Regarding the lacto-fermented beverages with strain A or B, the reduction in IL-6 levels was similar or even lower than of their corresponding non-fermented beverages, but still significant compared to untreated inflamed cells ($p < 0.05$). An exception was a greater reduction for both lacto-fermented thyme beverages compared to their corresponding non-fermented ones ($p < 0.05$) (Figure 1A). Regarding IL-8 values, lower but statistically significant ($p < 0.05$) values were detected for lacto-fermented thyme (using Strain A) and rosemary (using Strain B) beverages, but not for the other samples compared to their corresponding non-fermented beverages, although a non-significant trend was also observed for both lacto-fermented echinacea beverages and rosemary fermented with Strain A (Figure 1B). Finally, no statistically significant differences were observed in the reduction in PGE2 levels between the four lacto-fermented beverages and their corresponding non-fermented ones (Figure 1C).

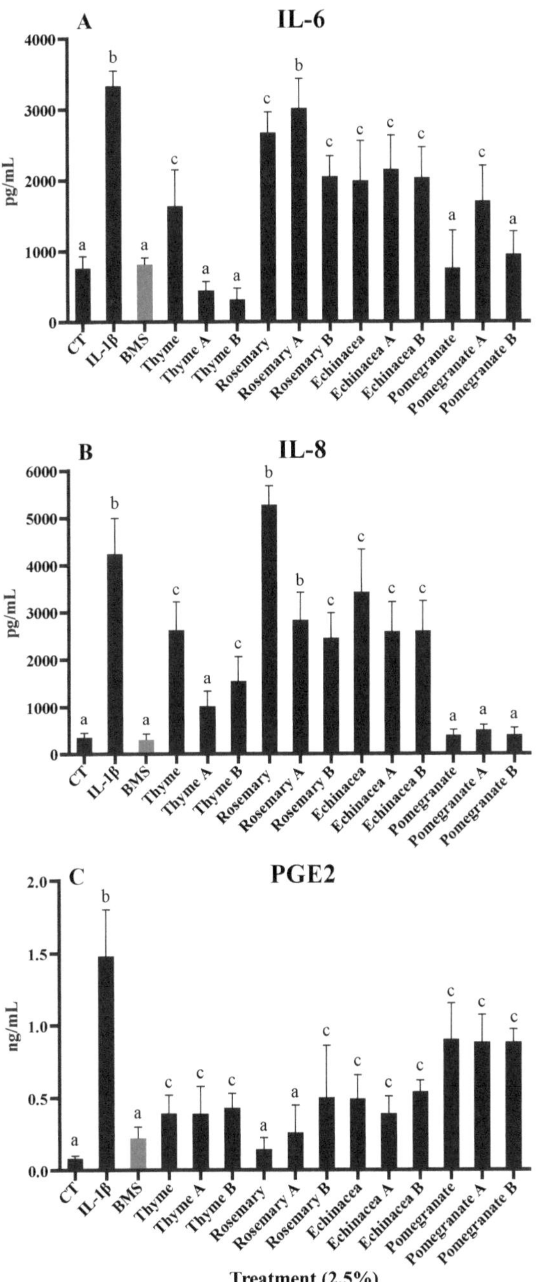

Figure 1. Pro-inflammatory cytokines (IL-6 (**A**) and IL-8 (**B**)) and PGE_2 levels (**C**) produced in the CCD18-Co culture media after 18 h of treatment, as measured by ELISA after exposure to IL-1β (1 ng/mL) alone or in combination with the four lacto-fermented herbal teas and their corresponding non-fermented beverage samples at a subtoxic dose of 2.5%. The selective IKK-2 inhibitor (BMS 345541; BMS) at 5 μM was assayed as a positive control of the anti-inflammatory effect. Results are shown as the mean ± SD of three independent experiments. Different letters indicate significant differences $p < 0.05$.

3.7. Correlation of (Poly)Phenolics with the Antioxidant and Anti-Inflammatory Effects

The correlation between the total polyphenol content (TPC) and individual (poly)phenol detected in each of the herbal infusions was analyzed, including data of the non-fermented infusions and those after two fermentations. The antioxidant activity (measured by DPPH and FRAP, and inhibition of XO) and the three pro-inflammatory markers (IL-6, IL-8, and PGE2) were also analyzed. Correlation and p values for all analyses in each herbal infusion are detailed in Supplementary Tables S3–S6. According to the data represented in Figure 2A, the Pearson correlation analysis revealed significantly negative correlations with the TPC of thyme with all pro-inflammatory markers and XO inhibition, while a positive correlation was found with DPPH activity. Individually, luteolin, chysoeriol glucoside, eriodictyol, and quercetin showed similar correlations, while rosmarinic acid only correlated positively with FRAP activity (r = 0.96, p = 0.002) (Figure 2A). For rosemary infusions, no statistically significant correlations were found with TPC for any marker evaluated. However, our analysis revealed significantly negative correlations with all individual (poly)phenols detected, except for rosmarinic acid, with all pro-inflammatory markers and XO inhibition. In addition, positive correlations were found with DPPH and FRAP activity (Figure 2B). Regarding echinacea infusions, the Pearson correlation analyses revealed significantly negative correlations between TPC with all pro-inflammatory markers and XO inhibition, while positive correlations were found between DPPH and FRAP activity. However, a positive correlation was also found for XO inhibition. Individually, a similar trend, although not statistically significant, was found for all phenolics detected in echinacea, with the exception of caffeic acid (Figure 2C). Finally, for pomegranate peel infusions, statistically significant inverse correlations between TPC and all pro-inflammatory markers were found. However, although a trend was observed, there were no statistically positive correlations with DPPH or FRAP activity nor an inverse correlation with XO inhibition. Among individual (poly)phenolics, most of them showed similar trends to TPC, with the exception of punicalin and punigluconin, which showed positive correlations with the three pro-inflammatory markers (Figure 2D).

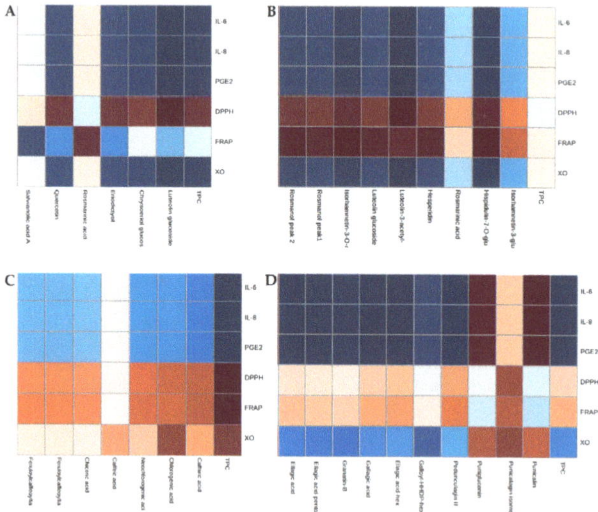

Figure 2. Heatmap analysis of the Pearson correlation between the total phenolic content (TPC) and individual (poly)phenolics detected with the three pro-inflammatory markers (IL-6, IL-8, and PGE$_2$) and antioxidant activities measured by DPPH, FRAP, and XO inhibition. Thyme (**A**), rosemary (**B**), echinacea (**C**), and pomegranate peel (**D**) infusions. Values higher than the averages (positive correlation) are indicated in red, and lower values in blue (negative correlation).

4. Discussion

The consumption of and demand for herbal teas and infusions continues to increase globally as caffeine-free beverages with attractive flavors, as well as many functional health benefits. Several aromatic herbs and plant extracts have traditionally been used for their antioxidant properties and for treating intestinal inflammation, indigestion, infections, etc., effects which are attributed to the predominant (poly)phenol compounds in these plant-based products [11,14,15]. However, although these bioactive compounds could be ingested in significant amounts, most (poly)phenols present in plant extracts have low bioavailability and reach the colon almost unaltered, where they can be metabolized by the gut microbiota to release numerous metabolites that are then absorbed and display biological effects [22,30]. In general, the biological action of (poly)phenols depend upon their bioaccessibility, which refers to the fraction of an ingested compound that is available for absorption in the gut and can also exert health benefits in the intestinal tract. In this line, different factors can affect the bioaccesibility of phenolics present in plant extracts, such as their water solubility, their presence mostly as glycosides, and their tight bond to food matrices [31]. Therefore, in recent decades, in order to enhance the bioaccessibility and bioavailability of the plant-food (poly)phenols and, therefore, their health benefits, different technological and biotechnological processes have been developed, including fermentation with lactic acid bacteria (LAB), which could be able to release phenolic compounds from the matrix and (or) lead to modifications and conversion to other phenolic compounds with improved bioavailability [8,31,32]. Herein, the present study has corroborated the impact of lacto-fermentation on several herbal infusions from thyme, rosemary, echinacea, and pomegranate peel on the (poly)phenolic composition and the improvement of their in vitro antioxidant and anti-inflammatory effects compared to their corresponding non-fermented infusions. The total (poly)phenol content (TPC) and the phenolic profile analyses by HPLC-MS/MS revealed that lacto-fermentation markedly increased the levels of most (poly)phenolics in all herbal infusions compared to their non-fermented counterparts. This is in agreement with many studies that have reported that LAB species and strains are able to increase the production of polyphenolic compounds in many foods and beverages [4,33–35]. Under our specific analysis conditions, we did not detect the formation of new compounds, as we followed a targeted strategy to compare fermented vs. non-fermented infusions. However, despite the significant differences observed between fermented and non-fermented infusions for many compounds, we cannot rule out the possibility of new bioactive compounds (e.g., phenolics) forming during the fermentation process.

In our study, we isolated from olive and black carrot two strains of *Lactiplantibacillus plantarum*, which are commonly used and found in foods, dairy products, and beverages and have the Generally Recognized as Safe (GRAS) status from the US Food and Drug Administration (US FDA) and the Qualified Presumption of Safety (QPS) status from the European Food Safety Authority (EFSA) [36,37]. Among herbal infusions, thyme infusion showed the highest increase in the (poly)phenolic content after fermentation, followed by the echinacea infusion. Surprisingly, although *L. plantarum* possesses β-glucosidase activities, the lacto-fermentation did not decrease, but rather increased the content of phenolic glycosides presents in thyme and rosemary, such as luteolin, chrysoeriol, isorhamnetin, or hispidulin. In contrast, previous studies have reported the high glycosidase activity of *L. plantarum*, with an improvement in the bioaccessibility and bioavailability of food phenolic compounds such as isoflavones and other flavonoid glycosides, such as quercetin or kaempferol. This is accompanied by an increase in their antioxidant activity [34,35,38,39]. However, according to these studies, differences were observed in the deglycosylation rate depending on the strains used (i.e., greater activity for *L. plantarum* 748T than for *L. plantarum* 9567), while in our study, the 16S rDNA sequencing indicated that the strains used for fermentation were *L. plantarum* 129 J1 and P1, which could be present lower β-glucosidase activity than others. Furthermore, another recent study described that some strains of *L. plantarum* do not possess any enzyme exhibiting β-glucosidase activity [40]. Therefore, our study corroborated that the impact of fermentation can be dependent both on LAB species with GRAS status and between strains of the same species. On the other hand, the release of phenolic compounds after lacto-fermentation could be related to the action of other enzymes that contribute to the breakdown of plant cell wall, thus improving their bioaccessibility. In this regard, other microbial enzymes apart from glucosidases, such as amylases, cellulases, chitinases, esterases, invertases, etc., produced during fermentation could degrade the cell walls of the plants and therefore improve the extraction of phenolic compounds [41].

Regarding the health benefits of lacto-fermentation, the antioxidant and anti-inflammatory effects of the lacto-fermented beverages and their corresponding non-fermented infusions were measured. We report here for the first time that thyme-, rosemary-, echinacea-, and pomegranate-based non-fermented infusions exhibit in vitro antioxidant effects similar to other plant-derived products (e.g., extracts or juices) [42–45]. According to the literature on the health benefits of lacto-fermentation, we observed enhanced antioxidant effects for most of the lacto-fermented infusions. The lacto-fermented thyme and rosemary infusions had an increased radical-scavenging activity (DPPH) and ferric reducing ability (FRAP, as well as to inhibit xanthine oxidase. Similarly, lacto-fermented echinacea infusions showed a significant increase in the DPPH and FRAP activities. Surprisingly, lacto-fermentations had controversial effects on pomegranate infusions, with a decrease in FRAP activity no matter the strain used and a higher DPPH activity when fermented with *L. plantarum* 129 J1 (strain A). These findings are in line with previous studies conducted on echinacea extract fermented with *L. plantarum* 1MR20 [46] and on pomegranate juice fermented with selected *L. plantarum* PU1 [47], which exhibited higher antioxidant effects than their respective unfermented controls. On the other hand, regarding the anti-inflammatory effects, all herbal infusion showed significant decreases in the production of pro-inflammatory markers (IL-6, IL-8, and PGE2) on IL-1β-induced CCD18-Co human colon myofibroblasts. This is in agreement with previous studies conducted in systemic and colon models where the anti-inflammatory effect was attributed to the phenolic fraction [21,48–50]. After lacto-fermentation with the two strains of *L. plantarum*, the reduction in all pro-inflammatory markers was preserved, with the exception of a greater reduction for lacto-fermented thyme compared to the corresponding non-fermented one and, to a lesser extent, for the fermented echinacea and rosemary infusions.

Overall, findings suggest that the lacto-fermentation of herbal infusions could enhance their antioxidant and anti-inflammatory effects, and this effect could be explained, at least partly, by the increase in and release of their phenolic fractions. Although no individual phenolics were evaluated, previous preclinical studies have found both antioxidant and anti-inflammatory activities in the gastrointestinal tract for many, such as rosmarinic acid, ellagic acid, chicoric acid, and caffeic acid, which are present in significant amounts in these herbs and plant extracts [51–54]. Furthermore, in order to explain whether the effect could be mediated by the (poly)phenolics, we ran a Pearson correlation analysis to evaluate a potential statistical correlation between the changes in the (poly)phenolics through lacto-fermentation and the antioxidant and anti-inflammatory effects. Through correlation analysis, we found that both the total phenolic content and most of the detected individual (poly)phenolics were positively correlated with the three evaluated pro-inflammatory markers and the XO inhibition. On the other hand, they were negatively correlated with the antioxidant activities (measured by DPPH and FRAP), mainly for thyme and echinacea, while those of pomegranate peel infusions correlated better with the inflammatory markers.

However, although we obtained some achievements, there are some limitations of the present study, and they should be further investigated. In this line, the antioxidant and anti-inflammatory properties could be also attributed to other fermentation-derived products such as vitamins, bioactive peptides, etc., or the synergy between different bioactives.

Furthermore, although the effect of the fermentation on these herbal infusions clearly improved the release of phenolics, the role of digestion and the interaction with the gut microbiota should be considered. Thus, firstly, during digestion, the structures of the (poly)phenolics can be hydrolyzed and modified, altering their bioavailability and potential health benefits [55]. Therefore, increasing the (poly)phenolic content through lacto-fermentation could preserve its health effects. In this line, a recent study carried out with extracts of different echinacea parts reported that their anti-inflammatory effects were preserved after in vitro gastrointestinal digestion despite a reduction in the concentration of their phenolics in those with higher contents [21]. Regarding the interaction with the microbiota, a higher content of phenolics, included glycosylated forms, which can be cleaved by endogenous or microbial enzymes from gut microbiota to release an absorbable and(or) metabolizable aglycone forms, could favor a higher conversion of phenolic compounds to more biologically active compounds (e.g., ellagic acid to urolithins) [56], as well as the prebiotic effect of the phenolics to positively modulate the gut microbiota [22,57].

5. Conclusions

Our results underscore the differential impact of lacto-fermentation on specific phenolic compounds in different herbal infusions, highlighting the complex interplay between microbial strains and phenolic composition during fermentation processes. The results showed that fermentation with *L. plantarum* markedly increased the content (accessibility) of phenolic compounds compared to the non-fermented infusions, with an enhancement of the antioxidant and anti-inflammatory activities. Among the fermented herbal infusions, those from thyme and rosemary showed more significant effects the corresponding non-fermented ones.

In conclusion, this research demonstrates that the health benefits of various herbal infusions can be improved with LAB fermentation based on greater accessibility of their phenolic compounds, thus being able to exert higher antioxidant and anti-inflammatory activities. Overall, this study suggests that lacto-fermentation could be used as tool to produce novel functional herbal infusions or beverages with higher antioxidant and anti-inflammatory activities. Furthermore, our results provide a scientific basis for highlighting the effective bioactivity of lacto-fermented herbal infusions that could be consumed in alleviating certain oxidative-stress-related diseases, intestinal inflammatory conditions, and related disorders.

Supplementary Materials: The following supporting information can be downloaded at: https://www.mdpi.com/article/10.3390/antiox13050562/s1, Table S1: pH values of lacto-fermented herbal teas; Table S2: Content of phenolics of the different plant material used in this study; Table S3: Pearson correlations (r value) and p values in thyme infusions between the total phenolic content (TPC) and individual (poly)phenolics detected with the three pro-inflammatory markers (IL-6, IL-8, and PGE2) and antioxidant activities measured by DPPH, FRAP, and XO inhibition; Table S4: Pearson correlations (r value) and *p* values in rosemary infusions between the total phenolic content (TPC) and individual (poly)phenolics detected with the three pro-inflammatory markers (IL-6, IL-8, and PGE2) and antioxidant activities measured by DPPH, FRAP, and XO inhibition; Table S5: Pearson correlations (r value) and *p* values in echinacea infusions between the total phenolic content (TPC) and individual (poly)phenolics detected with the three pro-inflammatory markers (IL-6, IL-8, and PGE2) and antioxidant activities measured by DPPH, FRAP, and XO inhibition; Table S6: Pearson correlations (r value) and *p* values in pomegranate peel infusions between the total phenolic content (TPC) and individual (poly)phenolics detected with the three pro-inflammatory markers (IL-6, IL-8, and PGE2) and antioxidant activities measured by DPPH, FRAP, and XO inhibition.

Author Contributions: Conceptualization, L.-E.M. and A.G.-S.; methodology, T.O., M.Á.Á.-G., S.M., F.V., A.B., D.F., E.P., L.-E.M. and A.G.-S.; writing—original draft preparation, L.-E.M. and A.G.-S.; writing—review and editing, all authors; funding acquisition, E.P., C.M. and A.G.-S. All authors have read and agreed to the published version of the manuscript.

Funding: This project was funded by the European Union's Horizon 2020 research and innovation programme under the PhenolAcTwin Project (Grant Agreement No: 951994). This work was supported by the grant TED2021-130962B-C22, funded by the MCIN/AEI/10.13039/501100011033 and by the "European Union NextGenerationEU/PRTR" program; the grant 22030/PI/22, funded by the Programa Regional de Fomento de la Investigación Científica y Técnica (Plan de Actuación 2022) de la Fundación Séneca-Agencia de Ciencia y Tecnología de la Región de Murcia, Spain; and the AGROAL-

NEXT programme, supported by MCIN with funding from the European Union NextGenerationEU (PRTR-C17.I1) and Fundación Séneca, Comunidad Autónoma Región de Murcia (CARM).

Institutional Review Board Statement: Not applicable.

Informed Consent Statement: Not applicable.

Data Availability Statement: All data included in this study are available upon request by contacting the corresponding authors.

Conflicts of Interest: The authors declare no conflicts of interest.

References

1. Shah, A.M.; Tarfeen, N.; Mohamed, H.; Song, Y. Fermented Foods: Their Health-Promoting Components and Potential Effects on Gut Microbiota. *Fermentation* **2023**, *9*, 118. [CrossRef]
2. Şanlier, N.; Gökcen, B.B.; Sezgin, A.C. Health Benefits of Fermented Foods. *Crit. Rev. Food Sci. Nutr.* **2019**, *59*, 506–527. [CrossRef] [PubMed]
3. Dimidi, E.; Cox, S.; Rossi, M.; Whelan, K. Fermented Foods: Definitions and Characteristics, Impact on the Gut Microbiota and Effects on Gastrointestinal Health and Disease. *Nutrients* **2019**, *11*, 1806. [CrossRef]
4. Filannino, P.; Bai, Y.; Di Cagno, R.; Gobbetti, M.; Gänzle, M.G. Metabolism of Phenolic Compounds by *Lactobacillus* spp. during Fermentation of Cherry Juice and Broccoli Puree. *Food Microbiol.* **2015**, *46*, 272–279. [CrossRef]
5. Saleem, G.N.; Gu, R.; Qu, H.; Bahar Khaskheli, G.; Rashid Rajput, I.; Qasim, M.; Chen, X. Therapeutic Potential of Popular Fermented Dairy Products and Its Benefits on Human Health. *Front. Nutr.* **2024**, *11*, 1328620. [CrossRef]
6. Zapaśnik, A.; Sokołowska, B.; Bryła, M. Role of Lactic Acid Bacteria in Food Preservation and Safety. *Foods* **2022**, *11*, 1283. [CrossRef] [PubMed]
7. Icer, M.A.; Özbay, S.; Ağagündüz, D.; Kelle, B.; Bartkiene, E.; Rocha, J.M.F.; Ozogul, F. The Impacts of Acidophilic Lactic Acid Bacteria on Food and Human Health: A Review of the Current Knowledge. *Foods* **2023**, *12*, 2965. [CrossRef]
8. Ashaolu, T.J. A Review on Selection of Fermentative Microorganisms for Functional Foods and Beverages: The Production and Future Perspectives. *Int. J. Food Sci. Technol.* **2019**, *54*, 2511–2519. [CrossRef]
9. Mathur, H.; Beresford, T.P.; Cotter, P.D. Health Benefits of Lactic Acid Bacteria (LAB) Fermentates. *Nutrients* **2020**, *12*, 1679. [CrossRef]
10. Zhu, H.; Guo, L.; Yu, D.; Du, X. New Insights into Immunomodulatory Properties of Lactic Acid Bacteria Fermented Herbal Medicines. *Front. Microbiol.* **2022**, *13*, 1073922. [CrossRef]
11. Liu, Y.; Guo, C.; Zang, E.; Shi, R.; Liu, Q.; Zhang, M.; Zhang, K.; Li, M. Review on Herbal Tea as a Functional Food: Classification, Active Compounds, Biological Activity, and Industrial Status. *J. Future Foods* **2023**, *3*, 206–219. [CrossRef]
12. Long, X.; Hu, C.; Geng, Y. Exploration and Enlightenment of Non-Camellia Tea Beverage Consumption Market for Health Care. *J. Green Sci. Technol.* **2023**, *25*, 275–280.
13. Atoui, A. Tea and Herbal Infusions: Their Antioxidant Activity and Phenolic Profile. *Food Chem.* **2005**, *89*, 27–36. [CrossRef]
14. Chandrasekara, A.; Shahidi, F. Herbal Beverages: Bioactive Compounds and Their Role in Disease Risk Reduction—A Review. *J. Tradit. Complement. Med.* **2018**, *8*, 451–458. [CrossRef] [PubMed]
15. Poswal, F.S.; Russell, G.; Mackonochie, M.; MacLennan, E.; Adukwu, E.C.; Rolfe, V. Herbal Teas and Their Health Benefits: A Scoping Review. *Plant Foods Hum. Nutr.* **2019**, *74*, 266–276. [CrossRef] [PubMed]
16. Parham, S.; Kharazi, A.Z.; Bakhsheshi-Rad, H.R.; Nur, H.; Ismail, A.F.; Sharif, S.; RamaKrishna, S.; Berto, F. Antioxidant, Antimicrobial and Antiviral Properties of Herbal Materials. *Antioxidants* **2020**, *9*, 1309. [CrossRef] [PubMed]
17. Martins, N.; Barros, L.; Santos-Buelga, C.; Silva, S.; Henriques, M.; Ferreira, I.C.F.R. Decoction, Infusion and Hydroalcoholic Extract of Cultivated Thyme: Antioxidant and Antibacterial Activities, and Phenolic Characterisation. *Food Chem.* **2015**, *167*, 131–137. [CrossRef] [PubMed]
18. Mena, P.; Cirlini, M.; Tassotti, M.; Herrlinger, K.; Dall'Asta, C.; Del Rio, D. Phytochemical Profiling of Flavonoids, Phenolic Acids, Terpenoids, and Volatile Fraction of a Rosemary (*Rosmarinus officinalis* L.) Extract. *Molecules* **2016**, *21*, 1576. [CrossRef] [PubMed]
19. Akhtar, S.; Ismail, T.; Fraternale, D.; Sestili, P. Pomegranate Peel and Peel Extracts: Chemistry and Food Features. *Food Chem.* **2015**, *174*, 417–425. [CrossRef]
20. Vieira, S.F.; Gonçalves, S.M.; Gonçalves, V.M.F.; Llaguno, C.P.; Macías, F.; Tiritan, M.E.; Cunha, C.; Carvalho, A.; Reis, R.L.; Ferreira, H.; et al. Echinacea Purpurea Fractions Represent Promising Plant-Based Anti-Inflammatory Formulations. *Antioxidants* **2023**, *12*, 425. [CrossRef]
21. Ávila-Gálvez, M.Á.; Giménez-Bastida, J.A.; Karadeniz, B.; Romero-Reyes, S.; Espín, J.C.; Pelvan, E.; González-Sarrías, A. Polyphenolic Characterization and Anti-Inflammatory Effect of In Vitro Digested Extracts of *Echinacea Purpurea* L. Plant Parts in an Inflammatory Model of Human Colon Cells. *Int. J. Mol. Sci.* **2024**, *25*, 1744. [CrossRef] [PubMed]
22. Espín, J.C.; González-Sarrías, A.; Tomás-Barberán, F.A. The Gut Microbiota: A Key Factor in the Therapeutic Effects of (Poly)Phenols. *Biochem. Pharmacol.* **2017**, *139*, 82–93. [CrossRef] [PubMed]

23. Ndoye, S.; Fraisse, D.; Akendengué, B.; Dioum, M.; Gueye, R.; Sall, C.; Seck, I.; Felgines, C.; Seck, M.; Senejoux, F. Antioxidant and Antiglycation Properties of Two Mango *(Mangifera Indica* L.) Cultivars from Senegal. *Asian Pac. J. Trop. Biomed.* **2018**, *8*, 137. [CrossRef]

24. Ávila-Gálvez, M.Á.; Romo-Vaquero, M.; González-Sarrías, A.; Espín, J.C. Kinetic Disposition of Dietary Polyphenols and Methylxanthines in the Rat Mammary Tissue. *J. Funct. Foods* **2019**, *61*, 103516. [CrossRef]

25. Fraisse, D.; Bred, A.; Felgines, C.; Senejoux, F. Screening and Characterization of Antiglycoxidant Anthocyanins from Vaccinium Myrtillus Fruit Using DPPH and Methylglyoxal Pre-Column HPLC Assays. *Antioxidants* **2020**, *9*, 512. [CrossRef]

26. Katalinić, V.; Generalić, I.; Skroza, D.; Ljubenkov, I.; Teskera, A.; Konta, I.; Boban, M. Insight in the Phenolic Composition and Antioxidative Properties of Vitis Vinifera Leaves Extracts. *Croat. J. Food Sci. Technol.* **2009**, *1*, 7–15.

27. Sowndhararajan, K.; Joseph, J.M.; Rajendrakumaran, D. In Vitro Xanthine Oxidase Inhibitory Activity of Methanol Extracts of Erythrina Indica Lam. Leaves and Stem Bark. *Asian Pac. J. Trop. Biomed.* **2012**, *2*, S1415–S1417. [CrossRef]

28. González-Sarrías, A.; Espín-Aguilar, J.C.; Romero-Reyes, S.; Puigcerver, J.; Alajarín, M.; Berná, J.; Selma, M.V.; Espín, J.C. Main Determinants Affecting the Antiproliferative Activity of Stilbenes and Their Gut Microbiota Metabolites in Colon Cancer Cells: A Structure–Activity Relationship Study. *Int. J. Mol. Sci.* **2022**, *23*, 15102. [CrossRef]

29. Odlaug, T.E.; Pflug, I.J. Clostridium Botulinum and Acid Foods. *J. Food Prot.* **1978**, *41*, 566–573. [CrossRef]

30. González-Sarrías, A.; Espín, J.C.; Tomás-Barberán, F.A. Non-Extractable Polyphenols Produce Gut Microbiota Metabolites That Persist in Circulation and Show Anti-Inflammatory and Free Radical-Scavenging Effects. *Trends Food Sci. Technol.* **2017**, *69*, 281–288. [CrossRef]

31. Polia, F.; Pastor-Belda, M.; Martínez-Blázquez, A.; Horcajada, M.-N.; Tomás-Barberán, F.A.; García-Villalba, R. Technological and Biotechnological Processes To Enhance the Bioavailability of Dietary (Poly)Phenols in Humans. *J. Agric. Food Chem.* **2022**, *70*, 2092–2107. [CrossRef] [PubMed]

32. Lee, N.-K.; Paik, H.-D. Bioconversion Using Lactic Acid Bacteria: Ginsenosides, GABA, and Phenolic Compounds. *J. Microbiol. Biotechnol.* **2017**, *27*, 869–877. [CrossRef]

33. Wang, L.-C.; Pan, T.-M.; Tsai, T.-Y. Lactic Acid Bacteria-Fermented Product of Green Tea and Houttuynia Cordata Leaves Exerts Anti-Adipogenic and Anti-Obesity Effects. *J. Food Drug Anal.* **2018**, *26*, 973–984. [CrossRef] [PubMed]

34. De Montijo-Prieto, S.; Razola-Díaz, M.d.C.; Barbieri, F.; Tabanelli, G.; Gardini, F.; Jiménez-Valera, M.; Ruiz-Bravo, A.; Verardo, V.; Gómez-Caravaca, A.M. Impact of Lactic Acid Bacteria Fermentation on Phenolic Compounds and Antioxidant Activity of Avocado Leaf Extracts. *Antioxidants* **2023**, *12*, 298. [CrossRef] [PubMed]

35. Letizia, F.; Fratianni, A.; Cofelice, M.; Testa, B.; Albanese, G.; Di Martino, C.; Panfili, G.; Lopez, F.; Iorizzo, M. Antioxidative Properties of Fermented Soymilk Using Lactiplantibacillus Plantarum LP95. *Antioxidants* **2023**, *12*, 1442. [CrossRef] [PubMed]

36. Rychen, G.; Aquilina, G.; Azimonti, G.; Bampidis, V.; Bastos, M.d.L.; Bories, G.; Chesson, A.; Cocconcelli, P.S.; Flachowsky, G.; Gropp, J.; et al. Guidance on the Characterisation of Microorganisms Used as Feed Additives or as Production Organisms. *EFSA J.* **2018**, *16*, e05206. [CrossRef]

37. Yilmaz, B.; Bangar, S.P.; Echegaray, N.; Suri, S.; Tomasevic, I.; Manuel Lorenzo, J.; Melekoglu, E.; Rocha, J.M.; Ozogul, F. The Impacts of Lactiplantibacillus Plantarum on the Functional Properties of Fermented Foods: A Review of Current Knowledge. *Microorganisms* **2022**, *10*, 826. [CrossRef] [PubMed]

38. Landete, J.M.; Curiel, J.A.; Rodríguez, H.; de las Rivas, B.; Muñoz, R. Aryl Glycosidases from Lactobacillus Plantarum Increase Antioxidant Activity of Phenolic Compounds. *J. Funct. Foods* **2014**, *7*, 322–329. [CrossRef]

39. María Landete, J.; Hernández, T.; Robredo, S.; Dueñas, M.; de las Rivas, B.; Estrella, I.; Muñoz, R. Effect of Soaking and Fermentation on Content of Phenolic Compounds of Soybean (*Glycine Max* Cv. Merit) and Mung Beans (*Vigna Radiata* [L] Wilczek). *Int. J. Food Sci. Nutr.* **2015**, *66*, 203–209. [CrossRef]

40. Plaza-Vinuesa, L.; Hernandez-Hernandez, O.; Sánchez-Arroyo, A.; Cumella, J.M.; Corzo, N.; Muñoz-Labrador, A.M.; Moreno, F.J.; de las Rivas, B.; Muñoz, R. Deciphering the Myrosinase-like Activity of *Lactiplantibacillus Plantarum* WCFS1 among GH1 Family Glycoside Hydrolases. *J. Agric. Food Chem.* **2022**, *70*, 15531–15538. [CrossRef]

41. Muñoz, R.; de las Rivas, B.; Rodríguez, H.; Esteban-Torres, M.; Reverón, I.; Santamaría, L.; Landete, J.M.; Plaza-Vinuesa, L.; Sánchez-Arroyo, A.; Jiménez, N.; et al. Food Phenolics and Lactiplantibacillus Plantarum. *Int. J. Food Microbiol.* **2024**, *412*, 110555. [CrossRef] [PubMed]

42. Pellati, F.; Benvenuti, S.; Magro, L.; Melegari, M.; Soragni, F. Analysis of Phenolic Compounds and Radical Scavenging Activity of Echinacea Spp. *J. Pharm. Biomed. Anal.* **2004**, *35*, 289–301. [CrossRef]

43. Vallverdú-Queralt, A.; Regueiro, J.; Martínez-Huélamo, M.; Rinaldi Alvarenga, J.F.; Leal, L.N.; Lamuela-Raventos, R.M. A Comprehensive Study on the Phenolic Profile of Widely Used Culinary Herbs and Spices: Rosemary, Thyme, Oregano, Cinnamon, Cumin and Bay. *Food Chem.* **2014**, *154*, 299–307. [CrossRef] [PubMed]

44. Chen, Y.-L.; Sung, J.-M.; Lin, S.-D. Effect of Extraction Methods on the Active Compounds and Antioxidant Properties of Ethanolic Extracts of Echinacea Purpurea Flower. *Am. J. Plant Sci.* **2015**, *6*, 201–212. [CrossRef]

45. Mantzourani, I.; Kazakos, S.; Terpou, A.; Alexopoulos, A.; Bezirtzoglou, E.; Bekatorou, A.; Plessas, S. Potential of the Probiotic Lactobacillus Plantarum ATCC 14917 Strain to Produce Functional Fermented Pomegranate Juice. *Foods* **2018**, *8*, 4. [CrossRef] [PubMed]

46. Rizzello, C.G.; Coda, R.; Macías, D.S.; Pinto, D.; Marzani, B.; Filannino, P.; Giuliani, G.; Paradiso, V.M.; Di Cagno, R.; Gobbetti, M. Lactic Acid Fermentation as a Tool to Enhance the Functional Features of *Echinacea* spp. *Microb. Cell Fact.* **2013**, *12*, 44. [CrossRef] [PubMed]

47. Pontonio, E.; Montemurro, M.; Pinto, D.; Marzani, B.; Trani, A.; Ferrara, G.; Mazzeo, A.; Gobbetti, M.; Rizzello, C.G. Lactic Acid Fermentation of Pomegranate Juice as a Tool to Improve Antioxidant Activity. *Front. Microbiol.* **2019**, *10*, 1550. [CrossRef] [PubMed]

48. Park, J.B. Identification and Quantification of a Major Anti-Oxidant and Anti-Inflammatory Phenolic Compound Found in Basil, Lemon Thyme, Mint, Oregano, Rosemary, Sage, and Thyme. *Int. J. Food Sci. Nutr.* **2011**, *62*, 577–584. [CrossRef] [PubMed]

49. Chohan, M.; Naughton, D.P.; Jones, L.; Opara, E.I. An Investigation of the Relationship between the Anti-Inflammatory Activity, Polyphenolic Content, and Antioxidant Activities of Cooked and In Vitro Digested Culinary Herbs. *Oxid. Med. Cell Longev.* **2012**, *2012*, 627843. [CrossRef]

50. Mastrogiovanni, F.; Mukhopadhya, A.; Lacetera, N.; Ryan, M.; Romani, A.; Bernini, R.; Sweeney, T. Anti-Inflammatory Effects of Pomegranate Peel Extracts on In Vitro Human Intestinal Caco-2 Cells and Ex Vivo Porcine Colonic Tissue Explants. *Nutrients* **2019**, *11*, 548. [CrossRef]

51. Hussein, O.E.; Hozayen, W.G.; Bin-Jumah, M.N.; Germoush, M.O.; Abd El-Twab, S.M.; Mahmoud, A.M. Chicoric Acid Prevents Methotrexate Hepatotoxicity via Attenuation of Oxidative Stress and Inflammation and Up-Regulation of PPARγ and Nrf2/HO-1 Signaling. *Environ. Sci. Pollut. Res.* **2020**, *27*, 20725–20735. [CrossRef] [PubMed]

52. Jin, B.-R.; Chung, K.-S.; Hwang, S.; Hwang, S.N.; Rhee, K.-J.; Lee, M.; An, H.-J. Rosmarinic Acid Represses Colitis-Associated Colon Cancer: A Pivotal Involvement of the TLR4-Mediated NF-KB-STAT3 Axis. *Neoplasia* **2021**, *23*, 561–573. [CrossRef] [PubMed]

53. Xiang, C.; Liu, M.; Lu, Q.; Fan, C.; Lu, H.; Feng, C.; Yang, X.; Li, H.; Tang, W. Blockade of TLRs-Triggered Macrophage Activation by Caffeic Acid Exerted Protective Effects on Experimental Ulcerative Colitis. *Cell Immunol.* **2021**, *365*, 104364. [CrossRef] [PubMed]

54. Li, X.; Xu, L.; Peng, X.; Zhang, H.; Kang, M.; Jiang, Y.; Shi, H.; Chen, H.; Zhao, C.; Yu, Y.; et al. The Alleviating Effect of Ellagic Acid on DSS-Induced Colitis *via* Regulating Gut Microbiomes and Gene Expression of Colonic Epithelial Cells. *Food Funct.* **2023**, *14*, 7550–7561. [CrossRef] [PubMed]

55. Wojtunik-Kulesza, K.; Oniszczuk, A.; Oniszczuk, T.; Combrzyński, M.; Nowakowska, D.; Matwijczuk, A. Influence of In Vitro Digestion on Composition, Bioaccessibility and Antioxidant Activity of Food Polyphenols—A Non-Systematic Review. *Nutrients* **2020**, *12*, 1401. [CrossRef]

56. García-Villalba, R.; Giménez-Bastida, J.A.; Cortés-Martín, A.; Ávila-Gálvez, M.Á.; Tomás-Barberán, F.A.; Selma, M.V.; Espín, J.C.; González-Sarrías, A. Urolithins: A Comprehensive Update on Their Metabolism, Bioactivity, and Associated Gut Microbiota. *Mol. Nutr. Food Res.* **2022**, *66*, e2101019. [CrossRef]

57. Marco, M.L.; Heeney, D.; Binda, S.; Cifelli, C.J.; Cotter, P.D.; Foligné, B.; Gänzle, M.; Kort, R.; Pasin, G.; Pihlanto, A.; et al. Health Benefits of Fermented Foods: Microbiota and Beyond. *Curr. Opin. Biotechnol.* **2017**, *44*, 94–102. [CrossRef]

 *antioxidants*

Article

Deciphering the Crucial Roles of the Quorum-Sensing Transcription Factor SdiA in NADPH Metabolism and (*S*)-Equol Production in *Escherichia coli* Nissle 1917

Zhe Wang [1,2], Yiqiang Dai [1,2], Fidelis Azi [3], Mingsheng Dong [1,*] and Xiudong Xia [1,2,4,5,*]

1   College of Food Science and Technology, Nanjing Agricultural University, Nanjing 210095, China; wangzhe@stu.njau.edu.cn (Z.W.)
2   Institute of Agro-Product Processing, Jiangsu Academy of Agricultural Sciences, Nanjing 210014, China
3   Department of Chemical Engineering, Guangdong Technion-Israel Institute of Technology, Shantou 515063, China
4   Jiangsu Key Laboratory for Food Quality and Safety-State Key Laboratory Cultivation Base, Ministry of Science and Technology, Nanjing 210014, China
5   School of Food and Biological Engineering, Jiangsu University, Zhenjiang 212013, China
*   Correspondence: dongms@njau.edu.cn (M.D.); 20140034@jaas.ac.cn (X.X.)

Abstract: The active metabolite (*S*)-equol, derived from daidzein by gut microbiota, exhibits superior antioxidative activity compared with its precursor and plays a vital role in human health. As only 25% to 50% of individuals can naturally produce equol when supplied with isoflavone, we engineered probiotic *E. coli* Nissle 1917 (EcN) to convert dietary isoflavones into (*S*)-equol, thus offering a strategy to mimic the gut phenotype of natural (*S*)-equol producers. However, co-fermentation of EcN-eq with fecal bacteria revealed that gut microbial metabolites decreased NADPH levels, hindering (*S*)-equol production. Transcriptome analysis showed that the quorum-sensing (QS) transcription factor SdiA negatively regulates NADPH levels and (*S*)-equol biosynthesis in EcN-eq. Screening AHLs showed that SdiA binding to C10-HSL negatively regulates the pentose phosphate pathway, reducing intracellular NADPH levels in EcN-eq. Molecular docking and dynamics simulations investigated the structural disparities in complexes formed by C10-HSL with SdiA from EcN or *E. coli* K12. Substituting *sdiA_EcN* in EcN-eq with *sdiA_K12* increased the intracellular NADPH/NADP$^+$ ratio, enhancing (*S*)-equol production by 47%. These findings elucidate the impact of AHL-QS in the gut microbiota on EcN NADPH metabolism, offering insights for developing (*S*)-equol-producing EcN probiotics tailored to the gut environment.

Keywords: isoflavone; (*S*)-equol; quorum sensing; *E. coli* Nissle 1917; NADPH; biosynthesis

 **check for updates**

Citation: Wang, Z.; Dai, Y.; Azi, F.; Dong, M.; Xia, X. Deciphering the Crucial Roles of the Quorum-Sensing Transcription Factor SdiA in NADPH Metabolism and (*S*)-Equol Production in *Escherichia coli* Nissle 1917. *Antioxidants* **2024**, *13*, 259. https://doi.org/10.3390/antiox13030259

Academic Editor: Myung-Ji Seo

Received: 16 January 2024
Revised: 17 February 2024
Accepted: 18 February 2024
Published: 20 February 2024

1. Introduction

(*S*)-equol is a gut microbial metabolite of daidzein and daidzin, which are the major isoflavones in soybean, and it exhibits heightened antioxidative and estrogenic activities compared with its precursors [1,2]. Its potential in alleviating oxidative stress, menopausal syndrome, cardiovascular disease, and osteoporosis, as well as reducing the risk of prostate, colon, and breast cancer, has garnered substantial clinical interest [2,3]. However, only a fraction (25 to 50%) of people harbor gut bacteria with an (*S*)-equol-producing pathway capable of metabolizing dietary daidzein into (*S*)-equol [4,5]. Fecal transplant therapy has demonstrated success in conferring an (*S*)-equol-producing phenotype in hosts by providing them with exogenous gut bacteria expressing the necessary enzymes for daidzein conversion into (*S*)-equol [6,7]. Recently, engineered bacteria-integrated (*S*)-equol-producing pathways have been utilized as efficient whole-cell biocatalysts for (*S*)-equol production [3,8]. Given that EcN is a probiotic widely used as a host for the development of engineered probiotics [9,10], it can be used to develop an engineered probiotic with the capability of converting dietary isoflavones into (*S*)-equol efficiently [11].

The (*S*)-equol-producing pathway involves four enzymatic steps: daidzein reductase (DZNR) catalyzes the conversion of daidzein to (*R*)-dihydrodaidzein, which can further undergo conversion to (*S*)-dihydrodaidzein facilitated by dihydrodaidzein racemase (DDRC) [3,4]. The subsequent reduction of (*S*)-dihydrodaidzein to *t*-tetrahydrodaidzein occurs through dihydrodaidzein reductase (DHDR), followed by the transformation of *t*-tetrahydrodaidzein into (*S*)-equol via tetrahydrodaidzein reductase (THDR) [5]. Notably, the enzymes DZNR and DHDR belong to the category of NADPH-dependent oxidoreductases, and the biosynthesis of 1 mol of (*S*)-equol necessitates the consumption of 2 mol of NADPH [8,12]. Therefore, the conversion of daidzein to (*S*)-equol will be impaired when intracellular NADPH is insufficient [8].

Quorum sensing (QS) is vital in bacterial physiological processes, allowing bacteria to adapt to environmental changes and explore new environmental niches [13]. This communication significantly impacts exogenous probiotic metabolism and phenotype, influencing traits such as acid resistance, adhesion, and intestinal colonization [14,15]. The LuxR/I-type QS systems are commonly seen in most Gram-negative proteobacteria, where the QS signal is provided by acyl homoserine lactone (AHL) [13]. Typically, these bacteria encode LuxI synthase, for AHL production, and the cognate LuxR transcription factor, which is regulated by AHL [13,16]. However, in EcN, only the LuxR-type transcription factor SdiA is present without the LuxI-type synthase, which functions in intraspecies and interspecies communication by detecting exogenous AHLs from other bacteria [16–18]. Recent studies found that AHL-QS could regulate NADPH metabolism in microorganisms by regulating the glucose-6-phosphate dehydrogenase and pentose phosphate pathways [19–22].

The impact of intestinal flora on engineered EcN and the biosynthesis of (*S*)-equol, particularly in relation to QS regulation, remains largely unexplored, despite the documented production of AHLs by intestinal flora [16]. Therefore, the aims of this study were (1) to investigate the influence of gut microbiota on NADPH metabolism, with a specific focus on the role of the QS transcription factor SdiA, and (2) to elucidate the molecular mechanism by which AHL-SdiA regulates NADPH metabolism, with the aim of optimizing the (*S*)-equol production of engineered EcN strains in the intestinal environment. These results could clarify the effect of AHL-QS of gut microbiota on the NADPH metabolism of EcN, which is instructive for developing engineered probiotics suited to the gut environment.

2. Materials and Methods

2.1. Strains and Chemicals

E. coli DH5α was utilized for plasmid amplification, while EcN was employed for gene expression. A comprehensive list of all strains used in this investigation is presented in Table 1. All genetic manipulations, including the use of restriction enzymes, DNA ligases, High-Fidelity DNA polymerase, and one-step directional cloning kits, were sourced from Yeasen Biotech (Shanghai, China). Standard and commercial soy isoflavones (SI; 80% *w/w* isoflavones) were purchased from Yuanye Biotech (Shanghai, China). The isoflavones in the SI used in this study comprised 522.35 μg/mg of daidzin, 184.72 μg/mg of glycitin, 77.51 μg/mg of genistin, 20.38 μg/mg of daidzein, 10.05 μg/mg of glycitein, and 6.17 μg/mg of genistein.

2.2. Plasmids and DNA Manipulation

The specifics of the primers and plasmids utilized in this study are outlined in Tables S1 and S2. The synthesis of *dznr* from *Asaccharobacter celatus*, as well as *ddrc*, *thdr*, and *dhdr* from *Slackia isoflavoniconvertens*, was performed by Sangon (Shanghai, China), incorporating codon optimization for *E. coli* [3,8]. For the construction of pETM6-*Pnar*, the *nar* promoter sequence (*Pnar*) was chemically synthesized, aligning with previously reported sequences [3]. The *Avr*II- and *Nde*I-digested *Pnar* fragment was subsequently cloned into the *Avr*II/*Nde*I-digested pETM6 plasmid, creating pETM6-*Pnar*. To obtain the *Pnar* sequence, the primers Pf_Pnar and Pr_Pnar were employed in a PCR reaction. The resulting PCR product mixture underwent *Dpn*I digestion and subsequent ligation using one-step cloning kits.

Table 1. Strains used in this study.

Strains	Relevant Properties	Source
E. coli DH5α (collection no.: DSM 6897)	F$^-$, φ80d *lacZ*ΔM15, Δ*(lacZYA-argF)*U169, *recA*1, *endA*1, *hsdR*17(rk$^-$, mk$^+$), *phoA*, *supE*44λ$^-$, *thi*$^-$, *gyrA*96, *relA*1	Invitrogen (Invitrogen, Carlsbad, CA, USA)
EcN (collection no.: DSM 115365)	Wild-type *E. coli* Nissle 1917	Lab stock
EcN-eq	EcN, *malK*::P$_{nar}$-*dznr*-P$_{nar}$-*ddrc*-P$_{nar}$-*dhdr*-P$_{nar}$-*thdr*, *exo/cea*::P$_{nar}$-*bglF*-P$_{nar}$-*bglB*, Δ*ptsG*::Kan^R	This study
EcN-eq Δ*decR*	EcN-eq, Δ*decR*	This study
EcN-eq Δ*HW372_01960*	EcN-eq, Δ*HW372_01960*	This study
EcN-eq Δ*yhjC*	EcN-eq, Δ*yhjC*	This study
EcN-eq Δ*HW372_03545*	EcN-eq, Δ*HW372_03545*	This study
EcN-eq Δ*sdiA*	EcN-eq, Δ*sdiA*	This study
EcN-eq Δ*yhaJ*	EcN-eq, Δ*yhaJ*	This study
EcN-eq pETM6-P$_{nar}$-*decR*	EcN-eq carrying pETM6-P$_{nar}$-*decR*	This study
EcN-eq pETM6-P$_{nar}$-*HW372_01960*	EcN-eq carrying pETM6-P$_{nar}$-*HW372_01960*	This study
EcN-eq pETM6-P$_{nar}$-*yhjC*	EcN-eq carrying pETM6-P$_{nar}$-*yhjC*	This study
EcN-eq pETM6-P$_{nar}$-*HW372_03545*	EcN-eq carrying pETM6-P$_{nar}$-*HW372_03545*	This study
EcN-eq pETM6-P$_{nar}$-*sdiA*	EcN-eq carrying pETM6-P$_{nar}$-*sdiA*	This study
EcN-eq pETM6-P$_{nar}$-*yhaJ*	EcN-eq carrying pETM6-P$_{nar}$-*yhaJ*	This study
EcN-eq Δ*sdiA*::*sdiA_K12*	EcN-eq, Δ*sdiA*::*sdiA_K12*	This study

The *malEK* and *exo/cea* locus of EcN were selected as suitable sites for transgene integration and expression [23,24]. Subsequently, EcN genomic DNA served as a template for amplifying the corresponding homology arm fragments via PCR. *Avr*II and *Sal*I restriction sites were introduced at the upstream homology arm to facilitate further manipulations during the PCR amplification. These resulting DNA fragments were efficiently cloned into the pUC57 vector using one-step cloning kits, creating pUC57-*malEK* and pUC57-*exo/cea*.

Primer pairs consisting of Pf_bglF/Pr_bglF and Pf_bglB/Pr_bglB were employed to clone *bglF* and *bglB* from *E. coli* K12 genomic DNA into pETM6-*Pnar*, resulting in the construction of pETM6-*Pnar-bglF-Pnar-bglB*. The plasmid pUC57-*exo/cea-Pnar-bglF-Pnar-bglB*, harboring donor DNA fragments, was established by integrating the appropriate fragments into the *Avr*II/*Sal*I-digested pUC57-*exo/cea*. The donor DNA was obtained through PCR amplification. To mediate the integration of the homologous arm into the chosen locus, transformation was carried out using the Cas9-recombinase-expressing plasmids pEcCas and pTarget [25,26]. Specific sgRNAs for the *malEK* and *exo/cea* loci were identified and ranked within the EcN genetic background using online software (https://chopchop.cbu.uib.no/# (accessed on 19 November 2023)) [27]. For the creation of the EcN-eq strain, *Nde*I/*Kpn*I-digested synthesized DNA fragments were cloned into the *Nde*I/*Kpn*I-digested pETM6-*Pnar*. Intermediate plasmids resulting from this process were subsequently digested with *Avr*II and *Sal*I, and the resulting DNA fragments were inserted into *Spe*I/*Sal*I sites of the corresponding plasmids using BioBrick cloning [28]. This led to the development of pETM6-*Pnar-dznr-Pnar-ddrc-Pnar-dhdr-Pnar-thdr*. The quadruple-gene cassette was integrated into pUC57-*exo/cea*, and the donor DNA was incorporated into the *exo/cea* locus following the previously described method. CRISPR/Cas9-mediated gene deletions were employed to implement the relevant gene knockouts, with a constitutively expressed kanamycin resistance gene introduced during the homologous recombination of the inactivated *ptsG*.

In order to facilitate the overexpression of *decR*, *HW372_01960*, *yhjC*, *HW372_03545*, *sdiA*, and *yhaJ*, each gene was individually amplified from the EcN genome using their respective primers. Subsequently, each gene was introduced into pETM6-*Pnar* using one-step cloning kits.

2.3. Culture Conditions

Luria–Bertani (LB) medium served as the cultivation medium for *E. coli* cells during gene cloning, plasmid propagation, and inoculum preparation. A single colony was selected and cultured in LB medium, followed by overnight incubation at 37 °C with shaking at 220 rpm. For in vitro fermentation experiments involving engineered EcN strains, a basal nutrient medium was utilized. The stock basal nutrient medium, with a pH of 7.0, consisted of peptone (10.00 g/L), yeast extract (10.00 g/L), $NaHCO_3$ (10.00 g/L), glucose (5.00 g/L), bile salts (2.50 g/L), L-cysteine hydrochloride (2.50 g/L), NaCl (0.150 g/L), K_2HPO_4 (0.2 g/L), KH_2PO_4 (0.2 g/L), heme (0.2 g/L), $MgSO_4$ (0.05 g/L), $CaCl_2$ (0.05 g/L), resazurin (5 mg/L), Tween 80 (10 mL), and vitamin K1 (50 μL) [29]. Before use, the stock medium was diluted at a ratio of 1:5.

Human fecal samples were voluntarily provided by eight healthy young donors (four females and four males) who refrained from antibiotic use for at least four months and who had no history of gastrointestinal disorders. Each fresh fecal sample was collected in stool tubes and combined with a 1:2 (*w/v*) ratio of modified sterile saline (9.0 g/L NaCl and 0.5 g/L cysteine-HCl) under anaerobic conditions in an anaerobic glove box (Xinmiao YQX-11, Shanghai, China). Fecal supernatants were obtained by centrifugation at 300 rpm for 20 min, followed by thorough mixing in equal proportions. Sterile fecal filtrates were then acquired by filtering the fecal supernatant through a sterile 0.2 μm membrane. Subsequently, the stock basal nutrient medium was blended with either fecal supernatants or sterile fecal filtrates at a volume ratio of 1:4. The initial inoculum of engineered EcN was set at 2×10^7 CFU/mL and incubated at 37 °C within an anaerobic glove box. To extract reaction samples, two volumes of ethyl acetate were used, and isoflavonoids were concentrated using a rotary evaporator for subsequent chromatographic separation.

2.4. SDS-PAGE Gel Electrophoresis

Following incubation, the cultured cells were retrieved through centrifugation at $9500\times g$ for 10 min at 4 °C and subsequently resuspended in 1 mL of PBS buffer. In order to prepare the samples for electrophoresis, 0.5 mL of the suspension was combined with 0.5 mL of $2\times$ SDS-PAGE sample loading buffer (Beyotime, Shanghai, China), boiled for 10 min, allowed to cool, and then subjected to centrifugation at $11,500\times g$ for 2 min. Then, 20 μL of the resulting samples were applied to pre-cast 12% gels from Yeasen (Shanghai, China). Proteins were separated by electrophoresis at 120 V for a duration of 60–90 min.

2.5. Determination of Intracellular NADPH Levels

The conversion of OD_{600} values to cell numbers was achieved through correlation with colony counting, and the cell densities were subsequently obtained by reading the OD_{600}. A total of 4–5 million cells were collected through centrifugation (2 min, $13,000\times g$). Following the manufacturer's instructions, the cell precipitates were washed using a mixed solution (70 mM HEPES, 60% MeOH) and resuspended in 0.9 mL of the acid extraction buffer (Beyotime, Shanghai, China) from the kit. The cell suspension underwent sonication for 1 min and was then centrifuged at $10,000\times g$ for 10 min at 4 °C. The supernatant (200 μL) was transferred to a new centrifuge tube, and an equal volume of alkaline buffer (Beyotime, Shanghai, China) was added to neutralize it. This mixture was then centrifuged at $12,000\times g$ at 4 °C for 10 min, and the resulting supernatant was collected. The quantification of $NADP^+$ and NADPH was performed using the $NADP^+$/NADPH Assay Kit from Beyotime, China.

2.6. Transcriptome Analysis by RNA-seq

The cultures underwent a 2 h incubation at 37 °C prior to RNA extraction. For this, 10 mL of collected cells were rapidly frozen in liquid nitrogen, and total bacterial RNA was extracted using the RNAprep pure Kit (TIANGEN, Beijing, China). RNA quality was checked by 1% agarose gel electrophoresis to verify the integrity of the RNA preparations. The extracted RNA was then transported on dry ice to Sangon Biotech (Shanghai, China) for transcriptome re-sequencing analysis. To obtain mRNA, the total RNA underwent rRNA removal, and the resultant mRNA was utilized as a template for DNA synthesis. Sequencing of the cDNA libraries was performed using the Illumina HiSeq platform. For annotation purposes, the EcN genome served as a reference. Differentially expressed genes between the sample and engineered strains were determined based on a false discovery rate (FDR) $\leq$ 0.05 and a fold change (log2Ratio) $\geq$ 2. To categorize genes at the KEGG_B_class level, the pathway-enrichment analysis tool Omicshare was employed.

2.7. Real-Time PCR Measurements for Transcriptional Analysis

Real-time PCR was employed to assess the transcriptional expression levels of the targeted genes. The initial isolation of total RNA from recombinant cells was performed using the MolPure Bacterial RNA Kit from Yeasen (Shanghai, China). The subsequent synthesis of cDNA was achieved using the Strand cDNA Synthesis SuperMix for qPCR, also from Yeasen. For the real-time PCR analysis, the LightCycler 480 Real-Time PCR System from Roche (Mannheim, Germany) was utilized, along with the qPCR SYBR Green Master Mix. Quantitative PCR amplification was conducted using 2 µL of diluted cDNA template. The final concentration of primers in the reaction solution was 0.2 µM, and the total volume was adjusted to 20 µL with ddH$_2$O. The amplification conditions included an initial predenaturation step at 95 °C for 2 min, followed by a two-step reaction (95 °C, 10 s; 60 °C, 30 s) for 40 cycles. The 16S gene was chosen as the endogenous reference gene to determine the relative mRNA expression levels of *zwf* (glucose-6-phosphate dehydrogenase) and *gnd* (6-phosphogluconate dehydrogenase), applying the $2^{-\Delta\Delta Ct}$ method [25].

2.8. Molecular Docking

The prediction of the SDH structure was conducted using AlphaFold2 through Google Colab [30]. Molecular docking procedures involving SdiAs with C10-HSL were carried out using the Discover Studio 2021 Libdock module. The docking pose selected adhered to the distance restraints of SdiAs with C10-HSL, emphasizing the highest docking score and the absence of adverse contacts. Analysis of the docking-complex structure was conducted utilizing the open-source PyMol.

2.9. MD Simulations

The MD simulations were carried out using GROMACS 2022 to analyze heat fluctuations. In brief, the protein structure was treated with the Charmm36 force field. Each system's proteins were solvated in the TIP3P water box. The MD simulations were conducted at 310 K for 100 ns. Energy minimization involved the use of the steepest descent for 1000 steps, followed by gradual heating of each system to 310 K and equilibration for 100 ps NVT and 100 ps NPT. The temperature was set to 310 K with a time constant of 0.1 ps, utilizing a V-rescale Berendsen thermostat. The pressure was maintained at 1 bar with a time constant of 2 ps, employing Parrinello–Rahman pressure coupling. Subsequently, the simulations were conducted for 100 ns, with results saved at intervals of 2 fs. The analysis of simulated proteins involved studying the root-mean-square deviation (RMSD), root-mean-square fluctuation (RMSF), and hydrogen-bond numbers using gmx-rms, gmx-rmsf, and gmx-gyrate, respectively.

2.10. Analytical Methods

Isoflavonoids were analyzed using the Agilent 1260 series HPLC system equipped with an Agilent ZORBAX SB-C18 column (4.6 × 250 mm, 5 µm). The separation of isoflavonoids was conducted with solvent A (containing 0.1% formic acid) and solvent B (methanol). The gradient elution program and detection methodology followed a previously established method [5]. For chiral analysis, the CHIRALPAK-IC column from Daicel Chemical Industries (4.6 mm × 150 mm, 5 µm) was utilized with an isocratic mobile phase of 1 mL/min (hexane/ethanol = 70:30) [3].

Reversed-phase solid-phase extraction was employed for the extraction of AHLs from feces [16]. In this process, each 100 µL sample was combined with 100 µL of a 100 nM

C6-HSL solution (in methanol) and 1600 µL of cold methanol (0.1% formic acid). Following vortex mixing for 15 s, the tubes were placed at -20 °C for 30 min and then centrifuged at $18,000\times g$ for 10 min. The supernatants were collected and subsequently speed-vacuum dried. Subsequently, 200 µL of methanol was added to the tube and vortex mixed for 30 s followed by the addition of 1.5 mL of water (0.1% formic acid). The resulting mixture was loaded onto an ISOLUTE C18 cartridge (100 mg/1 mL). After vacuum drying, the analytes were eluted with 1500 µL of acidified methanol (0.1% formic acid). The eluates were speed-vacuum-dried at 8 °C and then resuspended in 200 µL of 50% methanol (0.1% formic acid). Analysis was performed using a UHPLC system (Agilent, Santa Clara, CA USA) coupled with a Quantis triple quadrupole mass spectrometer via an electrospray ionization (ESI) source operated in the positive ion mode. The T3 (2.1 mm $\times$ 100 mm 1.8 µm) column from Waters (Milford, MA, USA) was used for chromatographic separation with mobile phase A consisting of 0.1% formic acid in water and mobile phase B consisting of 0.1% formic acid in acetonitrile. The AHLs were eluted at a flow rate of 200 µL/min using a gradient elution of the mobile phase starting at 5% B and then increasing to 10% B by 1 min, further increasing to 75% B by 3 min, and reaching 99% B by 6 min, which was held until 6.5 min; this was then decreased to 5% B by 13 min and held for 3.5 min, resulting in a total run time of 16.5 min. The injection volume was 10 µL, and the column temperature was maintained at 50 °C. Source parameters for the mass spectrometer were set as follows: spray voltage, 4500 V; ion-transfer-tube temperature, 350 °C; vaporizer temperature, 450 °C.

2.11. Statistical Analysis

We conducted statistical analyses using SPSS v26.0 (SPSS, Inc., Chicago, IL, USA). To assess statistical significance, we employed the Student's *t*-test, denoting significance levels as follows: * $p < 0.05$, ** $p < 0.01$, and *** $p < 0.001$. The error bars in the figures represent the standard deviation (SD).

3. Results and Discussion

3.1. Engineering E. coli Nissle 1917 for (S)-Equol Production from Daidzein and Daidzin through Chromosomal Integration

Isoflavones in soy products are in the form of glycoside isoflavones (e.g., daidzin), which need to be hydrolyzed to isoflavone aglycones (e.g., daidzein) to be utilized by *E. coli* [2,3,31,32]. Building upon a prior study where recombinant *E. coli* strains expressing both the β-glucoside influx transporter (BglF) and 6-phospho-β-glucosidase (BglB) from *E. coli* K12 showcased the ability to utilize daidzin, we opted for the introduction of *bglF* and *bglB* to enable daidzin utilization in EcN [3]. Subsequently, *dznr* from *Asaccharobacter celatus*, along with *ddrc*, *thdr*, and *dhdr* from *Slackia isoflavoniconvertens* were selected to convert daidzein into (S)-equol based on their efficient (S)-equol-production performance [8]. To regulate exogenous gene expression within the anaerobic gut environment, the anaerobically induced promoter *Pnar* was employed due to its functional adaptability across various growth phases of engineered *E. coli* strains [3,33]. These six genes underwent chromosomal integration and were transcriptionally regulated by the *Pnar* promoter, thereby generating the strain EcN-eq (Figure 1A). The (S)-equol-producing capacity of EcN-eq was examined in basal nutrient medium containing 100 mg/L soy isoflavone (SI; 80% *w/w* isoflavone), resulting in the generation of 129.3 µM of (S)-equol with a conversion rate of 96.4% (mol/mol) after 150 min of bioconversion (Figure 1B,C). The chiral HPLC analysis determined the stereochemical configuration of (S)-equol, with the enantiomeric excess values exceeding 99.0% (S) (Figure S1).

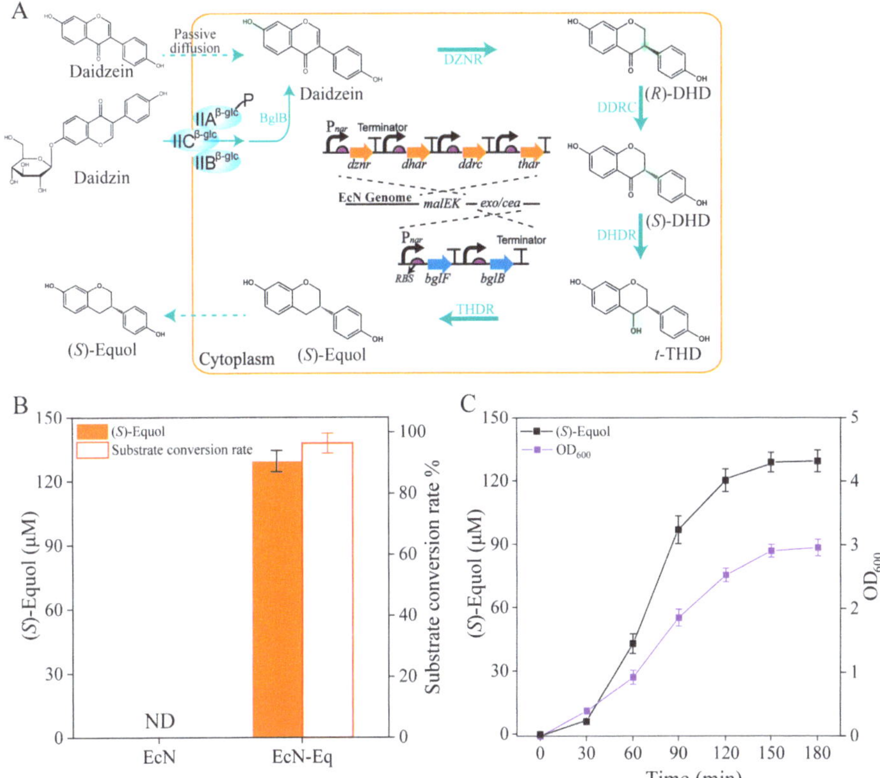

Figure 1. Construction of the EcN-eq strain for the (*S*)-equol conversion from daidzin and daidzein. (**A**) Illustration outlining the gene constructs incorporated into the EcN-eq strain, developed by gene insertion into the EcN genome. The green color represents the moiety of the structural formula that has changed. (**B**) Utilization of whole-cell cultures of the EcN-eq strain for the conversion of daidzin and daidzein into (*S*)-equol. A substrate of 100 mg/L of SI was introduced into the medium. (**C**) Representation of the growth curve of the EcN-eq strain and its temporal profile for (*S*)-equol production. All experiments were performed in triplicate, and error bars indicate the standard deviation (SD) with a 95% confidence interval (CI).

3.2. (S)-Equol Production by the EcN-eq Strain in Media with Fecal Supernatants or Sterile Fecal Filtrates

The efficient EcN-eq strain, developed for (*S*)-equol production from SI via the daidzin-utilization pathway and the (*S*)-equol biosynthesis pathway in EcN, was initially tested under optimal growth conditions in monoculture. However, the outcomes might not accurately represent real-life scenarios [34]. To assess the potential of EcN-eq as a cooperative blend of probiotics in conferring metabolic pathways for converting dietary SI into (*S*)-equol, we examined its (*S*)-equol production in a basal nutrient medium containing human fecal supernatant in vitro. Human fecal samples were voluntarily provided by eight healthy young donors (four females and four males) who had refrained from antibiotic use for at least four months and who had no history of gastrointestinal disorders. As depicted in Figure 2A, when EcN-eq was absent in the fecal supernatant cultures, (*S*)-equol was detected in only three samples, reaching a maximum titer of 18.7 μM. In contrast, the presence of EcN-eq in the fecal-supernatant cultures led to a significantly higher (*S*)-equol titer, measuring 77.5 μM compared with the control. However, the (*S*)-equol production by

EcN-eq when co-cultured with gut microorganisms was 40.1% lower than in monoculture, displaying substantial variability among individuals (Figures 1C and 2A). Intriguingly, the biomass of EcN-eq co-cultured with gut microorganisms was significantly higher than that of EcN-eq in monoculture (Figure 2A).

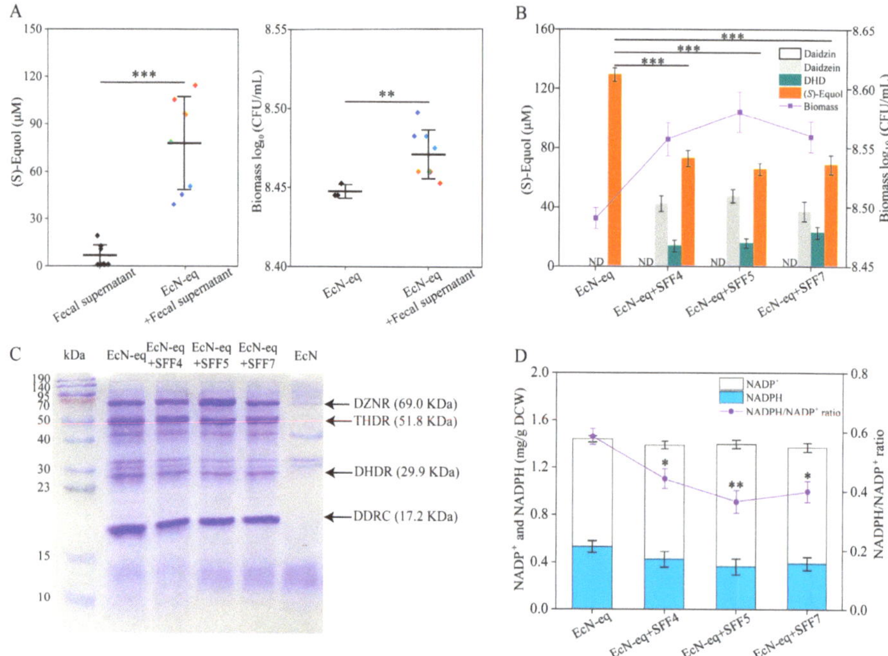

Figure 2. (*S*)-Equol production by the EcN-eq strain in media with fecal supernatants or sterile fecal filtrates. (**A**) Biomass and (*S*)-equol titer of the EcN-eq strain in media containing fecal supernatants. (**B**) Yield of (*S*)-equol and intermediates of the EcN-eq strain in media containing SFF4, SFF5, and SFF7, respectively. (**C**) Expression analysis of enzymes involved in the (*S*)-equol biosynthetic pathway by the EcN-eq strain in media containing SFF4, SFF5, and SFF7, respectively. (**D**) NADPH and NADP$^+$ content of the EcN-eq strain in media containing SFF4, SFF5, and SFF7, respectively. EcN-eq served as the control. Statistical analysis was conducted using a *t*-test, where * $p < 0.05$, ** $p < 0.01$, and *** $p < 0.001$. Error bars indicate the standard deviation (SD).

The interaction between gut microbiota and probiotics often occurs through microbial metabolites, exerting an influence on their metabolism and phenotype [14,35]. To identify the bottleneck in EcN-eq's performance when co-cultured with fecal bacteria, we scrutinized the intermediates and (*S*)-equol production during the whole-cell reaction of EcN-eq in cultures containing sterile fecal filtrates (i.e., SFF4, SFF5, and SFF7). Consistent with findings from the fecal-supernatant cultures, the presence of SFF4, SFF5, and SFF7 led to a notable reduction in (*S*)-equol production by EcN-eq compared with cultures without SFFs (Figure 2B). Remarkably, in these cultures, the major intermediates observed were daidzein and dihydrodaidzein (DHD), while daidzin was notably absent. Moreover, EcN-eq displayed a substantial increase in biomass in media containing SFF4, SFF5, and SFF7 relative to cultures without SFFs (Figure 2B). From these observations, we hypothesized that metabolites present in fecal-filtrate cultures potentially diminish the (*S*)-equol production titer of EcN-eq by impeding the conversion of daidzein into (*S*)-equol within EcN-eq.

Based on the above inference, we proceeded to examine the expression of pathway enzymes responsible for converting daidzein into (*S*)-equol in EcN-eq within the SSF-added media. As shown in Figures 2C and S2, the supplementation of SSF in the medium did not significantly impact the expression of DZNR, DDRC, DHDR, and THDR in EcN-eq. Given that DZNR and DHDR rely on NADPH as a coenzyme for their enzymatic activity [36,37], we further investigated the effects of SSF addition to the media on EcN-eq's intracellular NADPH levels and the NADPH/NADP$^+$ ratio (Figure 2D). The results revealed a significant decrease in EcN-eq's intracellular NADPH levels and NADPH/NADP$^+$ ratio upon the addition of SSF to the media. Notably, the declining trend in the NADPH/NADP$^+$ ratio corresponded consistently with the reduction observed in the (*S*)-equol titer (Figure 2B,D). NADPH is a pivotal component for biosynthesizing cellular components within the cell, while also serving as a crucial cofactor in producing various nutraceuticals and fine chemicals [38,39]. Notably, previous research has illustrated that engineered *E. coli* strains possess the capability to effectively convert daidzein (ranging between 197–500 μM) into (*S*)-equol using endogenous NADPH [3,40]. While enhancing intracellular NADPH levels can be achieved through the overexpression of enzymes responsible for its regeneration, this approach may disrupt the balance between oxidized and reduced forms of the cofactors [41]. Consequently, optimizing the endogenous NADPH regeneration pathway within the strain holds promise for enhancing the efficient synthesis of desired products.

3.3. Transcriptome Analysis of the EcN-eq Strain in Media with or without the Addition of Sterile Fecal Filtrates

To further elucidate the influence of metabolites present in fecal-filtrate cultures on the regulation of intracellular NADPH metabolism in EcN-eq, we conducted RNA-Seq analyses on EcN-eq samples that had been fermented for 60 min in media both with and without SSF5 addition. Genes exhibiting an absolute log2-fold change greater than 2 (with a *p*-value < 0.05) were categorized as differentially expressed genes (DEGs). Transcriptomic data indicated that 29 genes were upregulated while 17 genes were downregulated when comparing EcN-eq cultured in SSF5-containing medium to those in medium lacking SSF5 (Figure 3A). The identified DEGs were categorized by KEGG analysis (Figure 3B). The DEGs were predominantly associated with biological processes such as oxidoreductase activity, carbohydrate-derivative metabolism, and NADPH metabolism. Notably, genes involved in DNA binding, cell projection, and regulation of DNA-templated transcription showed increased expression, while those linked to the pentose phosphate pathway, oxidoreductase activity, transferase activity, and carbohydrate-derivative metabolic processes exhibited decreased expression (Figure 3B). As can be seen from Table 2, there were significant variations in the expression levels of six transcription factors (TFs) in EcN-eq when cultured in SSF5-containing medium. These TFs have the potential to modulate intracellular NADPH levels directly or indirectly.

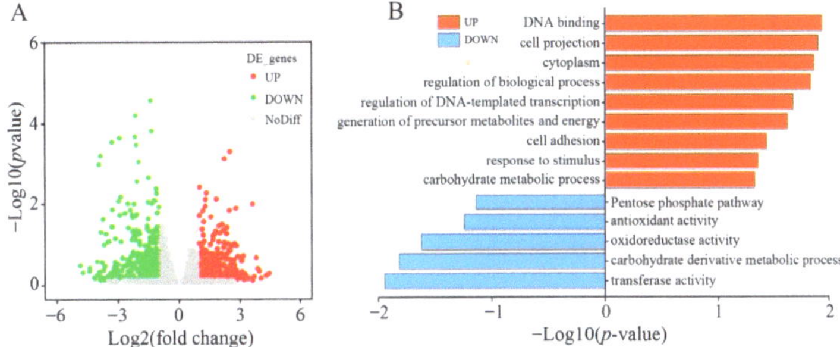

Figure 3. Transcriptome analysis of the EcN-eq strain in media with or without the addition of SFF5. (**A**) Volcano plot of differentially expressed genes in the EcN-eq strain in medium with or without added SFF5. (**B**) KEGG pathway-enrichment map, where the *X*-axis is the negative logarithmic transformation of *p*-values and the *Y*-axis is the enriched pathway; blue bars indicate pathways enriched by downregulated proteins, and red bars indicate pathways enriched by upregulated proteins.

Table 2. Differential gene expression analysis of the EcN-eq strain in medium without/with the addition of SSF5.

Gene	Annotation	log$_2$ (Fold Change)
Transporters		
yagG	sugar transporter	−2.9905
ydcS	polyamine transporter	−2.2575
fetB	iron-export ABC-transporter ATPase	2.5437
gatA	galactitol PTS	1.5344
Pentose phosphate pathway		
zwf	glucose-6-phosphate 1-dehydrogenase	−1.8321
gnd	6-phosphogluconate dehydrogenase	−2.6247
Acetate, anaplerotic, and other gluconeogenic pathways		
poxB	6-phosphofructokinase	2.7182
acs	acetyl-CoA synthetase	1.5124
pfkA	6-phosphofructokinase I	2.1273
gpmM	2,3-bisphosphoglycerate-independent phosphoglycerate mutase	1.5162
Stress-response pathway		
rsfA	ribosome-silencing factor	1.7628
yfcV	stress-response fimbriae	2.2365
ypjA	stress-response adhesin	1.9634
Cell to cell interaction		
hlyE	hemolysin E	2.6343
ypfA	adhesion-like autotransporter	1.7762
fimA	major type-I fimbrin	3.7689
yqjH	siderophore interaction protein	2.1547
Transcriptional regulator		
decR	AsnC-family transcriptional regulator	−1.9354
ascG	transcriptional regulator	3.2329
yhjC	LysR-family transcriptional regulator	2.7514
frvR	transcriptional regulator	−2.1672
sdiA	LuxI-family transcriptional regulator	3.4855
yhaJ	transcriptional regulator	2.7841

3.4. Quorum-Sensing Transcription Factor SdiA Modulates the Intracellular NADPH/NADP$^+$ Ratio

In order to elucidate the functional roles of the aforementioned transcriptional regulators, additional validation was conducted. This involved the construction of EcN-eq OE-*X* strains, characterized by the overexpression of the TFs, and EcN-eq ΔX strains, in which the TFs were knocked out. Subsequently, the intracellular NADPH/NADP$^+$ ratios were measured in media containing SSF5. As shown in Figure 4A, the intracellular NADPH/NADP$^+$ ratio of the EcN-eq strain $\Delta sdiA$ was significantly higher compared with the control, whereas the intracellular NADPH/NADP$^+$ ratio of the EcN-eq strain OE-*sdiA* was lower than that of the control. SdiA is a LuxR-type transcription factor that is recognized as a sensor for *E. coli* to detect AHLs produced by other bacteria in the environment [13]. These results strongly suggest the involvement of the QS transcription factor SdiA in regulating NADPH metabolism within EcN-eq.

Since SdiA recognizes a broad spectrum of AHLs, varying in acyl chain length and oxidation [17,18], we employed liquid mass spectrometry to quantify the species and concentrations of AHLs present in the SFFs. Table S2 exhibits the MRM transitions of the AHLs measured in this study. Figure 4B illustrates the substantial diversity in AHL types and concentrations across the eight SFF samples. Notably, AHLs with short side-chains (C$_{4-6}$-HSL) were absent in all samples, while AHLs featuring long side-chains (C$_{8-14}$-HSL) were detectable in each sample (Figure 4B). A previous study indicated distinct differences in SdiA complex formation involving different AHLs, resulting in the differential regulation of target genes within the *E. coli* genome [18]. In light of this, we proceeded to explore the impact of C8-HSL, C10-HSL, C12-HSL, and C14-HSL on EcN-eq's intracellular NADPH/NADP⁺ ratio and biomass. Notably, Figure 4C reveals that only the addition of C10-HSL in the medium elicited a significant reduction in EcN-eq's intracellular NADPH/NADP⁺ ratio and (*S*)-equol titer among the evaluated AHL candidates. Moreover, these AHLs contributed to a notable increase in the EcN-eq biomass compared with the control. However, no substantial differences were observed in EcN cultures supplemented with various AHLs.

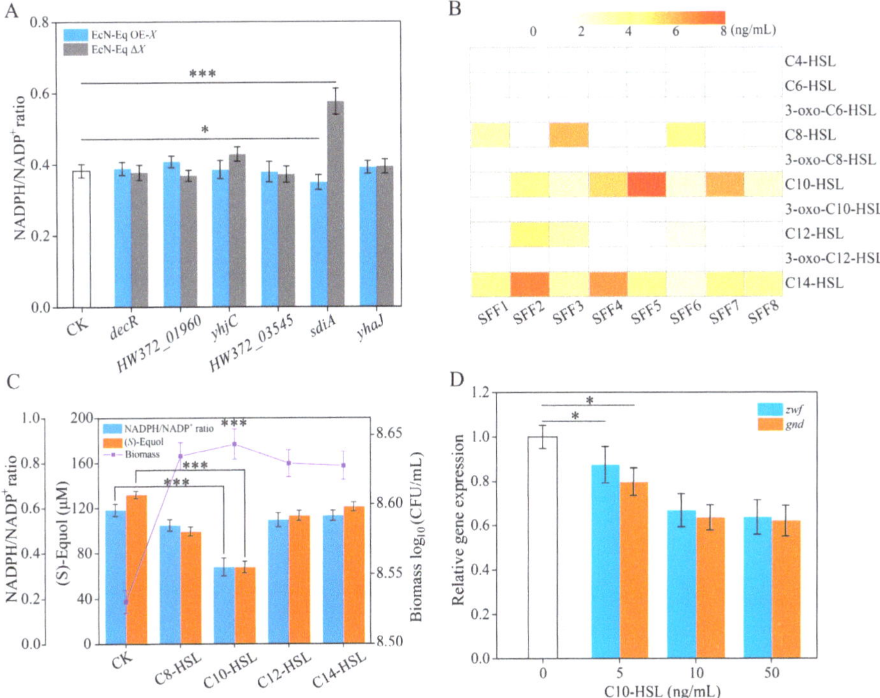

Figure 4. The SdiA quorum-sensing transcriptional regulator modulates the intracellular NADPH/NADP⁺ ratio in the EcN-eq strain. (**A**) Determination of NADPH/NADP⁺ ratios in transcription-factor-overexpression strains (EcN-eq OE-*X*) and knockout strains (EcN-eq Δ*X*). (**B**) Determination of N-acyl homoserine in SFF1 to SSF8 samples. (**C**) Effect of medium- and long-chain N-acyl homoserine lactones (C8-HSL to C14-HSL) on the NADPH/NADP⁺ ratio, (*S*)-equol titer, and biomass in the EcN-eq strain. The concentration of N-acyl homoserine lactones was 8 ng/mL. (**D**) Effect of different concentrations of C8-HSL on *zwf* and *gnd* gene expression in the EcN-eq strain. Statistical analysis was performed using a *t*-test; * *p* < 0.05, *** *p* < 0.001, and the error bars represent the standard deviation (SD).

In *E. coli* strains, NADPH production primarily occurs through the oxidative branch of the pentose phosphate pathway, wherein glucose-6-phosphate dehydrogenase (encoded by *zwf*) and 6-phosphogluconate dehydrogenase (encoded by *gnd*) play crucial roles in NADPH regeneration [8]. Transcriptomic analysis revealed a downregulation of both *zwf* and *gnd* expression in EcN-eq when supplemented with SSF5 (Table 2). To elucidate the regulatory role of the C10-HSL-SdiA complex on *zwf* and *gnd*, we assessed the impact of varying concentrations of C10-HSL (ranging from 0 to 50 ng/mL) in the culture medium on the expression of *zwf* and *gnd* in EcN-eq using RT-qPCR (Figure 4D). The results indicated that even at a low C10-HSL concentration of 5 ng/mL, there was a negative modulation of *zwf* and *gnd* expression in EcN-eq. Furthermore, higher concentrations of C10-HSL (10 and 50 ng/mL) led to a further decline in *zwf* and *gnd* expression in EcN-eq, reaching approximately 35% lower expression levels compared with the control (Figure 4D). In summary, the QS transcription factor SdiA appears to exert a negative regulatory effect on *zwf* and *gnd* in EcN-eq, potentially resulting in decreased intracellular NADPH levels.

3.5. Interaction Analysis of N-Decanoyl-L-Homoserine Lactone (C10-HSL) with SdiA_EcN

E. coli regulates growth and metabolic processes through SdiA in response to AHLs produced by other bacteria, enabling *E. coli* to navigate environmental variations and explore new ecological niches [18,42,43]. However, limited information exists on the role of SdiA in regulating intracellular NADPH metabolism in EcN. A previous study has shown that the N-terminal ligand-binding structural domain (LBD) of SdiA from the model organism *E. coli K12* has constrained accommodation for AHLs with acyl chain lengths beyond C8 [42]. Consequently, variations in the affinities of SdiA from different sources for long-side-chain AHLs could result in differential metabolic regulation by SdiA [18]. Given that EcN serves as a representative chassis for engineered probiotics, elucidating the disparities in C10-HSL interactions with SdiA_EcN and SdiA_K12 from *E. coli K12* could aid in developing engineered EcN strains tailored to the intestinal environment.

Given the unavailability of the crystal structure for SdiA_EcN, we employed AlphaFold2 for predictive modeling [30]. SdiA_EcN and SdiA_K12 exhibit structural resemblances, encompassing an AHL-binding ligand-binding domain (LBD) (Figure 5A). This domain adopts an alpha–beta–alpha sandwich configuration. Additionally, both share a C-terminal domain that functions as a DNA-binding structural domain, comprising four helices and featuring a helix–turn–helix motif for DNA binding (Figure 5A). Molecular-docking analysis revealed insights into the interactions between SdiA and C10-HSL at a molecular level (Figure 5A). The conformation of C10-AHL-SdiA_EcN is similar to that of C10-AHL-SdiA_K12, showcasing the stabilization of the acyl chain via hydrophobic residues [42]. However, notable distinctions arise in the hydrogen-bond interactions involving the lactone ring of C10-AHL and residues in SdiA_EcN versus SdiA_K12. In the docked conformation of SdiA_EcN with C10-AHL, evident hydrogen bonds form between W107 and the lactone ring, as well as between R111 and the lactone ring's carbonyl group. Conversely, in SdiA_K12, the hydrogen bond solely occurs between Y63 and the C1 carbonyl group. Literature reports indicate that mutations in specific distal or surface amino acid residues can alter the electrostatic charge distribution on the surface, affecting the LBD's interaction with ligands [44,45]. Therefore, the mutation of hydrophobic amino acid residue A66, situated in the alpha helix on the surface of SdiA_K12, to the hydrophilic amino acid S66 in SdiA_EcN may potentially induce variations in LBD volume, thereby influencing the differential binding of SdiA_EcN and SdiA_K12 to C10-HSL (Figure S3).

The molecular-docking findings were validated through an additional 100 ns of molecular-dynamics simulations. The root-mean-square deviation (RMSD), a pivotal parameter for assessing conformational fluctuations within the protein–ligand complex, inversely correlates with complex stability [46]. Notably, during the 60–100 ns period, the RMSD values of C10-HSL-SdiA_EcN exhibited lower values compared with C10-HSL-SdiA_K12, implying heightened thermal stability within the C10-HSL-SdiA_EcN complex (Figure 5B). Furthermore, examining the root-mean-square fluctuation (RMSF) values of residues 75–110 within the LBD revealed lower fluctuations in the C10-HSL-SdiA_EcN complex compared with C10-HSL-SdiA_K12, indicating reduced flexibility in this region for the former (Figure 5C). Quantification of hydrogen bonding at the binding site was performed to elucidate the nature of interactions between C10-HSL and SdiAs. Throughout the simulation, the number of stable hydrogen bonds between SdiA_EcN and C10-HSL remained consistently around one, with a maximum of three hydrogen bonds formed, indicative of a more stable binding between SdiA_EcN and C10-HSL (Figure 5D). Conversely, minimal stable hydrogen bonds were observed between SdiA_K12 and C10-HSL, suggesting the lack of a stable conformation within the C10-HSL-SdiA_K12 complex (Figure 5E).

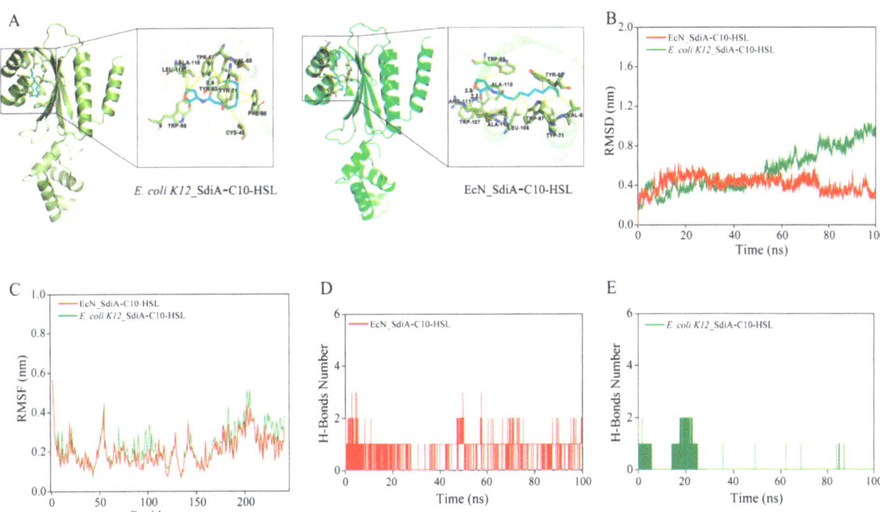

Figure 5. Molecular-docking analysis and MD simulation analysis of *E. coli K12*_SdiA and EcN_SdiA with C10-HSL. (**A**) Molecular-docking analysis of *E. coli K12*_SdiA and EcN_SdiA with C10-HSL. (**B**) RMSD analysis of *E. coli K12*_SdiA and EcN_SdiA with C10-HSL. (**C**) RMSF analysis of *E. coli K12*_SdiA and EcN_SdiA with C10-HSL. (**D**) Time profiles of hydrogen bonding of *E. coli K12*_SdiA with C10-HSL. (**E**) Time profiles of hydrogen bonding of EcN_SdiA with C10-HSL.

3.6. Replacing sdiA_EcN with sdiA_K12 Improved (S)-Equol Production and Avoided Biomass Reduction

Given the global gene regulation and growth-promoting role of *sdiA* in EcN, we replaced *sdiA_EcN* with *sdiA_K12* to elevate the intracellular NADPH levels within EcN-eq, aiming to enhance its (*S*)-equol biosynthesis. The construction of the EcN-eq strain Δ*sdiA::sdiA_K12* involved the replacement of *sdiA_EcN* coding sequences with *sdiA_K12* utilizing CRISPR/Cas9- and λ-Red-mediated recombination [3,26]. Subsequently, the intracellular NADP$^+$/NADPH ratios were assessed for both EcN-eq Δ*sdiA::sdiA_K12* and EcN-eq in media containing fecal filtrates. As shown in Figure 6A, the intracellular NADPH/NADP$^+$ ratio of EcN-eq Δ*sdiA::sdiA_K12* exhibited a significant increase compared with that of EcN-eq. Following a 150 min biotransformation period, the (*S*)-equol titer of EcN-eq Δ*sdiA::sdiA_K12* reached 113.7 μM with an 84.8% substrate conversion rate, indicating a 47% increase compared with EcN-eq (Figure 6B). Furthermore, the EcN-eq Δ*sdiA::sdiA_K12* group displayed a lower standard deviation in the (*S*)-equol titer than the EcN-eq group and showed no significant difference in biomass (Figure 6B,C).

Isoflavones are commonly utilized as antioxidative dietary supplements. However, (*S*)-equol, a secondary metabolite of isoflavone by the gut microbiota, exhibits significantly higher antioxidative capacity than its precursor [2,4]. Consequently, developing EcN strains with the ability to efficiently convert dietary isoflavones into (*S*)-equol may represent an effective strategy to confer the (*S*)-equol-producing phenotype of the host. Nevertheless, the intricacies of the gut microbiome may have contributed to the challenges encountered in in vivo studies involving engineered EcN strains [47]. This study offers insights into the development of (*S*)-equol-producing EcN probiotics tailored for the intestinal environment by elucidating the impact of AHL-SdiA on EcN NADPH metabolism.

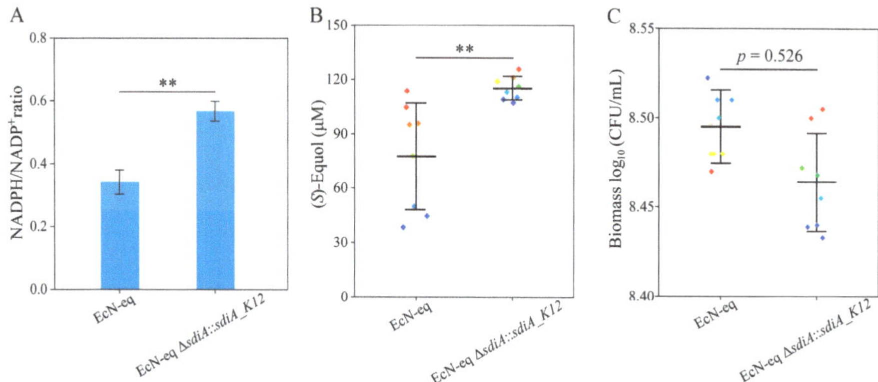

Figure 6. (**A**) NADPH/NADP$^+$ ratio of the EcN-eq and EcN-eq *sdiA::sdiA_K12* strains in media containing fecal filtrates. (**B,C**) (*S*)-Equol production titers (**B**) and biomass (**C**) of the EcN-eq strain *sdiA::sdiA_K12* in media containing fecal supernatants. Statistical analysis was performed using a *t*-test; ** $p < 0.01$, and the error bars represent the standard deviation (SD).

4. Conclusions

This study delves into how AHL-SdiA-mediated QS diminishes (*S*)-equol production by modulating NADPH metabolism in the engineered probiotic EcN. The results reveal that SdiA acts as a global transcription factor in EcN, regulating NADPH metabolism of the (*S*)-equol-producing EcN-eq strain in the intestinal environment. The complex of C10-HSL with SdiA was shown to reduce the intracellular NADPH/NADP$^+$ ratio by negatively regulating the expression of *zwf* and *gnd*. On this basis, the *sdiA* of EcN-eq was replaced with *sdiA_K12*, which has a low affinity for C10-HSL, resulting in the construction of an engineered EcN probiotic that efficiently produces (*S*)-equol in the intestinal environment. Considering the crucial role of NADPH as a cofactor in redox reactions, these findings offer valuable insights for engineering probiotics tailored to the gut environment. However, it is worth noting that this study was conducted in vitro; thus, in vivo experiments in an intestinal environment should be further implemented to confirm these findings.

Supplementary Materials: The following supporting information can be downloaded at: https://www.mdpi.com/article/10.3390/antiox13030259/s1, Table S1. Nucleotide sequences of primers. Table S2. Vectors used in this study. Figure S1. Chiral HPLC analysis of daidzin metabolites converted by strain EcN-eq. (A). Reference standards of (*S*)-Equol and (*R*)-Equol. (B). HPLC spectrum displaying the reaction sample from strain EcN-eq. Figure S2. Densitometric semi-quantifications of the SDS-PAGE in Figure 2C. Experiments in this study were conducted in triplicate, and error bars signify standard deviation (SD) with a 95% confidence interval (CI). Figure S3. Sequence alignment of *E. coli K12*_SdiA and *EcN*_SdiA.

Author Contributions: Methodology, Z.W. and X.X.; Software, Z.W., Y.D. and F.A.; Validation, Y.D.; Investigation, Z.W.; Writing–original draft, Z.W.; Writing–review & editing, F.A. and X.X.; Visualization, Z.W., F.A. and M.D.; Supervision, M.D.; Project administration, M.D.; Funding acquisition, Y.D. and X.X. All authors have read and agreed to the published version of the manuscript.

Funding: The financial support from the Jiangsu Agriculture Science and Technology Innovation Fund (Grant No. CX(21)2003), the Natural Science Foundation of Jiangsu Province (Grants No. BK20231392), and the Postgraduate Research & Practice Innovation Program of Jiangsu Province (Grant No. KYCX23_0773) are gratefully acknowledged.

Institutional Review Board Statement: The study was approved by the Ethics Committee of Jiangsu Academy of Agricultural Sciences (ethic code: Z202322137).

Informed Consent Statement: All subjects signed a written informed consent form to join the study.

Data Availability Statement: Data are available on request from the authors.

Conflicts of Interest: The authors declare no competing financial interests.

References

1. Bešlo, D.; Došlić, G.; Agić, D.; Rastija, V.; Šperanda, M.; Gantner, V.; Lučić, B. Polyphenols in ruminant nutrition and their effects on reproduction. *Antioxidants* **2022**, *11*, 970. [CrossRef] [PubMed]
2. Kim, I.-S. Current perspectives on the beneficial effects of soybean isoflavones and their metabolites for humans. *Antioxidants* **2021**, *10*, 1064. [CrossRef] [PubMed]
3. Wang, Z.; Li, X.; Azi, F.; Dai, Y.; Xu, Z.; Yu, L.; Zhou, J.; Dong, M.; Xia, X. Biosynthesis of (*S*)-equol from soy whey by metabolically engineered *Escherichia coli*. *J. Agric. Food Chem.* **2023**, *71*, 6110–6119. [CrossRef] [PubMed]
4. Wang, X.; Chen, B.; Fang, X.; Zhong, Q.; Liao, Z.; Wang, J.; Wu, X.; Ma, Y.; Li, P.; Feng, X. Soy isoflavone-specific biotransformation product S-equol in the colon: Physiological functions, transformation mechanisms, and metabolic regulatory pathways. *Crit. Rev. Food Sci.* **2022**, *62*, 1–29. [CrossRef] [PubMed]
5. Setchell, K.D.; Cole, S.J. Method of defining equol-producer status and its frequency among vegetarians. *J. Nutr.* **2006**, *136*, 2188–2193. [CrossRef] [PubMed]
6. Decroos, K.; Vanhemmens, S.; Cattoir, S.; Boon, N.; Verstraete, W. Isolation and characterisation of an equol-producing mixed microbial culture from a human faecal sample and its activity under gastrointestinal conditions. *Arch. Microbiol.* **2005**, *183*, 45–55. [CrossRef] [PubMed]
7. Liang, W.; Zhao, L.; Zhang, J.; Fang, X.; Zhong, Q.; Liao, Z.; Wang, J.; Guo, Y.; Liang, H.; Wang, L. Colonization potential to reconstitute a microbe community in pseudo germ-free mice after fecal microbe transplant from equol producer. *Front. Microbiol.* **2020**, *11*, 1221. [CrossRef]
8. Deng, H.; Gao, S.; Zhang, W.; Zhang, T.; Li, N.; Zhou, J. High titer of (*S*)-equol synthesis from daidzein in *Escherichia coli*. *ACS Synth. Biol.* **2022**, *11*, 4043–4053. [CrossRef]
9. Effendi, S.S.W.; Ng, I.S. Prospective and challenges of live bacterial therapeutics from a superhero *Escherichia coli* Nissle 1917. *Crit. Rev. Microbiol.* **2023**, *49*, 611–627. [CrossRef]
10. Yu, Y.; Gong, B.; Wang, H.; Yang, G.; Zhou, X. Chromosome evolution of *Escherichia coli* Nissle 1917 for high-level production of heparosan. *Biotechnol. Bioeng.* **2023**, *120*, 1081–1096. [CrossRef] [PubMed]
11. Kydd, L.; Shiveshwarkar, P.; Jaworski, J. Engineering *Escherichia coli* for conversion of dietary isoflavones in the gut. *ACS Synth. Biol.* **2022**, *11*, 3575–3582. [CrossRef] [PubMed]
12. Shimada, Y.; Yasuda, S.; Takahashi, M.; Hayashi, T.; Miyazawa, N.; Sato, I.; Abiru, Y.; Uchiyama, S.; Hishigaki, H. Cloning and expression of a novel NADP(H)-dependent daidzein reductase, an enzyme involved in the metabolism of daidzein, from equol-producing *Lactococcus* strain 20-92. *Appl. Environ. Microb.* **2010**, *76*, 5892–5901. [CrossRef] [PubMed]
13. Whiteley, M.; Diggle, S.P.; Greenberg, E.P. Progress in and promise of bacterial quorum sensing research. *Nature* **2017**, *551*, 313–320. [CrossRef] [PubMed]
14. Qian, X.; Tian, P.; Zhao, J.; Zhang, H.; Wang, G.; Chen, W. Quorum sensing of lactic acid bacteria: Progress and insights. *Food Rev. Int.* **2023**, *39*, 4781–4792. [CrossRef]
15. Spangler, J.R.; Dean, S.N.; Leary, D.H.; Walper, S.A. Response of *Lactobacillus plantarum* WCFS1 to the Gram-negative pathogen-associated quorum sensing molecule N-3-oxododecanoyl homoserine lactone. *Front. Microbiol.* **2019**, *10*, 715. [CrossRef] [PubMed]
16. Xue, J.; Chi, L.; Tu, P.; Lai, Y.; Liu, C.-W.; Ru, H.; Lu, K. Detection of gut microbiota and pathogen produced N-acyl homoserine in host circulation and tissues. *NPJ Biofilms Microbiomes* **2021**, *7*, 53. [CrossRef]
17. Nguyen, Y.; Nguyen Nam, X.; Rogers Jamie, L.; Liao, J.; MacMillan John, B.; Jiang, Y.; Sperandio, V. Structural and mechanistic roles of novel chemical ligands on the SdiA quorum-sensing transcription regulator. *mBio* **2015**, *6*, 10–1128. [CrossRef]
18. Shimada, T.; Shimada, K.; Matsui, M.; Kitai, Y.; Igarashi, J.; Suga, H.; Ishihama, A. Roles of cell division control factor SdiA: Recognition of quorum sensing signals and modulation of transcription regulation targets. *Genes Cells* **2014**, *19*, 405–418. [CrossRef]
19. Liu, L.; Zeng, X.; Zheng, J.; Zou, Y.; Qiu, S.; Dai, Y. AHL-mediated quorum sensing to regulate bacterial substance and energy metabolism: A review. *Microbiol. Res.* **2022**, *30*, 127102. [CrossRef]
20. Carneiro, D.G.; Almeida, F.A.; Aguilar, A.P.; Vieira, N.M.; Pinto, U.M.; Mendes, T.A.; Vanetti, M.C.D. *Salmonella enterica* optimizes metabolism after addition of acyl-homoserine lactone under anaerobic conditions. *Front. Microbiol.* **2020**, *11*, 1459. [CrossRef]
21. Yan, H.; Li, J.; Meng, J.; Li, J.; Jha, A.K.; Zhang, Y.; Wang, X.; Fan, Y. Insight into the effect of N-acyl-homoserine lactones-mediated quorum sensing on the microbial social behaviors in a UASB with the regulation of alkalinity. *Sci. Total Environ.* **2021**, *800*, 149413. [CrossRef] [PubMed]
22. Yan, C.; Li, X.; Zhang, G.; Zhu, Y.; Bi, J.; Hao, H.; Hou, H. Quorum sensing-mediated and growth phase-dependent regulation of metabolic pathways in *Hafnia alvei* H4. *Front. Microbiol.* **2021**, *12*, 567942. [CrossRef]
23. Yan, X.; Liu, X.-Y.; Zhang, D.; Zhang, Y.-D.; Li, Z.-H.; Liu, X.; Wu, F.; Chen, G.-Q. Construction of a sustainable 3-hydroxybutyrate-producing probiotic *Escherichia coli* for treatment of colitis. *Cell Mol. Immunol.* **2021**, *18*, 2344–2357. [CrossRef] [PubMed]
24. Leventhal, D.S.; Sokolovska, A.; Li, N.; Plescia, C.; Kolodziej, S.A.; Gallant, C.W.; Christmas, R.; Gao, J.-R.; James, M.J.; Abin-Fuentes, A.; et al. Immunotherapy with engineered bacteria by targeting the STING pathway for anti-tumor immunity. *Nat. Commun.* **2020**, *11*, 2739. [CrossRef] [PubMed]

25. Wang, Z.; Dai, Y.; Azi, F.; Wang, Z.; Xu, W.; Wang, D.; Dong, M.; Xia, X. Engineering *Escherichia coli* for cost-effective production of medium-chain fatty acids from soy whey using an optimized galactose-based autoinduction system. *Bioresour. Technol.* **2024**, *393*, 130145. [CrossRef]
26. Wang, Z.; Li, X.; Dai, Y.; Yin, L.; Azi, F.; Zhou, J.; Dong, M.; Xia, X. Sustainable production of genistin from glycerol by constructing and optimizing *Escherichia coli*. *Metab. Eng.* **2022**, *74*, 206–219. [CrossRef]
27. Labun, K.; Montague, T.G.; Krause, M.; Torres Cleuren, Y.N.; Tjeldnes, H.; Valen, E. CHOPCHOP v3: Expanding the CRISPR web toolbox beyond genome editing. *Nucleic Acids Res.* **2019**, *47*, W171–W174. [CrossRef]
28. Xu, P.; Vansiri, A.; Bhan, N.; Koffas, M.A.G. ePathBrick: A synthetic biology platform for engineering metabolic pathways in *E. coli*. *ACS Synth. Biol.* **2012**, *1*, 256–266. [CrossRef] [PubMed]
29. Xiao, L.; Zhang, C.; Zhang, X.; Zhao, X.; Chaeipeima Mahsa, G.; Ma, K.; Ji, F.; Azarpazhooh, E.; Ajami, M.; Rui, X.; et al. Effects of *Lacticaseibacillus paracasei* SNB-derived postbiotic components on intestinal barrier dysfunction and composition of gut microbiota. *Food Res. Int.* **2024**, *175*, 113773. [CrossRef]
30. Jumper, J.; Evans, R.; Pritzel, A.; Green, T.; Figurnov, M.; Ronneberger, O.; Tunyasuvunakool, K.; Bates, R.; Žídek, A.; Potapenko, A. Highly accurate protein structure prediction with AlphaFold. *Nature* **2021**, *596*, 583–589. [CrossRef]
31. Ionescu, V.S.; Popa, A.; Alexandru, A.; Manole, E.; Neagu, M.; Pop, S. Dietary phytoestrogens and their metabolites as epigenetic modulators with impact on human health. *Antioxidants* **2021**, *10*, 1893. [CrossRef]
32. Fiechter, G.; Raba, B.; Jungmayr, A.; Mayer, H.K. Characterization of isoflavone composition in soy-based nutritional supplements via ultra performance liquid chromatography. *Anal. Chim. Acta* **2010**, *672*, 72–78. [CrossRef]
33. Wichmann, J.; Behrendt, G.; Boecker, S.; Klamt, S. Characterizing and utilizing oxygen-dependent promoters for efficient dynamic metabolic engineering. *Metab. Eng.* **2023**, *77*, 199–207. [CrossRef]
34. Arcidiacono, S.; Spangler, J.R.; Litteral, V.; Doherty, L.A.; Stamps, B.; Walper, S.; Goodson, M.; Soares, J.W. In vitro fermentation evaluation of engineered sense and respond probiotics in polymicrobial communities. *ACS Biomater. Sci. Eng.* **2023**, *9*, 5176–5185. [CrossRef]
35. Ott, S.J.; Waetzig, G.H.; Rehman, A.; Moltzau-Anderson, J.; Bharti, R.; Grasis, J.A.; Cassidy, L.; Tholey, A.; Fickenscher, H.; Seegert, D. Efficacy of sterile fecal filtrate transfer for treating patients with Clostridium difficile infection. *Gastroenterology* **2017**, *152*, 799–811.e797. [CrossRef]
36. Tsuji, H.; Moriyama, K.; Nomoto, K.; Akaza, H. Identification of an enzyme system for daidzein-to-equol conversion in *Slackia* sp. strain NATTS. *Appl. Environ. Microbiol.* **2012**, *78*, 1228–1236. [CrossRef]
37. Shimada, Y.; Takahashi, M.; Miyazawa, N.; Ohtani, T.; Abiru, Y.; Uchiyama, S.; Hishigaki, H. Identification of two novel reductases involved in equol biosynthesis in *Lactococcus* strain 20-92. *J. Mol. Microbiol* **2011**, *21*, 160–172. [CrossRef]
38. Siziya, I.N.; Hwang, C.Y.; Seo, M.-J. Antioxidant potential and capacity of microorganism-sourced C30 carotenoids—A review. *Antioxidants* **2022**, *11*, 1963. [CrossRef]
39. Kim, M.; Jung, D.-H.; Hwang, C.Y.; Siziya, I.N.; Park, Y.-S.; Seo, M.-J. 4,4′-Diaponeurosporene production as C30 carotenoid with antioxidant activity in recombinant *Escherichia coli*. *Appl. Biochem.* **2023**, *195*, 135–151. [CrossRef]
40. Lee, P.G.; Kim, J.; Kim, E.J.; Jung, E.; Pandey, B.P.; Kim, B.G. P212A Mutant of dihydrodaidzein reductase enhances (*S*)-equol production and enantioselectivity in a recombinant *Escherichia coli* whole-cell reaction system. *Appl. Environ. Microbiol.* **2016**, *82*, 1992–2002. [CrossRef]
41. Wang, L.; Yu, H.; Xu, J.; Ruan, H.; Zhang, W. Deciphering the crucial roles of AraC-type transcriptional regulator Cgl2680 on NADPH metabolism and L-lysine production in *Corynebacterium glutamicum*. *World J. Microbiol. Biotechnol.* **2020**, *36*, 1–15. [CrossRef]
42. Kim, T.; Duong, T.; Wu, C.-a.; Choi, J.; Lan, N.; Kang, S.W.; Lokanath, N.K.; Shin, D.; Hwang, H.-Y.; Kim, K.K. Structural insights into the molecular mechanism of *Escherichia coli* SdiA, a quorum-sensing receptor. *Acta Crystallogr. D* **2014**, *70*, 694–707. [CrossRef]
43. Yang, Y.; Zhou, M.; Hardwidge, P.R.; Cui, H.; Zhu, G. Isolation and characterization of N-acyl homoserine lactone-producing bacteria from cattle rumen and swine intestines. *Front. Cell. Infect. Microbiol.* **2018**, *8*, 155. [CrossRef]
44. Gu, J.; Xu, Y.; Nie, Y. Role of distal sites in enzyme engineering. *Biotechnol. Adv.* **2023**, *63*, 108094. [CrossRef]
45. Xu, X.; Zeng, D.; Wu, D.; Lin, J. Single-point mutation near active center increases substrate affinity of alginate lyase AlgL-CD. *Appl. Biochem.* **2021**, *193*, 1513–1531. [CrossRef]
46. Wu, Q.; Zhang, C.; Dong, W.; Lu, H.; Yang, Y.; Li, W.; Xu, Y.; Li, X. Simultaneously enhanced thermostability and catalytic activity of xylanase from *Streptomyces rameus* L2001 by rigidifying flexible regions in loop regions of the N-terminus. *J. Agric. Food Chem.* **2023**, *71*, 12785–12796. [CrossRef]
47. Pedrolli, D.B.; Ribeiro, N.V.; Squizato, P.N.; de Jesus, V.N.; Cozetto, D.A. Engineering microbial living therapeutics: The synthetic biology toolbox. *Trends Biotechnol.* **2019**, *37*, 100–115. [CrossRef]

Article

Cultivar and Harvest Time of Almonds Affect Their Antioxidant and Nutritional Profile through Gut Microbiota Modifications

Adriana Delgado-Osorio [1], Beatriz Navajas-Porras [1], Sergio Pérez-Burillo [1], Daniel Hinojosa-Nogueira [1], Ángela Toledano-Marín [1], Silvia Pastoriza de la Cueva [1], Oleg Paliy [2] and José Ángel Rufián-Henares [1,3,*]

1 Departamento de Nutrición y Bromatología, Instituto de Nutrición y Tecnología de los Alimentos, Centro de Investigación Biomédica, Universidad de Granada, Av. del Hospicio, s/n, 18012 Granada, Spain; adrianadelgado@ugr.es (A.D.-O.); beatriznavajas@ugr.es (B.N.-P.); spburillo@uma.es (S.P.-B.); dhinojosa@ugr.es (D.H.-N.); antolemarin@correo.ugr.es (Á.T.-M.); spdelacueva@ugr.es (S.P.d.l.C.)
2 Department of Biochemistry and Molecular Biology, Boonshoft School of Medicine, Wright State University, Dayton, OH 45435, USA; oleg.paliy@wright.edu
3 Instituto de Investigación Biosanitaria ibs.GRANADA, Universidad de Granada, Avda. de Madrid 15, 2a Planta, 18012 Granada, Spain
* Correspondence: jarufian@ugr.es

Abstract: Almonds are a rich source of beneficial compounds for human health. In this work, we assessed the influence of almond cultivars and harvest time on their morphological (length, width and thickness) and nutritional (ash, moisture, proteins) profiles. We also evaluated the impact of an in vitro digestion and fermentation process on almonds' antioxidant and phenolic content, as well as their support of gut microbiota community and functionality, including the production of short-chain fatty acids (SCFAs), lactic and succinic acids. The length, width, and thickness of almonds varied significantly among cultivars, with the latter two parameters also exhibiting significant changes over time. Moisture content decreased with maturity, while protein and ash increased significantly. Total antioxidant capacity released by almonds after digestion and fermentation had different trends depending on the antioxidant capacity method used. The fermentation step contributed more to the antioxidant capacity than the digestion step. Both cultivar and harvest time exerted a significant influence on the concentration of certain phenolic compounds, although the total content remained unaffected. Similarly, fecal microbiota modulation depended on the cultivar and maturity stage, with the Guara cultivar and late maturity showing the largest effects. Cultivar type also exerted a significant impact on the concentration of SCFAs, with the Guara cultivar displaying the highest total SCFAs concentration. Thus, we conclude that cultivar and harvest time are key factors in shaping the morphological and nutritional composition of almonds. In addition, taking into account all the results obtained, the Guara variety has the best nutritional profile.

Keywords: almond; cultivar; harvest time; digestion; fermentation; gut microbiota; metabolites

Citation: Delgado-Osorio, A.; Navajas-Porras, B.; Pérez-Burillo, S.; Hinojosa-Nogueira, D.; Toledano-Marín, Á.; Pastoriza de la Cueva, S.; Paliy, O.; Rufián-Henares, J.Á. Cultivar and Harvest Time of Almonds Affect Their Antioxidant and Nutritional Profile through Gut Microbiota Modifications. *Antioxidants* **2024**, *13*, 84. https://doi.org/10.3390/antiox13010084

Academic Editor: Myung-Ji Seo

Received: 15 December 2023
Revised: 28 December 2023
Accepted: 29 December 2023
Published: 9 January 2024

1. Introduction

Almonds (*Prunus dulcis* (Mill.) D.A. Webb) are one of the most consumed nuts worldwide. These nuts are part of the Mediterranean diet and are eaten raw, blanched, roasted, fried, and caramelized; processed to flour; used in non-dairy beverages; or used as an ingredient for use in foods. In addition to their high gastronomic value, almonds stand out for their high nutritional profile. They represent an important source of lipids, mainly monounsaturated and polyunsaturated fatty acids, proteins, dietary fiber, and micronutrients such as vitamin E, potassium, phosphorus, calcium, and magnesium [1]. Almonds are also known to be a major source of phenolic compounds. These are responsible for the flavor and aroma of the almonds, as well as for their antioxidant, antiviral, and antibacterial properties [2,3].

The amount and type of nutrients present in almonds are influenced by different factors, including the cultivar, time of harvest, and the specific agroclimatic conditions in which they are grown [4]. Cultivar and harvest time are two of the main factors determining the morphological and nutritional profile of almonds. Numerous studies have confirmed the influence of these factors on morphological measurements (width, length, and thickness of the almond), as well as on the moisture, carbohydrate, lipid, protein, ash, antioxidant, mineral, phenolic, and tocopherol contents of almonds [5–10]. However, the results of these studies are partially contradictory, especially regarding the effect of harvest time, since there is no consistent trend on how the chemical profile of the almond evolves with time. These discrepancies could be related to the differences in those cultivars and harvest times considered, as well as due to other, yet unstudied, factors.

An additional key element in understanding how almond compounds impact human health is their bioaccessibility, which can be defined as the amount of a nutrient that is released from the food matrix in the gastrointestinal tract, and which becomes accessible for absorption. The bioaccessibility of nutrients is determined by multiple factors, including human digestion and fermentation by the gut microbiota [11]. In this context, dietary fiber and phenolic compounds are interesting to study because they are poorly digested in the upper gastrointestinal tract and enter the large intestine, where they are transformed by the gut microbiota into simpler metabolites with enhanced absorption and bioactivity [12]. The most common metabolites of dietary fiber, short-chain fatty acids (SCFAs), such as acetic, propionic and butyric acids, along with the metabolites of phenolic compounds have demonstrated multiple benefits in human health [13,14]. In addition to SCFAs, the gut microbiota produces other metabolites in the course of fermentation: lactic and succinic acids. These compounds have a significant impact on human health, as they are shared metabolites in human and microbial metabolic pathways and participate in microbiota–host cross-talk [15]. Other relevant actions of the gut microbiota include the release of antioxidant molecules from the food matrix and the production of antioxidant metabolites, leading to an increase in the antioxidant capacity of foods [16].

There is, however, insufficient information available on how human digestion and fermentation by the gut microbiota influence the antioxidant capacity of almonds and the bioaccessibility of their nutrients. A few studies have attempted to understand the effects of in vivo human digestion and fermentation by the gut microbiota on almonds, with a focus on analyzing gut microbiota-derived metabolites generated after the ingestion of almonds, such as the metabolites of phenolic compounds in urine and plasma [17–19] and the SCFAs in feces [20,21]. On the other hand, a group of studies focused on understanding the effects of in vitro digestion and fermentation and quantified SCFAs [22–24], phenolic compounds, and antioxidant capacity [24] generated after the process. Nevertheless, none of these studies took into account factors such as cultivar and harvest time, which could introduce potential variability in the results.

Thus, the aim of this work is to study the influence of harvest time and cultivar on morphological measurements (width, length and thickness of the almond) as well as the nutritional composition of almonds (ash, moisture and protein content) and evaluate the impact of an in vitro digestion and fermentation process in the antioxidant, phenolic, SCFAs, lactic and succinic acids profiles of almonds.

2. Materials and Methods

2.1. Chemicals

Firstly, (±)-6-Hydroxy-2,5,7,8-tetramethylchromane-2-carboxylic acid (Trolox), 2,2′-Azino-bis(3-ethylbenzothiazoline-6-sulfonic acid) diammonium salt (ABTS), 2,4,6-Tri(2-pyridyl)-s-triazine (TPTZ), 2,2 Diphenyl-1-1-picrythydrazul hydrate 95% (DPPH), potassium persulphate, 3,5-dicaffeoylquinic acid, caffeic acid, dimethyl caffeic, chlorogenic acid, ferulic acid, *p*-coumaric acid, gallic acid, tyrosol, *m*-hydroxyphenylacetic acid, Folin–Ciocalteu reagent, sodium hydroxide, methanol, hydrogen peroxide, hydrochloric acid, sulfuric acid, formic acid, succinic acid, propionic acid, acetic acid, isobutyric acid, lactic acid,

(+)-catechin, 3-(3-hydroxyphenyl)propionic acid, 3,4-dihydroxyphenylacetic acid, 3-(3,4-dihydroxyphenyl)propionic acid, (-)-epicatechin, kaempferol, phenol, quercetin, *p*-coumaric acid, naringenin, phloroglucinol, ferulic acid, urolithin A, urolithin B, iron (III) chloride hexahydrate, sodium acetate, potassium chloride, potassium di-hydrogen phosphate, sodium mono-hydrogen carbonate, sodium chloride, magnesium chloride hexahydrate, ammonium carbonate, calcium chloride dihydrate, sodium di-hydrogen phosphate, tryptone, cysteine, sodium sulfide, resazurin, salivary α-amylase, pepsin, bile acids (porcine bile extract), ethanol and Milli-Q® water were from Sigma-Aldrich. Pancreatin from porcine pancreas were purchased from Alpha Aesar. Diethyl ether and acetonitrile were from Honeywell. *N*-butyric acid was from Acros Organics, and 5-(3′, 4′-dihydroxyphenyl)-γ-valerolactone was purchased from TRC Canada. Moreover, (-)-Epigallocatechin and naringin were from Extrasynthese. Rutin was from PhytoLab. KjTabsTM VCM tablets, KjTabsTM VS Antifoam tablets, and boric acid solution (4% solution + indicator) were from Thermo Fisher Scientific (Waltham, MA, USA).

2.2. Samples

Almonds were cultivated in a private plot in Dúrcal (Granada, Spain, 36.962126, −3.560991). They belonged to the Spanish cultivars Guara (G), Vairo (V), Marta (MT), Marinada (MD), and Marcona (MC) and were harvested at three different times: time 1 (T1), time 2 (T2) and time 3 (T3). T1 and T2 for the five cultivars corresponded to 2 August 2023 and 24 August 2023, respectively, while T3 for Guara, Vairo and Marta cultivars corresponded to 26 September 2023 and T3 for Marinada and Marcona cultivars corresponded to 2 October 2023. After harvesting, the hulls and shells were removed, and almond kernels were stored at −80 °C until analysis.

2.3. Morphological Measurements

For each cultivar and harvest time, 10 almonds were randomly selected to measure kernel length, width and thickness by a caliper. Results were expressed in centimeters.

2.4. Determination of Ash, Moisture and Protein Content

Ash, moisture and protein content were determined in duplicate for each cultivar and harvest time by AOAC 923.03, 925.09 and 950.48 methods, respectively. Almond samples were grounded and homogenized, and 5 g of the mixture was used for each ash and moisture determination, while 0.5 g was employed for determining protein content. The results were expressed in percentages of ash, moisture, and protein content.

2.5. In Vitro Digestion and Fermentation

The samples were prepared by grinding each almond cultivar collected at each harvest time. Then, 100 mg of the resulting mixture was weighed. The samples were subjected to a previously described in vitro digestion and fermentation process [25] in order to simulate the human digestion–fermentation process without performing a nutritional intervention.

Briefly, the in vitro digestion consisted of an oral phase (2 min at 37 °C with 75 U/mL α-amylase), a gastric phase (2 h at 37 °C with 2000 U/mL pepsin at pH 3), and an intestinal phase (2 h at 37 °C with 13.37 mg/mL pancreatin and 10 mM bile acids at pH 7). The enzymatic reactions were stopped by immersing the tubes in ice. Once the three phases were finished, the samples were centrifuged at 6000 rpm for 10 min at 4 °C. The supernatant obtained represented the soluble and potentially absorbable fraction in the small intestine. Then, 10% of the supernatant was added to the solid residue, after which both mixed fractions were lyophilized and frozen at −80 °C. The remaining supernatant was also stored at −80 °C.

In vitro fermentation was performed with fresh feces from three healthy donors (people who had not taken antibiotics three months before the stool collection and with a body mass index within the 20–25 range). Stools were pooled to reduce inter-individual variability [26]. The sample submitted to fermentation was the combination of the solid residue obtained after in vitro digestion (100 mg) plus 10% of the digestion supernatant. Two control fermentations were run containing only fecal inoculum and buffer but no digested almonds (designated BL). In vitro fermentation took place at 37 °C for 24 h in an oxygen-depleted atmosphere. Upon completion, the samples were immersed in ice to stop microbial activity and centrifuged at 6000 rpm for 10 min. The supernatant, which represented the potentially absorbable fraction in the large intestine, as well as the pellet, were stored at −80 °C.

2.6. Antioxidant Assays

The antioxidant capacity was evaluated in the two fractions obtained after the in vitro digestion and fermentation of almonds: the supernatants obtained after digestion and fermentation. The assays were performed in duplicate for each sample. The sum of the two fractions accounts for the total antioxidant capacity that almonds can exert within the human body.

2.6.1. Trolox Equivalent Antioxidant Capacity against ABTS Radicals (TEAC$_{ABTS}$) Assay

TEAC$_{ABTS}$ was tested following a previously described method [27]. Briefly, ABTS was prepared by mixing ABTS stock solution (7 mM) with 2.45 mM potassium persulphate and storing the mixture in the dark for 12 h, after which it was diluted with a 50:50 ethanol:water solution. Then, 280 µL of diluted ABTS and 20 µL of sample or Trolox standard were added to a transparent 96-well polystyrene microplate (Biogen Científica, Madrid, Spain) and absorbance readings at 730 nm were monitored for 20 min on a Cytation 5 microplate reader (Agilent Technologies, Santa Clara, CA, USA) at 37 °C. Calibration was performed with a Trolox stock solution ranging from 0.01 to 1.00 mg/mL. Results were expressed as mmol Trolox equivalents per kg of sample.

2.6.2. Trolox Equivalent Antioxidant Capacity Referred to Reducing Capacity (TEAC$_{FRAP}$) Assay

TEAC$_{FRAP}$ was conducted following a previously described procedure [28]. The FRAP reagent was composed of 2.5 mL of 10 mM TPTZ solution, 2.5 mL of 20 mM FeCl$_3$·6H$_2$O and 25 mL of 0.3 M acetate buffer at pH 3.6. Moreover, 280 µL of FRAP reagent and 20 µL sample or Trolox standard were mixed in a 96-well microplate (Biogen Científica) and absorbance readings at 595 nm were monitored for 30 min with a Cytation 5 microplate reader (Agilent Technologies) at 37 °C. Calibration was performed with a Trolox stock solution ranging from 0.01 to 1.00 mg/mL. Results were expressed as mmol Trolox equivalents per kg of sample.

2.6.3. Trolox Equivalent Antioxidant Capacity against DPPH Radicals (TEAC$_{DPPH}$) Assay

TEAC$_{DPPH}$ was performed as in [29]. Then, 20 µL of the sample was mixed with 280 µL DPPH reagent (74 mg DPPH/L methanol) in a transparent 96-well polystyrene microplate (Biogen Científica) plate. Absorbance readings at 517 nm were monitored for 60 min using a Cytation 5 microplate reader (Agilent Technologies) at 37 °C. Calibration was performed with a Trolox stock solution ranging from 0.01 to 1.00 mg/mL. Results were expressed as mmol Trolox equivalents per kg of sample.

2.6.4. Folin–Ciocalteu Assay

For performing the Folin–Ciocalteu assay, the procedure described by [30] was adapted to a microplate reader. Furthermore, 30 µL of the sample was mixed in a well with 15 µL of Folin–Ciocalteu reagent and 255 µL of 2.35% sodium carbonate. Absorbance readings were monitored at 725 nm for 60 min at 37 °C in a Cytation 5 microplate reader (Agilent Technologies). Calibration was performed with a gallic acid stock solution ranging from 0.01 to 1.00 mg/mL. Results were expressed as mg of gallic acid equivalents/kg sample.

2.7. Ultra-High Performance Liquid Chromatography (UHPLC) Analysis

Chromatographic analyses were performed using a UHPLC Agilent Infinity II LC System equipped with a Diode Array Detector and a Refractive Index Detector.

2.7.1. Analysis of Phenolic Compounds

Phenolic compounds were extracted from the fermentation supernatant with diethyl ether [31]. The extraction procedure was carried out in duplicate for each sample. The fermentation supernatant samples were centrifuged at 13,000 rpm for 2 min, after which 800 µL of the supernatant was taken. Subsequently, 1 mL of diethyl ether was added to the 800 µL of supernatant and stored at 4 °C in the dark for 24 h. After that time, the organic phase corresponding to the diethyl ether with the extracted phenolic compounds was collected in a new tube, and two new extractions were carried out with 1 mL of diethyl ether at room temperature. Subsequently, a rotary evaporator at 30 °C was used to evaporate the diethyl ether, and the dry residue was redissolved in 1 mL of Milli-Q water/methanol 1:1 (v/v). Finally, the extracts were filtered with a 0.22 µm filter and collected in vials.

For the UHPLC analysis, the column used was an Agilent Poroshell 120 SB-Aq (4.6 × 100 mm, 2.7 µm). The mobile phase consisted of Solvent A (Milli-Q water/formic acid, 99.9:0.1) and Solvent B (acetonitrile/formic acid, 99.9:0.1). The method used was a gradient elution: 0–28 min (20% Solvent A, 80% Solvent B), 28–32 min (60% Solvent A, 40% Solvent B), 32–33 min (95% Solvent A, 5% Solvent B), 33–35 min (20% Solvent A, 80% Solvent B), 35–38 min (95% Solvent A, 5% Solvent B). The flow rate was maintained at 0.2 mL/min. The injection volume was 5 µL. The column temperature was set at 30 °C. The wavelength selected to measure the absorbance of the samples was 255 nm. The chromatographic analysis was performed in duplicate for each sample.

Quantification was carried out using calibration with the following external standards: (+)-catechin, (-)-epicatechin, (-)-epigallocatechin, rutin, 3,4-dihydroxyphenylacetic acid, 3-(3,4-dihydroxyphenyl)propionic acid, 3-(3-hydroxyphenyl)propionic acid, 5-(3′,4′-dihydroxyphenyl)-γ-valerolactone, phenol, phloroglucinol, ferulic acid, *p*-coumaric acid, naringenin, naringin, urolithin A, urolithin B, quercetin and kaempferol. Results were expressed in mg phenolic compound/kg of digestion residue.

2.7.2. Analysis of SCFAs, Lactic and Succinic Acids

The preparation of the sample for the chromatographic analysis was performed by centrifuging the fermentation supernatant samples at 13,300 rpm for 5 min, filtering them through a 0.22 µm filter and performing a 1:10 dilution with 1 M hydrochloric acid [32]. The column used was an Agilent Poroshell 120 SB-Aq (3 × 150 mm, 2.7 µm). The mobile phase was 5 mM sulfuric acid with isocratic elution at a flow rate of 0.5 mL/min. The injection volume was 5 µL. The column and RID temperature were set at 35 °C. The chromatographic analysis was performed in duplicate for each sample.

Quantification was carried out using calibration with the following external standards: lactic, acetic, succinic, propionic, *N*-butyric and isobutyric acids. *N*-butyric and isobutyric acids were quantified together. Total SCFAs were calculated as the sum of acetic, propionic and butyric acid. Results were expressed in mmol of each acid per L of the fermentation supernatant.

2.8. Microbial Genomic DNA Isolation and High Throughput Sequencing

Prokaryotic genomic DNA was isolated from the pellet obtained after sample fermentation using a ZR bacterial/fungal DNA kit (Zymo Research, Irvine, CA, USA) following the manufacturer's instructions. The V4 hypervariable region of the 16S rRNA gene was amplified using primers complementary to the flanking conserved sequences (forward primer complementary sequence GCCAGCMGCCGCGG and reverse primer complementary sequence GGACTACHVGGGTWTCTAAT). The forward primer incorporated a 6–7 nucleotide barcode to allow for sample multiplexing on the sequencer. In PCR amplifications, 25 ng of the starting gDNA material was first subjected to four cycles of linear elongation with the forward primer only in order to reduce sample-to-sample PCR bias [33] followed by twenty-five cycles of traditional exponential PCR. Amplicon sequencing was carried out on the Ion Torrent Personal Genome Machine (Thermo Fisher, USA) using a 318 Chip v2. After quality filtering, we obtained 32,485 reads per sample on average. All sequence reads were processed in QIIME [34] following our standard pipeline [35] to obtain the 16S rRNA gene-copy, number-adjusted, rarefied taxon counts. This final dataset was used for all further analyses.

2.9. Statistical Analyses

The statistical significance of the data was tested by the non-parametric Kruskal–Wallis analysis of variance test, followed by the pair-wise Games–Howell post-hoc tests to compare the samples that showed significant variation ($p < 0.05$). The harvest time and cultivar of the almonds were used as factors in the Kruskal–Wallis tests. All statistical analyses were performed using SPSS.

Multivariate statistical analyses were performed on the genus-level microbial abundance dataset generally following the approaches we described previously [36]. These included unconstrained principal coordinates analysis (PCoA) utilizing phylogenetic weighted UniFrac distance as a measure of sample dissimilarity and constrained canonical correspondence analysis (CCA). Logistic regression (LR) with Lasso regularization (threshold C = 0.2) was chosen to generate sample classification models, as was the case previously [37]. Lasso regularization allowed us to limit the number of discriminatory variables defining each sample type. Model performance was assessed by a 20-fold cross-validation algorithm. Statistical tests (one-way analysis of variance (ANOVA) unless otherwise stated) were carried out in SPSS.

3. Results and Discussion

3.1. Influence of Cultivar and Harvest Time on Ash, Moisture, Protein, and Morphology of Almonds

The harvest time of almonds had a significant impact on their width and thickness, while their length remained unaffected (Supplementary Materials, Table S1). Almond thickness tended to decrease significantly over time, while almond width initially increased and then returned to its original level. This general decrease can be attributed to a loss of moisture over time. However, the evolution of morphological measurements over time did not follow the same trend for all cultivars (Table 1). In fact, when comparing the morphological measurements between the five cultivars, significant differences were observed in all three morphological measurements (Supplementary Materials, Table S2). These results reinforce the idea that cultivar plays a major role in almond morphology.

A study conducted on 24 traditional almond cultivars in the central–western Iberian Peninsula revealed differences in the width, thickness, and length of these cultivars [38]. Another study involving 10 cultivars of diverse origins reported variations in the morphological measurements among them and highlighted distinct trends in almond length and width over time for each one [10]. In the same study, over the course of the two harvest times considered, almond length increased in five cultivars while decreasing in the other five. However, almond width increased in seven cultivars while decreasing in three. Additionally, the thickness decreased over time for the 10 cultivars under examination [10]. Thus, genotype has a strong influence on almond morphological characteristics.

Table 1. Morphological measurements, ash, moisture and protein profile of five almond cultivars collected at three harvest times and submitted to in vitro digestion-fermentation.

Cultivar	Harvest Time	Length (cm)	Width (cm)	Thickness (cm)	Ash (%)	Moisture (%)	Protein (%)
Guara	T1	2.47 ± 0.13	1.09 ± 0.10 [a]	0.70 ± 0.09 [a]	1.38 ± 0.12 [a]	65.1 ± 1.10	11.2 ± 1.13 [a]
	T2	2.51 ± 0.13	1.28 ± 0.08	0.70 ± 0.05 [a]	1.40 ± 0.21 [a]	47.9 ± 0.20	12.8 ± 0.62 [a]
	T3	2.41 ± 0.06	1.14 ± 0.14 [a]	0.40 ± 0.05	5.55 ± 0.27	7.00 ± 0.10	24.6 ± 4.82
Significance		NS	*	*	NS	*	*
Vairo	T1	2.24 ± 0.09	1.04 ± 0.09	0.56 ± 0.07 [a]	1.61 ± 0.01 [ab]	63.1 ± 0.6	14.8 ± 5.54 [ab]
	T2	2.18 ± 0.11	1.11 ± 0.07	0.55 ± 0.05 [a]	1.65 ± 0.14 [a]	44.7 ± 0.8	13.8 ± 1.03 [a]
	T3	2.26 ± 0.10	1.11 ± 0.12	0.36 ± 0.05	3.63 ± 0.25 [b]	6.00 ± 0.30	22.3 ± 0.72 [b]
Significance		NS	NS	*	NS	*	*
Marta	T1	2.34 ± 0.13	1.01 ± 0.12 [ab]	0.73 ± 0.05 [a]	1.37 ± 0.06	66.9 ± 1.1 [a]	12.5 ± 3.28
	T2	2.27 ± 0.12	1.10 ± 0.07 [a]	0.71 ± 0.07 [a]	1.64 ± 0.04	55.1 ± 2.00 [a]	9.36 ± 2.56
	T3	2.25 ± 0.12	0.93 ± 0.09 [b]	0.42 ± 0.06	4.56 ± 0.69	6.50 ± 0.00	22.6 ± 1.13
Significance		NS	*	*	NS	NS	*
Marinada	T1	2.07 ± 0.12 [a]	1.09 ± 0.10 [ab]	0.71 ± 0.03 [a]	1.14 ± 0.07	76.6 ± 0.30	6.96 ± 0.21
	T2	2.28 ± 0.18 [b]	1.05 ± 0.08 [a]	0.49 ± 0.06 [a]	1.55 ± 0.02	47.7 ± 1.30	12.8 ± 1.74
	T3	2.11 ± 0.14 [ab]	1.20 ± 0.13 [b]	0.69 ± 0.09	3.25 ± 1.08	8.80 ± 0.60	21.4 ± 1.54
Significance		*	*	*	NS	NS	*
Marcona	T1	1.80 ± 0.08	1.20 ± 0.08	0.84 ± 0.05	1.41 ± 0.09 [a]	54.6 ± 2.80 [a]	10.2 ± 0.82
	T2	1.70 ± 0.12	1.14 ± 0.11	0.75 ± 0.07	1.63 ± 0.09 [a]	50.3 ± 0.50 [a]	11.8 ± 1.74
	T3	1.74 ± 0.05	1.17 ± 0.05	0.62 ± 0.06	3.71 ± 0.15	9.30 ± 1.10	21.2 ± 5.33
Significance		NS	NS	*	NS	*	*

Statistical differences among samples were tested by the Kruskal–Wallis test at the 5% level of significance (NS: not significant. *: significant). For the samples that showed statistical differences, a common letter indicates that samples are not significantly different based on the pair-wise Games–Howell post-hoc test at the 5% level of significance.

In terms of the chemical composition, studied cultivars did not elicit a significant impact on the moisture, protein or ash content (Supplementary Materials, Table S3). Nevertheless, the cultivars with the highest levels of protein and ash were Vairo and Guara, respectively, while the lowest moisture was found in the Vairo cultivar.

For all cultivars, moisture content decreased significantly over time (Table 1), consistent with the previous reports [10,39–43]. A higher moisture level shortens the final product shelf life. We observed a significant increase in protein and ash content over time (Supplementary Materials, Table S4). There were significant differences between the first and the third harvest times: the ash content tripled, and the protein content doubled probably due to the decrease in water content. Regarding ash content, all five cultivars followed a consistent trend of increasing their content at each harvest time. However, in terms of protein content, all cultivars increased from T1 to T3, with Vairo and Marta cultivars experiencing a slight decrease from T1 to T2. These trends suggest that almonds accumulate protein and minerals over time, resulting in an increase in their nutritional value. The increase in the protein [10,43,44] and ash [10] content has been reported in the literature. However, although ash content serves as an indicator of mineral content, the trend in mineral levels can vary depending on the specific mineral being analyzed [6] and the almond cultivar [7,8,45].

In summary, these results indicate that the optimal time to harvest almonds is during the third period, as it leads to increased nutritional value and reduced moisture content (Supplementary Materials, Table S4), which in turn extends shelf life.

3.2. Antioxidant Capacity of the Samples Obtained after In Vitro Digestion and Fermentation

3.2.1. Evolution of the Total Antioxidant Capacity over Harvest Time

Total antioxidant capacity was calculated by adding antioxidant capacity released during in vitro digestion and in vitro fermentation. The evolution of total antioxidant capacity over harvest time differed among the assays (Figure 1).

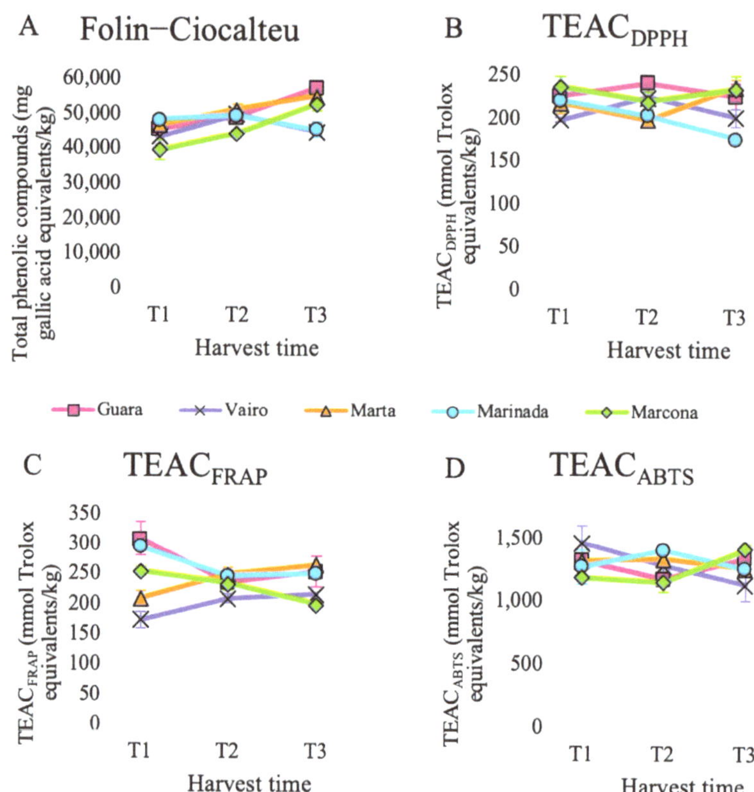

Figure 1. Total antioxidant capacity of five almond cultivars collected at three harvest times and submitted to in vitro digestion-fermentation measured by (**A**) Folin–Ciocalteu, (**B**) TEAC$_{DPPH}$, (**C**) TEAC$_{FRAP}$ and (**D**) TEAC$_{ABTS}$ assays.

The total phenolic content, measured via the Folin–Ciocalteu assay, exhibited significant differences over time, increasing at each time point, except for the Marinada cultivar (Supplementary Materials, Table S5) (Figure 1A). This general increase can be attributed to the accumulation of phenolic compounds during almond development. A study that analyzed the total phenolic content of almonds harvested at two different times also reported an increase for six out of the ten evaluated cultivars [10].

When analyzing the evolution of TEAC$_{ABTS}$, TEAC$_{FRAP}$, and TEAC$_{DPPH}$ over time, no significant differences were observed (see Supplementary Materials, Table S5). The general trend for TEAC$_{DPPH}$ was a decrease over time, observed in all cultivars from T1 to T3 except for Marta (Figure 1B). This contrasts with a study that reported an increase in TEAC$_{DPPH}$ over time for 9 out of the 10 cultivars [10].

On the other hand, the general tendency for TEAC$_{FRAP}$ and TEAC$_{ABTS}$ was an initial decrease followed by a slight increase (Supplementary Materials, Table S5). This phenomenon could be attributed to the fact that TEAC$_{FRAP}$ and TEAC$_{ABTS}$ tend to yield comparable results because the same compounds are reactive in the FRAP and ABTS assays. However, there was no common pattern in the evolution of $TEAC_{FRAP}$ and $TEAC_{ABTS}$ when considering different cultivars. While cultivars Marta and Vairo tended to increase their $TEAC_{FRAP}$ from T1 to T3, Guara, Marinada, and Marcona did the opposite (Figure 1C). Regarding $TEAC_{ABTS}$, Marinada, Marta and Vairo tended to suffer a decrease in antioxidant capacity from T1 to T3, while Guara and Marcona showed an increase (Figure 1D).

When we compared cultivars against each other, we observed significant differences among cultivars for $TEAC_{FRAP}$ and $TEAC_{DPPH}$ (see Supplementary Materials, Table S6). For FRAP and DPPH assays, Guara was the cultivar that exhibited the highest total antioxidant capacity. For ABTS, it was Marinada, and for Folin–Ciocalteu, it was Marta. Different studies from different regions in the world have analyzed the influence of the cultivar on the antioxidant capacity of almonds [46–48]. However, not many studies have focused on Spanish cultivars. In one study [10], which included the Spanish cultivar Marcona, featured in this experiment, DPPH and Folin–Ciocalteu assays showed a wide variability among the 10 almond cultivars studied, highlighting the substantial influence of almond genotype on the antioxidant capacity of almonds. In this study, Marcona was one of the cultivars that exhibited lower values in both assays, as was the case in our study for Folin–Ciocalteu but not for $TEAC_{DPPH}$. In another study that analyzed almond skin extracts [49], Guara reported a higher $TEAC_{FRAP}$ and total phenolic content than Marcona, while Marcona had higher $TEAC_{DPPH}$ values than Guara [50]. Nevertheless, Marcona almond oil was reported to have higher $TEAC_{DPPH}$ than Guara almond oil.

Altogether, these results highlight the notorious impact of the cultivar as well as the antioxidant assay used to measure the evolution of the total antioxidant capacity over time.

3.2.2. Contribution of In Vitro Digestion-Fermentation Fractions to Total Antioxidant Capacity

The contribution of each fraction to the total antioxidant capacity is shown as a percentage in Figure 2. Data from almonds harvested at T3 were used for this purpose since it is considered the ideal maturity for consumption, as stated above. Fermentation released higher antioxidant capacity than in vitro digestion in all four assays used (Figure 2). This contribution was most notable in the DPPH assay, followed by the FRAP, ABTS and Folin–Ciocalteu assays. This pattern has also been seen in other studies. According to Li et al. [24], the fermentation fraction exhibited higher $TEAC_{DPPH}$, $TEAC_{FRAP}$ and total phenolic content measured by Folin–Ciocalteu than oral, gastric, and intestinal digestion fractions. That study also found that, in general, colonic fermentation provided the highest bioaccessibility for most phenolic compounds, thereby contributing to the antioxidant capacity. In another investigation of almond bagasse [51], fermentation contributed more to $TEAC_{DPPH}$ and total phenolic content measured by Folin–Ciocalteu than gastric and intestinal digestions. In another study on nuts, the fermented fraction contributed more to $TEAC_{DPPH}$, $TEAC_{FRAP}$ and total phenolic content measured by the Folin–Ciocalteu assay in comparison with the digested fraction. This may be attributed to the gut microbiota releasing antioxidant compounds from the food matrix or generating antioxidant metabolites from compounds that have undergone incomplete digestion and absorption in the upper gastrointestinal tract. Almonds are a rich source of dietary fiber [23], which decreases the bioaccessibility of some nutrients during digestion and allows them to reach the large intestine [52]. There, the gut microbiota increases the bioaccessibility of these compounds as well as releases metabolites that contribute to the antioxidant capacity of the almond.

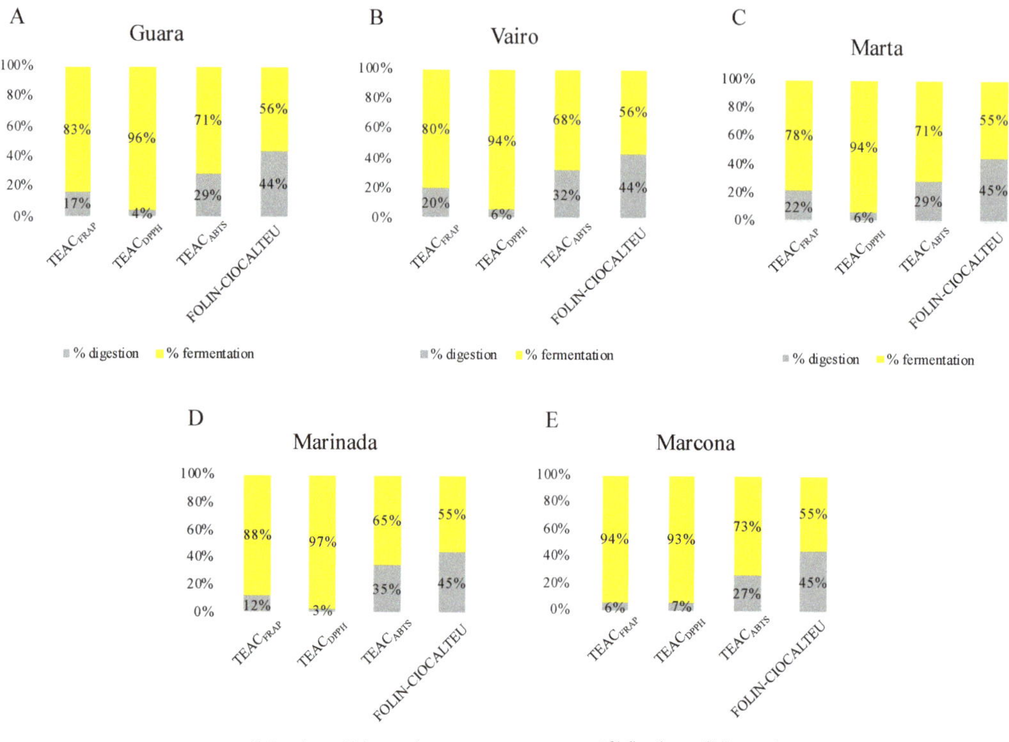

Figure 2. Contribution of the in vitro digestion and fermentation fractions to the total antioxidant capacity of (**A**) Guara (**B**) Vairo (**C**) Marta (**D**) Marinada (**E**) Marcona almonds harvested at T3.

3.3. Microbiota Community Structure Supported by Fermentation of Digested Almond Samples

We used the 16S rRNA gene amplicon sequencing to obtain profiles of human fecal microbial communities maintained on the digested almond samples. Ordination analyses indicated that the microbiota community structure overlapped among different cultivars and harvest time samples with few notable exceptions (Figure 3A). The microbiota community supported by the fermentation of the Guara almond cultivar was the most distinct among cultivars; many of the late-harvesting-time (T3) samples also clustered together. Both the cultivar and the harvest time had statistically significant contributions to the overall microbiota abundance dataset variance, as revealed by the constrained CCA analysis (Figure 3A). All communities were dominated by the genus *Escherichia/Shigella* in class Gammaproteobacteria, likely indicating the presence of residual oxygen in the fermentation vessels. An abundance of the genus *Phocaeicola* (formerly members of the Bacteroides genus, class Bacteroidia) was supported by all almond samples in comparison with blanks. Blank samples, on the other hand, maintained the genus *Enteroccus*. Examining the distribution of microbial classes and abundant genera among sample types, Guara almond fermentation expanded the abundance of the genus *Dorea* (see Figure 3B,C), known for its ability to degrade dietary fiber [53]. Considering differences among various harvesting times, T3 samples were significantly more abundant in the microbial phylotypes assigned to *Clostridium sensus stricto* (*SS*). Earlier harvesting time promoted a higher abundance of *Blautia* members (Figure 3C). Despite the presence of large amounts of unsaturated fatty acids in almond seeds, we have not detected appreciable amounts of previously noted "lipophilic" microbial genera such as *Bilophila* and *Alistipes* [54] in the fermented almond samples. The abundance of Gammaproteobacteria, another "lipophilic" taxon, did increase in the presence of digested almonds (Figure 3B).

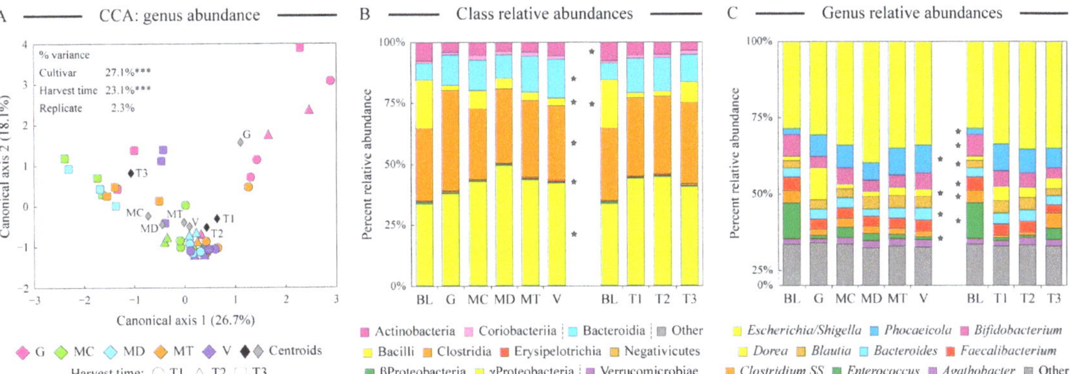

Figure 3. Comparison of microbiota composition among fermented samples. Similarity of microbial communities at the genus taxonomic level among all almond-fermented samples was assessed by the constrained canonical correspondence analysis (CCA, panel (**A**)). The percentage of total dataset variance explained by each axis is shown in parentheses. The relative contribution of explanatory variables to the overall variance of the dataset is shown with *** denoting $p < 0.001$. Panels (**B,C**) display the relative microbial abundances among sample types at the class (panel (**B**)) and genus (panel (**C**)) levels. Classes are ordered based on their phylum assignment; genera are ordered by the average abundance among all samples. Note the compression of the Y axis between 0% and 25% relative abundance values in panel (**C**). Star denotes the statistically significant difference (at $\alpha = 0.01$ level) in the taxon abundance among sample groups as calculated by the analysis of variance algorithm. Abbreviations are as follows: G (Guara), MC (Marcona), MD (Marinada), MT (Marta), V (Vairo), BL (fermentation blank).

We used a logistic regression algorithm to determine whether microbial communities can be classified into distinct classes based on either almond cultivar or harvest time. The classification procedure generated models predicting the classification of each sample based on its microbial composition, and the outcome of this analysis is shown in Figure 4A. Consistent with ordination analysis described above, the Guara cultivar together with Marinada was predicted the best among cultivars. Among harvest time classes, the late-harvested samples (T3) were the only class with great classification (93.3% correct prediction' area under the curve score of 0.950). Due to the applied Lasso regularization, only a handful of microbial genera designated each class, as displayed in the lists in Figure 4A. The abundance of *Dorea*, which defined the microbiota of the fermented Guara cultivar, is highlighted on the weighted UniFrac-based PCoA plot shown in Figure 4B. Similarly, *Clostridium SS* abundance, defining the T3 sample type, is overlaid onto the same PCoA plot in Figure 4C. The abundance of the genus *Erwinia* (class Gammaproteobacteria) was found to differ among all three harvest types—it had an average relative abundance of 0.61% in T1 samples; 0.48% abundance in T2 samples, and 0.14% in T3 samples.

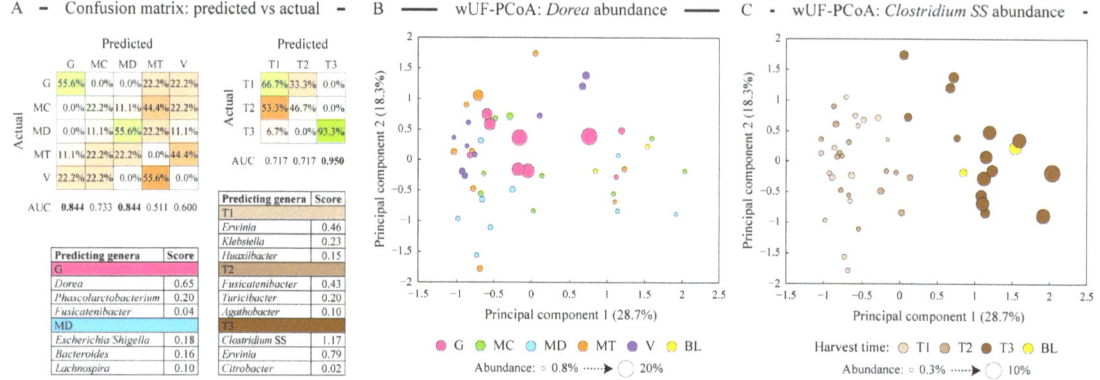

Figure 4. Modeling the differences in the microbiota profiles among sample types. Panel (**A**) displays the results of logistic regression (LR) based discriminant analysis of sample types. Separate models were generated for the microbiota distribution among the almond variety sample types and among the almond maturity sample types as shown. Confusion matrices reveal the concordance of the predicted vs actual class labels of all profiled samples, displayed as proportions of actual class labels assigned by the LR classifier to different classes. Green background highlights the correct prediction and orange background—misclassified cases. Area under the curve (AUC) values (range 0–1) illustrate the performance of each LR classifier. The top three genera predicting sample classes with high AUC values are shown in the tables; higher score represents higher contribution of the relative abundance of that genus to the classification model. Panels (**B**,**C**) depict the distribution of all samples in the weighted UniFrac distance-based principal coordinates analysis (PCoA). The percentage of total dataset variance explained by each axis is shown in parentheses. The size of each circle is proportionate to either *Dorea* (panel (**B**)) or *Clostridium sensus stricto* (SS, panel (**C**)) relative abundance in the corresponding sample as shown in the panel legend. Abbreviations are as follows: G (Guara), MC (Marcona), MD (Marinada), MT (Marta), V (Vairo), BL (fermentation blank).

3.4. Phenolic Compounds, SCFAs and Lactic and Succinic Acids Measured after In Vitro Digestion and Fermentation

3.4.1. Phenolic Compounds

In the analysis of the influence of the cultivar on the concentrations of individual phenolic compounds (Table 2), we observed significant differences among the cultivars for the concentration of (-)-epicatechin, (-)-epigallocatechin, ferulic acid, *p*-coumaric acid, rutin, 3-(3,4-dihydroxyphenyl)propionic acid, 3-(3-hydroxyphenyl)propionic acid, 5-(3′,4′-dihydroxyphenyl)-γ-valerolactone, phloroglucinol and urolithin B. The total concentration of individual phenolic compounds did not vary significantly among cultivars, similar to our results for the total phenolic content measured by the Folin–Ciocalteu assay. However, it is worth noting that the cultivar with the highest phenolic compound concentration was Guara. Other studies have reported variations in the individual profile of phenolic compounds among cultivars, with the extent of variation depending on the almond's origin [47]. In Spanish cultivars, Marcona and Guara almond skin extracts reported differences regarding individual phenolic concentrations. Guara had higher concentrations of (+)-catechin, (-)-epicatechin and naringin compared to Marcona, while Marcona had higher concentrations of naringenin than Guara [50].

Table 2. Phenolic compound concentrations (mg/kg) of five almond cultivars submitted to in vitro digestion-fermentation.

	Guara	Vairo	Marta	Marinada	Marcona	Significance
(+)-Catechin	104 ± 48.0	97.7 ± 55.6	73.4 ± 53.1	67.7 ± 63.3	83.0 ± 48.2	NS
(-)-Epicatechin	140 ± 65.8 [ab]	90.1 ± 35.7 [acd]	66.4 ± 18.9 [bce]	38.2 ± 15.3 [f]	49.0 ± 17.1 [def]	*
(-)-Epigallocatechin	49.9 ± 9.33 [abcd]	61.6 ± 9.71 [aef]	43.2 ± 14.5 [begh]	33.4 ± 15.8 [cgi]	52.0 ± 15.2 [dfhi]	*
Ferulic acid	0.002 ± 0 [abcd]	0.005 ± 0.003 [aefg]	0.002 ± 0 [behi]	0.002 ± 0.001 [cfhj]	0.002 ± 0 [dgij]	*
Kaempferol	0.006 ± 0.002	0.005 ± 0.001	0.006 ± 0.002	0.004 ± 0.001	0.008 ± 0.003	NS
Naringenin	0.003 ± 0.001	0.004 ± 0	0.003 ± 0.001	0.003 ± 0.001	0.003 ± 0.001	NS
Naringin	0.003 ± 0.002	0.003 ± 0.001	0.002 ± 0.001	0.002 ± 0	0.003 ± 0.001	NS
p-Coumaric acid	0.002 ± 0.001 [abcd]	0.004 ± 0.003 [aefg]	0.002 ± 0.002 [behi]	0.001 ± 0 [cfhj]	0.006 ± 0.006 [dgij]	*
Quercetin	54.4 ± 16.2	61.8 ± 8.64	51.9 ± 10.5	51.8 ± 15.0	53.6 ± 13.8	NS
Rutin	84.0 ± 11.6 [abcd]	80.7 ± 7.86 [aefg]	75.6 ± 5.03 [behi]	76.7 ± 2.28 [cfh]	81.9 ± 3.63 [dgi]	*
3,4-dihydroxyphenylacetic acid	38.4 ± 14.3	35.4 ± 10.3	31.5 ± 13.4	23.7 ± 13.5	28.7 ± 5.84	NS
3-(3,4-dihydroxyphenyl)propionic acid	269 ± 106 [ab]	146 ± 73.4 [acde]	84.3 ± 66.7 [cfg]	87.8 ± 42.0 [dfh]	161 ± 95.6 [begh]	*
3-(3-hydroxyphenyl)propionic acid	21.0 ± 7.28 [abcd]	4.97 ± 11.6 [aefg]	9.20 ± 8.17 [behi]	11.7 ± 12.8 [cfhj]	19.1 ± 11.7 [dgij]	*
5-(3′,4′-dihydroxyphenyl)-γ-valerolactone	54.6 ± 57.0 [ab]	183 ± 157 [acde]	274 ± 115 [cfg]	240 ± 113 [dfh]	213 ± 134 [begh]	*
Phenol	9.45 ± 2.79	8.41 ± 1.72	9.04 ± 4.11	10.6 ± 4.38	9.79 ± 4.42	NS
Phloroglucinol	207 ± 80.2 [abcd]	259 ± 56.2 [ae]	199 ± 34.4 [befg]	168 ± 46.7 [cfh]	186 ± 35.5 [dgh]	*
Urolithin A	0.022 ± 0.011	0.026 ± 0.009	0.021 ± 0.008	0.017 ± 0.006	0.025 ± 0.016	NS
Urolithin B	0.002 ± 0 [abc]	0.002 ± 0.001 [adef]	0.001 ± 0 [d]	0.002 ± 0 [be]	0.004 ± 0.002 [cf]	*
Total	1055 ± 39.2	1027 ± 40.9	921 ± 113	814 ± 24.0	944 ± 24.3	NS

Statistical differences among samples were tested by the Kruskal–Wallis test at the 5% level of significance (NS: not significant. *: significant). For the samples that showed statistical differences, a common letter indicates that samples are not significantly different based on the pair-wise Games–Howell post-hoc test at the 5% level of significance.

From the 18 phenolic compounds quantified in our study, 10 were phenolic compounds naturally found in almonds. Their order of abundance was, from highest to lowest: (+)-catechin, rutin, (-)-epicatechin, quercetin, (-)-epigallocatechin, kaempferol, naringenin, p-coumaric acid, ferulic acid, and naringin. The order of abundance of these phenolic compounds in almonds is similar to that reported in a review that brings together data from 61 studies [2]. Our results are generally within the ranges reported in that review, although we found some of the compounds at higher concentrations. This could be attributed to the increase in the bioaccessibility of phenolic compounds during in vitro fermentation, since a large proportion of these phenolic compounds are associated with dietary fiber, which remains intact after digestion and becomes a substrate for the gut microbiota that releases the bound phenolic compounds.

Harvest time did not have a significant impact on the concentration of 13 out of the 18 phenolic compounds analyzed (Table 3). Nevertheless, the concentrations of (-)-epigallocatechin, quercetin, and phenol exhibited a significant decrease over time, while the concentration of naringenin significantly increased. The total concentration of individual phenolic compounds showed no significant variation among harvest times but displayed a tendency to decrease. From T1 to T3, 13 phenolic compounds decreased their content, while 3 of them increased and 2 remained stable. The decrease in phenolic compound concentrations over time does not align with the significant increase observed in the total phenolic content determined by the Folin–Ciocalteu assay. This discrepancy may arise from factors such as not measuring the concentrations of all almond phenolic compounds or the reactivity of the Folin–Ciocalteu reagent with non-phenolic reducing compounds [55].

To the best of our knowledge, there are no studies evaluating the evolution of the individual phenolic profile of almonds during maturation. For other fruits, such as figs, the concentration of individual phenolic compounds decreased from T1 to T3 for 6 out of the 10 compounds determined [56]. In another study [57], the total concentration of individual phenolics decreased in various apple cultivars during ripening, although the trends were not consistent across all cultivars. Furthermore, it has been demonstrated that some in vitro functional properties of vegetables vary with harvest time, with hops showing higher phenolic content and functional properties in the early harvests due to modifications in the profile of phenolic compounds [58].

Table 3. Phenolic compound concentrations (mg/kg) of almonds collected at three harvest times and submitted to an in vitro digestion-fermentation.

	T1	T2	T3	Significance
(+)-Catechin	97.3 ± 55.6	71.5 ± 53.9	85.6 ± 50.5	NS
(-)-Epicatechin	97.3 ± 56.9	78.7 ± 59.2	53.6 ± 18.7	NS
(-)-Epigallocatechin	54.1 ± 18.5 [ab]	47.6 ± 13.3 [ac]	40.2 ± 12.11 [bc]	*
Ferulic acid	0.003 ± 0.002	0.002 ± 0.001	0.003 ± 0.002	NS
Kaempferol	0.006 ± 0.003	0.005 ± 0.002	0.007 ± 0.002	NS
Naringenin	0.003 ± 0.001 [a]	0.003 ± 0.001 [ab]	0.004 ± 0.001 [b]	*
Naringin	0.002 ± 0.001	0.003 ± 0.001	0.003 ± 0.001	NS
p-Coumaric acid	0.004 ± 0.005	0.002 ± 0.002	0.003 ± 0.002	NS
Quercetin	67.9 ± 5.92	48.0 ± 9.78 [a]	48.2 ± 10.3 [a]	*
Rutin	81.8 ± 10.0	81.1 ± 5.24	76.6 ± 4.77	NS
3,4-dihydroxyphenylacetic acid	36.6 ± 12.4	32.6 ± 7.41	26.5 ± 14.7	NS
3-(3,4-dihydroxyphenyl)propionic acid	162 ± 99.3	139 ± 70.3	142 ± 133	NS
3-(3-hydroxyphenyl)propionic acid	16.2 ± 12.6	10.7 ± 8.78	13.6 ± 13.2	NS
5-(3′,4′-dihydroxyphenyl)-γ-valerolactone	214 ± 117	180 ± 138	185 ± 163	NS
Phenol	12.5 ± 2.86	8.77 ± 2.86 [a]	7.08 ± 2.57 [a]	*
Phloroglucinol	220 ± 69.2	204.7 ± 47.1	186 ± 59.3	NS
Urolithin A	0.032 ± 0.011	0.018 ± 0.006 [a]	0.016 ± 0.005 [a]	*
Urolithin B	0.002 ± 0.002	0.002 ± 0.001	0.002 ± 0.001	NS
Total	1073 ± 49.4	921 ± 72.1	863 ± 159	NS

Statistical differences among samples were tested by the Kruskal–Wallis test at the 5% level of significance (NS not significant. *: significant). For the samples that showed statistical differences, a common letter indicates that samples are not significantly different based on the pair-wise Games–Howell post-hoc test at the 5% level of significance.

We also identified eight microbial phenolic metabolites, listed in order from highest to lowest concentrations: 5-(3′,4′)-dihydroxyphenyl-γ-valerolactone, phloroglucinol, 3-(3,4-dihydroxyphenyl)propionic acid, 3,4-dihydroxyphenylacetic acid, 3-(3-hydroxyphenyl)propionic acid, phenol, urolithin A, and urolithin B. This follows the trend of abundance reported by a clinical study [19] that measured phenolic microbial metabolites in urine after the ingestion of an almond skin extract. The most abundant metabolite, 5-(3′,4′-dihydroxyphenyl)-γ-valerolactone, could be a marker of the transformation of almond flavan-3-ols and proanthocyanidins. Moreover, 3-(3,4-Dihydroxyphenyl)propionic acid, 3-(3-hydroxyphenyl)propionic acid, and 3,4-dihydroxyphenylacetic acid are also metabolites of flavan-3-ols, as well as of flavonols, flavones, flavanones, and hydroxycinnamic acids. Phloroglucinol is a metabolite arising from the metabolization by the gut microbiota of a diverse group of phenolic compounds, among which we find flavonols, flavones, flavanones, flavan-3-ols, isoflavones, and anthocyanins. Phenol is one of the simplest metabolites that can be derived from the degradation of the parent phenolic compounds by the intestinal microbiota. Urolithins A and B are produced by the transformation of ellagic acid, a compound present in ellagitannins [59,60].

3.4.2. SCFAs, Lactic and Succinic Acids

We observed significant differences in propionic, butyric, and succinic acid content among maturation stages, with their levels increasing from T1 to T3 (Table 4). The amount of acetic and lactic acids also increased from T1 to T3, but the differences were not statistically significant. Furthermore, the total SCFA content did not vary significantly among harvest times, although there was a slight increase as nuts matured. This suggests that T3 is the optimal time to harvest almonds, as it produces the highest SCFA concentrations upon gut microbiota fermentation. This could be attributed to T3 being the time when almonds have accumulated the highest fiber content (probably due to moisture loss), serving as a substrate for SCFA production. It has been demonstrated that almonds are a rich source of total dietary fiber and soluble dietary fiber [23], and that levels of almond neutral detergent fiber increase with fruit development [43]. These results could explain the increase in total SCFAs over time seen in our study.

Table 4. SCFAs, lactic and succinic acids concentrations (mM) of five almond cultivars collected at three harvest times and submitted to in vitro digestion-fermentation.

Harvest Time	Acetic Acid	Propionic Acid	Butyric Acid	Lactic Acid	Succinic Acid	Total SCFAs
T1	12.0 ± 1.62	1.80 ± 0.38 [ab]	0.44 ± 0.25 [a]	7.13 ± 0.65	12.2 ± 0.22 [ab]	14.3 ± 2.22
T2	12.9 ± 2.61	3.51 ± 2.17 [ac]	0.50 ± 0.31 [a]	6.89 ± 0.58	8.62 ± 4.67 [ac]	15.6 ± 4.65
T3	14.5 ± 6.12	3.82 ± 3.05 [bc]	2.35 ± 1.92	7.36 ± 1.00	12.2 ± 2.64 [bc]	16.9 ± 10.9
Significance	NS	*	*	NS	*	NS
Cultivar	Acetic acid	Propionic acid	Butyric acid	Lactic acid	Succinic acid	Total SCFAs
Guara	19.5 ± 5.16 [abc]	6.16 ± 3.2 [abcd]	2.48 ± 2.29 [abcd]	6.07 ± 0.19 [a]	7.50 ± 3.76 [abcd]	28.2 ± 10.4 [abcd]
Vairo	12.6 ± 0.92 [ade]	2.21 ± 0.57 [aefg]	0.97 ± 1.00 [aefg]	7.58 ± 0.10 [bcd]	12.2 ± 0.62 [aefg]	15.7 ± 2.42 [aefg]
Marta	12.1 ± 0.38 [bd]	2.25 ± 0.44 [behi]	1.22 ± 1.35 [behi]	7.49 ± 0.33 [bef]	12.8 ± 0.56 [behi]	15.6 ± 1.46 [beh]
Marinada	11.1 ± 0.08 [ce]	2.97 ± 1.94 [cfhj]	0.33 ± 0.05 [cfhj]	7.23 ± 0.53 [ceg]	9.54 ± 5.10 [cfhj]	14.4 ± 2.03 [cfhi]
Marcona	10.6 ± 0.09	1.63 ± 0.18 [dgij]	0.48 ± 0.24 [dgij]	7.26 ± 1.09 [adfg]	13.0 ± 0.74 [dgij]	12.7 ± 0.36 [dgi]
Significance	*	*	*	*	*	*

Statistical differences among samples were tested by the Kruskal–Wallis test at the 5% level of significance (NS: not significant. *: significant). For the samples that showed statistical differences, a common letter indicates that samples are not significantly different based on the pair-wise Games–Howell post-hoc test at the 5% level of significance.

We also detected significant differences when comparing SCFA production among cultivars (Table 4). Vairo and Marcona cultivars had the highest content of lactic and succinic acids, respectively. In contrast, Guara exhibited the highest total SCFAs and acetic, propionic, and butyric acid levels. This could be attributed to Guara having the highest fiber content, matching the results of supported microbiota community analysis described above. In a study comparing three apple cultivars [61], the cultivar with the highest total and soluble dietary fiber content reported the highest total SCFAs compared to the rest. These findings suggest that Guara might be the best cultivar choice due to its higher proportion of beneficial SCFAs and lower lactic and succinic acid levels, which are intermediates in metabolic pathways and whose effects on health are not yet fully understood.

To the best of our knowledge, no studies have evaluated the production of SCFAs and lactic and succinic acids following the in vitro digestion and fermentation of almonds for the purpose of comparing different cultivars and harvest times. The studies available in the literature primarily focus on whether almond consumption in vitro or in vivo could be beneficial in terms of SCFA production. Previous studies on the in vitro digestion and fermentation of almonds have assessed SCFA production using different approaches. Some studies have compared the SCFA production of various nuts and have consistently ranked almonds as a significant source, placing them second out of five [62], first out of five [63], first out of six [24] and third out of five [23] among the nuts studied. Another study observed a substantial production of butyrate following an in vitro digestion and fermentation of finely ground almonds, which could be attributed to the disruption of cell wall fibers and increased bioaccessibility [22]. On the other hand, in vivo studies have not found substantial differences after the consumption of almonds in terms of SCFA production [20,21]. However, there is a need for more clinical studies to further explore the benefits of almond consumption [64].

4. Conclusions

This work confirms the influence of the cultivar and harvest time on some morphological and nutritional characteristics of almonds. Cultivar type had a significant impact on length, width and thickness, but not on moisture, protein and ash content. Width, thickness and moisture content significantly decreased over time, while the protein and ash increased. The total antioxidant capacity released by almonds after an in vitro digestion and fermentation process had different trends: total phenolic content measured by the Folin–Ciocalteu assay significantly increased over time, while $TEAC_{ABTS}$, $TEAC_{FRAP}$ and $TEAC_{DPPH}$ remained unaffected. Nevertheless, we reported significant differences for $TEAC_{FRAP}$ and $TEAC_{DPPH}$ among cultivars. The fermentation step contributed more to the total antioxidant capacity of almonds than the digestion step. The concentration of individual phenolic compounds was influenced more significantly by cultivar type than by harvest time. The same behavior was observed with total SCFAs. Nevertheless, neither cultivar nor harvest time had a significant impact on the total concentration of individual phenolic compounds.

The Guara cultivar showed the best nutritional profile, primarily due to its high levels of ash, protein, total SCFAs, $TEAC_{FRAP}$, $TEAC_{DPPH}$, total phenolic content measured by the Folin–Ciocalteu assay, and total concentration of individual phenolic compounds. In addition, the best harvest time was T3, characterized by a higher total phenolic content measured by the Folin–Ciocalteu assay along with increased ash, protein, total SCFAs, and reduced moisture. Consistent with these findings, the Guara cultivar and T3 harvest time supported human-derived gut microbial communities distinct from the rest of the almond samples.

Supplementary Materials: The following supporting information can be downloaded at: https://www.mdpi.com/article/10.3390/antiox13010084/s1, Table S1: Morphological measurements of almonds collected at three harvest times and subjected to in vitro digestion-fermentation; Table S2: Morphological measurements of five almond cultivars subjected to in vitro digestion-fermentation; Table S3: Ash, moisture and protein content of five almond cultivars subjected to in vitro digestion-fermentation; Table S4: Ash, moisture and protein content of almonds collected at three harvest times and subjected to in vitro digestion-fermentation; Table S5: Total antioxidant capacity of almonds collected at three harvest times and subjected to in vitro digestion-fermentation; Table S6: Total antioxidant capacity of five almond cultivars subjected to in vitro digestion-fermentation.

Author Contributions: Conceptualization, S.P.-B. and J.Á.R.-H.; methodology, A.D.-O., D.H.-N. and Á.T.-M.; formal analysis, A.D.-O., B.N.-P., Á.T.-M., S.P.-B. and O.P.; investigation, D.H.-N., S.P.d.l.C., O.P. and J.Á.R.-H.; resources, O.P. and J.Á.R.-H.; data curation, S.P.-B. and B.N.-P.; writing—original draft preparation, A.D.-O.; writing—review and editing, S.P.-B., S.P.d.l.C., O.P. and J.Á.R.-H.; supervision, J.Á.R.-H.; funding acquisition, J.Á.R.-H. and O.P. All authors have read and agreed to the published version of the manuscript.

Funding: This work was supported by the Plan propio de Investigación y Transferencia of the University of Granada under the program "Intensificación de la Investigación, modalidad B" and by the National Science Foundation award DBI-1335772.

Institutional Review Board Statement: The study was conducted according to the guidelines of the Declaration of Helsinki. It was approved by the Ethics Committee of the University of Granada (protocol code 1080/CEIH/2020).

Informed Consent Statement: Informed consent was obtained from all fecal donors involved in the study.

Data Availability Statement: Data are available to other researchers upon request to the corresponding authors (J.Á.R.-H.).

Acknowledgments: The authors thank the support of the Unit of Excellence 'UNETE' from the University of Granada (reference UCE-PP2017-05). This work is part of Adriana Delgado-Osorio's doctoral thesis, carried out as part of the "Nutrition and Food Sciences Programme" at the University of Granada. We also thank Gary Kash for his assistance with DNA amplification experiments.

Conflicts of Interest: The authors declare no conflicts of interest.

References

1. Barreca, D.; Nabavi, S.M.; Sureda, A.; Rasekhian, M.; Raciti, R.; Silva, A.S.; Annunziata, G.; Arnone, A.; Tenore, G.C.; Süntar, İ.; et al. Almonds (*Prunus dulcis* Mill. D. A. Webb): A Source of Nutrients and Health-Promoting Compounds. *Nutrients* **2020**, *12*, 672. [CrossRef]
2. Bolling, B.W. Almond Polyphenols: Methods of Analysis, Contribution to Food Quality, and Health Promotion. *Compr. Rev. Food Sci. Food Saf.* **2017**, *16*, 346–368. [CrossRef]
3. Musarra-Pizzo, M.; Ginestra, G.; Smeriglio, A.; Pennisi, R.; Sciortino, M.T.; Mandalari, G. The Antimicrobial and Antiviral Activity of Polyphenols from Almond (*Prunus dulcis* L.) Skin. *Nutrients* **2019**, *11*, 2355. [CrossRef]
4. Özcan, M.M. A Review on Some Properties of Almond: Impact of Processing, Fatty Acids, Polyphenols, Nutrients, Bioactive Properties, and Health Aspects. *J. Food Sci. Technol.* **2023**, *60*, 1493–1504. [CrossRef]
5. Matthäus, B.; Özcan, M.M.; Juhaimi, F.A.; Adiamo, O.Q.; Alsawmahi, O.N.; Ghafoor, K.; Babiker, E.E. Effect of the Harvest Time on Oil Yield, Fatty Acid, Tocopherol and Sterol Contents of Developing Almond and Walnut Kernels. *J. Oleo Sci.* **2018**, *67*, 39–45. [CrossRef]
6. Özcan, M.M.; Lemiasheuski, V. The Effect of Harvest Times on Mineral Contents of Almond and Walnut Kernels. *Erwerbs-Obstbau* **2020**, *62*, 455–458. [CrossRef]
7. Özcan, M.M.; Uslu, N. Effect of Variety on Bioactive Properties, Phytochemicals and Nutrients of Almond Kernels. *Erwerbs-Obstbau* **2023**, *65*, 981–988. [CrossRef]
8. Piscopo, A.; Romeo, F.V.; Petrovicova, B.; Poiana, M. Effect of the Harvest Time on Kernel Quality of Several Almond Varieties (*Prunus dulcis* (Mill.) D.A. Webb). *Sci. Hortic.* **2010**, *125*, 41–46. [CrossRef]
9. Beltrán Sanahuja, A.; Maestre Pérez, S.E.; Grané Teruel, N.; Valdés García, A.; Prats Moya, M.S. Variability of Chemical Profile in Almonds (*Prunus dulcis*) of Different Cultivars and Origins. *Foods* **2021**, *10*, 153. [CrossRef] [PubMed]
10. Summo, C.; Palasciano, M.; De Angelis, D.; Paradiso, V.M.; Caponio, F.; Pasqualone, A. Evaluation of the Chemical and Nutritional Characteristics of Almonds (*Prunus dulcis* (Mill). D.A. Webb) as Influenced by Harvest Time and Cultivar. *J. Sci. Food Agric.* **2018**, *98*, 5647–5655. [CrossRef] [PubMed]
11. Santos, D.I.; Saraiva, J.M.A.; Vicente, A.A.; Moldão-Martins, M. 2–Methods for Determining Bioavailability and Bioaccessibility of Bioactive Compounds and Nutrients. In *Innovative Thermal and Non-Thermal Processing, Bioaccessibility and Bioavailability of Nutrients and Bioactive Compounds*; Barba, F.J., Saraiva, J.M.A., Cravotto, G., Lorenzo, J.M., Eds.; Woodhead Publishing Series in Food Science, Technology and Nutrition; Woodhead Publishing: Sawston, UK, 2019; pp. 23–54. ISBN 978-0-12-814174-8.
12. Thomson, C.; Garcia, A.L.; Edwards, C.A. Interactions between Dietary Fibre and the Gut Microbiota. *Proc. Nutr. Soc.* **2021**, *80*, 398–408. [CrossRef]
13. Campos-Perez, W.; Martinez-Lopez, E. Effects of Short Chain Fatty Acids on Metabolic and Inflammatory Processes in Human Health. *Biochim. Biophys. Acta (BBA)—Mol. Cell Biol. Lipids* **2021**, *1866*, 158900. [CrossRef]
14. Cortés-Martín, A.; Selma, M.V.; Tomás-Barberán, F.A.; González-Sarrías, A.; Espín, J.C. Where to Look into the Puzzle of Polyphenols and Health? The Postbiotics and Gut Microbiota Associated with Human Metabotypes. *Mol. Nutr. Food Res.* **2020**, *64*, 1900952. [CrossRef]
15. Fernández-Veledo, S.; Vendrell, J. Gut Microbiota-Derived Succinate: Friend or Foe in Human Metabolic Diseases? *Rev. Endocr. Metab. Disord.* **2019**, *20*, 439–447. [CrossRef]
16. Navajas-Porras, B.; Pérez-Burillo, S.; Valverde-Moya, Á.J.; Hinojosa-Nogueira, D.; Pastoriza, S.; Rufián-Henares, J.Á. Effect of Cooking Methods on the Antioxidant Capacity of Plant Foods Submitted to In Vitro Digestion–Fermentation. *Antioxidants* **2020**, *9*, 1312. [CrossRef]
17. Garrido, I.; Urpi-Sarda, M.; Monagas, M.; Gómez-Cordovés, C.; Martín-álvarez, P.J.; Llorach, R.; Bartolomé, B.; Andrés-Lacueva, C. Targeted Analysis of Conjugated and Microbial-Derived Phenolic Metabolites in Human Urine After Consumption of an Almond Skin Phenolic Extract. *J. Nutr.* **2010**, *140*, 1799–1807. [CrossRef]
18. Llorach, R.; Garrido, I.; Monagas, M.; Urpi-Sarda, M.; Tulipani, S.; Bartolome, B.; Andres-Lacueva, C. Metabolomics Study of Human Urinary Metabolome Modifications After Intake of Almond (*Prunus dulcis* (Mill.) D.A. Webb) Skin Polyphenols. *J. Proteome Res.* **2010**, *9*, 5859–5867. [CrossRef]
19. Urpi-Sarda, M.; Garrido, I.; Monagas, M.; Gómez-Cordovés, C.; Medina-Remón, A.; Andres-Lacueva, C.; Bartolomé, B. Profile of Plasma and Urine Metabolites after the Intake of Almond [*Prunus dulcis* (Mill.) D.A. Webb] Polyphenols in Humans. *J. Agric. Food Chem.* **2009**, *57*, 10134–10142. [CrossRef]
20. Choo, J.M.; Tran, C.D.; Luscombe-Marsh, N.D.; Stonehouse, W.; Bowen, J.; Johnson, N.; Thompson, C.H.; Watson, E.-J.; Brinkworth, G.D.; Rogers, G.B. Almond Consumption Affects Fecal Microbiota Composition, Stool pH, and Stool Moisture in Overweight and Obese Adults with Elevated Fasting Blood Glucose: A Randomized Controlled Trial. *Nutr. Res.* **2021**, *85*, 47–59. [CrossRef] [PubMed]
21. Creedon, A.C.; Dimidi, E.; Hung, E.S.; Rossi, M.; Probert, C.; Grassby, T.; Miguens-Blanco, J.; Marchesi, J.R.; Scott, S.M.; Berry, S.E.; et al. The Impact of Almonds and Almond Processing on Gastrointestinal Physiology, Luminal Microbiology, and Gastrointestinal Symptoms: A Randomized Controlled Trial and Mastication Study. *Am. J. Clin. Nutr.* **2022**, *116*, 1790–1804. [CrossRef] [PubMed]
22. Mandalari, G.; Nueno-Palop, C.; Bisignano, G.; Wickham, M.S.J.; Narbad, A. Potential Prebiotic Properties of Almond (*Amygdalus communis* L.) Seeds. *Appl. Environ. Microbiol.* **2008**, *74*, 4264–4270. [CrossRef]

23. Şahin, M.; Arioglu-Tuncil, S.; Ünver, A.; Deemer, D.; Lindemann, S.R.; Tunçil, Y.E. Dietary Fibers of Tree Nuts Differ in Composition and Distinctly Impact the Fecal Microbiota and Metabolic Outcomes In Vitro. *J. Agric. Food Chem.* **2023**, *71*, 9762–9771. [CrossRef]

24. Li, M.; Lu, P.; Wu, H.; de Souza, T.S.P.; Suleria, H.A.R. In Vitro Digestion and Colonic Fermentation of Phenolic Compounds and Their Bioaccessibility from Raw and Roasted Nut Kernels. *Food Funct.* **2023**, *14*, 2727–2739. [CrossRef] [PubMed]

25. Pérez-Burillo, S.; Rufián-Henares, J.A.; Pastoriza, S. Towards an Improved Global Antioxidant Response Method (GAR+) Physiological-Resembling in Vitro Digestion-Fermentation Method. *Food Chem.* **2018**, *239*, 1253–1262. [CrossRef] [PubMed]

26. Pérez-Burillo, S.; Molino, S.; Navajas-Porras, B.; Valverde-Moya, Á.J.; Hinojosa-Nogueira, D.; López-Maldonado, A.; Pastoriza, S.; Rufián-Henares, J.Á. An in Vitro Batch Fermentation Protocol for Studying the Contribution of Food to Gut Microbiota Composition and Functionality. *Nat. Protoc.* **2021**, *16*, 3186–3209. [CrossRef]

27. Re, R.; Pellegrini, N.; Proteggente, A.; Pannala, A.; Yang, M.; Rice-Evans, C. Antioxidant Activity Applying an Improved ABTS Radical Cation Decolorization Assay. *Free Radic. Biol. Med.* **1999**, *26*, 1231–1237. [CrossRef]

28. Benzie, I.F.F.; Strain, J.J. The Ferric Reducing Ability of Plasma (FRAP) as a Measure of "Antioxidant Power": The FRAP Assay. *Anal. Biochem.* **1996**, *239*, 70–76. [CrossRef]

29. Yen, G.-C.; Chen, H.-Y. Antioxidant Activity of Various Tea Extracts in Relation to Their Antimutagenicity. *J. Agric. Food Chem.* **1995**, *43*, 27–32. [CrossRef]

30. Singleton, V.L.; Rossi, J.A. Colorimetry of Total Phenolics with Phosphomolybdic-Phosphotungstic Acid Reagents. *Am. J. Enol. Vitic.* **1965**, *16*, 144–158. [CrossRef]

31. Muñoz, A.E.; Álvarez, M.B.; Oliveras-López, M.-J.; Martínez, R.G.; Henares, J.Á.R.; Herrera, M.O. Determination of Polyphenolic Compounds by Ultra-Performance Liquid Chromatography Coupled to Tandem Mass Spectrometry and Antioxidant Capacity of Spanish Subtropical Fruits. *Agric. Sci.* **2018**, *9*, 180–199. [CrossRef]

32. Panzella, L.; Pérez-Burillo, S.; Pastoriza, S.; Martín, M.Á.; Cerruti, P.; Goya, L.; Ramos, S.; Rufián-Henares, J.Á.; Napolitano, A.; d'Ischia, M. High Antioxidant Action and Prebiotic Activity of Hydrolyzed Spent Coffee Grounds (HSCG) in a Simulated Digestion–Fermentation Model: Toward the Development of a Novel Food Supplement. *J. Agric. Food Chem.* **2017**, *65*, 6452–6459. [CrossRef]

33. Paliy, O.; Foy, B.D. Mathematical Modeling of 16S Ribosomal DNA Amplification Reveals Optimal Conditions for the Interrogation of Complex Microbial Communities with Phylogenetic Microarrays. *Bioinformatics* **2011**, *27*, 2134–2140. [CrossRef] [PubMed]

34. Caporaso, J.G.; Kuczynski, J.; Stombaugh, J.; Bittinger, K.; Bushman, F.D.; Costello, E.K.; Fierer, N.; Peña, A.G.; Goodrich, J.K.; Gordon, J.I.; et al. QIIME Allows Analysis of High-Throughput Community Sequencing Data. *Nat. Methods* **2010**, *7*, 335–336. [CrossRef] [PubMed]

35. Rajakaruna, S.; Freedman, D.A.; Sehgal, A.R.; Bui, X.; Paliy, O. Diet Quality and Body Mass Indices Show Opposite Associations with Distal Gut Microbiota in a Low-Income Cohort. *SDRP-JFST* **2019**, *4*, 846–851. [CrossRef]

36. Paliy, O.; Shankar, V. Application of Multivariate Statistical Techniques in Microbial Ecology. *Mol. Ecol.* **2016**, *25*, 1032–1057. [CrossRef]

37. Craig, M.P.; Rajakaruna, S.; Paliy, O.; Sajjad, M.; Madhavan, S.; Reddy, N.; Zhang, J.; Bottomley, M.; Agrawal, S.; Kadakia, M.P. Differential MicroRNA Signatures in the Pathogenesis of Barrett's Esophagus. *Clin. Transl. Gastroenterol.* **2020**, *11*, e00125. [CrossRef] [PubMed]

38. Pérez-Sánchez, R.; Morales-Corts, M.R. Agromorphological Characterization and Nutritional Value of Traditional Almond Cultivars Grown in the Central-Western Iberian Peninsula. *Agronomy* **2021**, *11*, 1238. [CrossRef]

39. Cherif, A.; Sebei, K.; Boukhchina, S.; Kallel, H.; Belkacemi, K.; Arul, J. Kernel Fatty Acid and Triacylglycerol Composition for Three Almond Cultivars during Maturation. *J. Am. Oil Chem. Soc.* **2004**, *81*, 901–905. [CrossRef]

40. Egea, G.; González-Real, M.M.; Baille, A.; Nortes, P.A.; Sánchez-Bel, P.; Domingo, R. The Effects of Contrasted Deficit Irrigation Strategies on the Fruit Growth and Kernel Quality of Mature Almond Trees. *Agric. Water Manag.* **2009**, *96*, 1605–1614. [CrossRef]

41. Kazantzis, I.; Nanos, G.D.; Stavroulakis, G.G. Effect of Harvest Time and Storage Conditions on Almond Kernel Oil and Sugar Composition. *J. Sci. Food Agric.* **2003**, *83*, 354–359. [CrossRef]

42. Nanos, G.D.; Kazantzis, I.; Kefalas, P.; Petrakis, C.; Stavroulakis, G.G. Irrigation and Harvest Time Affect Almond Kernel Quality and Composition. *Sci. Hortic.* **2002**, *96*, 249–256. [CrossRef]

43. Soler, L.; Canellas, J.; Saura-Calixto, F. Changes in Carbohydrate and Protein Content and Composition of Developing Almond Seeds. *J. Agric. Food Chem.* **1989**, *37*, 1400–1404. [CrossRef]

44. Hawker, J.S.; Buttrose, M.S. Development of the Almond Nut (*Prunus dulcis* (Mill.) D. A. Webb). Anatomy and Chemical Composition of Fruit Parts from Anthesis to Maturity. *Ann. Bot.* **1980**, *46*, 313–321. [CrossRef]

45. Levent, O. A Detailed Comparative Study on Some Physicochemical Properties, Volatile Composition, Fatty Acid, and Mineral Profile of Different Almond (*Prunus dulcis* L.) Varieties. *Horticulturae* **2022**, *8*, 488. [CrossRef]

46. Barreira, J.C.M.; Ferreira, I.C.F.R.; Oliveira, M.B.P.P.; Pereira, J.A. Antioxidant Activity and Bioactive Compounds of Ten Portuguese Regional and Commercial Almond Cultivars. *Food Chem. Toxicol.* **2008**, *46*, 2230–2235. [CrossRef] [PubMed]

47. Bolling, B.W.; Dolnikowski, G.; Blumberg, J.B.; Chen, C.-Y.O. Polyphenol Content and Antioxidant Activity of California Almonds Depend on Cultivar and Harvest Year. *Food Chem.* **2010**, *122*, 819–825. [CrossRef] [PubMed]

48. Oliveira, I.; Meyer, A.S.; Afonso, S.; Aires, A.; Goufo, P.; Trindade, H.; Gonçalves, B. Phenolic and Fatty Acid Profiles, α-Tocopherol and Sucrose Contents, and Antioxidant Capacities of Understudied Portuguese Almond Cultivars. *J. Food Biochem.* **2019**, *43*, e12887. [CrossRef] [PubMed]

49. Maestri, D.; Martínez, M.; Bodoira, R.; Rossi, Y.; Oviedo, A.; Pierantozzi, P.; Torres, M. Variability in Almond Oil Chemical Traits from Traditional Cultivars and Native Genetic Resources from Argentina. *Food Chem.* **2015**, *170*, 55–61. [CrossRef]

50. Valdés, A.; Vidal, L.; Beltrán, A.; Canals, A.; Garrigós, M.C. Microwave-Assisted Extraction of Phenolic Compounds from Almond Skin Byproducts (*Prunus amygdalus*): A Multivariate Analysis Approach. *J. Agric. Food Chem.* **2015**, *63*, 5395–5402. [CrossRef]

51. Duarte, S.; Puchades, A.; Jiménez-Hernández, N.; Betoret, E.; Gosalbes, M.J.; Betoret, N. Almond (*Prunus dulcis*) Bagasse as a Source of Bioactive Compounds with Antioxidant Properties: An In Vitro Assessment. *Antioxidants* **2023**, *12*, 1229. [CrossRef]

52. Wojtunik-Kulesza, K.; Oniszczuk, A.; Oniszczuk, T.; Combrzyński, M.; Nowakowska, D.; Matwijczuk, A. Influence of In Vitro Digestion on Composition, Bioaccessibility and Antioxidant Activity of Food Polyphenols—A Non-Systematic Review. *Nutrients* **2020**, *12*, 1401. [CrossRef] [PubMed]

53. D'hoe, K.; Conterno, L.; Fava, F.; Falony, G.; Vieira-Silva, S.; Vermeiren, J.; Tuohy, K.; Raes, J. Prebiotic Wheat Bran Fractions Induce Specific Microbiota Changes. *Front. Microbiol.* **2018**, *9*, 31. [CrossRef] [PubMed]

54. Agans, R.; Gordon, A.; Kramer, D.L.; Perez-Burillo, S.; Rufián-Henares, J.A.; Paliy, O. Dietary Fatty Acids Sustain the Growth of the Human Gut Microbiota. *Appl. Environ. Microbiol.* **2018**, *84*, e01525-18. [CrossRef] [PubMed]

55. Sánchez-Rangel, J.C.; Benavides, J.; Heredia, J.B.; Cisneros-Zevallos, L.; Jacobo-Velázquez, D.A. The Folin–Ciocalteu Assay Revisited: Improvement of Its Specificity for Total Phenolic Content Determination. *Anal. Methods* **2013**, *5*, 5990–5999. [CrossRef]

56. Ates, U. Harvest Time Influences Quality Attributes and Phenolic Composition of Fig Fruit: Insights from Physicochemical Analysis and Antioxidant Activity Assessment. *Erwerbs-Obstbau* **2023**, *65*, 1627–1632. [CrossRef]

57. Wojdyło, A.; Oszmiański, J. Antioxidant Activity Modulated by Polyphenol Contents in Apple and Leaves during Fruit Development and Ripening. *Antioxidants* **2020**, *9*, 567. [CrossRef]

58. Inui, T.; Okumura, K.; Matsui, H.; Hosoya, T.; Kumazawa, S. Effect of Harvest Time on Some in Vitro Functional Properties of Hop Polyphenols. *Food Chem.* **2017**, *225*, 69–76. [CrossRef]

59. Rowland, I.; Gibson, G.; Heinken, A.; Scott, K.; Swann, J.; Thiele, I.; Tuohy, K. Gut Microbiota Functions: Metabolism of Nutrients and Other Food Components. *Eur. J. Nutr.* **2018**, *57*, 1–24. [CrossRef]

60. Selma, M.V.; Espín, J.C.; Tomás-Barberán, F.A. Interaction between Phenolics and Gut Microbiota: Role in Human Health. *J. Agric. Food Chem.* **2009**, *57*, 6485–6501. [CrossRef]

61. Koutsos, A.; Lima, M.; Conterno, L.; Gasperotti, M.; Bianchi, M.; Fava, F.; Vrhovsek, U.; Lovegrove, J.A.; Tuohy, K.M. Effects of Commercial Apple Varieties on Human Gut Microbiota Composition and Metabolic Output Using an In Vitro Colonic Model. *Nutrients* **2017**, *9*, 533. [CrossRef]

62. Schlörmann, W.; Birringer, M.; Lochner, A.; Lorkowski, S.; Richter, I.; Rohrer, C.; Glei, M. In Vitro Fermentation of Nuts Results in the Formation of Butyrate and C9,T11 Conjugated Linoleic Acid as Chemopreventive Metabolites. *Eur. J. Nutr.* **2016**, *55*, 2063–2073. [CrossRef] [PubMed]

63. Lux, S.; Scharlau, D.; Schlörmann, W.; Birringer, M.; Glei, M. In Vitro Fermented Nuts Exhibit Chemopreventive Effects in HT29 Colon Cancer Cells. *Br. J. Nutr.* **2012**, *108*, 1177–1186. [CrossRef] [PubMed]

64. Zuelch, M.L.; Radtke, M.D.; Holt, R.R.; Basu, A.; Burton-Freeman, B.; Ferruzzi, M.G.; Li, Z.; Shay, N.F.; Shukitt-Hale, B.; Keen, C.L.; et al. Perspective: Challenges and Future Directions in Clinical Research with Nuts and Berries. *Adv. Nutr.* **2023**, *14*, 1005–1028. [CrossRef] [PubMed]

antioxidants

Article

Antioxidant Activities, Phenolic Compounds, and Sensory Acceptability of Kombucha-Fermented Beverages from Bamboo Leaf and Mulberry Leaf

Ruo-Gu Xiong [1] , Si-Xia Wu [1], Jin Cheng [1], Adila Saimaiti [1], Qing Liu [2] , Ao Shang [2], Dan-Dan Zhou [1], Si-Yu Huang [1], Ren-You Gan [3,*] and Hua-Bin Li [1,*]

[1] Guangdong Provincial Key Laboratory of Food, Nutrition and Health, Department of Nutrition, School of Public Health, Sun Yat-sen University, Guangzhou 510080, China; xiongrg@mail2.sysu.edu.cn (R.-G.X.); wusx6@mail2.sysu.edu.cn (S.-X.W.); chengj225@mail2.sysu.edu.cn (J.C.); saimaiti@mail2.sysu.edu.cn (A.S.); zhoudd6@mail2.sysu.edu.cn (D.-D.Z.); huangsy9@mail2.sysu.edu.cn (S.-Y.H.)
[2] School of Chinese Medicine, Li Ka Shing Faculty of Medicine, The University of Hong Kong, Hong Kong 999077, China; liuqing2@connect.hku.hk (Q.L.); shangao@connect.hku.hk (A.S.)
[3] Singapore Institute of Food and Biotechnology Innovation (SIFBI), Agency for Science, Technology and Research (A*STAR), 31 Biopolis Way, Singapore 138669, Singapore
* Correspondence: ganry@sifbi.a-star.edu.sg (R.-Y.G.); lihuabin@mail.sysu.edu.cn (H.-B.L.)

Abstract: Kombucha is traditional drink made from the fermentation of a black tea infusion, and is believed to offer a variety of health benefits. Recently, exploring kombucha made from alternative substrates has become a research hotspot. In this paper, two novel kombucha beverages were produced with bamboo leaf or mulberry leaf for the first time. Moreover, the effects of fermentation with leaf residues (infusion plus residues) or without leaf residues (only infusion) on the antioxidant properties of kombucha were compared. The ferric-reducing antioxidant power assay, Trolox equivalent antioxidant capacity assay, Folin–Ciocalteu method, and high-performance liquid chromatography were utilized to measure the antioxidant capacities, total phenolic contents, as well as some compound concentrations of the kombucha. The results showed that two types of kombucha had high antioxidant capacities. Moreover, kombucha fermented with bamboo leaf residues (infusion plus residues) significantly enhanced its antioxidant capabilities (maximum increase 83.6%), total phenolic content (maximum increase 99.2%), concentrations of some compounds (luteolin-6-C-glucoside and isovitexin), and sensory acceptability, compared to that without residues (only infusion). In addition, fermentation with leaf residues had no significant effect on mulberry leaf kombucha. Overall, the bamboo leaf was more suitable for making kombucha with residues, while the mulberry leaf kombucha was suitable for fermentation with or without residues.

Keywords: kombucha; mulberry leaf; bamboo leaf; fermentation; antioxidant activities

Citation: Xiong, R.-G.; Wu, S.-X.; Cheng, J.; Saimaiti, A.; Liu, Q.; Shang, A.; Zhou, D.-D.; Huang, S.-Y.; Gan, R.-Y.; Li, H.-B. Antioxidant Activities, Phenolic Compounds, and Sensory Acceptability of Kombucha-Fermented Beverages from Bamboo Leaf and Mulberry Leaf. *Antioxidants* **2023**, *12*, 1573. https://doi.org/10.3390/antiox12081573

Academic Editor: Myung-Ji Seo

Received: 5 July 2023
Revised: 4 August 2023
Accepted: 5 August 2023
Published: 6 August 2023

1. Introduction

Kombucha is a popular drink that is traditionally made by fermenting a mixture of black tea infusion and sugar with the symbiotic culture of bacteria and yeast (SCOBY) [1]. The fermentation process converts the sweetened tea infusion into an acidic taste and slightly fizzy drink. Moreover, several studies showed that kombucha contained a variety of bioactive components, and possessed abundant bioactivities, such as antioxidant activity, anti-inflammation, immune regulation, anti-diabetes, anti-obesity, hepatoprotection, blood pressure regulation, and anticancer effects [2–6]. Several tea and non-tea plants have been tested as alternative substrates for kombucha production, such as green tea, oolong tea, vegetables, fruits, and medicinal herbs, each with its unique chemical components that may influence the bioactivities of the final kombucha product [7,8].

The mulberry leaf contains many bioactive compounds, such as chlorogenic acid, rutin, and astragalin [9,10]. It is widely utilized in traditional medicines and functional foods due to its numerous health benefits, including antioxidant, anti-inflammatory, anti-diabetes, cardiovascular protective, and anticancer activities, and so on [11,12]. On the other hand, the bamboo leaf has been widely used in functional foods and traditional medicines, and provides multiple health benefits, including antioxidant properties, cardiovascular protection, anti-obesity, anti-diabetes, etc. [13,14]. Several studies suggested that the high content of phenolic compounds in the bamboo leaf were significant contributors to its health benefits [15,16]. There is no existing literature that has reported on the kombucha beverages made by the bamboo leaf or mulberry leaf. In addition, kombucha fermentation with tea residues, such as black tea, green tea, and sweet tea, could enhance the antioxidant properties as well as polyphenol contents of beverages according to our previous studies [8,17]. In this study, we investigated the antioxidant activities, polyphenol contents, as well as sensory acceptability of kombucha beverages made from mulberry leaf and bamboo leaf, and compared the impact of fermentation with or without leaf residues on these indicators. This study could be helpful for the valorization of the cheap mulberry and bamboo leaves.

2. Materials and Methods

2.1. Materials and Reagents

Mulberry leaf was bought from Anhui Tongjuntang Biotechnology Co., Ltd. (Bozhou, China). Fresh bamboo leaf was obtained in Guangzhou city (Guangzhou, China) and then was dried in air. The kombucha starter culture was purchased from Shandong Ruyun Edible Fungus Planting Co., Ltd. (Liaocheng, China).

Folin–Ciocalteu's phenol reagent, gallic acid, 2,20-azinobis (3-ethylbenothiazoline-6-sulfonic acid) diammonium salt (ABTS), 2,4,6-tri(2-pyridyl)-S-triazine (TPTZ), and 6-hydroxy-2,5,7,8-tetramethylchromane-2-carboxylic acid (Trolox) were acquired from Sigma-Aldrich (St. Louis, MO, USA). Acetic acid, ethanol, hydrochloric acid, iron(II) sulfate heptahydrate, iron(III) chloride hexahydrate, potassium persulfate, and sodium acetate were procured from Tianjin Chemical Factory (Tianjin, China). Sodium carbonate was bought from Shanghai Yuanye Biological Technology Co., Ltd. (Shanghai, China). Sucrose, methanol, and formic acid were purchased from Macklin Chemical Factory (Shanghai, China). The standard chemicals, involving astragalin, chlorogenic acid, caffeine, ellagic acid, epicatechin, gallic acid, kaempferol, quercetin, quercitrin, isovitexin, luteolin-6-C-glucoside, and rutin, were obtained from Derick Biotechnology Co., Ltd. (Chengdu, China).

2.2. Activation of Kombucha Starter Culture

This procedure was performed following the manufacturer's instructions. Briefly, 5 g teabag of black tea, 100 g sucrose and 1 L boiled water were added into a flask and mixed thoroughly. The teabag was removed after a 5 min soaking period and the mixture was cooled to room temperature (25 °C). The kombucha fungus, fermented broth, and cellulosic layer were then added to the mixture. The mixture was then placed into a clean and dark environment for 14 days at room temperature for subsequent inoculation.

2.3. Kombucha Preparation Based on Bamboo Leaf and Mulberry Leaf

The kombucha beverages were divided into 4 groups: (a) kombucha based on bamboo leaf with residues; (b) kombucha based on bamboo leaf without residues; (c) kombucha based on mulberry leaf with residues; (d) kombucha based on mulberry leaf without residues. In each glass flask, 200 mL of distilled water and 20 g of sucrose were added and heated in a boiling water bath. Then, 2 g of mulberry leaf or bamboo leaf was put into the flasks and kept for 5 min. After cooling to room temperature, the infusion was obtained by filtering the mixture through a sieve for fermentation without residues. In the case of fermentation with residues, filtration was skipped. Then, 20 mL of activated kombucha starter culture was added to each flask. These flasks were put into the clean, dark, and room-temperature environment for fermentation. The collected samples were filtered through 0.22 μm membranes for subsequent experiments.

2.4. Antioxidant Capacities and Total Phenolic Content Assessment

The antioxidant capacities assessment included ferric reducing antioxidant power (FRAP) and Trolox-equivalent antioxidant capacity (TEAC) assays [8,17].

The FRAP assay procedure followed the protocol described in the literature [17] and the results were reported in μmol Fe^{2+}/L. The TEAC assay procedure also followed the literature [8] and the results were reported in μmol Trolox/L.

The total phenolic content (TPC) was assessed using the Folin–Ciocalteu method as described in previous literature [8,17], and the results were reported in mg of gallic acid equivalent (GAE)/L.

2.5. Evaluation of Bioactive Compounds

The bioactive compounds in kombucha beverages were identified and quantified using high-performance liquid chromatography (HPLC) with a photodiode array detector (PDA) (Waters, Milford, MA, USA), following the previous studies with minor adjustment [8,17]. The Agilent Zorbax Eclipse XDB-C18 column (Santa Clara, CA, USA), with a size of 250 mm × 4.6 mm, 5 μm was used for separation. The gradient elution program was detailed in the literature [8,17].

2.6. Sensory Pilot Study

The sensory pilot study of kombucha beverages was conducted based on previous studies [8,17]. Different types of kombucha samples were evaluated by 8 participants (seven graduate students and one professor) from the Department of Nutrition, School of Public Health, Sun Yat-sen University, for their odor, color, flavor, sourness, and overall acceptability. These participants have participated in sensory evaluation of kombucha in our previous studies [8,17] and have rich experience. The hedonic scale ranged from 1–9 levels, where 9 indicated extremely like, 8 indicated greatly like, 7 indicated moderately like, 6 indicated slightly like, 5 indicated neither like nor dislike, 4 indicated slightly dislike, 3 indicated moderately dislike, 2 indicated greatly dislike, and 1 indicated extremely dislike.

2.7. Statistical Analysis

All experiments were conducted in triplicate and the results were shown as mean ± standard deviation (SD). Data processing and analysis were performed using Excel 2016 (Microsoft, Washington, DC, USA) and SPSS 25.0 (IBM Corp., Armonk, NY, USA). The one-way analysis of variance (ANOVA) plus post hoc Fisher least significant difference (LSD) was used to test the significance of multiple groups (same kombucha at different fermentation times) and the corresponding pairwise comparisons between groups. The one-way ANOVA was utilized to test the significance of kombucha fermented with residues group and kombucha fermented without residues group. Moreover, the Pearson correlation coefficient was utilized to determine the correlation between parameters and compound concentrations, and heat maps were plotted (https://www.chiplot.online (accessed on 17 May 2023)). In addition, partial least squares regression (PLSR) was also utilized to test the relationships between parameters and compound concentration. The statistical significance was set at $p < 0.05$.

3. Results and Discussion

In this study, kombucha beverages based on bamboo leaf and mulberry leaf were investigated because both leaves contain many bioactive compounds and possess various bioactivities. The appearances of kombucha beverages fermented by bamboo leaf and mulberry leaf are shown in Figure 1.

(a)   (b)

Figure 1. The appearances of kombucha beverages from bamboo leaf and mulberry leaf. (**a**) bamboo leaf kombucha fermented with or without residues; (**b**) mulberry leaf kombucha fermented with or without residues.

3.1. Antioxidant Activities

3.1.1. FRAP Values

The FRAP assay is commonly used to measure antioxidant capacity in food products by evaluating the reduction ability of substances on ferric ions [17]. The FRAP values of kombucha based on bamboo leaf as well as mulberry leaf are displayed in Figure 2.

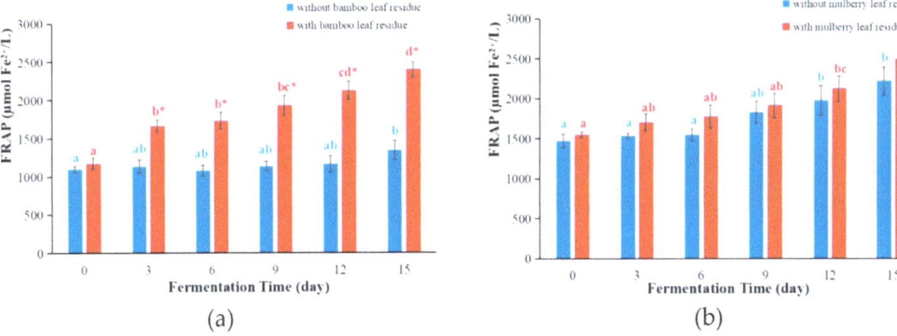

(a)   (b)

Figure 2. The changes in FRAP values during kombucha fermentation. (**a**) FRAP values of bamboo leaf kombucha; (**b**) FRAP values of mulberry leaf kombucha. In each color, the distinct letter represents significant differences ($p < 0.05$) in kombucha at different fermentation time. The * represents significant difference between kombucha fermented with leaf residues and that without leaf residues under the same fermentation time ($p < 0.05$).

For bamboo leaf kombucha fermented with leaf residues, the FRAP values increased over the fermentation period and reached the peak on day 15, which was 2.04 times that of day 0 (Figure 2a). For bamboo leaf kombucha fermented without leaf residues, the FRAP values hardly changed with the prolongation of the fermentation time. Moreover, the FRAP values of bamboo leaf kombucha fermented with leaf residues were remarkedly higher than those of kombucha fermented without leaf residues. This changing tendency was similar to that reported in the literature. For example, a study found that the FRAP value of green tea kombucha fermented with residues was significantly higher than that without residues [17]. Another study also found that fermentation with tea residues enhanced the FRAP values in kombucha based on vine tea and sweet tea compared with those without tea residues [8]. These results might be caused by many reasons. For example, boiling for only 5 min could not extract all the bioactive compounds from plant materials [18]

93

and the rest of the bioactive components could be extracted with the help of enzymes and microbiota during kombucha fermentation. In addition, the FRAP values of bamboo leaf kombucha (both fermented with and without leaf residues) were higher than that of kombucha-fermented soy whey (681.03 µM Fe^{2+}/L) [19].

For both mulberry leaf kombucha fermented with and without leaf residues, the FRAP values gradually raised over time and arrived at their peak on day 15, which were 1.62 times and 1.51 times compared with those of day 0, respectively (Figure 2b). Perhaps this was because some bioactive compounds (such as phenolic compounds and flavonoids) could be produced under the action of microorganisms, which could enhance the antioxidant activities of kombucha [20,21]. Moreover, the FRAP values of mulberry leaf kombucha fermented with leaf residues were slightly higher than those of kombucha fermented without leaf residues, although the difference was not statistically significant ($p > 0.05$). The FRAP values of mulberry leaf kombucha (both fermented with and without leaf residues) were higher than that of kombucha-fermented soy whey (681.03 µM Fe^{2+}/L) [19].

3.1.2. TEAC Values

The antioxidant capacity of substances could also be measured using the TEAC assay by comparing their ability to scavenge $ABTS^{\bullet+}$ radical cations with that of a standard reference compound Trolox [8]. The TEAC values of kombucha based on bamboo leaf residues and mulberry leaf residues are shown in Figure 3.

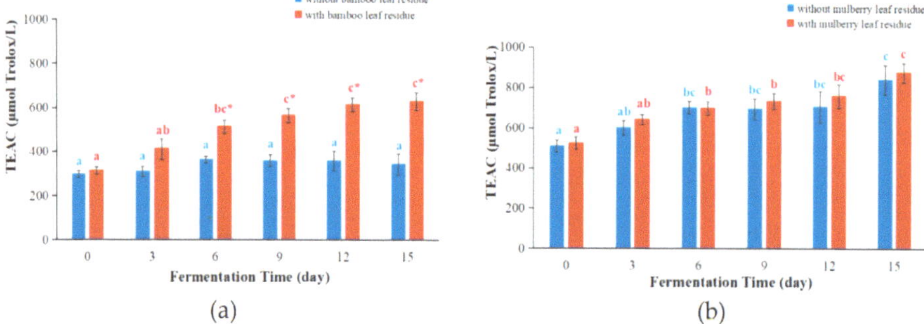

Figure 3. The changes in TEAC values during kombucha fermentation. (**a**) TEAC values of bamboo leaf kombucha; (**b**) TEAC values of mulberry leaf kombucha. In each color, the distinct letter represents significant differences ($p < 0.05$) in kombucha at different fermentation times. The * represents significant difference between kombucha fermented with leaf residues and that without leaf residues under the same fermentation time ($p < 0.05$).

For bamboo leaf kombucha fermented with leaf residues (Figure 3a), the TEAC values raised above fermentation and got to the top on day 15, which was 2.01-fold higher compared with that of day 0. For bamboo leaf kombucha fermented without leaf residues, the TEAC values remained relatively stable and did not change significantly as the fermentation time increased. The TEAC values of bamboo leaf kombucha fermented with leaf residues were significantly higher than those of kombucha fermented without leaf residues, by 1.84-fold on day 15. The reason might also be the same as the FRAP values. In addition, the TEAC values of bamboo leaf kombucha (both with and without leaf residues) were higher than that of black carrot kombucha (53.72 µmol Trolox/L) [22].

The TEAC values of kombucha based on mulberry leaf are shown in Figure 3b. For both mulberry leaf kombucha fermented with and without leaf residues, the TEAC values gradually increased as the fermentation time extended, and reached their maximum on day 15, with 1.68-fold and 1.66-fold increases compared with those of day 0, respectively. Several studies had reported similar results with kombucha prepared with black tea and green tea infusions [23]. Additionally, there was no statistical difference between the mulberry leaf kombucha fermented with leaf residues and that without leaf residues ($p > 0.05$). Moreover, the TEAC values of mulberry leaf kombucha (both fermented with and without residues) were higher than that of black carrot kombucha (53.72 μmol Trolox/L) [22].

For the bamboo leaf kombucha fermented with leaf residue and mulberry leaf kombucha (both fermented with and without leaf residues), the TEAC values were persistently elevated with the prolongation of fermentation time. Another study also obtained a similar tendency to the TEAC values [24]. However, a study found that the TEAC value of kombucha fermented by the green tea infusion was increased before day 7 and then decreased [25]. The differences in the radical scavenging activity observed in various kombucha might be attributed to the difference in the types of fermentation substrates and microbiota present in different kombucha beverages [26].

3.2. TPC Values

The Folin–Ciocalteu method is a spectrophotometric assay that measures TPC by reacting with phenolic compounds in the sample and is widely used in many studies [17,27]. The TPC values of kombucha based on bamboo leaf and mulberry leaf are shown in Figure 4.

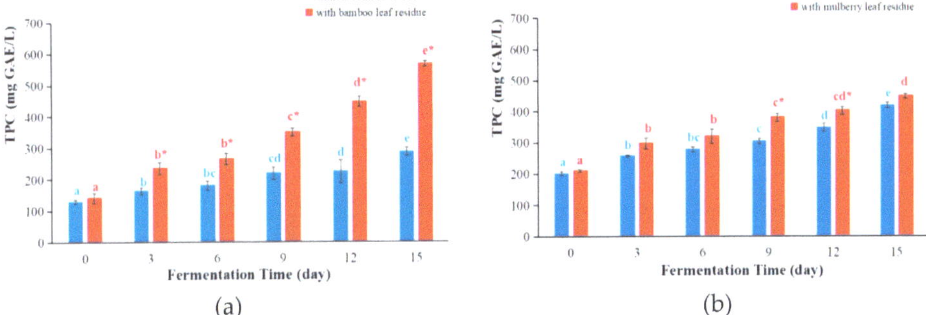

Figure 4. The changes in TPC values during kombucha fermentation. (**a**) TPC values of bamboo leaf kombucha; (**b**) TPC values of mulberry leaf kombucha. In each color, the distinct letter represents significant differences ($p < 0.05$) in kombucha at different fermentation times. The * represents significant difference between kombucha fermented with leaf residues and that without leaf residues under the same fermentation time ($p < 0.05$).

For bamboo leaf kombucha fermented with leaf residues (Figure 4a), the TPC values increased as the fermentation time progressed and reached its maximum on day 15, which was a 4.03-fold increase compared with that of day 0. For bamboo leaf kombucha fermented without leaf residues, the TPC values increased with the extension of the fermentation time and peaked on day 15, which was 2.07 times that of day 0. The tendency was similar to the previous studies [28,29]. Obviously, the TPC values of bamboo leaf kombucha fermented with leaf residues were higher compared with those of kombucha fermented without leaf residues, with a 1.99- and 1.97-fold increase on days 12 and 15. Moreover, the TPC value of bamboo leaf kombucha fermented with leaf residues was higher than those of kombucha fermented with the black tea infusion (412.25 mg GAE/L) and strawberry tree (*Arbutus unedo*) fruits (9.0 mg GAE/100 mL) [30,31].

The TPC values of kombucha based on the mulberry leaf are shown in Figure 4b. For both mulberry leaf kombucha fermented with or without leaf residues, the TPC values gradually raised over time and got to the top on day 15, which were 2.07 times and 2.13 times those of day 0, respectively. Additionally, the TPC values of mulberry leaf kombucha fermented with leaf residues were slightly higher than that of kombucha fermented without leaf residues. Furthermore, the TPC values of mulberry leaf kombucha (both fermented with and without leaf residues) were higher than that fermented with strawberry tree (*Arbutus unedo*) fruits (9.0 mg GAE/100 mL) [31].

In brief, fermentation with leaf residues increased the TPC values in both bamboo leaf and mulberry leaf kombucha. As mentioned before, boiling for only 5 min could not extract all the bioactive compounds from the plant materials [22] and the rest of the polyphenols in the leaves could be extracted over the fermentation period. Moreover, the microbial hydrolysis could increase the degradation of complex polyphenols to small molecules [28,29], which could further increase the TPC values in kombucha fermented with leaf residues.

3.3. Concentrations of Bioactive Compounds in Kombucha

The bioactive compounds in kombucha beverages were separated and quantified using HPLC-PDA, as displayed in Figure 5. In bamboo leaf kombucha, four compounds were identified as gallic acid, chlorogenic acid, luteolin-6-C-glucoside, and isovitexin (Figure 5b,c). In mulberry leaf kombucha, four compounds were identified as gallic acid, chlorogenic acid, rutin, and astragalin (Figure 5d,e).

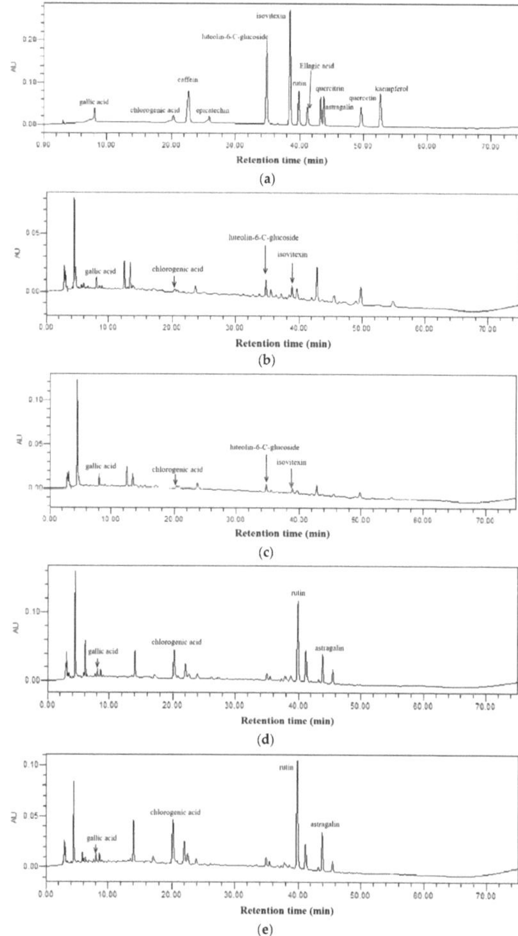

Figure 5. The representative HPLC chromatograms of standards and kombucha beverages at 270 nm. (**a**) standards; (**b**) bamboo leaf kombucha fermented without leaf residues; (**c**) bamboo leaf kombucha fermented with leaf residues; (**d**) mulberry leaf kombucha fermented without leaf residues; (**e**) mulberry leaf kombucha fermented with leaf residues.

The compounds in kombucha were quantified using the peak area under the maximum absorption wavelength, and the corresponding results are presented in Figure 6.

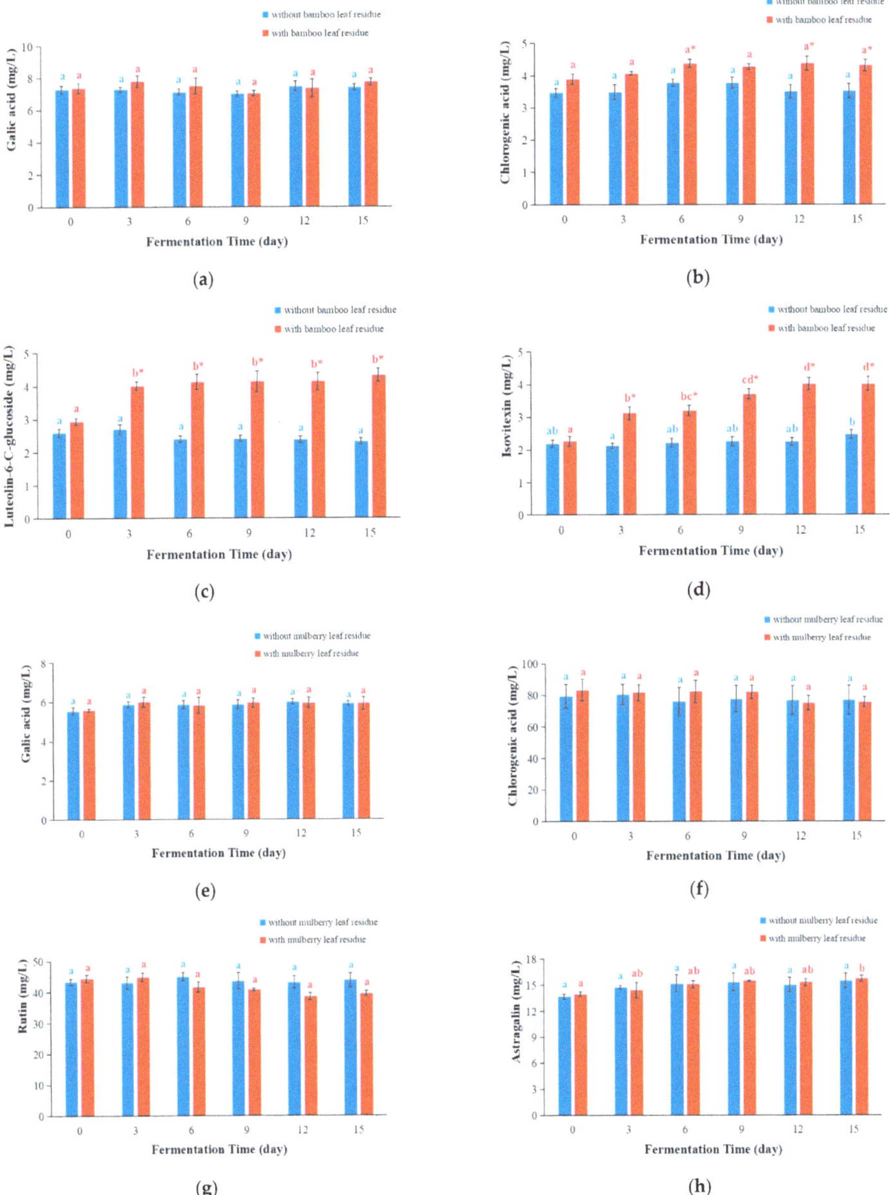

Figure 6. Changes in the contents of main phenolic compounds during kombucha fermentation. (**a–d**) bamboo leaf kombucha and (**e–h**) mulberry leaf kombucha. In each color, the distinct letter represents significant differences ($p < 0.05$) in kombucha at different fermentation times. The * represents significant difference between kombucha fermented with leaf residues and that without leaf residues under the same fermentation time ($p < 0.05$).

For bamboo leaf kombucha fermented without leaf residues, the concentrations of gallic acid, chlorogenic acid, luteolin-6-C-glucoside, and isovitexin remained relatively stable and did not change significantly as the fermentation time increased. This might be due to the absence of microorganisms or enzymes in kombucha that can degrade these compounds. For bamboo leaf kombucha fermented with leaf residues, there was no significant change observed in the concentrations of gallic acid and chlorogenic acid with the progression of the fermentation time (Figure 6a,b), which showed differences from kombucha based on black tea and green tea [8,17]. A study found that the content of gallic acid significantly increased through the enzymatic degradation of green tea extract with tannase [32]. Therefore, it could be due to the absence of compounds that could be degraded to produce gallic acid and chlorogenic acid in bamboo leaf residues. The concentrations of luteolin-6-C-glucoside increased on day 3 and then stayed stable ((Figure 6c), which might be because luteolin-6-C-glucoside continued to dissolve from leaf residues during the first three days and almost completely dissolved at day 3. The concentrations of isovitexin were increased over the prolonged fermentation period and reached the peak on day 12 (Figure 6d), which might be because the isovitexin kept dissolving from leaf residues until it was completely dissolved. Compared with the bamboo leaf kombucha without leaf residues, the concentrations of chlorogenic acid were higher in the bamboo leaf kombucha fermented with leaf residues, with 1.16, 1.25, and 1.23 times on days 6, 12, and 15, respectively; the concentrations of luteolin-6-C-glucoside and isovitexin were higher in the kombucha fermented with bamboo leaf residues, being 1.78 and 1.86 times higher on days 12 and 15, respectively. Additionally, the concentration of gallic acid in bamboo leaf kombucha fermented with leaf residues was similar to those of kombucha fermented without leaf residues. Generally speaking, these results suggested that fermentation with leaf residues could be helpful for the full extraction of bioactive compounds in bamboo leaf.

For mulberry leaf kombucha fermented without leaf residues, the concentrations of gallic acid, chlorogenic acid, rutin, and astragalin did not show significant changes with the prolongation of the fermentation time (Figure 6e–h). For mulberry leaf kombucha fermented with leaf residues, the concentration of gallic acid, chlorogenic acid, and rutin did not change with the prolongation of the fermentation time. Moreover, the concentration of astragalin slightly raised over time and got the highest level on day 15. In addition, there was no statistically significant difference in the concentrations of these compounds between the mulberry leaf kombucha fermented with leaf residues and that without leaf residues.

Furthermore, although the FRAP, TEAC, and TPC values of the bamboo leaf or mulberry leaf kombucha beverages were lower than those of black tea or green tea kombucha beverages [17], the different components in these beverages could result in varying health benefits, and the consumers could choose different beverages according to their demand. Especially because of the absence of caffeine in bamboo leaf and mulberry leaf kombucha beverages, for individuals sensitive to caffeine, bamboo leaf or mulberry leaf kombucha beverages could be better choices than black tea and green tea kombucha beverages, which contain high concentrations of caffeine.

3.4. Correlations between Parameters and Concentrations of Compounds

The correlations between FRAP, TEAC, TPC, and concentrations of compounds are obtained through heat map visualization and are shown in Figure 7.

For FRAP values vs. TEAC values, significant correlations were found in bamboo leaf kombucha fermented with leaf residues (R = 0.93), mulberry leaf kombucha fermented without leaf residues (R = 0.85), and mulberry leaf kombucha fermented with leaf residues (R = 0.89), which suggested that the antioxidant compounds in these beverages not only had the ability to reduce ferric ions, but also scavenge $ABTS^{\bullet+}$ radical cations. For FRAP values vs. TPC values, they were highly correlated in bamboo leaf kombucha without (R = 0.64) or with (R = 0.95) leaf residues, and mulberry leaf kombucha without (R = 0.88) or with (R = 0.91) leaf residues. The findings suggested that the phenolic compounds could be contributed to the ability to reduce Fe^{3+}. For TEAC values vs. TPC values, there were significant correlations in bamboo leaf kombucha fermented with leaf residues (R = 0.88), mulberry leaf kombucha fermented without leaf residues (R = 0.87), and mulberry leaf kombucha fermented with leaf residues (R = 0.94), which indicated that phenolic compounds might be major contributors to the free radical scavenging function of kombucha beverages.

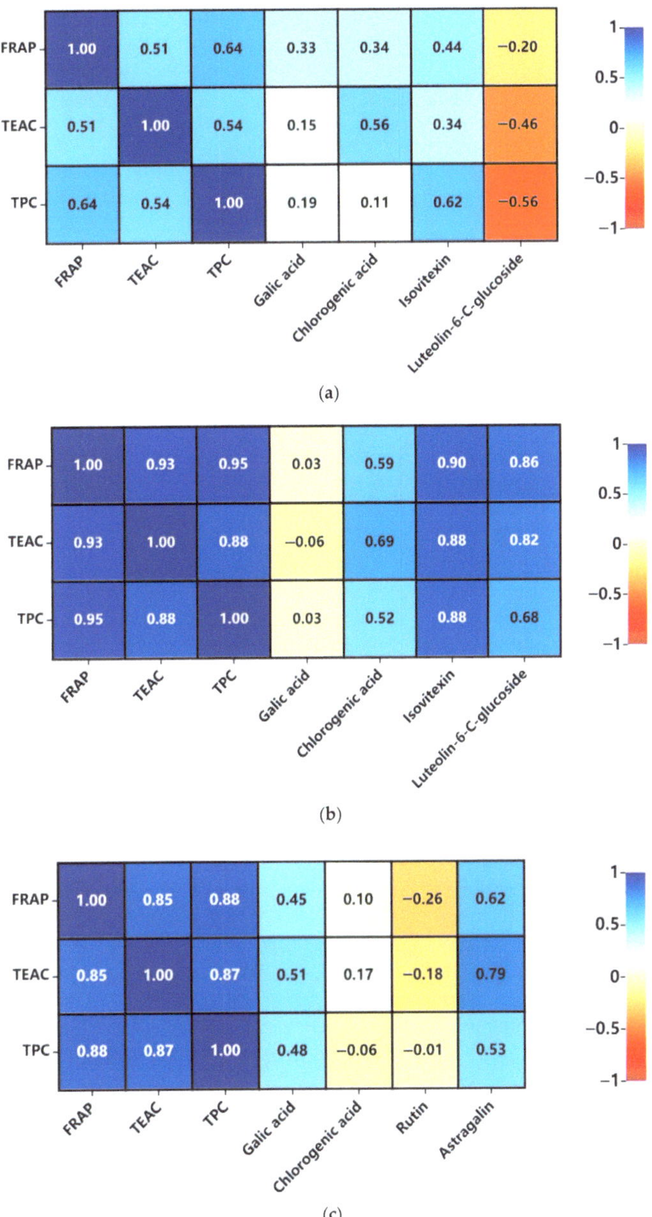

(a)

(b)

(c)

Figure 7. *Cont.*

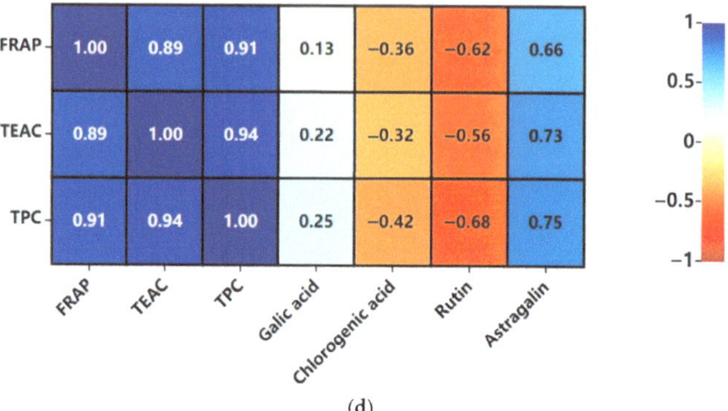

(d)

Figure 7. Heat map analysis of compound concentrations and associated parameters. (**a**) bamboo leaf kombucha fermented without leaf residues; (**b**) bamboo leaf kombucha fermented with leaf residues; (**c**) mulberry leaf kombucha fermented without leaf residues; (**d**) mulberry leaf kombucha fermented with leaf residues.

With regards to the correlation between FRAP values and bioactive compounds in bamboo leaf kombucha, significant correlations were found between the FRAP values and concentration of isovitexin (R = 0.90), as well as the FRAP values and the concentration of luteolin-6-C-glucoside (R = 0.86) in bamboo leaf kombucha fermented with leaf residues, suggesting that the isovitexin and luteolin-6-C-glucoside could potentially contribute to the antioxidant capacity. Moreover, the studies reported that isovitexin and luteolin were major flavonoid carbon glycosides in bamboo leaf that contributed to its antioxidant activity [33–35], which aligned with the outcomes obtained in this study. For the correlation between the FRAP values and bioactive compounds in mulberry leaf kombucha, significant correlations were found between the FRAP values and the concentration of astragalin in mulberry leaf kombucha fermented without leaf residues (R = 0.62), as well as the FRAP values and the concentration of astragalin in mulberry leaf kombucha fermented with leaf residues (R = 0.66).

With regards to the correlation between TEAC values and bioactive compounds in bamboo leaf kombucha, significant correlations were found between TEAC values and several compounds, including chlorogenic acid (R = 0.69), isovitexin (R = 0.88), and luteolin-6-C-glucoside (R = 0.82), in bamboo leaf kombucha fermented with leaf residues. For the correlation between the TEAC values and bioactive compounds in mulberry leaf kombucha, significant correlations existed between the TEAC values and astragalin (R = 0.79) in mulberry leaf kombucha fermented without leaf residues, and between the TEAC values and rutin (R = 0.73) in mulberry leaf kombucha fermented with leaf residues.

With regards to the correlation between TPC values and bioactive compounds in bamboo leaf kombucha, significant correlations were found between the TPC values and several compounds, including isovitexin (R = 0.88) and luteolin-6-C-glucoside (R = 0.68), in bamboo leaf kombucha fermented with leaf residues. Moreover, TPC values were related to isovitexin (R = 0.62) in bamboo leaf kombucha fermented without leaf residues. For the correlation between the TPC values and bioactive compounds in mulberry leaf kombucha, the TPC values were related to astragalin (R = 0.75) in mulberry leaf kombucha fermented with leaf residues.

On the other hand, the correlations between the concentrations of compounds and the parameters were also studied using the PLSR model and the optimum number of PLS-factors required for the models was three. The results are displayed in Figures 8 and 9.

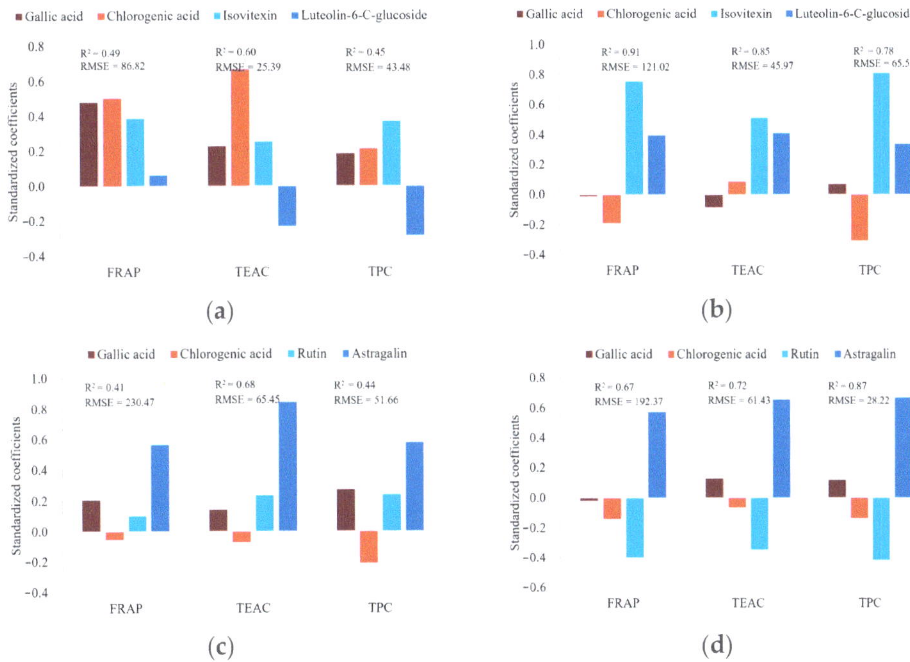

Figure 8. Standardized coefficients in PLSR models. (**a**) bamboo leaf kombucha fermented without leaf residues; (**b**) bamboo leaf kombucha fermented with leaf residues; (**c**) mulberry leaf kombucha fermented without leaf residues; (**d**) mulberry leaf kombucha fermented with leaf residues. RMSE, root mean square error.

For bamboo leaf kombucha fermented without leaf residues (Figures 8a and 9a–c), gallic acid, chlorogenic acid, and isovitexin generally had positive impacts on the FRAP, TEAC, and TPC values. Moreover, the prediction errors were small in the FRAP, TEAC, and TPC values, suggesting high prediction accuracy. For bamboo leaf kombucha fermented with leaf residues (Figures 8b and 9d–f), isovitexin and luteolin-6-C-glucoside had positive impacts on the FRAP, TEAC, and TPC values, which were consisted with those of the Pearson correlation coefficients. Moreover, the R^2 were high (0.91, 0.85, and 0.78, respectively), as well as the prediction errors being small in the FRAP, TEAC, and TPC values. The results showed that the model performed well. For mulberry leaf kombucha (both fermented without and with leaf residues), astragalin had strong positive impacts on the FRAP, TEAC, and TPC values, which were similar to those of the Pearson correlation coefficients. In addition, the R^2 were generally high in these models, which indicated models performed well. Furthermore, the prediction errors were small in the TEAC and TPC values, suggesting high prediction accuracy (Figure 8c,d and Figure 9g–l).

Although the results for the correlations between the concentrations of compounds and parameters from the Pearson correlation coefficients and PLSR model had little differences, they were generally similar.

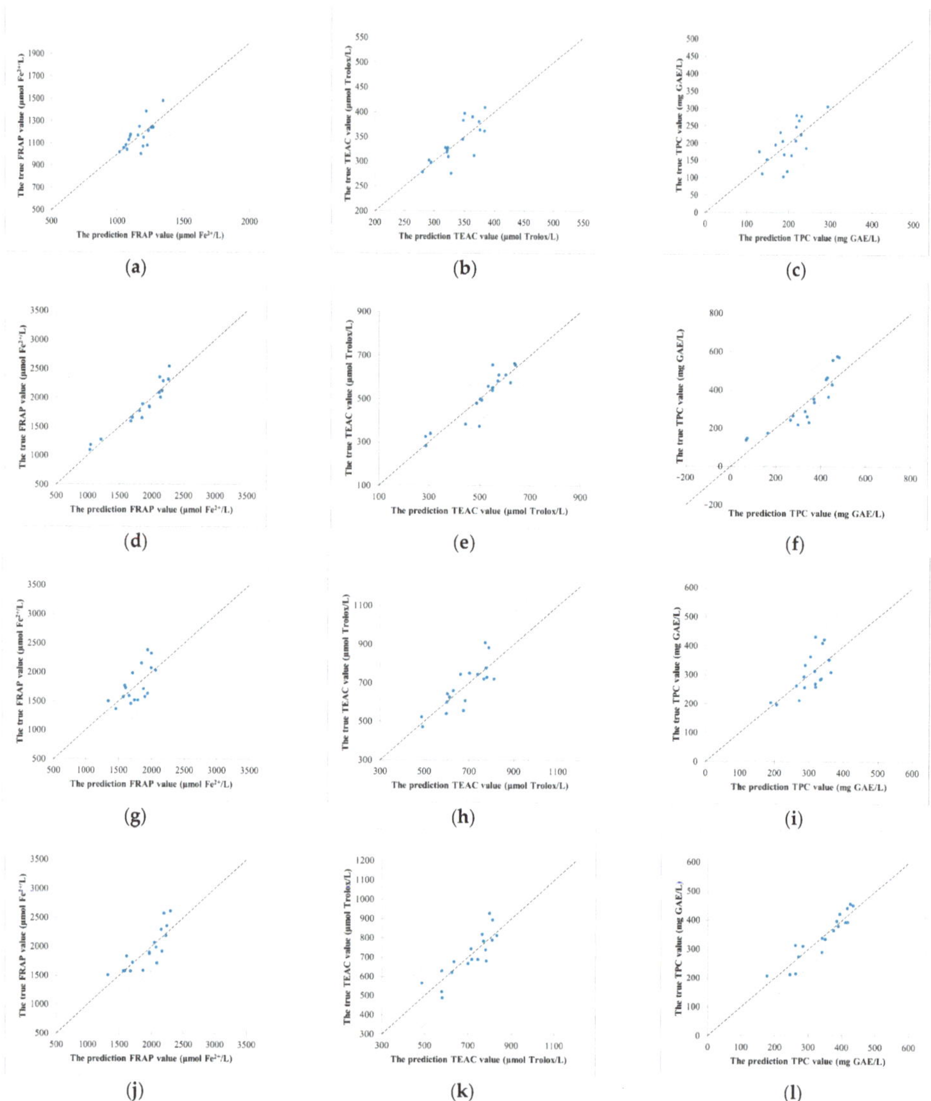

Figure 9. The correlation between the predicted values and true values. (**a–c**) bamboo leaf kombucha fermented without leaf residues; (**d–f**) bamboo leaf kombucha fermented with leaf residues; (**g–i**) mulberry leaf kombucha fermented without leaf residues; (**j–l**) mulberry leaf kombucha fermented with leaf residues.

3.5. Sensory Pilot Study

The sensory pilot study included odor, color, flavor, sourness, and overall acceptability of the kombucha beverages. The findings are displayed in Figure 10.

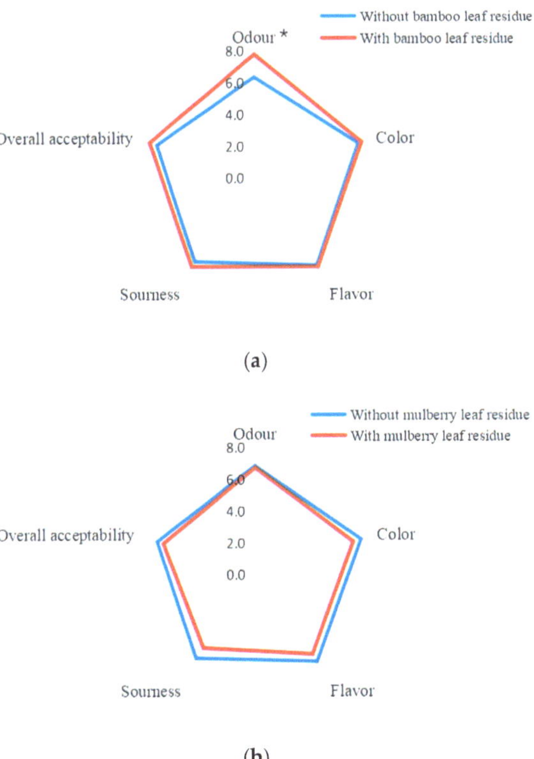

Figure 10. Sensory pilot study of kombucha beverages. (**a**) bamboo leaf kombucha fermented with or without leaf residues; (**b**) mulberry leaf kombucha fermented with or without leaf residues. The * represents significant difference between kombucha fermented with leaf residues and that without leaf residues ($p < 0.05$).

For bamboo leaf kombucha, the odor of kombucha fermented with bamboo leaf residues was better than those without leaf residues ($p < 0.05$). The scores of sourness and overall acceptability of kombucha fermented with bamboo leaf residues were slightly higher than those without leaf residues, although the differences showed not statistically significant ($p > 0.05$). Moreover, fermentation with or without residues had no difference on the color and flavor of the bamboo leaf kombucha.

For mulberry leaf kombucha, the color, flavor, sourness, and overall acceptability of the kombucha fermented without bamboo leaf residues were slightly better than those with leaf residues, but the differences were not statistically significant ($p > 0.05$). Moreover, fermentation with or without residues had no difference on the odor of the mulberry leaf kombucha.

In general, fermentation with leaf residues could enhance the hedonic scores of bamboo leaf kombucha to some extent, while it could not enhance the sensory acceptability of mulberry leaf kombucha. Moreover, among the four kombucha beverages in this study, bamboo leaf kombucha fermented with leaf residues got the highest scores in each indicator.

4. Conclusions

In this study, mulberry leaf kombucha and bamboo leaf kombucha have been explored. The results suggested that fermentation with leaf residues could markedly increase the FRAP, TEAC, and TPC values in bamboo leaf kombucha beverages with an enhancement of the sensory acceptability, but the enhancement in mulberry leaf kombucha beverages was not statistically significant, except for the TPC values. Moreover, several bioactive compounds in these kombucha beverages were separated and quantified using HPLC-PDA, and they might contribute to the antioxidant capacities of these beverages. Fermentation with leaf residues increased the concentrations of certain bioactive compounds in bamboo leaf kombucha beverages, but not in mulberry leaf kombucha beverages. Overall, fermentation with leaf residues might be a more appropriate option for making bamboo leaf kombucha beverages, while both fermentation with and without leaf residues showed no difference for making mulberry leaf kombucha beverages. In total, this study found that bamboo leaf and mulberry leaf could be used as alternative substrates to produce kombucha beverages with strong antioxidant activities and various bioactive compounds, which could be used to prevent and manage certain oxidative-stress-related diseases. In addition, this study could be helpful for the value-added utilization of mulberry and bamboo leaves.

Author Contributions: Conceptualization, R.-G.X., R.-Y.G. and H.-B.L.; Investigation, R.-G.X., S.-X.W., J.C., A.S. (Adila Saimaiti), Q.L., A.S. (Ao Shang), D.-D.Z. and S.-Y.H.; Methodology, R.-G.X., S.-X.W., J.C., A.S. (Adila Saimaiti), Q.L., A.S. (Ao Shang), D.-D.Z. and S.-Y.H.; Supervision, R.-Y.G. and H.-B.L.; Writing—original draft, R.-G.X.; Writing—review and editing, R.-Y.G. and H.-B.L.; Funding acquisition, H.-B.L. All authors have read and agreed to the published version of the manuscript.

Funding: This research was funded by the Key Project of Guangdong Provincial Science and Technology Program (No. 2014B020205002).

Institutional Review Board Statement: The study was conducted according to the guidelines of the Declaration of Helsinki, and approved by the Ethics Committee of Sun Yat-sen University (protocol code 2022-001; January 2022).

Informed Consent Statement: Informed consent was obtained from all subjects involved in the study.

Data Availability Statement: Data are contained within the article.

Acknowledgments: We thank Lin Zheng for the technical support.

Conflicts of Interest: The authors declare that they have no known competing financial interests or personal relationships that could have appeared to influence the work reported in this paper.

References

1. De Miranda, J.F.; Ruiz, L.F.; Silva, C.B.; Uekane, T.M.; Silva, K.A.; Gonzalez, A.G.M.; Fernandes, F.F.; Lima, A.R. Kombucha: A review of substrates, regulations, composition, and biological properties. *J. Food Sci.* **2022**, *87*, 503–527. [CrossRef] [PubMed]
2. Bortolomedi, B.M.; Paglarini, C.S.; Brod, F.C.A. Bioactive compounds in kombucha: A review of substrate effect and fermentation conditions. *Food Chem.* **2022**, *385*, 132719. [CrossRef] [PubMed]
3. Kapp, J.M.; Sumner, W. Kombucha: A systematic review of the empirical evidence of human health benefit. *Ann. Epidemiol.* **2019**, *30*, 66–70. [CrossRef]
4. Vargas, B.K.; Fabricio, M.F.; Ayub, M.A.Z. Health effects and probiotic and prebiotic potential of Kombucha: A bibliometric and systematic review. *Food Biosci.* **2021**, *44*, 101332. [CrossRef]
5. Abaci, N.; Deniz, F.S.S.; Orhan, I.E. Kombucha—An ancient fermented beverage with desired bioactivities: A narrowed review. *Food Chem. X* **2022**, *14*, 100302. [CrossRef] [PubMed]
6. Da Silva, J.C.; Mafaldo, I.M.; Brito, I.D.; Cordeiro, A. Kombucha: Formulation, chemical composition, and therapeutic potentialities. *Curr. Res. Food Sci.* **2022**, *5*, 360–365. [CrossRef]
7. Emiljanowicz, K.E.; Malinowska-Panczyk, E. Kombucha from alternative raw materials—The review. *Crit. Rev. Food Sci. Nutr.* **2020**, *60*, 3185–3194. [CrossRef]
8. Saimaiti, A.; Huang, S.Y.; Xiong, R.G.; Wu, S.X.; Zhou, D.D.; Yang, Z.J.; Luo, M.; Gan, R.Y.; Li, H.B. Antioxidant capacities and polyphenol contents of kombucha beverages based on vine tea and sweet tea. *Antioxidants* **2022**, *11*, 1655. [CrossRef]
9. Chan, E.W.C.; Wong, S.K.; Tangah, J.; Inoue, T.; Chan, H.T. Phenolic constituents and anticancer properties of *Morus alba* (white mulberry) leaves. *J. Integr. Med.-JIM* **2020**, *18*, 189–195. [CrossRef]
10. He, X.Y.; Chen, X.; Ou, X.Q.; Ma, L.Y.; Xu, W.T.; Huang, K.L. Evaluation of flavonoid and polyphenol constituents in mulberry leaves using HPLC fingerprint analysis. *Int. J. Food Sci. Technol.* **2020**, *55*, 526–533. [CrossRef]
11. Shin, S.O.; Seo, H.J.; Park, H.; Song, H.J. Effects of mulberry leaf extract on blood glucose and serum lipid profiles in patients with type 2 diabetes mellitus: A systematic review. *Eur. J. Integr. Med.* **2016**, *8*, 602–608. [CrossRef]

12. Zhang, R.Y.; Zhang, Q.; Zhu, S.; Liu, B.Y.; Liu, F.; Xu, Y. Mulberry leaf (*Morus alba* L.): A review of its potential influences in mechanisms of action on metabolic diseases. *Pharmacol. Res.* **2022**, *175*, 106029. [CrossRef] [PubMed]

13. Ying, C.; Mao, Y.; Chen, L.; Wang, S.; Ling, H.; Li, W.; Zhou, X. Bamboo leaf extract ameliorates diabetic nephropathy through activating the AKT signaling pathway in rats. *Int. J. Biol. Macromol.* **2017**, *105*, 1587–1594. [CrossRef] [PubMed]

14. Yu, Y.; Li, Z.; Cao, G.; Huang, S.; Yang, H. Bamboo leaf flavonoids extracts alleviate oxidative stress in HepG2 cells via naturally modulating reactive oxygen species production and Nrf2-mediated antioxidant defense responses. *J. Food Sci.* **2019**, *84*, 1609–1620. [CrossRef]

15. Cheng, Y.Q.; Wan, S.Q.; Yao, L.N.; Lin, D.; Wu, T.; Chen, Y.J.; Zhang, A.L.; Lu, C.F. Bamboo leaf: A review of traditional medicinal property, phytochemistry, pharmacology, and purification technology. *J. Ethnopharmacol.* **2023**, *306*, 116166. [CrossRef]

16. Zhu, D.Y.; Wang, C.H.; Zhang, Y.S.; Yang, Y.; Shang, Y.F.; Niu, X.L.; Sun, L.Y.; Ma, Y.L.; Wei, Z.J. Insight into solvent effects on phenolic content and antioxidant activity of bamboo leaves extracts by HPLC analysis. *J. Food Meas. Charact.* **2018**, *12*, 2240–2246. [CrossRef]

17. Zhou, D.D.; Saimaiti, A.; Luo, M.; Huang, S.Y.; Xiong, R.G.; Shang, A.; Gan, R.Y.; Li, H.B. Fermentation with tea residues enhances antioxidant activities and polyphenol contents in kombucha beverages. *Antioxidants* **2022**, *11*, 155. [CrossRef] [PubMed]

18. Zhao, C.N.; Tang, G.Y.; Cao, S.Y.; Xu, X.Y.; Gan, R.Y.; Liu, Q.; Mao, Q.Q.; Shang, A.; Li, H.B. Phenolic profiles and antioxidant activities of 30 tea infusions from green, black, oolong, white, yellow and dark teas. *Antioxidants* **2019**, *8*, 215. [CrossRef] [PubMed]

19. Tu, C.; Tang, S.; Azi, F.; Hu, W.; Dong, M. Use of kombucha consortium to transform soy whey into a novel functional beverage. *J. Funct. Foods* **2019**, *52*, 81–89. [CrossRef]

20. Erskine, E.; Ozkan, G.; Lu, B.Y.; Capanoglu, E. Effects of fermentation process on the antioxidant capacity of fruit byproducts. *ACS Omega* **2023**, *8*, 4543–4553. [CrossRef]

21. Hur, S.J.; Lee, S.Y.; Kim, Y.C.; Choi, I.; Kim, G.B. Effect of fermentation on the antioxidant activity in plant-based foods. *Food Chem.* **2014**, *160*, 346–356. [CrossRef]

22. Yildiz, E.; Guldas, M.; Gurbuz, O. Determination of in-vitro phenolics, antioxidant capacity and bio-accessibility Kombucha tea produced from black carrot varieties grown in Turkey. *Food Sci. Technol.* **2021**, *41*, 180–187. [CrossRef]

23. Zou, C.; Li, R.Y.; Chen, J.X.; Wang, F.; Gao, Y.; Fu, Y.Q.; Xu, Y.Q.; Yin, J.F. Zijuan tea- based kombucha: Physicochemical, sensorial, and antioxidant profile. *Food Chem.* **2021**, *363*, 130322. [CrossRef] [PubMed]

24. Lopes, D.R.; Santos, L.O.; Prentice-Hernandez, C. Antioxidant and antibacterial activity of a beverage obtained by fermentation of yerba-mate (Ilex paraguariensis) with symbiotic kombucha culture. *J. Food Process. Preserv.* **2021**, *45*, e15101. [CrossRef]

25. Li, S.Y.; Zhang, Y.; Gao, J.R.; Li, T.; Li, H.Z.; Mastroyannis, A.; He, S.; Rahaman, A.; Chang, K. Effect of fermentation time on physiochemical properties of kombucha produced from different teas and fruits: Comparative study. *J. Food Qual.* **2022**, *2022*, 2342954. [CrossRef]

26. Lee, K.R.; Jo, K.; Ra, K.S.; Suh, H.J.; Hong, K.B. Kombucha fermentation using commercial kombucha pellicle and culture broth as starter. *Food Sci. Technol.* **2022**, *42*, e70020. [CrossRef]

27. Değirmencioğlu, N.; Yıldız, E.; Sahan, Y.; Güldas, M.; Gürbüz, O. Impact of tea leaves types on antioxidant properties and bioaccessibility of kombucha. *J. Food Sci. Technol.* **2021**, *58*, 2304–2312. [CrossRef]

28. Chakravorty, S.; Bhattacharya, S.; Chatzinotas, A.; Chakraborty, W.; Bhattacharya, D.; Gachhui, R. Kombucha tea fermentation: Microbial and biochemical dynamics. *Int. J. Food Microbiol.* **2016**, *220*, 63–72. [CrossRef] [PubMed]

29. Jakubczyk, K.; Kałduńska, J.; Kochman, J.; Janda, K. Chemical profile and antioxidant activity of the kombucha beverage derived from white, green, black and red tea. *Antioxidants* **2020**, *9*, 447. [CrossRef]

30. Ivanišová, E.; Meňhartová, K.; Terentjeva, M.; Harangozo, Ľ.; Kántor, A.; Kačániová, M. The evaluation of chemical, antioxidant, antimicrobial and sensory properties of kombucha tea beverage. *J. Food Sci. Technol.* **2020**, *57*, 1840–1846. [CrossRef]

31. Tejedor-Calvo, E.; Morales, D. Chemical and aromatic changes during fermentation of kombucha beverages produced using strawberry tree (*Arbutus unedo*) fruits. *Fermentation* **2023**, *9*, 326. [CrossRef]

32. Xu, X.Y.; Meng, J.M.; Mao, Q.Q.; Shang, A.; Li, B.Y.; Zhao, C.N.; Tang, G.Y.; Cao, S.Y.; Wei, X.L.; Gan, R.Y.; et al. Effects of tannase and ultrasound treatment on the bioactive compounds and antioxidant activity of green tea extract. *Antioxidants* **2019**, *8*, 362. [CrossRef] [PubMed]

33. Li, X.B.; Tao, W.Q.; Xun, H.; Yao, X.; Wang, J.; Sun, J.; Yue, Y.D.; Tang, F. Simultaneous determination of flavonoids from bamboo leaf extracts using liquid chromatography-tandem mass spectrometry. *Rev. Bras. Farmacogn.-Braz. J. Pharmacogn.* **2021**, *31*, 347–352. [CrossRef]

34. Ma, N.H.; Guo, J.; Chen, S.H.X.; Yuan, X.R.; Zhang, T.; Ding, Y. Antioxidant and compositional HPLC analysis of three common bamboo leaves. *Molecules* **2020**, *25*, 409. [CrossRef] [PubMed]

35. Ye, S.; Pan, F.; Yao, L.; Fang, H.; Cheng, Y.; Zhang, Z.; Chen, Y.; Zhang, A. Isolation, characterization of bamboo leaf flavonoids by size exclusion chromatography and their antioxidant properties. *Chem. Biodivers.* **2022**, *19*, e202200506. [CrossRef]

Article

Up-Cycling Grape Pomace through Sourdough Fermentation: Characterization of Phenolic Compounds, Antioxidant Activity, and Anti-Inflammatory Potential

Andrea Torreggiani [1], Chiara Demarinis [2], Daniela Pinto [3], Angela Papale [3], Graziana Difonzo [2], Francesco Caponio [2], Erica Pontonio [2], Michela Verni [1,*] and Carlo Giuseppe Rizzello [1]

[1] Department of Environmental Biology, "Sapienza" University of Rome, 00185 Rome, Italy; andrea.torreggiani@uniroma1.it (A.T.); carlogiuseppe.rizzello@uniroma1.it (C.G.R.)
[2] Department of Soil, Plant and Food Science, University of Bari Aldo Moro, 70126 Bari, Italy; chiara.demarinis@uniba.it (C.D.); graziana.difonzo@uniba.it (G.D.); francesco.caponio@uniba.it (F.C.); erica.pontonio@uniba.it (E.P.)
[3] Human Microbiome Advanced Project, 20129 Milan, Italy; dpinto@giulianipharma.com (D.P.); apapale@giulianipharma.com (A.P.)
* Correspondence: michela.verni@uniroma1.it

Abstract: Despite its appealing composition, because it is rich in fibers and polyphenols, grape pomace, the major by-product of the wine industry, is still discarded or used for feed. This study aimed at exploiting grape pomace functional potential through fermentation with lactic acid bacteria (LAB). A systematic approach, including the progressively optimization of the grape pomace substrate, was used, evaluating pomace percentage, pH, and supplementation of nitrogen and carbon sources. When grape pomace was used at 10%, especially without pH correction, LAB cell viability decreased up to 2 log cycles. Hence, the percentage was lowered to 5 or 2.5% and supplementations with carbon and nitrogen sources, which are crucial for LAB metabolism, were considered aiming at obtaining a proper fermentation of the substrate. The optimization of the substrate enabled the comparison of strains performances and allowed the selection of the best performing strain (*Lactiplantibacillus plantarum* T0A10). A sourdough, containing 5% of grape pomace and fermented with the selected strain, showed high antioxidant activity on DPPH and ABTS radicals and anti-inflammatory potential on Caco2 cells. The anthocyanins profile of the grape pomace sourdough was also characterized, showing qualitative and quantitative differences before and after fermentation. Overall, the grape pomace sourdough showed promising applications as a functional ingredient in bread making.

Keywords: grape marc; lactic acid bacteria; fermentation; phenolic compounds; Caco2 cells

Citation: Torreggiani, A.; Demarinis, C.; Pinto, D.; Papale, A.; Difonzo, G.; Caponio, F.; Pontonio, E.; Verni, M.; Rizzello, C.G. Up-Cycling Grape Pomace through Sourdough Fermentation: Characterization of Phenolic Compounds, Antioxidant Activity, and Anti-Inflammatory Potential. *Antioxidants* **2023**, *12*, 1521. https://doi.org/10.3390/antiox12081521

Academic Editor: Myung-Ji Seo

Received: 14 July 2023
Revised: 26 July 2023
Accepted: 27 July 2023
Published: 29 July 2023

1. Introduction

Wine industry by-products amount to roughly 5–7 million tons each year worldwide [1], with France, Italy, Spain, and the United States, in that order, being the biggest producers [2]. Within the wine brewing process, grape pomace is the major by-product and consists of peels, seeds, and a small amount of grape pulp, remaining after the wort is separated. Traditionally, part of the grape pomace generated is distilled to produce different local spirits; otherwise, it is dumped or used for animal feed or compost [3]. Nevertheless, grape pomace is rich in fibers and polyphenols; it is estimated that during wine brewing, roughly 60–70% of the total phenols, which include proanthocyanidins and a diversity of anthocyanin glycosides, remain in the pomace [4].

Such compounds could be extracted and recovered, although this would lead to more by-products. Hence, for its valorization, strategies that take into account a zero-waste approach, considering a qualitative and quantitative recovery optimization, should be preferred.

Several studies evaluated the effect of grape pomace addition to bread, pasta, cookies, muffins, dairy, and meat and fish products, yet all concluded that the fortification, especially if at high percentages, causes notable undesirable rheological and organoleptic changes [5]. Among the valorization strategies, fermentation, which is often used to improve the nutritional, functional, and technological properties of food and food by-products [6], might fit sustainability criteria. Few studies explored the possibility to ferment grape pomace with yeasts or edible fungi, yet very little can be found about the valorization strategies that use lactic acid bacteria (LAB) [7,8]. As a matter of fact, grape pomace, similar to grapes, can generate a hostile environment for the development of lactic acid bacteria, due to the low pH and high number of phenolic compounds, which are toxic for LAB [9,10]. Indeed, in winemaking, LAB species belonging to the genera *Pediococcus*, *Leuconostoc*, and the former *Lactobacillus* genus dominate grape must at the early stages of the fermentation, yet as the process goes on, most of them do not tolerate such conditions, and *Oenococcus* species mostly carry out the rest of the fermentation [9]. For this reason, it is imperative, when fermenting grape pomace, that an appropriate starter selection is performed, and the fermentation conditions optimized.

In this context, our study aimed at exploiting grape pomace functional potential through fermentation with LAB strains selected for their pro-technological and antioxidant features [11–15]. A systematic approach, which progressively guided us towards the proper strain and process conditions, was implemented to enhance pomace antioxidant and anti-inflammatory features with the prospect of its application as a functional ingredient in a sourdough bread.

2. Materials and Methods

2.1. Grape Pomace

The red grape pomace (*Vitis vinifera* L., cultivar Primitivo) used in this work was provided by a winery in Santeramo in Colle (Apulia, Southern Italy) following a seven-day maceration phase and collected after pressing. The grape pomace was dried at 70 °C for 60 min in a ventilated oven (Argolab-TCF120, Carpi, Italy) and finely ground with a laboratory mill Ika-Werke M20 (GMBH, and Co. KG, Staufen, Germany) to obtain a powder (grape pomace powder, GPP), further sieved with a 150 μm mesh. The proximate composition of grape pomace was the following: proteins, 11% of GPP dry matter (d.m.); lipids, 7% on d.m.; ashes, 10% on d.m.; and total dietary fiber, 42% on d.m.

Moisture was determined at 105 °C with a thermobalance MA35 (Sartorius, Gottinga, Germany); the activity water (a_w) was determined with a Humimeter RH2 (Schaller Messtechnik, St. Ruprecht an der Raab, Austria).

2.2. Microrganisms

Eleven lactic acid bacteria strains belonging to the Culture Collection of the Department of Soil, Plant and Food Sciences, University of Bari Aldo Moro, were used as starters for GPP fermentation: *Furfurilactobacillus rossiae* T0A16 and LB5, *Lactiplantibacillus plantarum* T6B10, T0A10, 18S9, H18, H64, and LB1, *Pediococcus acidilactici* 10MM0, *Leuconostoc mesenteroides* 12MM1, and *P. pentosaceus* H22. LAB strains were singly cultivated in De Man, Rogosa and Sharpe (MRS, Oxoid, Basingstoke, Hampshire, UK) at 30 °C until the late exponential phase of growth was reached (ca. 10 h). Before the inoculum, cells were harvested by centrifugation ($10,000\times g$, 10 min, 4 °C), washed twice in 50 mM phosphate buffer, pH 7.0, and re-suspended in tap water.

2.3. Starter Selection

2.3.1. Grape Pomace-Derived Substrates

Aiming at investigating LAB performances, different conditions to render grape pomace to a suitable substrate for their growth were considered: (i) GPP/water ratio of 10%, 5%, and 2.5% (*w/v*); (ii) pH correction at pH 6.0 with food grade sodium bicarbonate (E500 as coded by European Food Safety Authority EFSA) (Solvay, Brussel, Belgium);

(iii) supplementation with glucose (1% w/v) (Oxoid) and/or yeast extract (0.5% w/v) (Oxoid). Cells, collected as previously described, were resuspended in tap water, and used to singly inoculate the GPP substrates. Inoculum corresponded to 7 log cfu/mL; all fermentations were carried out at 30 °C for 24 h. For each condition considered, a not-inoculated control was prepared and analyzed before (Ct-T0) and after incubation at 30 °C for 24 h (Ct-T24).

2.3.2. Growth and Kinetics of Acidification

With the aim of selecting the best adapting strain, LAB growth and acidification was monitored. A FiveEasy Plus (Mettler-Toledo, Columbus, OH, USA) pH-meter was used for monitoring the fermentation processes. TTA (total titratable acidity) was determined according to the official AACC method 02-31.01 [16] and expressed as the amount of 0.1 M NaOH required to adjust the pH of 10 g sample in sterile water to 8.3. LAB was enumerated using MRS agar (Oxoid) agar medium, supplemented with cycloheximide (0.1 g/L). Plates were incubated, under anaerobiosis (AnaeroGen and AnaeroJar, Oxoid), at 30 °C for 48 h.

The kinetics of acidification were modeled according to the Gompertz equation as modified by Zwietering [17]: $y = k + A \exp\{-\exp[(Vmaxe/A)(\lambda - t) + 1]\}$, where y is the acidification extent expressed as ΔpH at the time t, k is the initial level of the depend variable to be modeled, A is the difference pH between inoculation and stationary phase, Vmax is the maximum acidification rate, and λ is the length of the latency phase expressed in hours.

2.3.3. In Vitro Antioxidant Activity

Methanolic extracts were obtained from GPP-derived substrates, as reported by Rizzello et al. [15], to determine the scavenging activity against the DPPH (2,2-diphenyl-1-picrylhydrazyl) radical [14]. The scavenging effect was expressed as shown in the following equation: DPPH scavenging activity (%) = [(blank absorbance − sample absorbance)/blank absorbance] × 100. The synthetic antioxidant butylated hydroxytoluene (BHT) was included as a reference (75 ppm) in the analysis.

2.4. Sourdough Fermentation and Characterization

Three type-II sourdoughs were produced by using type 0 wheat flour (Coop, Casalecchio di Reno, Italia) having the following proximal composition: moisture, 14.5% (w/w); proteins, 11.75% (w/w) on d.m.; carbohydrates, 83.4% (w/w) on d.m.; dietary fibers 2.9% (w/w) on d.m.; fat, 1.2% (w/w) on d.m.; ash, 0.7% (w/w) on d.m. GPP was mixed to wheat flour at percentages of 0, 2.5, and 5% (w/w), respectively, for sourdoughs SD0, SD2.5, and SD5.

Sourdoughs, having a dough yield (DY, dough weight × 100/flour weight) of 160, were produced using *L. plantarum* T0A10 as starter and fermented for 24 h at 30 °C.

The pH and TTA of the sourdoughs were determined before (t0) and after (t24) fermentation, as described above. Water/salt-soluble extracts (WSE) of the sourdoughs were prepared [18] and used to determine the content of total free amino acids (TFAA) by a Biochrom 30+ series Amino Acid Analyzer (Biochrom Ltd., Cambridge Science Park, Cambridge, UK) with a Li-cation-exchange column (4.6 × 200 mm internal diameter), as described by Verni et al. [19]. The WSE were also used to analyze lactic and acetic acids, respectively, with K-DLATE and K-ACET kits (Megazyme International Ireland Limited, Bray, Ireland). The quotient of fermentation (QF) was determined as the molar ratio between lactic and acetic acids. Radical scavenging activity on DPPH radical was determined on methanolic extracts, as described before. The evaluation of the scavenging activity against the ABTS+ radical (2,20-azino-di-(3-ethylbenzthiazoline sulfonate)) was evaluated using the CSO790 Kit (Sigma-Aldrich, Darmstadt, Germany), following the manufacturer's instructions on both methanol and aqueous extracts. The scavenging activity was expressed as Trolox equivalents.

2.5. Analysis of Anthocyanins by UHPLC-DAD-MS/MS

Aiming at characterizing the phenolic profile of the SD5 sourdough, methanolic extracts were obtained before (SD5-T0) and after 24 h of fermentation (SD5-T24) at 30 °C. SD0 sourdough, not containing GPP, was analyzed as control.

A UHPLC Ultimate 3000RS Dionex interfaced by H-ESI II probe with a LTQ Velos pro linear ion trap mass spectrometer (Thermo Fisher Scientific, Waltham, MA, USA) was used for the analysis of anthocyanins compounds. The UHPLC system was composed by qua-ternary pump, autosampler, column compartment, and detector. The analytical separation was achieved as previously reported [20] with some modification. A Hypersil GOLD aQ C18 column was used (100 mm of length, 2.1 mm internal diameter and 1.9 μm of particle size), held at 30 °C and at a constant flow of 0.3 mL min^{-1} with water-formic acid (90:10 v/v) (solvent A) and acetonitrile-formic acid (99.9:0.1 v/v) (solvent B). The gradient program of solvent A was as follows: 0–20 min from 98% to 30%; 20–24 min isocratic at 30%. Then, equilibration was performed at the initial conditions for 9 min. The PDA detector was set to scan from 220 to 600 nm of wavelength managed by a 3D field.

The MS parameter conditions were as follows: capillary temperature 320 °C; source heater temperature 280 °C; nebulizer gas N2; sheath gas flow 33 psi; auxiliary gas flow 5 arbitrary units; S-Lens RF Level 60%. Data were acquired in positive ionization mode. Samples were analyzed with two methods: a full scan method from 100 to 1000 m/z and a data-dependent experiment to collect MS2 data. The data-dependent settings were a full scan from 200 to 1000 for positive ionization, activation level 500 counts, isolation width 2 Da, default charge state 2, and CID energy 35.

The samples were filtered using syringe filters (LLG Labware, Meckenheim, Germany) in RC by 0.22 μm before injection into the equipment. All data were acquired and processed using Xcalibur v.2 (Thermo Fisher Scientific, Waltham, MA, USA). The injection volume was 5 μL. Tentative identification of compounds was performed using mass spectra (MS2), λmax, and retention time according to the literature [21,22]. Quantitative analysis was performed according to the external standard method based on calibration curves obtained by injecting different concentrations of standard solutions (R2 = 0.9972). Specifically, the standard used was malvidin-3-O-glucoside, which was purchased from phyproof® (PhytoLab, Dutendorfer, Germany).

2.6. Antioxidant Activity on Caco2 Cells

2.6.1. Caco2 Cells Culture

Human colon carcinoma Caco2 cells (ICLC HTL97023) provided from the National Institute for Cancer Research of Genoa (Italy), were routinely cultured in Dulbecco's modified Eagle's medium (DMEM) GLUTAMAX medium supplemented with 10% (v/v) heat-inactivated fetal bovine serum (FBS), 1% (v/v) HEPES 1M, 1% (v/v) non-essential amino acids (Gibco), 1% L-glutamine 200 mM, penicillin (100 U/mL), and streptomycin (100 mg/mL) (Aurogene, Italy) and maintained in 25 cm^2 culture flasks (BD Biosciences, Franklin Lakes, NJ, USA) at 37 °C in a 5% CO_2 and 95% air-humidified atmosphere. Confluent cultures were split 1:3–1:6 every two days, after washing with PBS 1X (without Ca^{2+} and Mg^{2+}), using Trypsin/EDTA and seed at 2–5 × 10^4 cell/cm2, 37 °C, 5% CO_2. Cell quantification was made through trypan blue assay.

Cell treatments were carried out by using freeze-dried methanolic extracts from SD5 (before and after 24 h of fermentation), resuspended in DMEM (10 mg/mL, stock solution), and sterilized through a 0.22 mm filter membrane (Millipore Corporation, Bedford, MA, USA).

2.6.2. Citotoxicity

Cell viability was measured according to the MTT assay [23]. After 24 h seeding on a 96-well plate, 80% confluent Caco2 cells were exposed to 10 μg/mL of freeze dried methanolic extracts from SD5. The control was the basal medium. Plates were incubated at 37 °C, 5% CO_2, for 24 h. After each treatment, the medium was aspirated and replaced with

100 μL per well of MTT solution. MTT was dissolved (5 mg/mL) in FBS and diluted 1:10 in the cell culture medium without phenol red. After 3 h of incubation, the basal medium was aspirated and 100 μL per well of DMSO were added to dissolve purple formazan product. The solution was shacked in the dark for 15 min at room temperature. The absorbance of the solutions was read at 570 nm in a microplate reader (BioTek Instruments Inc., Bad Friedrichshall, Germany). Each experiment was carried out in triplicate. Data were expressed as the mean percentage of viable cells compared to the culture in basal medium.

2.6.3. RNA Extraction and Real-Time-PCR

After treatment with the extracts from SD5, the expression of *TNF-α* and *IL-1β* from Caco2 cells was investigated through RT-PCR. When ca. 80% confluence was reached, Caco2 cells were harvested with trypsin/EDTA, seeded, at a density of 1×10^6 cells per well, into 12-well (Becton Dickinson France S.A., Meylan Cedex, France) plates and incubated at 37 °C, 5% CO_2, for 24 h. Cells in DMEM GLUTAMAX medium and DMEM GLUTAMAX with lipopolysaccharide (LPS) (10 μg/mL) were used as the controls. Freeze-dried methanolic extracts from SD5 at a concentration of 10 μg/mL were added to 80% confluent Caco2 cells with the same LPS concentration (10 μg/mL) and incubated at 37 °C for 24 h. For quantitative real-time PCR (RT-PCR), total RNA from Caco2 cells was extracted using Tri Reagent (Sigma Aldrich), as described by Chomczynski and Mackey [24]. The cDNA was synthesized from 2 μg RNA template in a 20 μL reaction volume, using the PrimeScript RT-PCR kit (Takara, Japan).

The cDNA was amplified and detected by the Stratagene Mx3000P Real-Time PCR System (Agilent Technologies Italia S.p.A., Milan, Italy). The amplification of cDNA was conducted using the following Taqman gene expression assays: TNF-α Hs00174128_m1 (tumor necrosis factor alpha), IL-1β Hs01555410_m1 (interleukin-1-beta), and Hs999999 m1 (human glyceraldehyde-3-phosphate dehydrogenase, GAPDH). GAPDH was used as a housekeeping gene. PCR amplifications were carried out in a 20 μL of total volume. The mixture of reaction contained 10 μL of 2× Premix Ex Taq (Takara, Kusatsu, Japan), 1 μL of 20× TaqMan gene expression assay, 0.4 μL of RoX Reference Dye II (Takara, Japan), 4.6 μL of water, and 4 μL of DNA. PCR conditions were the following: 95 °C for 30 s followed by 40 cycles of 95 °C for 5 s and 60 °C for 20 s. PCR reactions were performed in duplicate using an MX3000p PCR machine (Stratagene, La Jolla, CA, USA). Analyses were carried out in triplicate. The average value of target gene was normalized using *GAPDH* gene and the relative quantification of the levels of gene expression was determined by comparing the Δ cycle threshold (ΔCt) value [25]. Results were expressed as percent ratio to cells only treated with LPS.

2.7. Breadmaking

Three experimental breads were manufactured by using the bread machine Ariete 132 Panexpress 750 (De Longhi Appliances Srl, Campi Bisenzio, Italy) and the type 0 flour above mentioned. Experimental breads were as follows: cY-B, a control bread, leavened with 2% *w/w* baker's yeast; SD0-B, a control sourdough bread, containing 25% *w/w* of the SD0 sourdough, and leavened with 2% *w/w* baker's yeast; SD5-B, a sourdough bread containing 25% *w/w* of the SD5 sourdough, and leavened with 2% *w/w* baker's yeast.

All breads were obtained from doughs with DY 160 corresponding to a flour/water ratio of 62.5/37.5% (*w/w*), and all were added with commercial baker's yeast (AB Mauri Italy S.p.a., Casteggio, Italia). Proofing was performed at 28 °C for 1.5 h and baking at 180 °C for 50 min. Recipes are reported in Table S1.

2.8. Bread Characterization

2.8.1. Biochemical and Nutritional Characterization

The analysis of pH, TTA, organic acids, and TFAA of the dough after proofing were carried out as reported above. The proximal composition and energy value of experimental breads were determined following the AACC approved methods [16]. In detail, protein (total nitrogen × 5.7), fats, ash, dietary fibers, and moisture were determined according to the 46-11A, 30–10.01, 08–01, 32-05.01, and 44-15A methods, respectively. Carbohydrates were calculated as the difference [100 − (proteins + lipids + ash + total dietary fibers + starch)].

2.8.2. Technological Characterization

The dough leavening performance of the different samples was evaluated determining the volume increase of a 15 mL-dough placed in a graduated cylinder top, covered with a piece of Parafilm® and allowed to ferment at 28 °C. Results were expressed as the difference between the initial and the final volume/min (ΔV(mL)/min). Texture profile analysis was performed by using an FRTS-100N Texture Analyzer (Imada, Toyohashi, Japan) equipped with a 3 cm cylinder probe FR-HA-30J on boule-shaped loaves (200 g) stored for 2 h at room temperature after baking. The instrument settings were test speed 1 mm/s, 30% deformation of the sample, and two compression cycles, and the parameters evaluated were hardness, cohesiveness, springiness, and chewiness.

The chromaticity coordinates of the crust and crumb of the bread were obtained by a CS-10 colorimeter (CHN Spec Technology, Hangzhou, China) and reported as color difference, $\Delta E * ab$, calculated by the following equation:

$$\Delta E * ab = \sqrt{(\Delta L)^2 + (\Delta a)^2 + (\Delta b)^2}$$

where ΔL, Δa, and Δb are the differences for L, a^*, and b^* values between sample and reference (a white ceramic plate having L = 92.2, a^* = 0.15, and b^* = 0.85).

2.8.3. Sensory Analysis

Breads sensory analysis was performed by a trained panel group composed of 10 assessors (5 male and 5 females, mean age: 30 years, range: 25–54 years) with proven skills and previous experience in sensory evaluation of bread, pasta, and other cereal-based products. The sensory attributes, scored with a scale from 0 to 10 (with 10 being the highest score), were discussed with the assessors during the introductory 2 h-training session. Sensory attributes included: visual and tactile perception (crust and crumb color, elasticity, friability); taste (sweetness, bitterness, astringency, salty taste, herbaceous taste, acidic taste, overall flavor intensity); and scent perception (acidic odor). Bread slices (1.5 cm thick) were coded and served in a randomized order 4 h after baking. A glass of water was drunk by the panelists after each sample tasting.

2.9. Statistical Analysis

All the data were reported as the means of the data collected in three independent analyses. Data were subjected to one-way ANOVA; pair-comparison of treatment means was achieved by Tukey's procedure at $p < 0.05$, using the statistical software Statistica 12.5 (StatSoft Inc., Tulsa, OK, USA). The results of anthocyanins quantification were subjected to one-way analysis of variance followed by Tukey's procedure at $p < 0.05$, using OriginPro 2020 (OriginLab Corporation, Northampton, MA, USA).

3. Results
3.1. LAB Strain Selection

The GGP proximal composition (expressed as % on d.m.) was as follows: proteins, 12.35 ± 0.51; lipids 8.93 ± 0.21; dietary fibers, 53.14 ± 1.45; carbohydrates, 17.73 ± 0.14; ash, 7.84 ± 0.76. Moisture and a_w were, respectively, 5.76 ± 0.51% (w/w) and 0.3169 ± 0.0012.

Aiming at verifying the suitability of GPP as substrate for LAB growth, the 11 strains were inoculated in suspensions obtained with 10% GPP in water, with or without pH adjustment to 6.00 (initial pH of the GPP in water was 3.88 ± 0.02, while TTA was 10.4 ± 0.2 mL), with or without supplementations with glucose and/or yeast extract. No significant changes in pH and TTA were found for any of the conditions tested, while LAB cell density during the 24 h of incubation remained stable in conditions in which pH was corrected, while it decreased from 1 to 2 log cycles in all the others.

Assuming a strong inhibition of microbial growth linked to the abundance of polyphenols and pH, the test was repeated in a substrate containing lower GPP concentration, corresponding to 5% and 2.5% (*w/v*) in water. In substrates containing 5% *w/v* GPP, with the initial correction of pH to 6.00, a slight pH decrease (up to 0.5) was found only for three of the eleven strains inoculated (*L. plantarum* T0A10, H64, and *F. rossiae* T0A16). When the percentage of pomace in the substrate was reduced to 2.5%, the ΔpH value exceeded 0.5 units for all the strains inoculated.

In particular, acidification was intense when glucose and yeast extract were used simultaneously as supplements. Under these conditions, the decrease in pH was, for *L. plantarum* T0A10, *F. rossiae* T0A16, and *L. plantarum* H64, higher than 2.5 units.

Based on these results, the substrate containing 2.5% of GPP, adjusted to pH 6.00, and with added glucose and yeast extract was considered suitable for the subsequent comparison of the fermentative performances of the LAB strains.

The parameters obtained from the modeling of the acidification kinetics of the 11 strains are shown in Table 1. A, corresponding to the pH difference between the adaptation phase and the stationary phase, ranged from 2.878 and 2.361 (pH units), respectively for *F. rossiae* T0A16 and *L. plantarum* 18S9. The strains having the highest A values were *F. rossiae* T0A16 and *L. plantarum* H64 and T0A10 (Table 1). The maximum acidification rate was between 0.475 and 0.311 ΔpH/h. In particular, *P. pentosaceus* H22, *F. rossiae* T0A16 and *L. plantarum* H64 showed Vmax values higher than 0.45. The adaptation phase was longer than 4 h for *L. plantarum* T6B10, H18, H64, *P. pentosaceus* H22, and *F. rossiae* T0A16 and shorter than 3.4 h for all others.

Table 1. Parameters of the acidification kinetics. A, difference of the pH between inoculation and the stationary phase (ΔpH); Vmax, the maximum acidification rate (ΔpH/h); λ, length of the latency phase (h).

Strain	A	Vmax	λ
L. plantarum T0A10	2.83 ± 0.08 [a]	0.41 ± 0.02 [b]	3.39 ± 0.06 [d]
Ln. mesenteroides 12MM1	2.47 ± 0.05 [b]	0.41 ± 0.03 [b]	3.46 ± 0.05 [d]
L. plantarum LB1	2.41 ± 0.06 [c]	0.39 ± 0.01 [b]	3.31 ± 0.02 [d]
L. plantarum T6B10	2.40 ± 0.09 [c]	0.40 ± 0.03 [b]	4.39 ± 0.05 [c]
L. plantarum 18S9	2.36 ± 0.08 [d]	0.31 ± 0.02 [c]	3.31 ± 0.02 [d]
L. plantarum H18	2.46 ± 0.05 [b]	0.44 ± 0.02 [a]	5.19 ± 0.07 [a]
P. pentosaceus H22	2.44 ± 0.02 [b]	0.47 ± 0.03 [a]	5.44 ± 0.07 [a]
F. rossiae T0A16	2.88 ± 0.02 [a]	0.46 ± 0.03 [a]	4.94 ± 0.02 [b]
L. plantarum H64	2.87 ± 0.04 [a]	0.47 ± 0.02 [a]	4.92 ± 0.03 [b]

The data are the means of three independent experiments ± standard deviations (n = 3). [a–d] Values in the same column with different superscript letters differ significantly (*p* < 0.05).

Cell densities reached by LAB strains in the pomace substrate are shown in Figure 1A. At the end of the 24 h incubation at 30 °C, the starters that presented the highest cell density were *L. plantarum* T0A10 (9.57 ± 0.12 log CFU/mL), *L. plantarum* LB1 (9.47 ± 0.11 log CFU/mL), and *L. plantarum* H18 (9.41 ± 0.10 log CFU/mL). Slightly lower values were found for *L. plantarum* H64, 18S9 and T6B10, and *Ln. mesenteroides* 12MM1. Significantly (*p* < 0.05) lower cell densities were found for all the others, with *P. acidilactici* 10MM0 showing the lowest final cell density (7.00 ± 0.01 log CFU/mL) (Figure 1A).

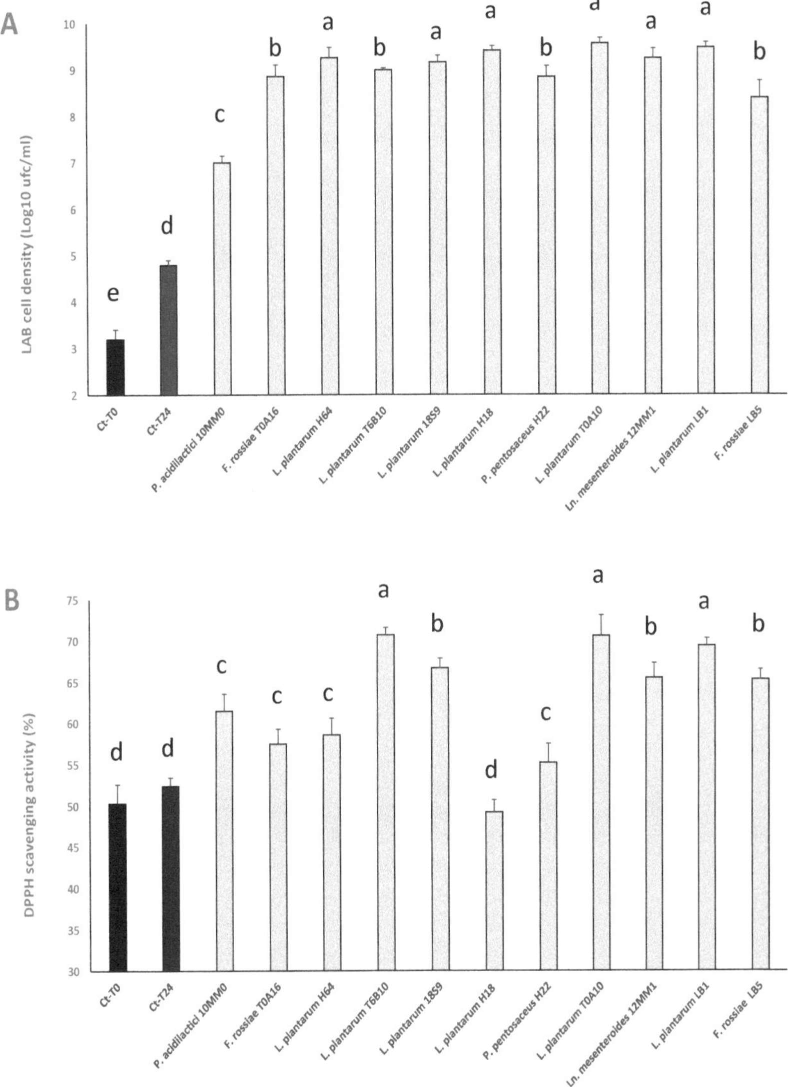

Figure 1. LAB cell density (**A**) and radical scavenging activity on DPPH radical (**B**) in grape pomace substrate (GPP 2.5% *w/v*, glucose 1% *w/v*; yeast extract 0.5% *w/v*, pH 6.0) inoculated (7 log cfu/mL) with *Furfurilactobacillus rossiae* T0A16 and LB5, *Lactiplantibacillus plantarum* T6B10, T0A10, 18S9, H18, H64, and LB1, *Pediococcus acidilactici* 10MM0, *Leuconostoc mesenteroides* 12MM1, and *P. pentosaceus* H22, and fermented at 30 °C for 24 h. A not-inoculated control was included and analyzed before (Ct-T0) and after incubation (Ct-T24). Data ± are the means of three independent analyses. [a–d] Values with different superscript letters differ significantly ($p < 0.05$). Bars of standard deviations are also represented.

With the aim of investigating the effect of fermentation on the antioxidant activity of the substrate, the radical scavenging activity was determined for all fermented substrates (methanolic extracts), comparing the results to that of a non-inoculated control, before and after incubation in the same conditions of the inoculated samples. In particular, the antioxidant activity was expressed in terms of percentage of DPPH radical stabilized after 10 min of reaction. The samples with the highest antioxidant activity were those fermented by *L. plantarum* T6B10 (70.7 ± 0.9%), T0A10 (70.6 ± 2.5%), LB1 (69.4 ± 0.9%), and 18S9 (66.7 ± 1.2%), *Ln. mesenteroides* 12MM1 (65.5 ± 1.8%), and *L. rossiae* LB5 (65.3 ± 1.2%) (Figure 1B). The radical scavenging activity observed for the substrates fermented by *L. plantarum* T6B10, T0A10, and LB1 was comparable ($p > 0.05$) to that of the synthetic antioxidant BHT (75 ppm), also included in the analysis (71.21 ± 0.50%).

Markedly and significantly ($p < 0.05$) lower antioxidant activity was found for *P. acidilactici* 10MM0 (61.5 ± 2.1%), *L. plantarum* H64 (58.6 ± 2.0%), *F. rossiae* T0A16 (57.5 ± 1.8%), *P. pentosaceus* H22 (55.2 ± 2.3%), and *L. plantarum* H18 (49.20 ± 1.5%) (Figure 1B).

Based on the results obtained thus far, *L. plantarum* T0A10, showing a short λ phase (Table 1), high final cell density (Figure 1A) and high antioxidant activity (Figure 1B), was selected for further experiments and used as starter for sourdough fermentation of wheat flour containing GPP.

3.2. Sourdough Fermentation

Three type-II sourdoughs, differing only for the percentage of GPP added (0%, 2.5%, and 5% of flour weight), were produced and characterized. In all cases, intense acidification was observed over the 24 h of incubation. A final pH of ca. 3.60 was found for the three sourdoughs (without significant differences) (Table 2). However, the addition of GPP affected the pH of the dough before fermentation; indeed, both the initial pH and the ΔpH decreased as the percentage of added pomace increased (Table 2). The addition of GPP, by virtue of the high concentration of organic acids contained therein, also modified TTA values, which were higher for SD2.5 and SD5 compared to SD0. The selected *L. plantarum* strain grew by two logarithmic cycles in all three doughs, although the final cell density was significantly ($p < 0.05$) higher in SD0 (Table 2).

FAA was analyzed, before and at the end of the fermentation process, to evaluate the proteolytic activity of the strain. In SD0, the release of FAA was particularly consistent (+83% compared to t0), but in SD2.5 and SD5, the increases corresponded to 57% and 29%, respectively (Table 2).

Lactic acid concentration was significantly ($p < 0.05$) lower in SD2.5 and SD5 compared to SD0, and the same trend, albeit with markedly lower concentrations, was observed for acetic acid (Table 2). The fermentation quotient was lower than 5 for all three sourdoughs, but significantly ($p < 0.05$) higher for SD5 (4.97) compared to SD2.5 and SD0 (Table 2).

The antioxidant activity of the three sourdoughs was analyzed as radical scavenging activity on DPPH, before and after the fermentation (Table 2). The mere addition of GPP increased the antioxidant activity from about 7% (SD0) to 34 and 74% (for SD2.5 and SD5, respectively). Fermentation further increased the antioxidant activity in all three sourdoughs up to 40%, with values of radical scavenging activity of 48 and 95%, respectively, for SD2.5 and SD5 at the end of the 24 h of incubation (Table 2).

An additional assay aimed at quantifying the total antioxidant activity (on the ABTS radical) showed antioxidant activities of 0.48 and 1.02 mM Trolox equivalent for SD0 and SD5, respectively (Table 2). ABTS test also confirmed increases of the ABTS radical scavenging activity after sourdough fermentation, ranging from 60 to 70%. During sourdough fermentation, the pH kinetic was modeled (Table 2). The parameter *A* significantly ($p < 0.05$) decreased as the percentage of GPP increased (Table 2). The Vmax was lower in the two supplemented sourdoughs (although without significant differences between SD2.5 and SD5). The adaptation phase, increased together with the GPP supplementation entity, varying from 3.72 h in SD0 to 4.28 h in SD5 (Table 2).

Table 2. Characterization of the type II sourdoughs inoculated with *L. plantarum* T0A10 (7 log cfu/g) and fermented at 30 °C for 24 h. GPP was mixed to the wheat flour at percentages of 0, 2.5, and 5% (*w/w*), respectively, for sourdoughs SD0, SD2.5, and SD5. Dough yield of the sourdoughs was 160.

	SD0		SD2.5		SD5	
	t0	tf	t0	tf	t0	tf
pH	5.69 ± 0.04 [a]	3.61 ± 0.03 [d]	5.12 ± 0.02 [b]	3.63 ± 0.01 [d]	4.73 ± 0.01 [c]	3.59 ± 0.02 [d]
TTA	1.82 ± 0.01 [e]	10.2 ± 0.18 [b]	4.21 ± 0.03 [d]	12.8 ± 0.26 [a]	6.22 ± 0.05 [c]	14.03 ± 0.32 [a]
Lactic acid bacteria (Log ufc/g)	2.03 ± 0.10 [c]	9.96 ± 0.09 [a]	2.74 ± 0.30 [c]	9.24 ± 0.07 [b]	2.80 ± 0.25 [c]	9.05 ± 0.13 [b]
Lactic acid (mmol/kg)	nd	90.41 ± 2.31 [a]	nd	85.23 ± 4.01 [b]	nd	80.21 ± 3.87 [c]
Acetic acid (mmol/kg)	nd	24.64 ± 1.05 [a]	1.22 ± 0.07 [b]	19.21 ± 2.27 [b]	1.98 ± 0.10 [b]	16.12 ± 1.54 [c]
QF	-	3.67 ± 0.16 [c]	-	4.43 ± 0.29 [b]	-	4.97 ± 0.17 [a]
Total free amino acids (mg/kg)	314 ± 12 [d]	576 ± 32 [a]	309 ± 16 [d]	485 ± 27 [b]	290 ± 10 [e]	375 ± 21 [c]
DPPH Radical Scavenging Activity (%)	7.0 ± 0.3 [f]	13.1 ± 0.7 [e]	34.3 ± 0.9 [d]	58.4 ± 0.8 [c]	74.1 ± 1.3 [b]	95.2 ± 1.6 [a]
ABTS Radical scavenging (mM Trolox eq)	0.11 ± 0.03 [d]	0.18 ± 0.05 [d]	0.30 ± 0.08 [c]	0.48 ± 0.15 [b]	0.62 ± 0.10 [b]	1.02 ± 0.03 [a]
Kinetics of acidification parameters						
A	2.04 ± 0.12 [a]		1.48 ± 0.07 [b]		1.17 ± 0.04 [c]	
Vmax (ΔpH/Δh)	0.42 ± 0.02 [a]		0.25 ± 0.01 [b]		0.22 ± 0.01 [b]	
λ (h)	3.72 ± 0.18 [b]		3.89 ± 0.13 [b]		4.28 ± 0.10 [a]	

The data are the means of three independent analysis $\pm$ standard deviations (n = 3). [a-e] Values in the same row with different superscript letters differ significantly ($p < 0.05$).

3.3. Anthocyanins Identification and Quantification

Being the sourdough formulation exhibiting the most promising functional potential, SD5 was chosen for further characterization of the anthocyanins. Table 3 presents the concentration of individual anthocyanins identified in each sample based on their molecular ions and the corresponding anthocyanidin fragments produced in the MS^2 experiment.

Table 3. Retention time, concentration (mg/kg), mass spectral, and UV-Vis data of anthocyanins identified in the type II sourdoughs with 5% (*w/w*) of GPP, before (SD5-T0) and after (SD5-T24) fermentation with *L. plantarum* T0A10.

Peak	RT (min)	Concentration (mg/kg)		MH^+ (*m/z*)	Fragments	λ_{max}	Anthocyanins
		SD5-T0	**SD5-T24**				
1.	6.68	41.7 ± 1.6 [a]	28.4 ± 0.9 [b]	479	317	278,524	Petunidin 3-*O*-glucoside
2.	7.16	-	23.3 ± 0.1	655	331,493	276,530	Malvidin 3,5-*O*-diglucoside
3.	7.37	40.3 ± 3.7 [a]	11.8 ± 0.8 [b]	463	301	282,522	Peonidin 3-*O*-glucoside
4.	7.66	847.5 ± 13.5 [a]	755.3 ± 22.0 [b]	493	331	258,528	Malvidin 3-*O*-glucoside
5.	8.12	12.6 ± 0.9 [b]	21.9 ± 0.6 [a]	561	399	280,512	Carboxypyranomalvidin-3-*O*-glucoside
6.	9.49	31.4 ± 2.3 [a]	21.0 ± 1.5 [b]	535	331	280,522	Malvidin-3-*O*-acetylglucoside
7.	9.99	30.2 ± 1.5 [a]	25.6 ± 0.3 [b]	655	331	280,530	Malvidin-3-*O*-caffeoylglucoside
8.	10.13	34.9 ± 0.01 [a]	32.3 ± 1.4 [a]	625	317	282,530	Petunidin-3-*O*-coumaroylglucoside
9.	10.68	41.5 ± 0.2 [a]	34.7 ± 1.3 [b]	609	301	282,522	Peonidin 3-*O*-coumaroylglucoside
10.	10.84	351.3 ± 13.4 [b]	410.8 ± 10.1 [a]	639	331	282,534	Malvidin-3-*O*-trans-coumaroylglucoside

The data are the means of two independent analysis ± standard deviations (n = 3). [a–b] Values in the same row with different superscript letters differ significantly ($p < 0.05$).

Overall, significant ($p < 0.05$) decreases, from 11 to 71%, were found in SD5-T24 compared to SD5-T0, for 6 out of 10 identified compounds. Whereas higher concentrations were found for carboxypyranomalvidin-3-*O*-glucoside and malvidin-3-*O*-trans-coumaroylglucoside after fermentation (74% and 17%, respectively). Another compound, malvidin 3,5-*O*-diglucoside, which was not found in SD5-T0, exceeded 20 mg/kg after fermentation.

3.4. Cytotoxicity and Anti-Inflammatory Effects

The MTT method was used as an indirect measure of cytotoxicity, through the determination of cell viability after treatments with the freeze-dried extracts from SD0 and SD5 (at concentrations of 0.01–0.1 mg/mL). After 24 h of incubation, none of the tested samples showed a significant ($p > 0.05$) cytotoxic effect on Caco2 cell line, compared to the control.

qRT-PCR was used to assess the anti-inflammatory potential of the samples, after cell treatment at 10 μg/mL. TNF-α and IL-1β, two cytokines involved in intestinal inflammation, were used as markers. Compared to positive control (medium + LPS), the treatment with SD5-T24 caused a marked and significant ($p < 0.05$) anti-inflammatory effect, with a decrease of the TNF-α gene expression of the 63% (Figure 2A), which was comparable ($p < 0.05$) to the expression observed in absence of LPS stimulation (Figure 2A). A lower, but also significant ($p < 0.05$) decrease (-17%) was also observed for SD5-T0 (Figure 2A). Compared to positive control (medium + LPS), only the SD5-T24 led to a significant ($p < 0.05$) decrease (-60%) of the *IL-1β* gene expression after 24 h of incubation (Figure 2B).

3.5. Breads
3.5.1. Biochemical and Nutritional Characterization

For the production of the breads, three different doughs were obtained. While cY-B was produced only with the use of baker's yeast, two others were added with 25% (of the total weight) of SD0 and SD5. During leavening (1.5 h at 28 °C), the increase in volume of the loaves (Table 4) was found to be similar ($p > 0.05$) for the three doughs. The final pH, on the other hand, was obviously lower for the loaves containing sourdough (4.59 and 4.48, respectively, for SD0-B and SD5-B). Accordingly, TTA had an opposite trend and was the lowest for cY-B (Table 4). The concentration of organic acids was affected by the sourdough presence. Indeed, lactic acid of SD0-B and SD5-B ranged from 23 to

24 mmol/kg, with no significant ($p < 0.05$) differences, while acetic acid was significantly higher in SD0-B. Organic acids were found only in traces in cY-B. The QF was lower in SD0-B compared to SD5-B (3.66 vs. 5.21). As expected, the concentration of FAA was higher in the sourdough breads than cY-B, obtained with baker's yeast. The antioxidant activity on SD5-B corresponded to 46% and was markedly higher than the two bread formulations not containing GPP.

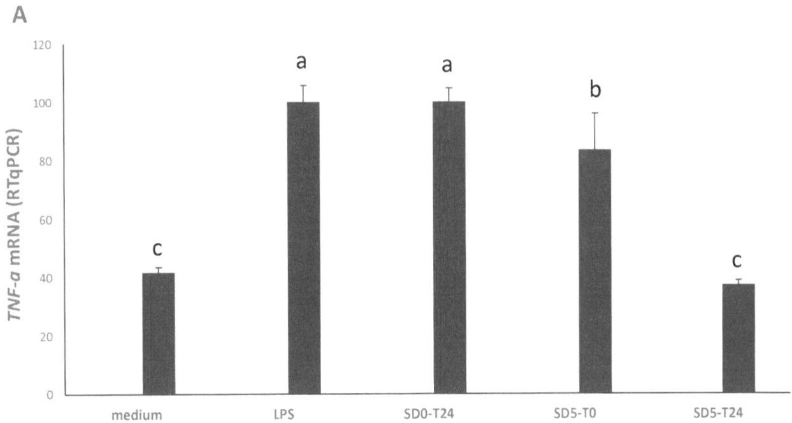

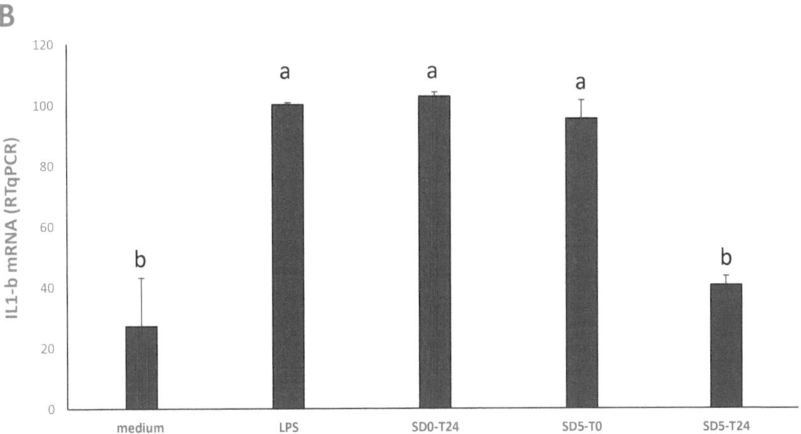

Figure 2. Expression of the tumor necrosis factor alpha (TNF-α) (panel **A**) and interleukin1 beta (Il-1β) (panel **B**) genes in Caco2 cells as determined by RT-PCR. Caco2 cells were treated at 37 °C for 24 h, under 5% CO_2, with: basal medium (negative control); basal medium with LPS (10 µg/mL) (positive control, LPS); and 10 µg/mL of the freeze-dried extracts from SD0 and SD5, before (T0) and after (T24) sourdough fermentation. Results are expressed as percent ratio to LPS treatment. Values are means of two experiments in triplicate. [a-c] Values with different superscript letters differ significantly ($p < 0.05$). Bars of standard deviations are also represented.

Table 4. Bread characterization data: cY-B, a control bread, leavened with 2% *w/w* baker's yeast; SD0-B, control sourdough bread, containing 25% *w/w* of the SD0 sourdough, and leavened with 2% *w/w* baker's yeast; SD5-B, a sourdough bread containing 25% *w/w* of the SD5 sourdough, and leavened with 2% *w/w* baker's yeast.

	cY-B	SD0-B	SD5-B
Volume increase (mL/min)	0.212 ± 0.013 [a]	0.198 ± 0.009 [a]	0.205 ± 0.008 [a]
pH	5.61 ± 0.02 [a]	4.59 ± 0.01 [b]	4.48 ± 0.02 [b]
TTA (mL NaOH 0.1 M)	2.5 ± 0.13 [b]	4.82 ± 0.24 [a]	5.10 ± 0.16 [a]
Lactic acid (mmol/kg)	1.32 ± 0.11 [b]	23.13 ± 0.49 [a]	24.18 ± 0.62 [a]
Acetic acid (mmol/kg)	1.40 ± 0.21 [c]	6.32 ± 0.28 [a]	4.64 ± 0.36 [b]
QF	-	3.66 ± 0.19 [b]	5.21 ± 0.14 [a]
Total Free Amino acid (mg/kg)	111 ± 4 [b]	279 ± 8 [a]	267 ± 6 [a]
DPPH Radical Scavenging Activity (%)	7.3 ± 0.2 [c]	12.0 ± 1.3 [b]	46.2 ± 2.3 [a]

The data are the means of three independent analysis ± standard deviations (n = 3). [a–c] Values in the same row with different superscript letters differ significantly ($p < 0.05$).

After cooking, the dough undergoes a drop in weight due to the evaporation of part of the water contained in the dough, hence the final moisture of the loaves was 27% (*w/w*).

Proximal composition of the breads was also determined (Table S2). The main differences related to the presence of GPP (SD5-B) compared to the other two breads mainly concerned dietary fibers content, which was 30% higher in SD5-B compared to cY-B (2.72%). Consequentially, carbohydrates were slightly but significantly ($p < 0.05$) lower in SD5-B. The differences in the nutritional label did not determine significant variations in the energy value of the three experimental breads.

The analysis of free amino acids performed with the Biochrom 30 automatic amino acid analyser (Figure 3) showed that sourdough fermentation caused the increase of Ala, Val, GABA (γ-aminobutyric acid), Lys, Trp, Arg and Pro concentrations (SD5-B and SD0-B vs. cY-B), whereas the presence of pomace generated an increase in Cys and Tyr (SD5-B vs. cY-B).

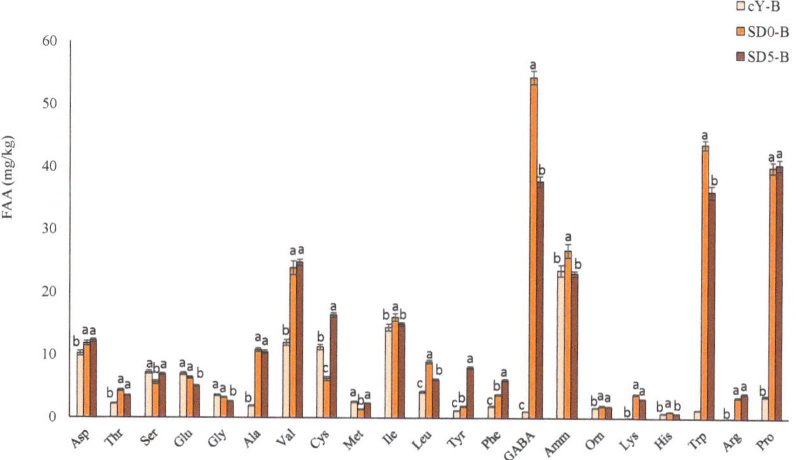

Figure 3. Free amino acids (mg/kg) in experimental breads. cY-B, a control bread, leavened with 2% *w/w* baker's yeast; SD0-B, control sourdough bread, containing 25% *w/w* of the SD0 sourdough, and leavened with 2% *w/w* baker's yeast; SD5-B, a sourdough bread containing 25% *w/w* of the SD5 sourdough, and leavened with 2% *w/w* baker's yeast. Data are the means of three independent analyses. [a–c] Values with different superscript letters differ significantly ($p < 0.05$). Bars of standard deviations are also represented.

3.5.2. Technological Characterization

Breads were subjected to a Textural Profile Analysis (Table 5). The specific volume of the bread containing SD5 was significantly ($p < 0.05$) lower than that of bread obtained with only baker's yeast (Table 5). The loaves had comparable ($p > 0.05$) hardness although the use of sourdough, especially SD5, gave lower cohesiveness to breads compared to baker's yeast control. Unlike what is commonly found for fiber-enriched breads, the springiness parameter, which is related to elasticity, was the highest for SD5-B. Nevertheless, chewiness of SD5-B was higher compared to other bread. The addition of GPP caused considerable color differences, expressed overall by the parameter ΔE (Table 5). The colorimetric coordinates showed a significant ($p < 0.05$) decrease in brightness (L) for both the crust and the crumb and, for the crust, a value of the parameter "a" (green/red index) considerably higher compared to that observed for the other two breads (Table 5).

Table 5. Volume, textural and color features of the experimental breads. cY-B, control bread, leavened with 2% *w/w* baker's yeast; SD0-B, control sourdough bread, containing 25% *w/w* of the SD0 sourdough, and leavened with 2% *w/w* baker's yeast; SD5-B, a sourdough bread containing 25% *w/w* of the SD5 sourdough, and leavened with 2% *w/w* baker's yeast.

	cY-B	SD0-B	SD5-B
Specific volume (cm³/g)	3.12 ± 0.06 [a]	3.06 ± 0.03 [ab]	2.92 ± 0.04 [b]
Hardness (N)	52.51 ± 1.20 [a]	51.76 ± 2.05 [a]	52.76 ± 0.14 [a]
Cohesiveness	0.57 ± 0.00 [a]	0.46 ± 0.05 [b]	0.35 ± 0.02 [c]
Springiness	0.84 ± 0.02 [b]	0.84 ± 0.01 [b]	1.15 ± 0.10 [a]
Chewiness (N)	27.06 ± 3.37 [b]	19.95 ± 0.95 [c]	44.30 ± 0.07 [a]
Crust color			
L	62.86 ± 1.43 [a]	57.51 ± 0.34 [a]	46.15 ± 2.79 [b]
a	−1.61 ± 0.24 [b]	−1.90 ± 0.34 [b]	1.20 ± 0.27 [a]
b	18.93 ± 0.38 [a]	16.54 ± 1.57 [a]	12.18 ± 0.89 [b]
ΔE	33.99 ± 1.34 [b]	38.07 ± 3.56 [b]	47.97 ± 2.61 [a]
Crumb color			
L	60.95 ± 3.63 [a]	56.61 ± 0.20 [b]	45.37 ± 2.05 [c]
a	−4.3 ± 0.52 [a]	−3.63 ±0.20 [a]	−0.15 ± 0.15 [b]
b	13.54 ± 0.4 [a]	12.59 ± 0.67 [a]	8.75 ± 0.14 [b]
ΔE	34.02 ± 3.47 [c]	37.88 ± 0.11 [b]	48.24 ± 2.02 [a]

The data are the means of three independent analysis ± standard deviations (n = 3). [a–c] Values in the same row with different superscript letters differ significantly ($p < 0.05$).

3.6. Sensory Profile

The sensory profile of the breads was assessed through a panel test. The average scores obtained from the evaluations of 10 assessors showed that the most evident differences of the SD5-B compared to the breads not containing GPP were related to crust and crumb color, which was markedly more intense (Figure 4).

The use of sourdough in SD0-B and SD5-B moreover conferred perceivable acidic taste and odor compared to cY-B (Figure 4). The friability of SD5-B was judged similar to the baker's yeast control, despite the higher presence of fibers. Astringency and herbaceous flavors, attributes included because typical of grape pomace, were clearly recognized by panelists, but not perceived as defect. Among the three breads, SD5-B received the highest score for the overall flavor intensity.

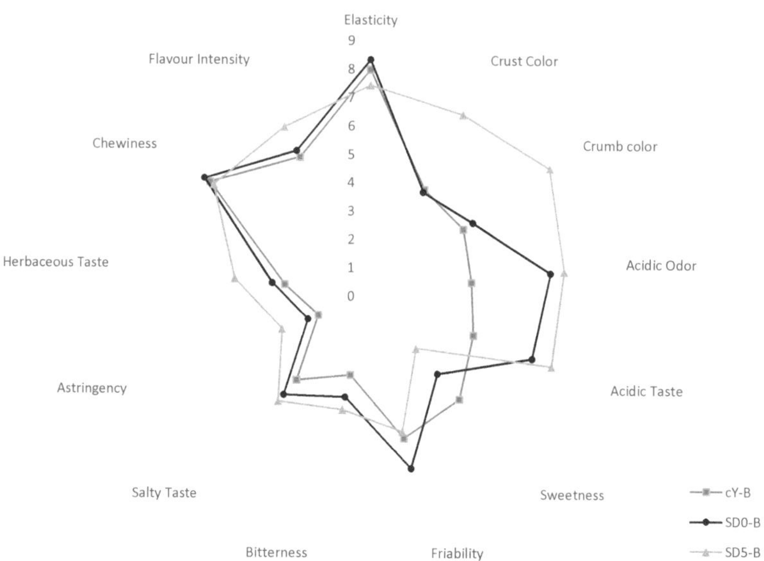

Figure 4. Spider web chart of the results obtained in the sensory analysis of the experimental breads. cY-B, control bread, leavened with 2% *w/w* baker's yeast; SD0-B, control sourdough bread, containing 25% *w/w* of the SD0 sourdough, and leavened with 2% *w/w* baker's yeast; SD5-B, a sourdough bread containing 25% *w/w* of the SD5 sourdough, and leavened with 2% *w/w* baker's yeast.

4. Discussion

As one of the major by-products of the wine industry, grape pomace valorization strategies that consider both quantitative and qualitative recovery optimizations, should be sought, especially in world-leading countries for wine production such as Italy. Exploiting grape pomace potential could be beneficial not only for the sustainability of the whole food system, but also to boost the functionality of the product made thereof. Several studies have demonstrated the correlation between grape pomace phenolics, particularly procyanidins, and antioxidant, antimutagenic, and anticarcinogenic activities [26]. Nevertheless, despite their health-promoting properties, it is well known that phenolic compounds can be ester bound or trapped within proteins or polysaccharides, on cell walls of food matrices [27]. Being bound to dietary fibers, the fraction potentially bio-accessible and absorbed in the gastro-intestinal tract can change [27]. Over the years, numerous studies have found that bioprocessing treatments, including fermentation, can modify the free/bound phenolics ratio. Indeed, several enzymes, among which feruloyl esterases, glucosidases, and tannases, have been described in LAB, mostly of the former *Lactobacillus* species, in fungi of the genera *Aspergillus* and *Penicillium*, as well as in *Saccharomyces cerevisiae* [28].

Based on the above considerations, this study aimed at exploiting grape pomace functional potential through fermentation with 11 lactic acid bacteria strains previously found to enhance the nutritional features of the matrices from which they were isolated [11–15]. Regardless, an obstacle that should be considered when fermenting grape pomace is its suitability as growth substrate. This is why a first experimental phase was aimed at identifying the percentage of pomace which, resuspended in water, would allow LAB to grow and acidify the medium. In fact, it was expected that, due to (i) the conspicuous removal of soluble nutrients (fermentable sugars and nitrogenous substances) during the winemaking process, (ii) the low pH [9], and (iii) the high concentration of polyphenols, having marked antimicrobial activity [10], the substrate was not optimal for microbial growth. Although to grape pomace polyphenols have been ascribed several functional properties, they can directly inhibit LAB growth by (i) interacting with cell membrane, thus causing membrane permeabilization and leakage of intracellular components, (ii) inhibiting enzymes essential for their metabolism, or (iii) chelating metal ions and nutrients [29]. For these reasons, when GPP was used at 10%, especially without pH correction, LAB cell density decreased up to 2 log cycles in 24 h. Hence, GPP percentage was lowered to 5 or 2.5% and supplementations with carbon and nitrogen sources, which are crucial for LAB metabolism, acting as an energy source and playing an important role in the growth [30], were considered. The optimization of the substrate enabled the comparison of strains performances and allowed the selection of the best performing strain (*L. plantarum* T0A10) based on growth, acidification, and ability to increase the DPPH radical scavenging activity. Indeed, among lactic acid bacteria, *L. plantarum* is the most utilized species for plant substrate fermentation due to its remarkable, although strain-dependent, metabolic activity against phenolic compounds [29]. Such activities allow the detoxification of phenolic compounds into metabolites having less antimicrobial activity and often higher antioxidant potential [28,29] justifying the best adaptation of the selected strain and the increased DPPH radical scavenging activity (up to 20%) compared to unfermented GPP (Figure 1B).

Hence, further experiments, aimed at including GPP in sourdough production, only involved *L. plantarum* T0A10. Based on the substrate optimization, only 2.5% and 5% of GPP supplementation were considered, whereas the modification of the pH was naturally provided by the buffering effect of the flour (Table 2). Despite all, LAB cell density, ΔpH, concentrations of organic acids, and release of FAA suggested an inhibitory effect of the pomace on the microorganism, albeit rather limited. Yet, considerable ($p < 0.05$) increases of the DPPH and ABTS radical scavenging activity were observed in sourdoughs (SD2.5 and SD5) proportional to the percentage of addition. Indeed, as expected, the sole addition of GPP enhanced the in vitro antioxidant activity (SD2.5-T0 and SD5-T0 compared to SD0), yet fermentation with *L. plantarum* T0A10 determined a further and significant increase (Table 2), confirming the results obtained in the substrate optimization process. Similar results in terms of antioxidant activity were obtained when grape pomace was fermented with edible fungi *Phanerochaete chrysosporium* and *Trametes gibbosa* [24]. Being the percentage providing the most functional potential, the sourdough containing the highest GPP content (SD5) was chosen for further characterization.

Studies have shown that anthocyanins and their glycosides are the most abundant phenolic compounds in grape pomace [4], which is why the characterization of the phenolic profile focused on this class of compounds. Overall, the total amount of anthocyanins did not differ before and after fermentation, yet differences in the quantities of some of the identified compounds were found. In particular, malvidin 3,5-*O*-diglucoside, which was not detected in SD5-T0, was found in SD5-T24. As explained above, it is likely that in GPP, malvidin 3,5-*O*-diglucoside was part of the bound fraction of anthocyanins, and as a consequence of fermentation, either due to the pH or the enzymatic activity [28,29,31] of *L. plantarum* T0A10, it was released in free form. Although malvidin 3,5-*O*-diglucoside is not commonly found in *V. vinifera*, traces of it were detected when grape pomace was inoculated with *Lacticaseibacillus casei* [32], suggesting a possible release as a consequence of fermentation. The same could be said for carboxypyranomalvidin-3-*O*-glucoside and malvidin-3-*O*-trans-coumaroylglucoside, which also increased with the fermentation. It should also be highlighted that several factors, among which the concentration and the structure of these compounds, as well as the pH of the environment are directly related to anthocyanins antioxidant capacity [31]. Indeed, it was observed that the radical scavenging activity of anthocyanins extracts significantly increases at acidic pH, which results in a more stable form of reduced anthocyanins by free radicals [33], explaining why, despite the total amount not differing between SD5-T0 and SD5-T24, higher DPPH and ABTS radical scavenging activity was found after fermentation (Table 2). Moreover, it was found by Han et al. [34] that malvidin-3-*O*-trans-coumaroylglucoside, which was higher in SD5-T24, has higher antioxidant capacity compared to other anthocyanins [34].

Along with substrate optimization and starter selection, it is worth underlining the importance of the methods used for the determination of food functional properties. In fact, when *L. plantarum* PU1 or *Bifidobacterium breve* 15A were used to ferment grape marc no significant variation of the antioxidant activity, estimated by ABTS, was observed compared to the unfermented control, whereas high inhibition of the oxidation of linoleic acid as well as protective effect against Caco2 cell oxidative stress was observed [7]. Thus, since in vitro assays can only be considered predictive tools of the functional activity in vivo and testing substances directly on animals or human is not an easy approach, ex vivo assays offer the right compromise to better understand food functional properties. In our study, the MTT assay on the Caco2 cell line was used to determine the cytotoxicity of SD5, before and after fermentation, and none was found. Conversely, the expression of *TNF-α* and *IL-1β* from Caco2 cells was investigated through RT-PCR after treatment with the extracts from SD0 and SD5. In vitro and in vivo antioxidant and anti-inflammatory properties of grape pomace, mainly ascribed to its phenolic compounds, have been already reported (for a review, see Bocsan et al. [35]). Yet, in our study, fermentation led to a further down-regulation of the expression of the pro-inflammatory proteins, such as cytokines (*IL-1β* and *TNF-α*), showing a significant anti-inflammatory effect of SD5-T24 compared to the other sourdoughs (before fermentation or without GPP addition). This could be due to a series of factors, among which the liberation into a free form of GPP anthocyanins which were not in the SD5-T0 and the effect of the pH. Moreover, in the conditions of this study, only anthocyanins were detected, whereas, even though limited, a contribution of the phenolic compounds provided by the wheat flour used for sourdough, as well as the effect of *L. plantarum* metabolism on them, cannot be excluded. Indeed, a series of hydroxycinnamic and hydroxybenzoic acids constitute most of the free phenolic compounds of the flour [36] and lactic acid bacteria are known for the ability to decarboxylate phenolic acids to the corresponding phenol or vinyl derivatives or hydrogenated them through phenolic acid reductases [28]. Metabolites of phenolic acids conversion, compared to their precursors, have reduced antimicrobial activity, but higher biological activities than their precursors [28], thus explaining the overall higher functional features of SD5-T24 compared to SD5-T0.

The functionality and health benefits of foods, especially fermented ones, are becoming increasingly important for consumers and in tourn food industries, which is why more research should focus on these aspects, particularly if a step towards the overall sustainability of the process is to be taken [29]. Therefore, SD5 was used as ingredient in bread making. In terms of nutritional composition, the main difference among breads was the higher fiber content in SD5-B compared to SD0-B and Cy-B. It should be noted that the increase in fiber is not only due to the addition of pomace, but we can hypothesize a contribution from resistant starch, which is generated following the biological acidification [37]. During sourdough fermentation, although negatively affected by GPP addition, a copious increase of GABA (up to 50-fold) was found compared to Cy-B. Such increase is of particular interest since γ-aminobutyric acid, produced by LAB as response mechanism to the low pH, is a functional compound with hypotensive, diuretic, and tranquilizing effects [38] which could further enhance bread functionality. The addition of GPP in the bread formulation, as expected, slightly impacted its technological properties; nevertheless, due to the small fortification and sourdough fermentation, known for its ability to modify rheology attributes [39], such effects were mitigated. These rheological attributes mainly rely on the physicochemical changes of the protein network, as a consequence of the acidic pH, which facilitate the larger dough expansion during fermentation [39]. Finally, the addition of SD5 to bread broadened its aroma profile with herbaceous and acid tastes due to the GPP itself and the sourdough fermentation, respectively. Overall, higher and pleasant flavor intensity was observed by panelists for SD5-B, which could be due to a variety of volatile (e.g., alcohols, acids, ketones, hydrocarbons, aldehydes, and esters) and non-volatile (e.g., amino acids) compounds developed during GPP fermentation [29]. Indeed, among non-volatile compounds proline and aspartic acid, which were all significantly higher (up to 11-fold) in SD5-B compared to Cy-B, are the ones that affect flavor the most [29].

5. Conclusions

This study demonstrated that lactic acid bacteria, specifically selected to ferment grape pomace, the major by-product of the wine industry, can be used to exploit its functional potential, increasing its antioxidant and anti-inflammatory features. The combination of sourdough fermentation and grape pomace proved to be an interesting ingredient for sourdough bread, with prospects of application that can boost both the sustainability of the food system and the intake of healthier and functional foods.

Supplementary Materials: The following supporting information can be downloaded at: https://www.mdpi.com/article/10.3390/antiox12081521/s1, Table S1: Bread recipes. cY-B, control bread; SD0-B, control sourdough bread, containing sourdough without grape pomace (SD0); SD5-B, bread containing sourdough added of 5% grape pomace (SD5); Table S2. Nutritional characteristics of the breads: cY-B, control bread, leavened with 2% w/w baker's yeast; SD0-B, control sourdough bread, containing 25% w/w of the SD0 sourdough, and leavened with 2% w/w baker's yeast; SD5-B, a sourdough bread containing 25% w/w of the SD5 sourdough, and leavened with 2% w/w baker's yeast.

Author Contributions: Conceptualization, C.G.R.; formal analysis, A.T., C.D., M.V., A.P. and G.D.; investigation, A.T.; resources, C.G.R., F.C., D.P. and E.P.; data curation, A.T., M.V. and C.G.R.; writing—original draft preparation, M.V. and C.G.R.; writing—review and editing, M.V. and C.G.R.; supervision, C.G.R. All authors have read and agreed to the published version of the manuscript.

Funding: This research received no external funding.

Institutional Review Board Statement: Not applicable.

Informed Consent Statement: Not applicable.

Data Availability Statement: The data presented in this study are available upon request from the corresponding author.

Acknowledgments: The authors gratefully thank Giammaria Giuliani and Fabio Rinaldi, respectively, member of the board of director and consultant of the Human Microbiome Advanced Project, Milan, Italy.

Conflicts of Interest: The authors declare no conflict of interest.

References

1. Machado, N.F.; Domínguez-Perles, R. Addressing facts and gaps in the phenolics chemistry of winery by-products. *Molecules* **2017**, *22*, 286. [CrossRef]
2. Beres, C.; Costa, G.N.; Cabezudo, I.; da Silva-James, N.K.; Teles, A.S.; Cruz, A.P.; Mellinger-Silva, C.; Tonon, R.V.; Cabral, L.M.C.; Freitas, S.P. Towards integral utilization of grape pomace from winemaking process: A review. *Waste Manag.* **2017**, *68*, 581–594. [CrossRef] [PubMed]
3. Perra, M.; Bacchetta, G.; Muntoni, A.; De Gioannis, G.; Castangia, I.; Rajha, H.N.; Manca, M.L.; Manconi, M. An outlook on modern and sustainable approaches to the management of grape pomace by integrating green processes, biotechnologies and advanced biomedical approaches. *J. Funct. Foods* **2022**, *98*, 105276. [CrossRef]
4. Yang, C.; Han, Y.; Tian, X.; Sajid, M.; Mehmood, S.; Wang, H.; Li, H. Phenolic composition of grape pomace and its metabolism. *Crit. Rev. Food Sci. Nutr.* **2022**, 1–17. [CrossRef] [PubMed]
5. Antonić, B.; Jančíková, S.; Dordević, D.; Tremlová, B. Grape pomace valorization: A systematic review and meta-analysis. *Foods* **2020**, *9*, 1627. [CrossRef]
6. Shiferaw Terefe, N.; Augustin, M.A. Fermentation for tailoring the technological and health related functionality of food products. *Crit. Rev. Food Sci. Nutr.* **2020**, *60*, 2887–2913. [CrossRef]
7. Campanella, D.; Rizzello, C.G.; Fasciano, C.; Gambacorta, G.; Pinto, D.; Marzani, B.; Scarano, N.; De Angelis, M.; Gobbetti, M. Exploitation of grape marc as functional substrate for lactic acid bacteria and bifidobacteria growth and enhanced antioxidant activity. *Food Microbiol.* **2017**, *65*, 25–35. [CrossRef]
8. Mišković Špoljarić, K.; Šelo, G.; Pešut, E.; Martinović, J.; Planinić, M.; Tišma, M.; Bucić-Kojić, A. Antioxidant and antiproliferative potentials of phenolic-rich extracts from biotransformed grape pomace in colorectal Cancer. *BMC Complement. Med.* **2023**, *23*, 29. [CrossRef]
9. Gil-Sánchez, I.; Bartolomé Suáldea, B.; Victoria Moreno-Arribas, M. Malolactic Fermentation. In *Red Wine Technology*; Morata, A., Ed.; Academic Press: Cambridge, MA, USA, 2018; pp. 85–98. ISBN 9780128144008.

10. Sabel, A.; Bredefeld, S.; Schlander, M.; Claus, H. Wine phenolic compounds: Antimicrobial properties against yeasts, lactic acid and acetic acid bacteria. *Beverages* **2017**, *3*, 29. [CrossRef]

11. Nionelli, L.; Montemurro, M.; Pontonio, E.; Verni, M.; Gobbetti, M.; Rizzello, C.G. Pro-technological and functional characterization of lactic acid bacteria to be used as starters for hemp (*Cannabis sativa* L.) sourdough fermentation and wheat bread fortification. *Int. J. Food Microbiol.* **2018**, *279*, 14–25. [CrossRef]

12. Nionelli, L.; Pontonio, E.; Gobbetti, M.; Rizzello, C.G. Use of hop extract as antifungal ingredient for bread making and selection of autochthonous resistant starters for sourdough fermentation. *Int. J. Food Microbiol.* **2018**, *266*, 173–182. [CrossRef] [PubMed]

13. Pontonio, E.; Verni, M.; Dingeo, C.; Diaz-de-Cerio, E.; Pinto, D.; Rizzello, C.G. Impact of enzymatic and microbial bioprocessing on antioxidant properties of hemp (*Cannabis sativa* L.). *Antioxidants* **2020**, *9*, 1258. [CrossRef] [PubMed]

14. Rizzello, C.G.; Nionelli, L.; Coda, R.; De Angelis, M.; Gobbetti, M. Effect of sourdough fermentation on stabilisation, and chemical and nutritional characteristics of wheat germ. *Food Chem.* **2010**, *119*, 1079–1089. [CrossRef]

15. Rizzello, C.G.; Lorusso, A.; Russo, V.; Pinto, D.; Marzani, B.; Gobbetti, M. Improving the antioxidant properties of quinoa flour through fermentation with selected autochthonous lactic acid bacteria. *Int. J. Food Microbiol.* **2017**, *241*, 252–261. [CrossRef]

16. AACC. *Approved Methods of Analysis*, 11th ed.; AOAC: St. Paul, MN, USA, 2010.

17. Zwietering, M.H.; Jongenburger, I.; Rombouts, F.M.; Van't Riet, K.J.A.E.M. Modeling of the bacterial growth curve. *Appl. Environ. Microbiol.* **1990**, *56*, 1875–1881. [CrossRef]

18. Weiss, W.; Vogelmeier, C.; Görg, A. Electrophoretic characterization of wheat grain allergens from different cultivars involved in bakers' asthma. *Electrophoresis* **1993**, *14*, 805–816. [CrossRef]

19. Verni, M.; Dingeo, C.; Rizzello, C.G.; Pontonio, E. Lactic acid bacteria fermentation and endopeptidase treatment improve the functional and nutritional features of *Arthrospira platensis*. *Front. Microbiol.* **2021**, *12*, 744437. [CrossRef]

20. Trani, A.; Verrastro, V.; Punzi, R.; Faccia, M.; Gambacorta, G. Phenols, volatiles and sensory properties of Primitivo wines from the" Gioia Del Colle" PDO Area. *S. Afr. J. Enol. Vitic.* **2016**, *37*, 139–148. [CrossRef]

21. Oliveira, J.; Alhinho da Silva, M.; Teixeira, N.; De Freitas, V.; Salas, E. Screening of anthocyanins and anthocyanin-derived pigments in red wine grape pomace using LC-DAD/MS and MALDI-TOF techniques. *J. Agric. Food Chem.* **2015**, *63*, 7636–7644. [CrossRef]

22. Kammerer, D.; Claus, A.; Carle, R.; Schieber, A. Polyphenol screening of pomace from red and white grape varieties (*Vitis vinifera* L.) by HPLC-DAD-MS/MS. *J. Agric. Food Chem.* **2004**, *52*, 4360–4367. [CrossRef]

23. Mosmann, T. Rapid colorimetric assay for cellular growth and survival: Application to proliferation and cytotoxicity assays. *J. Immunol. Methods* **1983**, *65*, 55–63. [CrossRef] [PubMed]

24. Chomczynski, P.; Mackey, K. Short technical reports. Modification of the TRI reagent procedure for isolation of RNA from polysaccharide-and proteoglycan-rich sources. *Biotechniques* **1995**, *19*, 942–945.

25. Vigetti, D.; Viola, M.; Karousou, E.; Rizzi, M.; Moretto, P.; Genasetti, A.; Clerici, M.; Hascall, V.C.; De Luca, G.; Passi, A. Hyaluronan-CD44-ERK1/2 regulate human aortic smooth muscle cell motility during aging. *J. Biol. Chem.* **2008**, *283*, 4448–4458. [CrossRef]

26. Yu, J.; Ahmedna, M. Functional components of grape pomace: Their composition, biological properties and potential applications. *Int. J. Food Sci. Technol.* **2013**, *48*, 221–237. [CrossRef]

27. Rocchetti, G.; Gregorio, R.P.; Lorenzo, J.M.; Barba, F.J.; Oliveira, P.G.; Prieto, M.A.; Simal-Gandara, J.; Mosele, J.I.; Motilva, M.J.; Tomas, M.; et al. Functional implications of bound phenolic compounds and phenolics–food interaction: A review. *Compr. Rev. Food Sci. Food Saf.* **2022**, *21*, 811–842. [CrossRef]

28. Verni, M.; Verardo, V.; Rizzello, C.G. How fermentation affects the antioxidant properties of cereals and legumes. *Foods* **2019**, *8*, 362. [CrossRef] [PubMed]

29. Liu, Z.; de Souza, T.S.P.; Wu, H.; Holland, B.; Dunshea, F.R.; Barrow, C.J.; Suleria, H.A.R. Development of Phenolic-Rich Functional Foods by Lactic Fermentation of Grape Marc: A Review. *Food Rev. Int.* **2023**, 1–20. [CrossRef]

30. Singh, V.; Haque, S.; Niwas, R.; Srivastava, A.; Pasupuleti, M.; Tripathi, C. Strategies for fermentation medium optimization: An in-depth review. *Front. Microbiol.* **2017**, *7*, 2087. [CrossRef]

31. Tena, N.; Martín, J.; Asuero, A.G. State of the art of anthocyanins: Antioxidant activity, sources, bioavailability, and therapeutic effect in human health. *Antioxidants* **2020**, *9*, 451. [CrossRef] [PubMed]

32. Anghel, L.; Milea, A.S.; Constantin, O.E.; Barbu, V.; Chitescu, C.; Enachi, E.; Râpeanu, G.; Mocanu, G.D.; Stănciuc, N. Dried grape pomace with lactic acid bacteria as a potential source for probiotic and antidiabetic value-added powders. *Food Chem.* **2023**, *X*, 100777. [CrossRef]

33. Kungsuwan, K.; Singh, K.; Phetkao, S.; Utama-ang, N. Effects of pH and anthocyanin concentration on color and antioxidant activity of *Clitoria ternatea* extract. *Food Appl. Biosc. J.* **2014**, *2*, 31–46. [CrossRef]

34. Han, F.; Ju, Y.; Ruan, X.; Zhao, X.; Yue, X.; Zhuang, X.; Quin, M.; Fang, Y. Color, anthocyanin, and antioxidant characteristics of young wines produced from spine grapes (*Vitis davidii* Foex) in China. *Food Nutr. Res.* **2017**, *61*, 1339552. [CrossRef] [PubMed]

35. Bocsan, I.C.; Măgureanu, D.C.; Pop, R.M.; Levai, A.M.; Macovei, S.O.; Pătrașca, I.M.; Chedea, V.S.; Buzoianu, A.D. Antioxidant and anti-inflammatory actions of polyphenols from red and white grape pomace in ischemic heart diseases. *Biomedicines* **2022**, *10*, 2337. [CrossRef] [PubMed]

36. Klepacka, J.; Fornal, Ł. Ferulic acid and its position among the phenolic compounds of wheat. *Crit. Rev. Food Sci. Nutr.* **2006**, *46*, 639–647. [CrossRef]

37. Sajilata, M.G.; Singhal, R.S.; Kulkarni, P.R. Resistant starch–a review. *Compr. Rev. Food Sci. Food Saf.* **2006**, *5*, 639–647. [CrossRef]
38. Yogeswara, I.B.A.; Maneerat, S.; Haltrich, D. Glutamate decarboxylase from lactic acid bacteria—A key enzyme in GABA synthesis. *Microorganisms* **2020**, *8*, 1923. [CrossRef]
39. Arora, K.; Ameur, H.; Polo, A.; Di Cagno, R.; Rizzello, C.G.; Gobbetti, M. Thirty years of knowledge on sourdough fermentation: A systematic review. *Trends Food Sci. Technol.* **2021**, *108*, 71–83. [CrossRef]

 *antioxidants*

Article

Antioxidative Properties of Fermented Soymilk Using *Lactiplantibacillus plantarum* LP95

Francesco Letizia , Alessandra Fratianni, Martina Cofelice , Bruno Testa , Gianluca Albanese , Catello Di Martino * , Gianfranco Panfili, Francesco Lopez and Massimo Iorizzo

Department of Agriculture, Environmental and Food Sciences, University of Molise, Via De Sanctis, 86100 Campobasso, Italy; f.letizia@studenti.unimol.it (F.L.); fratianni@unimol.it (A.F.); martina.cofelice@unimol.it (M.C.); bruno.testa@unimol.it (B.T.); g.albanese@studenti.unimol.it (G.A.); panfili@unimol.it (G.P.); lopez@unimol.it (F.L.)
* Correspondence: lello.dimartino@unimol.it

Abstract: In recent times, there has been a growing consumer interest in replacing animal foods with alternative plant-based products. Starting from this assumption, for its functional properties, soymilk fermented with lactic acid bacteria is gaining an important position in the food industry. In the present study, soymilk was fermented with *Lactiplantibacillus plantarum* LP95 at 37 °C, without the use of stabilizers as well as thickeners and acidity regulators. We evaluated the antioxidant capacity of fermented soymilk along with its enrichment in aglycone isoflavones. The conversion of isoflavone glucosides to aglycones (genistein, glycitein, and daidzein) was analyzed together with antioxidant activity (ABTS) measurements, lipid peroxidation measurements obtained by a thiobarbituric acid reactive substance (TBARS) assay, and apparent viscosity measurements. From these investigations, soymilk fermentation using *Lp. plantarum* LP95 as a starter significantly increased isoflavones' transformation to their aglycone forms. The content of daidzein, glycitein, and genistein increased after 24 h of fermentation, reaching levels of 48.45 ± 1.30, 5.10 ± 0.16, and 56.35 ± 1.02 μmol/100 g of dry weight, respectively. Furthermore, the antioxidant activity increased after 6 h with a reduction in MDA (malondialdehyde). The apparent viscosity was found to increase after 24 h of fermentation, while it slightly decreased, starting from 21 days of storage. Based on this evidence, *Lp. plantarum* LP95 appears to be a promising candidate as a starter for fermented soymilk production.

Keywords: functional food; soymilk; fermentation; isoflavones; *Lactiplantibacillus plantarum*

check for
updates

Citation: Letizia, F.; Fratianni, A.; Cofelice, M.; Testa, B.; Albanese, G.; Di Martino, C.; Panfili, G.; Lopez, F.; Iorizzo, M. Antioxidative Properties of Fermented Soymilk Using *Lactiplantibacillus plantarum* LP95. *Antioxidants* **2023**, *12*, 1442. https://doi.org/10.3390/antiox12071442

Academic Editor: Myung-Ji Seo

Received: 5 June 2023
Revised: 13 July 2023
Accepted: 15 July 2023
Published: 18 July 2023

1. Introduction

The growing interest of consumers in nutritionally enriched and health-promoting foods provokes interest in the eventual development of fermented functional foods [1]. In this view, the consumption of soybean (*Glycine max*) and its products has received attention from researchers and consumers both for the nutritional value and for its potential health benefits, including reducing the risk of cardiovascular diseases, diabetes, breast cancer, and chronic inflammation, due to the presence of bioactive compounds, such as isoflavones [2,3]. The water extract of soybeans, namely, soymilk, is an important oriental traditional beverage [4], considered a colloidal system as it contains protein and minerals [5]. This beverage can be a good substitute for animal-based milk and an ideal food for vegans diets and lactose-intolerant people [6]. Despite the listed properties of soymilk, its consumption in Western countries is still limited due to its unpleasant taste and its antinutritional factor, such as phytates that compromise the natural availability of minerals, as well as the presence of indigestible oligosaccharides, such as raffinose and stachyose, that cause flatulence [7,8]. Considering these limitations, the fermentative biotransformation approach using lactic acid bacteria (LAB) is an effective strategy to improve the nutritional, textural, and rheological properties of soymilk [3]. Isoflavones are phytochemical compounds naturally present in soymilk, and their enteric absorption is related to their chemical structure. The concentrations of the different forms of these bioactive compounds in soymilk are related to the soybean cultivar, time of soaking, the ratio of soybean to water, filtration, and thermal treatments [9,10]. The principal isoflavones of soy are daidzein, genistein, and glycitein, which exist in plants conjugated with sugars [11,12].

Glycosides can be hydrolyzed in the oral cavity [13] and, mainly, in the intestine to their aglycones, which are then absorbed in the intestine [14]. The flavonoid glycosides are considered biologically inactive, and their bioavailability requires the initial hydrolysis of the sugar moiety by intestinal β-glucosidases for uptake into the peripheral circulation [15].

Several studies have shown that it is possible to reverse the glucoside/aglycone ratio in soy food through LAB strains with β-glucosidase activity [16,17]. Therefore, the use of these bacteria as a starter, with the aforementioned enzymatic activity, in the fermentation of soymilk could contribute to increasing bioavailable isoflavones. [18].

Moreover, some LAB, when assumed through the consumption of fermented foods, show antioxidant effects, which differ among species due to different concentrations and types of antioxidants, and therefore could play a pivotal role in human health [19,20].

Among LAB species, *Lactiplantibacillus plantarum* has the Qualified Presumption of Safety (QPS) status from the European Food Safety Authority (EFSA) and the Generally Recognized as Safe (GRAS) status from the US Food and Drug Administration (US FDA) [21]. In addition, this bacterial species has a documented history as a microbial food culture [22–24]. Based on previous studies [25], in this work, we used the *Lp. plantarum* LP95 strain which possesses antimicrobial, β-glucosidase activities and exopolysaccharide (EPS) production capacity, as a starter to obtain fermented organic soymilk without the use of stabilizers, thickeners, and acidity regulators.

Therefore, the main aim of our study was to evaluate the antioxidant capacity of fermented soymilk and its enrichment in isoflavone aglycone.

Furthermore, the fermented soymilk was characterized by its flow behavior and apparent viscosity.

2. Materials and Methods

2.1. Bacterial Strain

In this study, *Lp. plantarum* LP95 (Genbank accession number: OM033654), isolated from fermented pollen (bee bread), was cultured at 37 °C in MRS medium (de Man, Rogosa and Sharpe Broth, Oxoid Ltd., Basingstoke, Hampshire, UK). After 12 h, bacterial cells were harvested by centrifugation (10,000 rpm, 5 min, 4 °C) and washed with sterile saline solution (0.9% *w/v* NaCl). The supernatant was discarded, and the cell pellet was recovered, resuspended in saline solution and properly diluted to adjust cell concentration before being used as the starter.

2.2. Chemicals

All chemicals were purchased from Sigma (Sigma-Aldrich, Merck KGaA, Darmstadt, Germany). Eluents and water for analytical analysis were purified using a Milli-Q system (Millipore, Burlington, MA, USA).

2.3. Preparation of Soymilk

Handcrafted soymilk was prepared according to Wang et al. [26] with some modification by soaking 400 g of mature yellow organic soybeans (*Glycine max*, marketed by BV&FdL S.r.l., Bentivoglio, Italy) in 1200 mL of sterile distilled water for 12 h.

Subsequently, soybeans were drained, dried, and blended with 1400 mL of water. The resulting product was subjected to blanching (90 °C, 10 min), afterward filtered and sterilized at 121 °C for 15 min, and finally cooled for the inoculum phase. Next, the entire mass was homogenized by stirring before adding the microbial starter. After the starter inoculation, soymilk was portioned into three different sterile glass containers (200 mL) for subsequent experimental and analytical steps.

2.4. Screening for Optimum Temperature

To optimize the fermentation phase, preliminary tests were carried out inoculating the obtained soymilk with *Lp. plantarum* LP95 at a concentration of 10^6 CFU/mL at three temperatures (20 °C, 28 °C, and 37 °C). The pH and viable cell counts were monitored at 0, 3, 6, 9, 24, 32, 48, and 72 h.

2.5. Production of Fermented Soymilk

According to the preliminary screening for the optimum temperature, the fermentation phase was conducted at 37 °C, inoculating the soymilk with *Lp. plantarum* LP95 (10^8 CFU/mL). After 24 h, the clot was broken, and the final product was stored at 4 °C. Samples were analyzed immediately and every 7 days until the 49th day.

2.6. Determination of Water-Holding Capacity (WHC)

WHC is defined as the capacity of soymilk to retain all or part of its water, and its determination was assayed according to the method described by Li et al. [5]. Briefly, fermented soymilk (30 g) was centrifuged at 8000 rpm at 4 °C for 15 min. The supernatant was collected, weighed, and the WHC was calculated using Equation (1):

$$WHC\ (\%) = \left(1 - \frac{W_1}{W_2}\right) \times 100 \tag{1}$$

W_1: weight of supernatant after centrifugation (g); W_2: fermented soymilk weight (g).

2.7. Flow Behavior and Apparent Viscosity

The flow behavior of the fermented soymilk was evaluated by a rotational rheometer (Haake Mars III, Thermo Scientific, Karlsruhe, DE) with a plate–plate measuring system (20 mm in diameter, 1 mm gap distance). Flow curves were generated by varying the shear rate from 0.001 to 300 s^{-1} over 2 min, recording the shear stress and viscosity values at the temperature of 37 °C (during the fermentation step) and 4 °C (during storage). The temperature was controlled by a heating and cooling system (Phoenix II, Thermo Scientific, Karlsruhe, Germany) combined with a Peltier system [27,28]. In addition, the apparent viscosity was obtained from flow curves at a shear rate of 50 s^{-1} at 4 °C [29]. Non-inoculated soymilk was used as the control.

2.8. Chemical Analysis

For subsequent chemical analyses, soymilk samples were frozen at −40 °C and then freeze-dried under vacuum at 15 Pa in a freeze-drier Genesis 25 ES dryer (VirTis Genesis 25ES, SP Industries Inc., Gardiner, NY, USA) for 48 h (maximum shelf temperature +20 °C). After freeze-drying, the samples were stored at room temperature. The freeze-drying treatment was necessary to guarantee the stability of the soymilk for the subsequent analytical steps. The moisture content (%) of the freeze-dried soymilk was determined according to the method described by Jung et al. [30].

2.8.1. Determination of Isoflavones

The content of isoflavones in soymilk during the fermentation stage was determined by HPLC, using a Dionex (Dionex, Sunnyvale, CA, USA) Ultimate 3000 quaternary pump equipped with a degasser, thermostated column compartment, and a diode array detector (DAD). Soymilk was sampled at 0, 3, 6, 9, and 24 h from the beginning of fermentation and subjected to freeze-drying. Then, an aliquot of 500 mg freeze-dried soymilk was subjected to a methanolic extraction, according to the method described by Fahmi et al. [31]. Briefly, 25 µL of methanolic extract was injected into a reversed-phase column (Phenomenex Luna 5 µm C18 100 Å 250 × 4.6 mm) protected with Security Guard Phenomenex (Phenomenex

Inc., Torrance, CA, USA). Elution was performed at a flow rate of 1 mL/min at 35 °C, using a solution of phosphoric acid 0.1% (eluent A) and absolute methanol (eluent B), setting the following gradient elution profile expressed as %B: 0 min 28%, 2 min 37%, 30 min 73%. Isoflavones were detected at a wavelength of 254 nm. The identification and quantification were performed using Chromeleon software ver. 6.80 through the retention times and calibration curves of daidzein, daidzin, genistein, genistin, glycitein, and glycitin, as external standards. Results were expressed as mg/100 g dry weight (D.W.).

2.8.2. Antioxidant Activity (ABTS Assay)

The 2,2′-Azino-bis(3-ethylbenzothiazoline-6-sulfonic acid) diammonium salt radical cation (ABTS$^{\bullet}$+) was used for estimating the total antioxidant activity (TAA), according to the method described by Re et al. [32]. Briefly, ABTS was dissolved in water to a 7 mM concentration. ABTS radical cations (ABTS$^{\bullet}$+) were produced by reacting the ABTS methanol solution with 2.45 mM potassium persulfate. The ABTS$^{\bullet}$+ solution was diluted with cold pure methanol up to an optical density (OD) of 0.700 at 745 nm. An aliquot of 20 mg of freeze-dried soymilk sample was subjected to methanolic extraction. After 2 h, 100 µL of methanolic extract was mixed with 900 µL of the ABTS$^{\bullet}$+ solution, and the decrease in absorbance was recorded using a BioSpectrometer (Eppendorf, Hamburg, Germany). Trolox was used as standard for the calibration curve. The TAA in soymilk was assayed at 0, 3, 6, 9, and 24 h after inoculum and expressed as the Trolox equivalent antioxidant capacity (TEAC; mg Trolox Eq./100 g D.W.).

2.8.3. Thiobarbituric Acid Reactive Substance (TBARS) Assay

The TBARS assay was used to estimate lipid peroxidation using malondialdehyde (MDA) as a marker. To 40 mg of freeze-dried soymilk, 1 mL of ethanol/water solution (80:20 *v/v*) was added, and the suspension was vigorously shaken. Then, samples were centrifuged at 13,000 rpm, and the supernatant was used to generate the malondialdehyde–thiobarbituric acid (MDA-TBA) complex, according to the method described by Hodges et al. [33], with some modifications. Absorbances were detected at 532, 440, and 600 nm, using a BioSpectrometer (Eppendorf, Hamburg, Germany). An aqueous solution of malondialdehyde bis(dimethyl acetal) was used as a standard for the calibration curve. The quantities of MDA during the fermentation phase were calculated using the modified equation reported by Landi [34].

2.9. Statistical Analysis

Statistical analysis was performed by analysis of variance (ANOVA), followed by Tukey's multiple comparisons, while a *t*-test was used for the analysis of the significance of the apparent viscosity during the fermentation stage, using SPSS Statistics 21 software (IBM Corp., Armonk, NY, USA). The calculations were performed on data obtained from the analysis carried out on three independent aliquots (n = 3) obtained from a single initial mass of soymilk after starter microbial inoculation. All data were expressed as a mean $\pm$ standard deviation ($\pm$SD).

3. Results
3.1. Preliminary Fermentation Trial

Although the fermentation kinetics have a similar pattern, preliminary tests revealed, as shown in Figure 1 and Tables S1 and S2 (Supplementary Materials), that 37 °C was the optimal temperature for the fermentation of soymilk using *Lp. plantarum* LP95. In detail, at this temperature, the viable cell counts increased progressively from 7.32 log CFU/mL to a value greater than 9 log CFU/mL, after 9 h of fermentation. Moreover, as shown in Figure 1, after 9 h at 37 °C, the fermented soymilk reached values below pH 5.0, lower than those recorded in tests at 20 °C (pH 5.98) and 28 °C (pH 5.31).

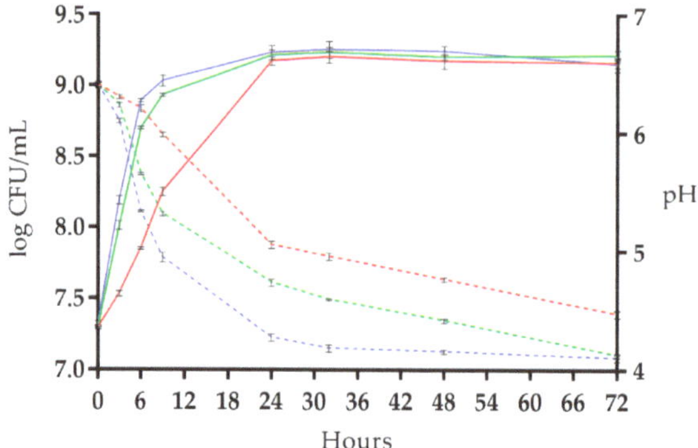

Figure 1. Viable cell count (log CFU/mL) of *Lp. plantarum* LP95 (continuous line) and pH values (dashed line) during the fermentation of soymilk at 20 °C (red line), 28 °C (green line), and 37 °C (blue line). Error bars indicate the standard deviation of the means (*n* = 3).

3.2. Microbiological and Physicochemical Parameters during Fermentation and Storage

Figure 2 shows the viable cell count and pH values using *Lp. plantarum* LP95 in soymilk during the fermentation and storage stages (see also Table S3). As inferred from the viable cell count of *Lp. plantarum* LP95, it reached a value of approximately 9 log CFU/mL during the fermentation phase, which remained constant over the following seven days. After 7 days, a fast decrease in viable cell count occurred, followed by a slower decrease, starting on the 28th day. However, the viable cell count remained above 7 log CFU/mL for up to 49 days of storage. Regarding the pH value of fermented soymilk, an acidification process was evident during fermentation (from 6.5 to 4.2), while during storage, the pH value remained unchanged. In addition, the WHC of fermented soymilk was determined after 24 h, and it reached a value of 43.1 ± 0.3%.

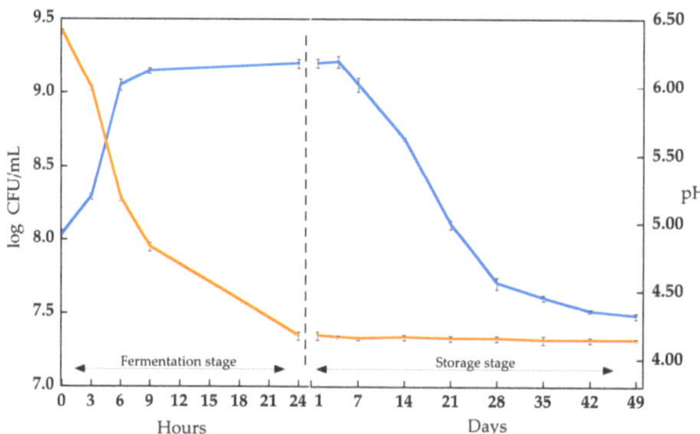

Figure 2. Viable cell count (log CFU/mL) of *Lp. plantarum* LP95 (blue line) and pH values (orange line) during soymilk fermentation (37 °C; 24 h) and fermented soymilk storage (4 °C; 49 days). Error bars indicate the standard deviation of the means (*n* = 3).

3.3. Flow Behavior and Apparent Viscosity of Fermented Soymilk

It is well-known that the correlation between the shear stress (τ) and the shear rate ($\dot{\gamma}$) defines the flow behavior of a sample, which can be classified as Newtonian behavior if the viscosity (η) does not vary with deformation or non-Newtonian behavior when the viscosity varies as a function of the applied shear rate or shear stress. In this study, the apparent viscosity of fermented soymilk was measured to give preliminary rheological information concerning fermented soymilk. The flow behavior of fermented soymilk during both fermentation (37 °C) and storage (4 °C) stages was evaluated, and the obtained curves are depicted in Figure 3A,B, respectively. In detail, all samples showed a non-Newtonian shear-thinning behavior, so the apparent viscosity decreased when the shear rate increased. In Figure 3A, it is possible to observe how fermentation led to changes in the soymilk, with a visible increase in viscosity after 24 h of fermentation. During the storage stage (Figure 3B), a slight decrease in the viscosity of the sample was observed, starting after 21 days. The consistency coefficient (K) and flow behavior index (n) were obtained as described by Hamet et al. [35], fitting the flow curves with the Ostwald–de Waele model, and the results are reported in Table S4. Moreover, the apparent viscosities of the fermented soymilk at a shear rate of 50 s^{-1} during the fermentation and storage stages are shown in Figure 4A,B, respectively, while the numerical data are reported in Tables S5 and S6. In detail, after 24 h of fermentation, the values of apparent viscosity were different compared to the sample at 0 h. However, after clot rupture at 7 days of storage (4 °C), the highest level of apparent viscosity occurred, while at 21 days, a rapid decline was observed.

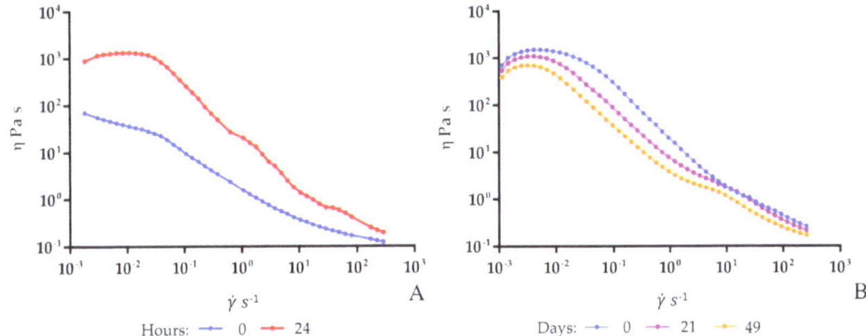

Figure 3. Viscosity curves of soymilk during fermentation using *Lp. plantarum* LP95 as a starter at 37 °C (**A**) and fermented soymilk during storage at 4 °C (**B**) at different times: viscosity (η) as a function of the shear rate (γ̇).

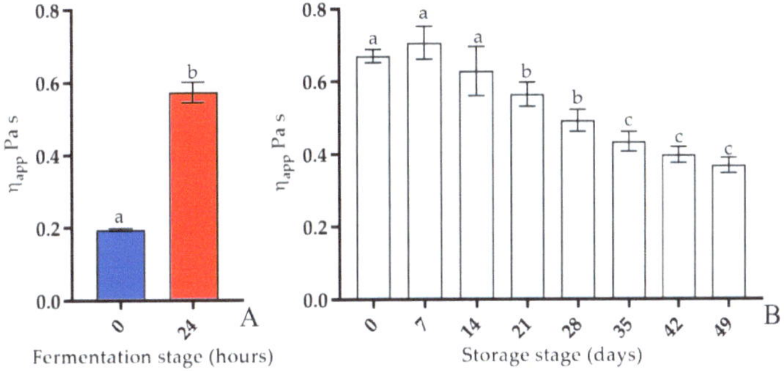

Figure 4. Apparent viscosity at 50 s^{-1} of soymilk during fermentation using *Lp. plantarum* LP95 as a starter at 37 °C (**A**) and fermented soymilk during storage at 4 °C (**B**) at different times. All values are expressed as the mean ± standard deviation ($n = 3$). Different lowercase letters in each bar indicate significant differences ($p < 0.05$).

3.4. Bioconversion of Isoflavones during Fermentation

Soymilk fermentation, using *Lp. plantarum* LP95 as a starter, significantly increased isoflavones in the aglycone-form. In detail, the results reported in Table 1 showed that the glycosylated isoflavones in soymilk, immediately after the inoculum (0 h), contained daidzin, glycitin, and genistin (11.26 ± 0.27, 39.34 ± 1.00, 40.39 ± 1.27 μmol/100 g D.W.), while the aglycone derivatives were daidzein, glycitein, and genistein (25.64 ± 0.02,

1.58 ± 0.18, and 21.58 ± 0.39 μmol/100 g D.W.). The content of daidzein, glycitein, and genistein increased after 24 h of fermentation, reaching the highest levels of 48.45 ± 1.30, 5.10 ± 0.16, and 56.35 ± 1.02 μmol/100 g D.W., respectively. The moisture content (%) of freeze-dried soymilk was 0.912 ± 0.008.

Table 1. Deglycosylation of soymilk isoflavones into aglycones during the fermentation stage of soymilk using *Lp. plantarum* LP95 as a starter. All values are expressed as the mean ± standard deviation (*n* = 3).

Isoflavones (μmol/100 g D.W.)	Fermentation Stage (hours)				
	0	3	6	9	24
Daidzin	11.26 ± 0.27 [a]	11.08 ± 0.26 [a]	0.57 ± 0.02 [b]	0.00 ± 0.00 [c]	0.00 ± 0.00 [c]
Glycitin	39.34 ± 1.00 [a]	37.78 ± 0.55 [b]	1.45 ± 0.10 [c]	0.00 ± 0.00 [d]	0.00 ± 0.00 [d]
Genistin	40.39 ± 1.27 [a]	40.11 ± 0.02 [a]	4.60 ± 0.09 [b]	0.00 ± 0.00 [c]	0.00 ± 0.00 [c]
Daidzein	25.64 ± 0.02 [b]	25.76 ± 1.11 [b]	47.71 ± 0.72 [a]	48.45 ± 1.74 [a]	48.45 ± 1.30 [a]
Glycitein	1.58 ± 0.18 [d]	1.65 ± 0.04 [d]	2.61 ± 0.17 [c]	3.69 ± 0.01 [b]	5.10 ± 0.16 [a]
Genistein	21.58 ± 0.39 [c]	21.77 ± 0.80 [c]	51.14 ± 0.84 [b]	54.75 ± 2.11 [a]	56.35 ± 1.02 [a]

Different lowercase letters (a–d) in each row indicate significant differences ($p < 0.05$).

3.5. Changes in Antioxidant Activity during Fermentation

During the fermentation stage, the measured values of TAA were significantly different compared to the soymilk immediately after inoculum (0 h). In particular, the results illustrated in Figure 5A and the numerical data reported in Table S7 have shown that the TEAC at 0, 3, 6, 9, and 24 h was 123.35 ± 1.58, 110.30 ± 6.03, 121.42 ± 5.26, 105.79 ± 2.34, and 108.76 ± 3.25 mg Trolox Eq./100 g D.W., respectively. Specifically, the percentage decrease in TEAC was already about 10% at 3 h; at 6 h, it was only 2%, and at 9 and 24 h, it was around 13% compared to the sample at the beginning of fermentation.

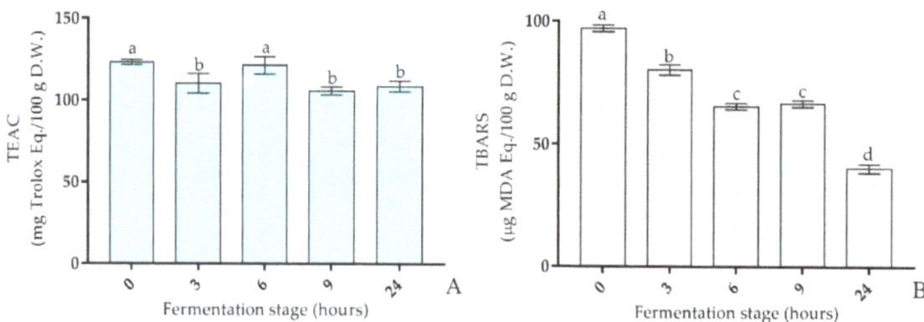

Figure 5. Variations in TEAC expressed as mg Trolox Eq./100 g D.W. (**A**) and TBARS expressed as μg MDA Eq./100 g D.W. (**B**) in soymilk during fermentation using *Lp. plantarum* LP95 as a starter. All values are expressed as the mean ± standard deviation (*n* = 3). Different lowercase letters in each bar indicate significant differences ($p < 0.05$).

3.6. Variation in MDA during Fermentation

The evolution of MDA levels during fermentation was assayed. In particular, the detected values showed significant differences during the fermentation stage. As reported in Figure 5B and Table S8, the highest level of MDA concentration was detected at 0 h (97.11 ± 1.33 μg MDA Eq./100 g D.W.); however, it significantly decreased at 3 h (80.26 ± 2.12 μg MDA Eq./100 g D.W.), while between 6 and 9 h (65.53 ± 1.30; 66.68 ± 1.35 μg MDA Eq./100 g D.W., respectively), there were no significant differences between the detected values. Finally, at 24 h, the MDA value reached its lowest level (40.17 ± 1.88 μg MDA Eq./100 g D.W.).

4. Discussion

The formulation of functional foods based on the use of fermentative starters that also possess probiotic properties is becoming an aspect of growing interest in the field of food biotechnology. In this investigation, *Lp. plantarum* LP95 showed good fermentation performance and some properties that overall improved the biofunctional characteristics of unfermented soymilk. It was also inferred that carbohydrates naturally present in soymilk, such as stachyose, raffinose, sucrose, glucose, and fructose, act as substrates for fermentation by LAB, leading to a lowering of pH due to the release of lactic acid [36,37]. The used starter LP95, after 24 h of fermentation at 37 °C, produced a compact clot. After clop rupture, the determined WHC was similar to data obtained by other authors using different LAB strains in soybean-based products [38–40].

The observed viscosity changes that occurred during fermentation were also attributable to protein coagulation phenomena [41]. In fact, as described by Xu et al. [42], the protease activity of *Lp. plantarum* might partially hydrolyze soy proteins by lowering soymilk pH. This negatively charges the denatured soy protein groups, leading to progressive coagulation. In addition, the ability to produce EPSs from *Lp. plantarum* LP95 [25] could have a positive influence on the coagulation phase [5]. To clarify this aspect, comparative studies will be conducted in the future on soymilk fermented by *Lp. plantarum* LP95 using non-EPS-producing LAB strains or coagulation with calcium chloride [43,44] as references, with and without the addition of EPSs [45].

The study of the apparent viscosity of samples is important to evaluate how fermentation could affect the sensorial quality of the final product. In general, the properties of fermented products are influenced by different factors, such as the type of microorganisms used and the duration and temperature at which fermentation takes place [46].

Samples analyzed in this study possessed a non-Newtonian shear-thinning behavior, where the apparent viscosity decreased when the shear rate increased. This behavior was also observed for other vegetable-fermented types of milk [35,39,47]. The flow behavior of the fermented product was then followed during the refrigerated storage phase. The same behavior observed during fermentation was also found during storage for up to the 14th day, with a drop in viscosity values in the following days. The physicochemical changes could also affect the texture and mouthfeel of the final product, so the apparent viscosity at a shear rate of 50 s^{-1} was obtained, as it can be related to the oral perception during the consumption of semisolid products [48,49].

The ability of this *Lactobacillus* to increase the viscosity of soymilk could also be related to the probable presence of EPSs, in agreement with previous studies [35,50–52]. In this respect, we can report that during the storage of samples containing fermented soymilk stored at 4 °C, the apparent viscosity remained substantially constant.

The bioavailability of glycosidic isoflavones first requires enzymatic hydrolysis of the sugar moiety, which occurs in the microvillous zone of enterocytes in the intestine [53]. Several LAB strains show specific β-glucosidase enzyme activity that hydrolyzes isoflavone glycosides to aglycones [54]. In detail, the fermentation of soymilk using *Lp. plantarum* LP95 as a starter culture confirmed the complete biotransformation of isoflavone glucosides into their aglyconic forms, which occurred after 6 h during the fermentation. Vegetable polyphenols introduced into a diet with soymilk act as an exogenous antioxidant system. In fact, they have an important antioxidant activity against the negative effects of free radicals, preventing oxidative stress and cell damage [55]. Data obtained on the total antioxidant capacity of the fermented soymilk samples were determined by the ABTS radical scavenging. In the first 3 h of the fermentation process, there was a slight reduction in the antioxidant activity, probably due to a depletion of antioxidant molecules by the bacterial cells used to mitigate oxidative stress [20]. After six hours, there was an increase in the antioxidant activity of the fermented soymilk, probably attributable to the deglycosylation by bacterial bioconversion of isoflavone glycosides. After, the antioxidant activity returns to a decrease, probably due to the scavenging of isoflavonoids [56] produced during bacterial metabolism.

Furthermore, an important parameter to monitor during food processing is the oxidative state of the final product. The oxidation of lipids, which has been extensively studied, proceeds via a free-radical chain mechanism, giving rise to numerous by-products. The main one of these is MDA, which is, in fact, a marker of oxidative damage in biological systems [57]. Monitoring lipid oxidation is important, as this process leads to the loss of nutritional factors and the development of toxic compounds in food, causing various diseases, such as atherosclerosis [58]. The detection of TBARS is frequently used to determine lipid oxidation. However, as reported by Ghani et al. [59], this measure does not indicate the antioxidant mechanism of action. Despite this potential limitation, there are features of the TBARS test that make it useful as a complement to other widely used screening tests, such as the TAA. In our study, the TBARS assay showed a progressive reduction in MDA levels during the fermentation of soymilk using *Lp. plantarum* LP95. Additionally, this trend coincides with the increase in the amount of detected isoflavone aglycones, and this may suggest a correlation between the two events.

5. Conclusions

In this article, we presented a study on the *Lp. plantarum* LP95 strain, which has good potential for improving the quality of fermented soymilk to develop functional food products. The enhancement of the antioxidant activity of soymilk has been shown.

The increase in isoflavone aglycone content during fermentation was a result of β-glucosidase activity toward isoflavone glucosides. This process allowed the obtaining of a soy beverage with enhanced antioxidant capacity able to contribute to the improvement in the health and nutritional status of consumers.

Furthermore, we also report that without using chemical additives, the soymilk obtained and stored at 4 °C has a good shelf life of up to 14 days.

Based on this evidence, *Lp. plantarum* LP95 appears to be a promising candidate as a starter culture to obtain a fermented soymilk rich in isoflavone aglycones and with enhanced health-promoting properties.

To validate this application, future investigations will be conducted on the shelf life and on the organoleptic and nutritional qualities of soymilk fermented by this bacterium.

Supplementary Materials: The following supporting information can be downloaded at https://www.mdpi.com/article/10.3390/antiox12071442/s1, Table S1: Viable cell count (log CFU/mL) of *Lp. plantarum* LP95 during the preliminary fermentation trial of soymilk at 20 °C, 28 °C, and 37 °C, Table S2: Changes in pH during the preliminary fermentation trial of soymilk with *Lp. plantarum* LP95 at 20 °C, 28 °C, and 37 °C, Table S3: Changes in viable cell count (log CFU/mL) and pH in soymilk during the fermentation (37 °C) and storage stage (4 °C) using *Lp. plantarum* LP95 as a starter, Table S4: Values of consistency coefficient (K) and flow behavior index (n) obtained by fitting the flow curves with the Ostwald–de Waele model, Table S5: Apparent viscosity (Pa s) at $50\,\mathrm{s}^{-1}$ of soymilk during the fermentation stage at 37 °C using *Lp. plantarum* LP95 as a starter, Table S6: Apparent viscosity (Pa s) at $50\,\mathrm{s}^{-1}$ of fermented soymilk during the storage stage at 4 °C using *Lp. plantarum* LP95 as a starter, Table S7: Variations in total antioxidant activity (TEAC; mg Trolox Eq. $100\,\mathrm{g}^{-1}$ D.W.) in soymilk during the fermentation stage using *Lp. plantarum* LP95 as a starter, Table S8: Variations in thiobarbituric acid reactive substance (TBARS; µg MDA Eq. $100\,\mathrm{g}^{-1}$ D.W.) in soymilk during the fermentation stage using *Lp. plantarum* LP95 as a starter.

Author Contributions: Conceptualization, C.D.M. and M.I.; methodology, C.D.M., M.I., M.C., F.L. (Francesco Lopez), G.P., A.F., and B.T.; software, M.C. and F.L. (Francesco Letizia); validation, G.A., F.L. (Francesco Lopez), and M.C.; formal analysis, F.L. (Francesco Letizia), M.C., C.D.M., F.L. (Francesco Lopez), G.P., B.T., and A.F.; writing—original draft preparation, F.L. (Francesco Letizia), C.D.M., and M.I.; writing—review and editing; F.L. (Francesco Letizia), M.I., A.F., G.P., C.D.M., and M.C.; project administration, C.D.M. and M.I. All authors have read and agreed to the published version of the manuscript.

Funding: This research received no external funding.

Institutional Review Board Statement: Not applicable.

Informed Consent Statement: Not applicable.

Data Availability Statement: The data are contained within the article or in the Supplementary Material.

Conflicts of Interest: The authors declare no conflict of interest.

References

1. Shah, A.M.; Tarfeen, N.; Mohamed, H.; Song, Y. Fermented Foods: Their Health-Promoting Components and Potential Effects on Gut Microbiota. *Fermentation* **2023**, *9*, 118. [CrossRef]
2. Messina, M. Soy and Health Update: Evaluation of the Clinical and Epidemiologic Literature. *Nutrients* **2016**, *8*, 754. [CrossRef] [PubMed]
3. Kumari, M.; Kokkiligadda, A.; Dasriya, V.; Naithani, H. Functional Relevance and Health Benefits of Soymilk Fermented by Lactic Acid Bacteria. *J. Appl. Microbiol.* **2022**, *133*, 104–119. [CrossRef] [PubMed]
4. Van Dyck, S. 3-The Impact of Singlet Oxygen on Lipid Oxidation in Foods. In *Oxidation in Foods and Beverages and Antioxidant Applications*; Decker, E.A., Ed.; Woodhead Publishing: Sawston, UK, 2010; pp. 57–75. ISBN 978-1-84569-648-1.
5. Li, C.; Li, W.; Chen, X.; Feng, M.; Rui, X.; Jiang, M.; Dong, M. Microbiological, Physicochemical and Rheological Properties of Fermented Soymilk Produced with Exopolysaccharide (EPS) Producing Lactic Acid Bacteria Strains. *LWT-Food Sci. Technol.* **2014**, *57*, 477–485. [CrossRef]
6. Subrota, H.; Shilpa, V.; Brij, S.; Vandna, K.; Surajit, M. Antioxidative Activity and Polyphenol Content in Fermented Soy Milk Supplemented with WPC-70 by Probiotic Lactobacilli. *Int. Food Res. J.* **2013**, *20*, 2125.
7. Theodoropoulos, V.C.T.; Turatti, M.A.; Greiner, R.; Macedo, G.A.; Pallone, J.A.L. Effect of Enzymatic Treatment on Phytate Content and Mineral Bioacessability in Soy Drink. *Food Res. Int.* **2018**, *108*, 68–73. [CrossRef]
8. Zhao, D.; Shah, N.P. Changes in Antioxidant Capacity, Isoflavone Profile, Phenolic and Vitamin Contents in Soymilk during Extended Fermentation. *LWT-Food Sci. Technol.* **2014**, *58*, 454–462. [CrossRef]
9. Chun, J.; Kim, G.M.; Lee, K.W.; Choi, I.D.; Kwon, G.; Park, J.; Jeong, S.; Kim, J.; Kim, J.H. Conversion of Isoflavone Glucosides to Aglycones in Soymilk by Fermentation with Lactic Acid Bacteria. *J. Food Sci.* **2007**, *72*, M39–M44. [CrossRef]
10. Yu, X.; Meenu, M.; Xu, B.; Yu, H. Impact of Processing Technologies on Isoflavones, Phenolic Acids, and Antioxidant Capacities of Soymilk Prepared from 15 Soybean Varieties. *Food Chem.* **2021**, *345*, 128612. [CrossRef]
11. Hu, C.; Wong, W.-T.; Wu, R.; Lai, W.-F. Biochemistry and Use of Soybean Isoflavones in Functional Food Development. *Crit. Rev. Food Sci. Nutr.* **2020**, *60*, 2098–2112. [CrossRef]
12. Zaheer, K.; Humayoun Akhtar, M. An Updated Review of Dietary Isoflavones: Nutrition, Processing, Bioavailability and Impacts on Human Health. *Crit. Rev. Food Sci. Nutr.* **2017**, *57*, 1280–1293. [CrossRef]
13. Walle, T.; Browning, A.M.; Steed, L.L.; Reed, S.G.; Walle, U.K. Flavonoid Glucosides Are Hydrolyzed and Thus Activated in the Oral Cavity in Humans. *J. Nutr.* **2005**, *135*, 48–52. [CrossRef]
14. Otieno, D.; Shah, N. Endogenous β-glucosidase and β-galactosidase Activities from Selected Probiotic Micro-organisms and Their Role in Isoflavone Biotransformation in Soymilk. *J. Appl. Microbiol.* **2007**, *103*, 910–917. [CrossRef]
15. Setchell, K.D.; Brown, N.M.; Zimmer-Nechemias, L.; Brashear, W.T.; Wolfe, B.E.; Kirschner, A.S.; Heubi, J.E. Evidence for Lack of Absorption of Soy Isoflavone Glycosides in Humans, Supporting the Crucial Role of Intestinal Metabolism for Bioavailability. *Am. J. Clin. Nutr.* **2002**, *76*, 447–453. [CrossRef]
16. Baú, T.; Garcia, S.; Ida, E. Changes in Soymilk during Fermentation with Kefir Culture: Oligosaccharides Hydrolysis and Isoflavone Aglycone Production. *Int. J. Food Sci. Nutr.* **2015**, *66*, 845–850. [CrossRef]
17. Hur, H.-G.; Lay Jr, J.O.; Beger, R.D.; Freeman, J.P.; Rafii, F. Isolation of Human Intestinal Bacteria Metabolizing the Natural Isoflavone Glycosides Daidzin and Genistin. *Arch. Microbiol.* **2000**, *174*, 422–428. [CrossRef]
18. Donkor, O.; Shah, N.P. Production of β-Glucosidase and Hydrolysis of Isoflavone Phytoestrogens by *Lactobacillus acidophilus*, *Bifidobacterium lactis*, and *Lactobacillus casei* in Soymilk. *J. Food Sci.* **2008**, *73*, M15–M20. [CrossRef]
19. Li, C.; Fan, Y.; Li, S.; Zhou, X.; Park, K.-Y.; Zhao, X.; Liu, H. Antioxidant Effect of Soymilk Fermented by *Lactobacillus plantarum* HFY01 on D-Galactose-Induced Premature Aging Mouse Model. *Front. Nutr.* **2021**, *8*, 667643. [CrossRef]
20. Feng, T.; Wang, J. Oxidative Stress Tolerance and Antioxidant Capacity of Lactic Acid Bacteria as Probiotic: A Systematic Review. *Gut Microbes* **2020**, *12*, 1801944. [CrossRef]
21. EFSA Panel on Additives and Products or Substances Used in Animal Feed (FEEDAP); Rychen, G.; Aquilina, G.; Azimonti, G.; Bampidis, V.; de Bastos, M.L.; Bories, G.; Chesson, A.; Cocconcelli, P.S.; Flachowsky, G.; et al. Guidance on the Characterisation of Microorganisms Used as Feed Additives or as Production Organisms. *EFSA J.* **2018**, *16*, e05206. [CrossRef]
22. Behera, S.S.; Ray, R.C.; Zdolec, N. *Lactobacillus plantarum* with Functional Properties: An Approach to Increase Safety and Shelf-Life of Fermented Foods. *BioMed Res. Int.* **2018**, *2018*, 9361614. [CrossRef] [PubMed]
23. Iorizzo, M.; Lombardi, S.J.; Macciola, V.; Testa, B.; Lustrato, G.; Lopez, F.; De Leonardis, A. Technological Potential of *Lactobacillus* Strains Isolated from Fermented Green Olives: In Vitro Studies with Emphasis on Oleuropein-Degrading Capability. *Sci. World J.* **2016**, *2016*, 1917592. [CrossRef] [PubMed]
24. Iorizzo, M.; Testa, B.; Lombardi, S.J.; García-Ruiz, A.; Muñoz-González, C.; Bartolomé, B.; Moreno-Arribas, M.V. Selection and Technological Potential of *Lactobacillus plantarum* Bacteria Suitable for Wine Malolactic Fermentation and Grape Aroma Release. *LWT* **2016**, *73*, 557–566. [CrossRef]
25. Letizia, F.; Albanese, G.; Testa, B.; Vergalito, F.; Bagnoli, D.; Di Martino, C.; Carillo, P.; Verrillo, L.; Succi, M.; Sorrentino, E. In Vitro Assessment of Bio-Functional Properties from *Lactiplantibacillus plantarum* Strains. *Curr. Issues Mol. Biol.* **2022**, *44*, 2321–2334. [CrossRef]

26. Wang, Y.-C.; Yu, R.-C.; Chou, C.-C. Antioxidative Activities of Soymilk Fermented with Lactic Acid Bacteria and Bifidobacteria. *Food Microbiol.* **2006**, *23*, 128–135. [CrossRef]
27. Cofelice, M.; Cinelli, G.; Lopez, F.; Di Renzo, T.; Coppola, R.; Reale, A. Alginate-Assisted Lemongrass (*Cymbopogon nardus*) Essential Oil Dispersions for Antifungal Activity. *Foods* **2021**, *10*, 1528. [CrossRef]
28. Cofelice, M.; Messia, M.C.; Marconi, E.; Cuomo, F.; Lopez, F. Effect of the Xanthan Gum on the Rheological Properties of Alginate Hydrogels. *Food Hydrocoll.* **2023**, *142*, 108768. [CrossRef]
29. Koksoy, A.; Kilic, M. Use of Hydrocolloids in Textural Stabilization of a Yoghurt Drink, Ayran. *Food Hydrocoll.* **2004**, *18*, 593–600. [CrossRef]
30. Jung, S.; Murphy, P.A.; Sala, I. Isoflavone Profiles of Soymilk as Affected by High-Pressure Treatments of Soymilk and Soybeans. *Food Chem.* **2008**, *111*, 592–598. [CrossRef]
31. Fahmi, R.; Khodaiyan, F.; Pourahmad, R.; Emam-Djomeh, Z. Effect of Ultrasound Assisted Extraction upon the Genistin and Daidzin Contents of Resultant Soymilk. *J. Food Sci. Technol.* **2014**, *51*, 2857–2861. [CrossRef]
32. Re, R.; Pellegrini, N.; Proteggente, A.; Pannala, A.; Yang, M.; Rice-Evans, C. Antioxidant Activity Applying an Improved ABTS Radical Cation Decolorization Assay. *Free Radic. Biol. Med.* **1999**, *26*, 1231–1237. [CrossRef]
33. Hodges, D.M.; DeLong, J.M.; Forney, C.F.; Prange, R.K. Improving the Thiobarbituric Acid-Reactive-Substances Assay for Estimating Lipid Peroxidation in Plant Tissues Containing Anthocyanin and Other Interfering Compounds. *Planta* **1999**, *207*, 604–611. [CrossRef]
34. Landi, M. Commentary to:"Improving the Thiobarbituric Acid-Reactive-Substances Assay for Estimating Lipid Peroxidation in Plant Tissues Containing Anthocyanin and Other Interfering Compounds" by Hodges et al., Planta (1999) 207: 604–611. *Planta* **2017**, *245*, 1067. [CrossRef]
35. Hamet, M.F.; Piermaria, J.A.; Abraham, A.G. Selection of EPS-Producing *Lactobacillus* Strains Isolated from Kefir Grains and Rheological Characterization of the Fermented Milks. *LWT-Food Sci. Technol.* **2015**, *63*, 129–135. [CrossRef]
36. Garro, M.S.; Font de Valdez, G.; Savoy de Giori, G. Determination of Oligosaccharides in Fermented Soymilk Products by High-Performance Liquid Chromatography. In *Environmental Microbiology: Methods and Protocols*; Walker, J.M., Spencer, J.F.T., Ragout de Spencer, A.L., Eds.; Humana Press: Totowa, NJ, USA, 2004; pp. 135–138. ISBN 978-1-59259-765-9.
37. Brereton, K.R.; Green, D.B. Isolation of Saccharides in Dairy and Soy Products by Solid-Phase Extraction Coupled with Analysis by Ligand-Exchange Chromatography. *Talanta* **2012**, *100*, 384–390. [CrossRef]
38. Baygut, H.; Cais-Sokolińska, D.; Bielska, P.; Teichert, J. Fermentation Kinetics, Microbiological and Physical Properties of Fermented Soy Beverage with Acai Powder. *Fermentation* **2023**, *9*, 324. [CrossRef]
39. Mei, J.; Feng, F.; Li, Y. Effective of Different Homogeneous Methods on Physicochemical, Textural and Sensory Characteristics of Soybean (*Glycine max* L.) Yogurt. *CyTA-J. Food* **2017**, *15*, 21–26.
40. Huang, K.; Liu, Y.; Zhang, Y.; Cao, H.; Luo, D.; Yi, C.; Guan, X. Formulation of Plant-Based Yoghurt from Soybean and Quinoa and Evaluation of Physicochemical, Rheological, Sensory and Functional Properties. *Food Biosci.* **2022**, *49*, 101831. [CrossRef]
41. O' Flynn, T.D.; Hogan, S.A.; Daly, D.F.; O' Mahony, J.A.; McCarthy, N.A. Rheological and Solubility Properties of Soy Protein Isolate. *Molecules* **2021**, *26*, 3015. [CrossRef]
42. Xu, Y.; Ye, Q.; Zhang, H.; Yu, Y.; Li, X.; Zhang, Z.; Zhang, L. Naturally Fermented Acid Slurry of Soy Whey: High-Throughput Sequencing-Based Characterization of Microbial Flora and Mechanism of Tofu Coagulation. *Front. Microbiol.* **2019**, *10*, 1088. [CrossRef]
43. Shun-Tang, G.; Ono, T.; Mikami, M. Incorporation of Soy Milk Lipid into Protein Coagulum by Addition of Calcium Chloride. *J. Agric. Food Chem.* **1999**, *47*, 901–905. [CrossRef] [PubMed]
44. Liu, Z.-S.; Chang, S.K.-C. Effect of Soy Milk Characteristics and Cooking Conditions on Coagulant Requirements for Making Filled Tofu. *J. Agric. Food Chem.* **2004**, *52*, 3405–3411. [CrossRef] [PubMed]
45. Zannini, E.; Waters, D.M.; Coffey, A.; Arendt, E.K. Production, Properties, and Industrial Food Application of Lactic Acid Bacteria-Derived Exopolysaccharides. *Appl. Microbiol. Biotechnol.* **2016**, *100*, 1121–1135. [CrossRef] [PubMed]
46. Peng, X.; Liao, Y.; Ren, K.; Liu, Y.; Wang, M.; Yu, A.; Tian, T.; Liao, P.; Huang, Z.; Wang, H. Fermentation Performance, Nutrient Composition, and Flavor Volatiles in Soy Milk after Mixed Culture Fermentation. *Process Biochem.* **2022**, *121*, 286–297. [CrossRef]
47. Yovanoudi, M.; Dimitreli, G.; Raphaelides, S.; Antoniou, K. Flow Behavior Studies of Kefir Type Systems. *J. Food Eng.* **2013**, *118*, 41–48. [CrossRef]
48. Pang, Z.; Xu, R.; Zhu, Y.; Li, H.; Bansal, N.; Liu, X. Comparison of Rheological, Tribological, and Microstructural Properties of Soymilk Gels Acidified with Glucono-δ-Lactone or Culture. *Food Res. Int.* **2019**, *121*, 798–805. [CrossRef]
49. Arancibia, C.; Bayarri, S.; Costell, E. Effect of Hydrocolloid on Rheology and Microstructure of High-Protein Soy Desserts. *J. Food Sci. Technol.* **2015**, *52*, 6435–6444. [CrossRef]
50. Masiá, C.; Geppel, A.; Jensen, P.E.; Buldo, P. Effect of *Lactobacillus rhamnosus* on Physicochemical Properties of Fermented Plant-Based Raw Materials. *Foods* **2021**, *10*, 573. [CrossRef]
51. Jurášková, D.; Ribeiro, S.C.; Silva, C.C. Exopolysaccharides Produced by Lactic Acid Bacteria: From Biosynthesis to Health-Promoting Properties. *Foods* **2022**, *11*, 156. [CrossRef]
52. Zhao, X.; Liang, Q. EPS-Producing *Lactobacillus plantarum* MC5 as a Compound Starter Improves Rheology, Texture, and Antioxidant Activity of Yogurt during Storage. *Foods* **2022**, *11*, 1660. [CrossRef]
53. Hendrich, S. Bioavailability of Isoflavones. *J. Chromatogr. B Analyt. Technol. Biomed. Life. Sci.* **2002**, *777*, 203–210. [CrossRef]

54. Rekha, C.; Vijayalakshmi, G. Isoflavone Phytoestrogens in Soymilk Fermented with β-Glucosidase Producing Probiotic Lactic Acid Bacteria. *Int. J. Food Sci. Nutr.* **2011**, *62*, 111–120. [CrossRef]

55. De Mello Andrade, J.M.; Fasolo, D. Polyphenol Antioxidants from Natural Sources and Contribution to Health Promotion. In *Polyphenols in Human Health and Disease*; Elsevier: Amsterdam, The Netherlands, 2014; pp. 253–265.

56. Cetinkaya, H.; Kulak, M.; Karaman, M.; Karaman, H.S.; Kocer, F. Flavonoid Accumulation Behavior in Response to the Abiotic Stress: Can a Uniform Mechanism Be Illustrated for All Plants? In *Flavonoids*; Justino, G.C., Ed.; IntechOpen: Rijeka, Croatia, 2017; Chapter 8; ISBN 978-953-51-3424-4.

57. Fernández, J.; Pérez-Álvarez, J.A.; Fernández-López, J.A. Thiobarbituric Acid Test for Monitoring Lipid Oxidation in Meat. *Food Chem.* **1997**, *59*, 345–353. [CrossRef]

58. Spickett, C.; Fedorova, M.; Hoffmann, R.; Forman, H. An Introduction to Redox Balance and Lipid Oxidation. In *Lipid Oxidation in Health and Disease*; CRC Press: Boca Raton, FL, USA, 2015; pp. 1–24.

59. Ghani, M.A.; Barril, C.; Bedgood, D.R., Jr.; Prenzler, P.D. Measurement of Antioxidant Activity with the Thiobarbituric Acid Reactive Substances Assay. *Food Chem.* **2017**, *230*, 195–207. [CrossRef]

 *antioxidants*

Article

The Radical Scavenging Activities and Anti-Wrinkle Effects of Soymilk Fractions Fermented with *Lacticaseibacillus paracasei* MK1 and Their Derived Peptides

Sulhee Lee [1], Sang-Pil Choi [1], Huijin Jeong [2], Won Kyu Yu [3], Sang Won Kim [3] and Young-Seo Park [2],*

[1] Kimchi Functionality Research Group, World Institute of Kimchi, Gwangju 61755, Republic of Korea; slee@wikim.re.kr (S.L.); spchoi@wikim.re.kr (S.-P.C.)

[2] Department of Food Science and Biotechnology, Gachon University, Seongnam 13120, Republic of Korea; gmlwls2218@gachon.ac.kr

[3] Yonsei University Dairy, Asan 31419, Republic of Korea; wonkyu@yonseidairy.com (W.K.Y.); ksw@yonseidairy.com (S.W.K.)

* Correspondence: ypark@gachon.ac.kr; Tel.: +82-31-750-5378

Abstract: Soybean-derived peptides exert several beneficial effects in various experimental models. However, only a few studies have focused on the radical scavenging and anti-wrinkle effects of soymilk-derived peptides produced via different processes, such as fermentation, enzymatic treatment, and ultrafiltration. Therefore, in this study, we investigated the radical scavenging and antiwrinkle effects of soymilk fractions produced using these processes. We found that 50SFMKUF5, a 5 kDa ultrafiltration fraction fermented with *Lacticaseibacillus paracasei* MK1 after flavourzyme treatment, exhibited the highest radical scavenging activity using the 2,2-diphenyl-1-picrylhydrazyl radical scavenging assay as well as potent anti-wrinkle effects assessed by type 1 procollagen production and tumor necrosis factor-α production in ultraviolet B (UVB)-treated human dermal fibroblasts and HaCaT keratinocytes. To identify potential bioactive peptides, candidate peptides were synthesized, and their anti-wrinkle effects were assessed. APEFLKEAFGVN (APE), palmitoyl-APE, and QIVTVEGGLSVISPK peptides were synthesized and used to treat UVB-irradiated fibroblasts, HaCaT keratinocytes, and α-melanocyte-stimulating hormone-induced B16F1 melanoma cells. Among these peptides, Pal-APE exerted the strongest effect. Our results highlight the potential of soymilk peptides as anti-aging substances.

Keywords: lactic acid bacteria; radical scavenging activity; anti-wrinkle; anti-aging; soybean milk; fermentation; peptide

Citation: Lee, S.; Choi, S.-P.; Jeong, H.; Yu, W.K.; Kim, S.W.; Park, Y.-S. The Radical Scavenging Activities and Anti-Wrinkle Effects of Soymilk Fractions Fermented with *Lacticaseibacillus paracasei* MK1 and Their Derived Peptides. *Antioxidants* **2023**, *12*, 1392. https://doi.org/10.3390/antiox12071392

Academic Editors: Stanley Omaye and Dimitrios Stagos

Received: 4 June 2023
Revised: 2 July 2023
Accepted: 4 July 2023
Published: 6 July 2023

1. Introduction

Increasing demand for plant-based protein sources and non-dairy milk alternatives has brought soybean (*Glycine max*) and soymilk to the forefront of nutritional research. Soybeans are valuable protein sources that have been extensively studied for their nutritional benefits. They are considered viable alternatives to animal products, such as milk and meat, and are one of the most consumed protein sources worldwide [1]. Soymilk, a soybean product, is a popular non-dairy milk alternative owing to its low cost, improved functional and nutritional properties, and potential health benefits. Soymilk is a good nutritional medium for the growth and proliferation of microorganisms, indicating its potential as a carrier for probiotics and other bioactive compounds [2].

Soybean-derived peptides are protein fragments produced from soybean proteins via different processes, such as hydrolysis, fermentation, food processing, and gastrointestinal digestion. These methods enable the production of many small peptides with radical scavenging properties from soy protein [3]. Some of these peptides exert numerous beneficial effects, such as antidiabetic, anticancer, hypotensive, anti-inflammatory, and radical scavenging effects, in various experimental models. Soybean peptides have an average molecular weight of approximately 3–10 kDa and high glutamic acid content [4]. Enzymes, bacteria, and types of soy protein affect the composition and production of peptides [5]. The use of microorganisms to ferment substances is a cost-effective approach for the production of bioactive peptides that is extensively employed in the dairy industry to improve the functionality of milk products and byproducts. Several fermented dairy products have functional health benefits linked to bioactive peptides. The majority of these products are obtained using lactic acid bacteria [6]. Furthermore, fermented soybean meal, soy foods, and soymilk exhibit high potency in combating oxidative stress [7–9].

Several studies have highlighted the association between the radical scavenging capacity and the anti-aging effects of different substances [10,11]. Aging is a natural process affecting all living organisms. Skin ages more rapidly than other parts of the body because of direct exposure to environmental factors, such as chemicals and solar ultraviolet (UV) radiation [12]. Excessive exposure to UVB radiation (290–320 nm) causes photoaging, which is characterized by dry and rough skin with wrinkles, irregular pigmentation, and reduced elasticity [13]. Exposure of the skin to UV radiation for an extended period stimulates the production of reactive oxygen species (ROS) in the epidermis, leading to oxidative stress [14]. ROS can activate keratinocytes to produce pro-inflammatory cytokines, such as interleukin (IL)-1, -6, and -8, and tumor necrosis factor (TNF)-α [15]. Subsequently, these cytokines induce the expression of matrix metalloproteinases (MMPs) in dermal fibroblasts [16]. MMPs degrade collagen and other proteins in the connective tissue. Moreover, increased MMP activity combined with decreased procollagen production impairs the skin structure, ultimately leading to wrinkle formation during skin photoaging [17].

In this study, we aimed to investigate the radical scavenging and anti-wrinkle effects of soymilk-derived peptides produced via different processes, namely fermentation, enzymatic treatment, and ultrafiltration (UF). The radical scavenging activities of soymilk-derived peptides were measured using the 2,2-diphenyl-1-picrylhydrazyl (DPPH) radical scavenging assay, and their anti-wrinkle effects were evaluated based on type 1 procollagen synthesis and TNF-α production in UVB-radiation-exposed human dermal fibroblasts and HaCaT keratinocytes. Furthermore, liquid chromatography–tandem mass spectrometry (LC-MS/MS) analysis was used to identify their potential bioactive peptides. Finally, candidate peptides were synthesized based on the analysis results, and their anti-wrinkle effects were evaluated.

2. Materials and Methods

2.1. Lactic Acid Bacteria and Fermentation of Soymilk

The strain and optimal conditions used for soymilk fermentation were as previously reported [18]. *Lacticaseibacillus paracasei* MK1 was isolated from aged kimchi and soymilk was obtained from Yonsei University Milk (Seoul, Republic of Korea). Sterilized soymilk was inoculated with *L. paracasei* MK1 to a viable count of 9.0 log CFU/mL and fermented at 30 °C for 18 h for subsequent use in this study.

2.2. Soymilk Sample Preparation

2.2.1. Commercial Protease Sample

Commercial protease-treated soymilk was prepared using protamex (Novozyme, Bagsværd, Denmark), a blend of microbial endo-proteases from *Bacillus subtilis*, or flavourzyme (Novozyme), a high-quality blend of endo- and exo-peptidases from *Aspergillus oryzae*, to compare *L. paracasei* MK1-fermented and commercial protease-treated soymilk. Protamex or flavourzyme (2 g of enzyme per kg of protein) was added to sterilized soymilk preheated to 50 °C, shaken for 4 h at the same temperature, and inactivated by heating at 80 °C for 10 min.

2.2.2. Fractionation of Soymilk Samples

UF was performed to fractionate the MK1-fermented and commercial protease-treated soymilk. MK1-fermented or protease-treated soymilk was centrifuged at 12,000× g for 20 min, and the supernatant was filtered via membrane filtration using a nitrocellulose filter membrane (Sigma, St. Louis, MO, USA) with 0.22 µm pore size, and the filtrate was used for UF. The Labscale TFF system (Millipore Corporation, Bedford, MA, USA) was used as a UF device, and cartridges were performed using BIOMAX 30 K and 5 K polyethersulfone (50 cm^2; Pellicon XL filter, Millipore Corporation). All soymilk samples prepared using this method are listed in Table 1.

Table 1. Preparation methods of various samples used in this study.

Sample Name	Flow Chart of Fermentation or/and Fractionation
50S	50% Soymilk
50SMK	50% Soymilk→Fermented with *L. paracasei* MK1
50SMKUF30	50% Soymilk→Fermented with *L. paracasei* MK1→UF 30 kDa
50SMKUF5	50% Soymilk→Fermented with *L. paracasei* MK1→UF 30 kDa→UF 5 kDa
SF	100% Soymilk→Flavourzyme
SP	100% Soymilk→Protamex
50SMKF	50% Soymilk→Fermented with *L. paracasei* MK1→Flavourzyme
50SMKFUF30	50% Soymilk→Fermented with *L. paracasei* MK1→Flavourzyme→UF 30 kDa
50SMKFUF5	50% Soymilk→Fermented with *L. paracasei* MK1→Flavourzyme→UF 30 kDa→UF 5 kDa
50SMKP	50% Soymilk→Fermented with *L. paracasei* MK1→Protamex
50SMKPUF30	50% Soymilk→Fermented with *L. paracasei* MK1→Protamex→UF 30 kDa
50SMKPUF5	50% Soymilk→Fermented with *L. paracasei* MK1→Protamex→UF 30 kDa→UF 5 kDa
50SFMK	50% Soymilk→Flavourzyme→Fermented with *L. paracasei* MK1
50SFMKUF30	50% Soymilk→Flavourzyme→Fermented with *L. paracasei* MK1→UF 30 kDa
50SFMKUF5	50% Soymilk→Flavourzyme→Fermented with *L. paracasei* MK1→UF 30 kDa→UF 5 kDa
50SPMK	50% Soymilk→Protamex→Fermented with *L. paracasei* MK1
50SPMKUF30	50% Soymilk→Protamex→Fermented with *L. paracasei* MK1→UF 30 kDa
50SPMKUF5	50% Soymilk→Protamex→Fermented with *L. paracasei* MK1→UF 30 kDa→UF 5 kDa
100S	100% Soymilk
100SMK	100% Soymilk→Fermented with *L. paracasei* MK1
100SMKUF30	100% Soymilk→Fermented with *L. paracasei* MK1→UF 30 kDa
100SMKUF5	100% Soymilk→Fermented with *L. paracasei* MK1→UF 30 kDa→UF 5 kDa
100SMKF	100% Soymilk→Fermented with *L. paracasei* MK1→Flavourzyme
100SMKFUF30	100% Soymilk→Fermented with *L. paracasei* MK1→Flavourzyme→UF 30 kDa
100SMKFUF5	100% Soymilk→Fermented with *L. paracasei* MK1→Flavourzyme→UF 30 kDa→UF5 kDa
100SMKP	100% Soymilk→Fermented with *L. paracasei* MK1→Protamex
100SMKPUF30	100% Soymilk→Fermented with *L. paracasei* MK1→Protamex→UF 30 kDa
100SMKPUF5	100% Soymilk→Fermented with *L. paracasei* MK1→Protamex→UF 30 kDa→UF 5 kDa
100SFMK	100% Soymilk→Flavourzyme→Fermented with *L. paracasei* MK1
100SFMKUF30	100% Soymilk→Flavourzyme→Fermented with *L. paracasei* MK1→UF 30 kDa
100SFMKUF5	100% Soymilk→Flavourzyme→Fermented with *L. paracasei* MK1→UF 30 kDa→UF 5 kDa
100SPMK	100% Soymilk→Protamex→Fermented with *L. paracasei* MK1
100SPMKUF30	100% Soymilk→Protamex→Fermented with *L. paracasei* MK1→UF 30 kDa
100SPMKUF5	100% Soymilk→Protamex→Fermented with *L. paracasei* MK1→UF 30 kDa→UF 5 kDa

2.3. Determination of the Radical Scavenging Activities of Soymilk Samples Using the DPPH Radical Assay

The radical scavenging activities of soymilk samples were determined using the DPPH radical scavenging assay as described by Liu et al. [19], with slight modifications. Briefly, the soymilk samples were mixed with 500 µM DPPH (Sigma) dissolved in ethanol in a ratio of 1:1 and kept in the dark at 20 °C for 20 min. Absorbance was measured at 540 nm using a microplate reader (Epoch, BioTek Instruments, Inc., Winooski, VT, USA). Ascorbic acid (Sigma) at a concentration of 1 mM was used as a positive control. The DPPH radical scavenging activity was calculated as follows:

$$\text{DPPH radical scavenging activity (\%)} = (A_{blank} - A_{sample})/A_{blank} \times 100 \qquad (1)$$

where A_{blank} is the absorbance of the blank, and A_{sample} is the absorbance of the sample.

2.4. Treatment of Cell Lines with Soymilk Fractions

2.4.1. Cells and Reagents

Normal human dermal fibroblast (fibroblast) and non-tumorigenic human keratinocyte (HaCaT) cell lines from the American Type Culture Collection (Manassas, VA, USA) and melanoma murine B16F1 cell line from the Korean Cell Line Bank (Seoul, Republic of Korea) were cultured in the Dulbecco's modified Eagle's medium (DMEM; Welgene, Daegu, Republic of Korea) supplemented with 10% fetal bovine serum (FBS; Welgene) and 1% penicillin/streptomycin (Welgene). Cells were grown at 37 °C with 5% CO_2 in an incubator.

2.4.2. Cell Viability Assay

Cell viability was assessed using 3-(4,5-dimethylthiazol-2-yl)-2,5-diphenyltetrazolium bromide (MTT; Sigma) as described by Kumar et al. [20]. Fibroblasts, HaCaT cells (7×10^3 cells/well), and B16F1 melanoma cells (5×10^3 cells/well) were treated with various concentrations of soymilk samples for 24 h. After adding 100 µL/well of MTT solution, the cells were incubated at 37 °C. After 90 min, the MTT solution was removed, 100 µL/well of dimethyl sulfoxide (Sigma) was added, and cells were incubated at 20 °C. The plates were gently shaken, and the absorbance at 570 nm was determined using a microplate reader (Perkin Elmer, Waltham, MA, USA).

2.4.3. UVB Exposure of Fibroblasts and HaCaT Cells

Human fibroblasts were seeded in a 6-well plate at a density of 2×10^5 cells/well. After 24 h of incubation, the medium was replaced with a serum-free medium for starvation. The UVB irradiation method consisted of replacing phosphate-buffered saline (Amresco LLC, Solon, OH, USA) after washing the cells, which were then exposed to UVB irradiation at 312 nm and 25 mJ/cm^2 using a VLX-3 W research radiometer (Vilber Loumet, Collégien, France). Then, cells were treated with various concentrations of soymilk in DMEM without FBS for 48 h. Then, the procollagen content in the culture supernatant was measured using the procollagen type-I C-Peptide EIA kit (Takara Bio, Inc., Shiga, Japan). Ascorbic acid (50 µM; Sigma) was used as a positive control, cells not treated with UV light were used as negative controls, and untreated soymilk samples were used as controls.

HaCaT keratinocytes were seeded in a 12-well plate at a density of 1×10^5 cells/well and incubated for 24 h. Then, HaCaT keratinocytes were treated with various concentrations of soymilk samples under UVB irradiation (at 312 nm and 12.5 mJ/cm^2) and incubated for 24 h. After incubation, the culture supernatant was analyzed for TNF-α secreted by the soymilk samples using the TNF alpha human enzyme-linked immunosorbent assay (ELISA) kit (#KHC3012; Invitrogen, Carlsbad, CA, USA). We used 1 µM/mL of 50 µM dexamethasone (Sigma) as a positive control, and cells not treated with UV as negative controls.

2.4.4. Melanin Content of B16F1 Cells

Melanin production in cells was measured using a modified version of the method described by Tsuboi et al. [21]. B16F1 melanoma cells were treated with samples derived from the soymilk fractions in media for three days. After incubation, the cell pellets were harvested and dissolved in 1 mL of 1 N NaOH at 100 °C for 30 min. The dissolved samples were centrifuged, and the absorbance of the supernatant was measured at 400 nm.

2.5. Chemical Parameters

The saccharide, carbohydrate, crude protein, crude lipid, and solid contents of the original soymilk were analyzed by the Korea Standards Test and Analysis Institute (Gyeonggi, Republic of Korea). Crude protein and lipid contents of the fermented soymilk fractions were determined by the Korea Food Research Institute (Wanju, Republic of Korea). Free amino acids were analyzed by the Korea Basic Science Institute (Daejeon, Republic of Korea) The high-performance liquid chromatography conditions are presented in Table S1. Analysis of peptides in soymilk samples was performed by the National Instrumentation Center for Environmental Management (Seoul, Republic of Korea), and the analysis conditions are presented in Table S2. Total nitrogen content was analyzed following the macro-Kjeldahl method [22] using the Kjeltec 8100 distilling and 2508 digestion units (Foss, Eden Prairie, MN, USA). Trichloroacetic acid (TCA)-soluble nitrogen content was determined as described by Rowland [23]. Briefly, 10 mL of the sample and the same amount of 24% (*w/v*) TCA solution were mixed well, incubated at room temperature for 30 min, and centrifuged at 8000 rpm for 15 min to measure the TCA-soluble nitrogen content in the supernatant. The amount of reducing sugar released was determined following the Somogyi–Nelson method [24], with slight modifications [25].

2.6. Statistical Analyses

Statistical analyses were conducted using the GraphPad 6.01 software (GraphPad Software Inc., San Diego, CA, USA). Data were analyzed using analysis of variance, followed by Dunnett's test for pairwise comparisons. Data are presented as the mean $\pm$ standard deviation. Statistical significance was set at * $p < 0.05$, ** $p < 0.01$, *** $p < 0.001$, and **** $p < 0.0001$ for all tests.

3. Results and Discussion

3.1. DPPH Radical Scavenging Activities of Soymilk Fractions

Fermentation using lactic acid bacteria produces bioactive antioxidants, which maintain food stability and are beneficial to humans [5]. In this study, after lactic acid bacterial fermentation and/or enzymatic treatment of soymilk, the radical scavenging activities of various fractions were determined via UF. DPPH radical scavenging activities were determined to measure the radical scavenging activities of the soymilk fractions. The radical scavenging activities of the soymilk fractions (Table 1) were measured, and ascorbic acid was used as a positive control. DPPH radical scavenging activity increased in a dose-dependent manner (Figure S1), and 50SMK with 50% soymilk fermented with *L. paracasei* MK1 exhibited the highest radical scavenging activity, with 66.11% scavenging activity (Figure 1). By contrast, the fractions without fermentation or enzyme treatment (50S) exhibited the lowest radical scavenging activity of 45.11%, with 50–60% DPPH radical scavenging activity.

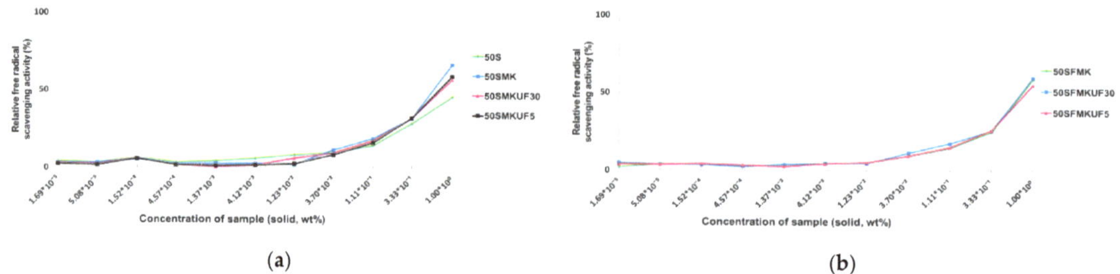

(a)   (b)

Figure 1. 2,2-diphenyl-1-picrylhydrazyl (DPPH) radical scavenging activities of various soymilk fractions. (**a**) 50S, 50SMK, 50SMKUF30, and 50SMKUF5. (**b**) 50SFMK, 50SFMKUF30, and 50SFMKUF5.

Next, we determined the half-maximal inhibitory concentration (IC_{50}), the point at which the radical scavenging activity is reduced by half, of the samples (Table 2). Except for 50S and 50SMK, no significant differences were observed among the other fractions. The 50S fraction had an IC_{50} of 1.24%, which was much higher than that of the other fractions, and 50SMK had an IC_{50} of 0.67%, which was lower than that of the other fractions. The DPPH radical scavenging activities and IC_{50} values of different fractions were similar, but the radical scavenging activity was the highest for 50SMK.

Table 2. Half-maximal inhibitory concentration (IC_{50}) values of soymilk fractions.

Sample	IC_{50} (Solid, wt%)	Sample	IC_{50} (Solid, wt%)
50S	1.24	100S	ND [1]
50SMK	0.67	100SMK	0.86
50SMKUF30	0.81	100SMKUF30	0.81
50SMKUF5	0.75	100SMKUF5	0.91
SF	ND1	100SMKF	0.91
SP	ND	100SMKFUF30	0.86
50SMKF	0.86	100SMKFUF5	0.95
50SMKFUF30	0.78	100SMKP	0.86
50SMKFUF5	0.86	100SMKPUF30	0.87
50SMKP	0.78	100SMKPUF5	0.91
50SMKPUF30	0.86	100SFMK	0.75
50SMKPUF5	0.91	100SFMKUF30	0.99
50SFMK	0.81	100SFMKUF5	0.86
50SFMKUF30	0.78	100SPMK	0.78
50SFMKUF5	0.86	100SPMKUF30	0.91
50SPMK	0.78	100SPMKUF5	0.91
50SPMKUF30	0.99		
50SPMKUF5	0.87		

[1] ND, not detected.

The fermentation of soymilk by lactic acid bacteria produces isoflavones and peptides [26–29]. Soymilk fermented with *L. paracasei* KUMBB005 reduces the DPPH radical scavenging activity by 27% [30], whereas that fermented with *L. paracasei* CD4 improves the free radical scavenging activity by 67% [31]. Soymilk fermented with *L. paracasei* MK1 exhibited high radical scavenging activity, possibly due to the peptides or isoflavones present in fermented soymilk.

3.2. Cytotoxicities of Soymilk Fractions in Cell Lines

The cytotoxicities of various soymilk fractions with radical scavenging activity were determined using a skin cell line. MTT reagent was used to measure the cytotoxicities of the soymilk fractions. The MTT assay measures the absorbance of formazan (purple) obtained after the reduction of MTT (yellow) by succinate dehydrogenase in the mitochondria to reflect the concentration of active cells [32].

3.2.1. Cytotoxicities of Soymilk Fractions in Fibroblasts

Cytotoxicity was evaluated by treating cells with various concentrations of soymilk fractions and comparing them to a control group with untreated cells (Figure S2). In particular, the highest growth rate of 129.67% was observed when fibroblasts were treated with ultrafiltered 100SPMKUF5, which was obtained by fermenting protamex-treated 100% soymilk with *L. paracasei* MK1 (Figure 2). This growth rate was observed at a concentration of 2.50×10^2. No significant differences were observed among all treated soymilk fractions. However, when fibroblast cells were treated with a concentration $\geq 1.25 \times 10^2$, the survival rate was higher than that of the control. Glycitin, a soybean isoflavone, exhibited non-cytotoxicity and increased cell proliferation in dermal fibroblasts in a dose-dependent manner [33]. Treating fibroblasts with soy peptides results in a non-toxic outcome similar to that in the positive control, AA2G (vitamin C derivative) [34]. Increased proliferation occurs as a result of extracting low-molecular-weight, physiologically active peptides from soybean protein fermented with *L. rhamnosus* and processing them in fibroblasts at different concentrations [35]. Therefore, the main functional ingredient derived from soybeans exhibits low toxicity and increases the proliferation of human dermal fibroblast cells.

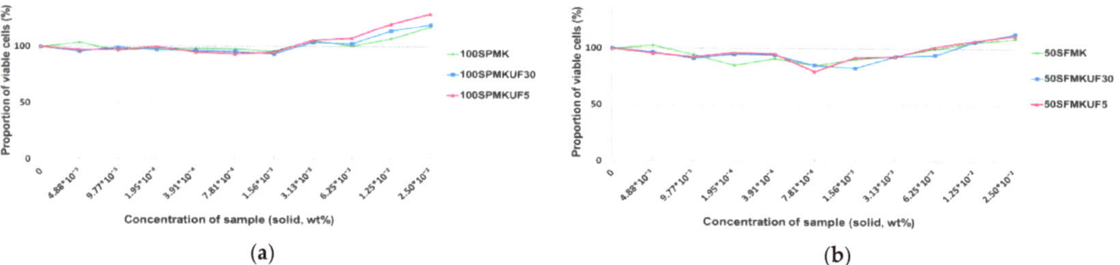

Figure 2. Effects of various soymilk fractions on the viability of human dermal fibroblasts (a) 100SPMK, 100SPMKUF30, and 100SPMKUF5. (b) 50SFMK, 50SFMKUF30, and 50SFMKUF5.

3.2.2. Cytotoxicities of Soymilk Fractions in HaCaT Keratinocytes

To determine their cytotoxicities, immortalized HaCaT keratinocytes were treated with various concentrations of the soymilk fractions (Figure S3). Even at higher concentrations of the sample, the viability of cells remained high at 0.1% concentration, with 50SFMKUF5 at 103.71%, 50SMKP at 102.02%, 50SFMKUF30 at 101.83%, 50SPMK at 101.45%, 100SMK at 101.41%, 50SFMK at 100.7%, and 100SPMK at 100.23% (Figure 3). Additionally, 100SMKFUF5, which showed the highest toxicity, had a cell viability of 59.63%, indicating that approximately 40% of the cell population was killed compared to that in the control. Kim et al. [35] reported no cytotoxicity in HaCaT keratinocytes treated with various concentrations of soybean peptides fermented with *L. rhamnosus*, even at a high concentration of 0.2%.

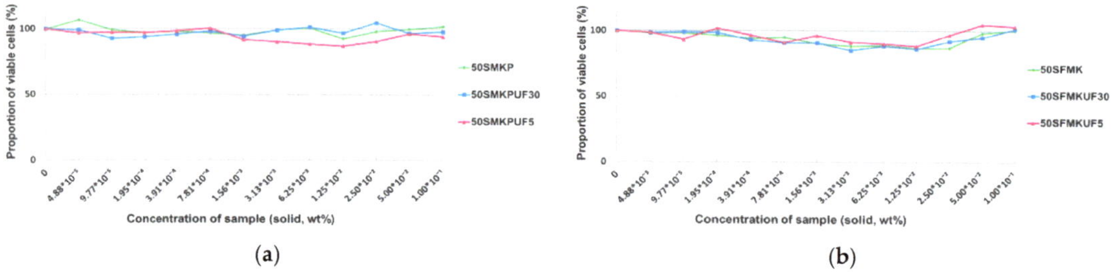

Figure 3. Cytotoxicities of various soymilk fractions in HaCaT keratinocytes. (a) 50SMKP, 50SMKPUF30, and 50SMKPUF5. (b) 50SFMK, 50SFMKUF30, and 50SFMKUF5.

Overall, proliferation increased when samples were treated with fibroblast cells, but toxicity was higher when HaCaT cells were treated with the 100% soymilk fraction than when treated with 50% soymilk. Based on these results, the anti-aging effects of the fractions prepared with 50% soymilk were determined by measuring the amount of type 1 procollagen and TNF-α produced.

3.3. Production of Type 1 Procollagen in Fibroblasts

UV exposure causes collagen degradation in the dermis, which is one of the main causes of wrinkles [36]. To measure the inhibitory effect of the sample on photoaging, human dermal fibroblasts were exposed to UVB, and the amount of newly generated procollagen produced during cell growth was measured. After UVB irradiation of the fibroblasts, the untreated control was set to 100%, ascorbic acid was used as the positive control, and the soymilk fractions were used to treat cells at concentrations of 0.1%, 0.01%, and 0.001% (Figure 4). When fibroblasts were treated with ascorbic acid, the production of procollagen was 197.76%. Procollagen production in cells treated with 50% soymilk was significantly lower than that in cells treated with ascorbic acid (Figure 4a). When 50SFMKUF5, the soymilk fraction obtained by fermenting 50% flavourzyme-treated soymilk with lactic acid bacteria and fractionating by 5 kDa UF, was treated with fibroblast cells, the production of procollagen was confirmed to be 164.23%. No significant differences were observed in the results of the treatment with ascorbic acid and 50SFMKUF5 at different concentrations, as shown in Figure 4b. By contrast, when the fibroblasts were treated with 0.1% 50SPMKUF5, a fermented soymilk fraction treated with protamex, the production of procollagen was the lowest, at 94.97%, which was a two-fold difference from the positive control (Figure S4). When collagen hydrolysate was prepared by treatment with commercial proteases (flavourzyme and protamex), the protein recovery and free amino group production of the collagen hydrolysate treated with flavourzyme were higher than those treated with protamex [37]. Low-molecular-weight fraction of soybean concentrate treated with flavourzyme exerts antioxidant activity [38]; however, there have been no reports on the production of procollagen in the low-molecular-weight fraction obtained via fermentation with commercial proteases and lactic acid bacteria.

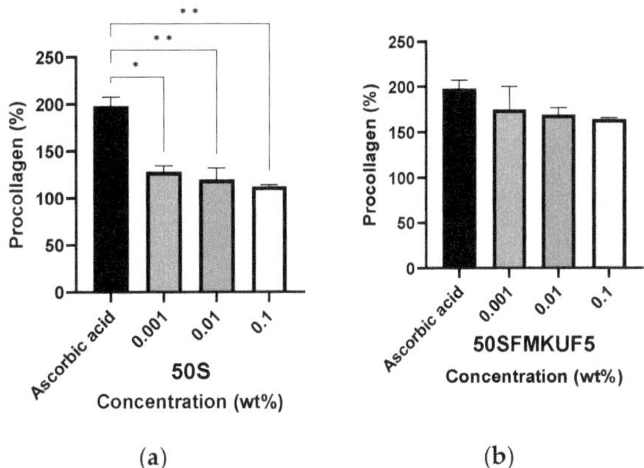

(a)   (b)

Figure 4. Production of type 1 procollagen in ultraviolet B (UVB)-irradiated human dermal fibroblasts treated with (**a**) 50S and (**b**) 50SFMKUF5. The mean ± standard deviation (SD) values of three independent experiments were analyzed using a two-way analysis of variance (ANOVA) and Dunnett's test (* $p < 0.05$ and ** $p < 0.01$).

3.4. Inhibition of TNF-α Production in HaCaT Keratinocytes

Exposure to UV radiation induces the release of pro-inflammatory mediators from skin cells, causing the infiltration and activation of immune cells [11]. The photoprotection of the skin is likely attributable to the antioxidant, anti-inflammatory, and anticarcinogenic properties of certain compounds [39]. In this study, the inflammation-alleviating effect of soymilk fractions on UVB-irradiated HaCaT keratinocytes was measured by assessing the amount of TNF-α, an indicator of inflammation, produced in response to treatment with soymilk fractions. After UVB irradiation of HaCaT keratinocytes, the untreated control was set to 100%, dexamethasone was used as the positive control, and the soymilk fractions were treated with 0.1, 0.02, 0.004, and 0.0008% cells (Figures 5 and S5). The production of TNF-α in HaCaT keratinocytes treated with dexamethasone was measured at 65.58%. The amount of TNF-α produced in the cells treated with 50SMKFUF5, 50SMKPUF5, 50SFMKUF30, 50SFMKUF5, and 50SPMKUF5 was lower than that in the positive control. This indicated that these soymilk fractions were more effective than dexamethasone in inhibiting TNF-α production. When HaCaT keratinocytes were treated with 50SPMKUF5 and 50SFMKUF5 concentrations of 0.2%, 23.23% and 23.44% of TNF-α were measured, respectively, indicating that inflammation was effectively inhibited. Although 50SPMKUF5 inhibited the production of TNF-α the most, it was excluded from the final selection because of its cytotoxicity, as determined in the cytotoxicity evaluation. The production of TNF-α is decreased in UVB-irradiated HaCaT keratinocytes treated with persimmon oligomeric proanthocyanidins [40] or the *Rhododendron weyrichii* flower extract [41]. Here, 50SFMKUF5 exerted anti-inflammatory effects on UVB-irradiated cells by reducing the production of TNF-α.

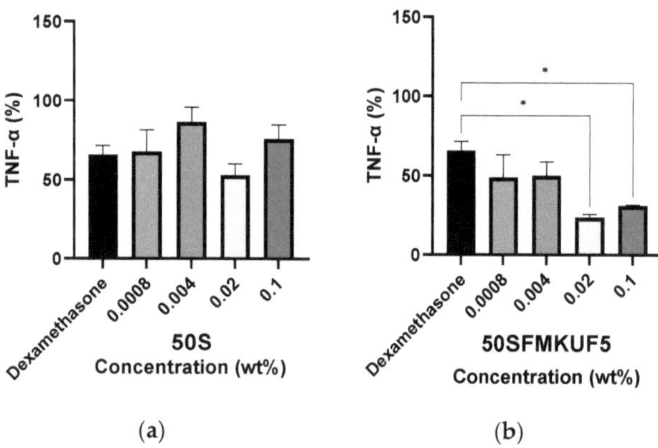

(a)   (b)

Figure 5. Inhibition of tumor necrosis factor (TNF)-α production in UVB-irradiated HaCaT keratinocytes treated with (**a**) 50S and (**b**) 50SFMKUF5. The mean ± SD values of three independent experiments were analyzed using two-way ANOVA and Dunnett's test (* $p < 0.05$).

In this study, 50SFMKUF5 was finally selected as the soymilk fraction with radical scavenging and anti-wrinkle effects based on efficacy evaluations, such as the DPPH radical scavenging activity, cell differentiation, MTT assays, and measurements of procollagen and TNF-α.

3.5. Characteristics of Functional Soymilk Fractions

3.5.1. General and Reducing Sugars in Soymilk Fractions

Next, we confirmed the differences between the components of soymilk and the selected 50SFMKUF5. Table 3 shows the saccharide, carbohydrate, crude protein, crude lipid, and solid contents in soymilk. The crude protein, crude lipid, and solid contents of 50SFMKUF5 were 0.30 g, 0.00 g, and 1.37%, respectively. Compared to 100% soymilk, all general components showed lower values, possibly due to the effects of fermentation, enzyme treatment, and UF.

Table 3. General components of 100% soymilk and 50SMKUF5.

Component	Content	
	100% Soymilk	**50SMKUF5**
Saccharide (g/100 g)	0.71	ND [1]
Carbohydrate (g/100g)	3.40	ND
Crude protein (g/100 g)	4.30	0.30
Crude lipid (g/100 g)	2.40	0.00
Solid (%)	10.80	1.37

[1] ND, not detected.

The results of reducing sugar analysis of the soymilk fractions are shown in Figure 6 There were no differences among the reducing sugars of the five fractions. However, 50SFMKUF5 exhibited a higher reducing sugar content of 1.67 µmol/mL than other soymilk fractions. The reducing sugar content of LAB-fermented soymilk decreases over time [42]. Additionally, in tofu residue treated with amylase, insoluble polysaccharides are converted to reducing sugars after enzyme treatment [43]. The results of this study showed that a significant portion of insoluble polysaccharides was converted by flavourzyme in 50SFMKUF5. The second highest amount of reducing equivalents was observed in 50SMKFUF5 cells treated with flavourzyme after fermentation.

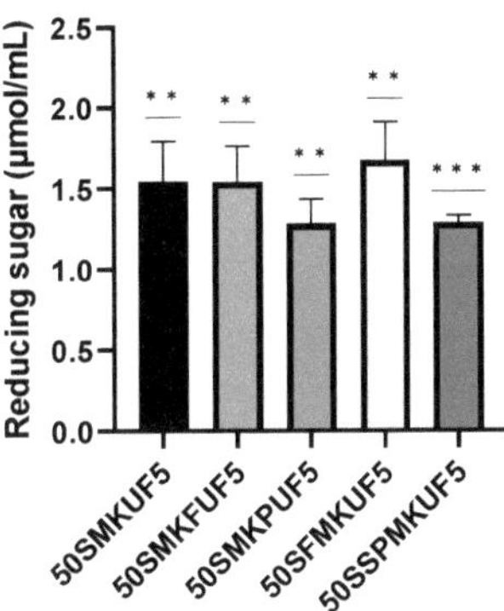

Figure 6. Reducing sugar contents of various soymilk fractions. The mean ± SD values of three independent experiments were analyzed using a *t*-test (** $p < 0.01$ and *** $p < 0.001$).

3.5.2. Free Amino Acid and Nitrogen Contents in Soymilk Fractions

During fermentation, the growth of lactic acid bacteria in soymilk relies on proteolysis to release amino acids [44]. In Table 4, the free amino acids of the fractions of less than 5 kDa obtained via UF were analyzed (50SMKUF5, 50SMKFUF5, 50SMKPUF5, 50SFMKUF5, and 50SPMKUF5). The total free amino acid (TA) content of 50SMKUF5 without enzyme treatment was the lowest, measuring 40.2 mg/100 g. On the other hand, the TA contents of 50SFMKUF5 and 50SPMKUF5, which were fermented after enzyme treatment, were higher than those of 50SMKFUF5 and 50SMKPUF5 treated with enzyme after fermentation. The TA content of 50SFMKUF5 was the highest at 67.3 mg/100 g, whereas that of 50SMKFUF5 was 46.8 mg/100 g. The free amino acid content of soymilk increases by 234.8% when fermented with five types of lactic acid bacteria and hydrolyzed with proteases [45]. Fermenting soymilk with *Lactobacillus delbrueckii* subsp. *Delbrueckii* increases the overall content of free amino acids [46]. These results showed that the TA content of the fraction treated with the enzyme later was higher than that of the fraction treated with the enzyme first because protease treatment of soymilk protein prior to fermentation allowed *L. paracasei* MK1 to cleave the protein more effectively.

Here, glutamic acid content accounted for 10.7% of the total free amino acids in 50SFMKUF5, which was lower compared to that reported in a previous study [47] on soymilk where the glutamic acid content was found to be 733 mg/100 mL, representing 19.1% of the total amino acids. Although the leucine and arginine contents in soymilk were 7.7 and 7.9%, respectively, in 50SFMKUF5, their amino acid contents were 16.6 and 17.7%, respectively. However, because the overall amino acid content was significantly reduced, many free amino acids were lost through various processes. The nutritionally important essential amino acid content was similar to that of TA. According to these results, when *L. paracasei* MK1 used soymilk protein in soymilk fermentation, enzymatic treatment resulted in better degradation of soymilk protein, suggesting that flavourzyme is more effective than the other enzymes.

Table 4. Compositions of free amino acids in various soymilk fractions.

Amino Acid	Concentration of Free Amino Acid (mg/100 g, %)				
	50SMKUF5	**50SMKFUF5**	**50SMKPUF5**	**50SFMKUF5**	**50SPMKUF5**
Aspartic acid	1.8 (4.5)	1.9 (4.1)	1.9 (4.5)	2.6 (3.9)	1.2 (2.8)
Glutamic acid	12.5 (31.0)	13.3 (28.4)	11.2 (27.0)	7.2 (10.7)	4.6 (10.5)
Asparagine	0.4 (1.0)	0.6 (1.2)	0.4 (0.9)	1.0 (1.5)	0.2 (0.4)
Serine	0.2 (0.6)	0.5 (1.1)	0.3 (0.6)	0.2 (0.3)	0.1 (0.2)
Glutamine	0.0 (0.0)	0.5 (1.1)	0.0 (0.0)	0.0 (0.0)	0.0 (0.0)
Glycine	0.8 (2.0)	0.7 (1.5)	0.7 (1.7)	0.6 (0.8)	1.2 (2.7)
Histidine	1.6 (3.9)	1.6 (3.4)	1.7 (4.2)	1.3 (2.0)	1.2 (2.8)
Arginine	10.9 (27.0)	10.2 (21.8)	10.7 (25.8)	11.9 (17.7)	9.6 (22.1)
Threonine	0.3 (0.6)	0.7 (1.6)	0.4 (1.0)	0.5 (0.8)	0.5 (1.1)
Alanine	2.4 (6.0)	2.5 (5.3)	2.5 (6.1)	2.9 (4.3)	3.1 (7.1)
Proline	2.9 (7.2)	2.7 (5.9)	3.8 (9.1)	1.5 (2.2)	1.0 (2.3)
Tyrosine	1.5 (3.7)	0.7 (1.6)	1.5 (3.6)	3.1 (4.5)	1.4 (3.2)
Valine	0.3 (0.7)	0.7 (1.4)	0.3 (0.7)	2.2 (3.3)	1.7 (4.0)
Methionine	0.0 (0.0)	0.7 (1.6)	0.0 (0.0)	0.3 (0.5)	0.5 (1.2)
Cysteine	0.0 (0.0)	0.1 (0.3)	0.1 (0.2)	0.1 (0.2)	0.0 (0.0)
Isoleucine	0.0 (0.0)	0.6 (1.2)	0.2 (0.4)	2.4 (3.6)	2.1 (4.8)
Leucine	0.4 (0.9)	2.1 (4.5)	0.6 (1.4)	11.2 (16.6)	4.7 (10.8)
Phenylalanine	0.6 (1.5)	2.4 (5.1)	1.0 (2.4)	15.0 (22.3)	6.4 (14.6)
Tryptophan	2.7 (6.7)	2.6 (5.6)	3.0 (7.3)	2.7 (4.0)	3.9 (8.9)
Lysine	1.0 (2.5)	1.7 (3.6)	1.2 (3.0)	0.6 (0.8)	0.2 (0.4)
TA [1]	40.2 (100)	46.8 (100)	41.4 (100)	67.3 (100)	43.6 (100)
EA [2]	6.9 (16.8)	13.1 (28.0)	8.4 (20.4)	36.2 (53.9)	21.2 (43.6)

[1] TA, total free amino acid. [2] EA, essential amino acid (Thr + Val + Met + Ile + Leu + Phe + Lys + Trp + His).

Total nitrogen content provides an estimate of the number of peptides or proteins present in the sample, whereas the TCA-soluble nitrogen content estimates the nitrogen compound content when proteins undergo hydrolysis by proteases [48]. The total and TCA-soluble nitrogen contents of the soymilk fractions were measured to evaluate the changes in nitrogen content due to fermentation and enzyme treatment. The total nitrogen content of 50% soymilk was the highest at 2.82 mg N/100 mL, but its TCA-soluble nitrogen content was the lowest at 0.12 mg N/100 mL (Table 5). The total nitrogen content of the samples treated with the enzyme was higher than that of the untreated samples. Free amino acids, oligopeptides, amines, and other compounds comprising non-protein nitrogen (NPN) are an important group of cosmetic components owing to their technological and functional significance. These compounds play a significant role in cosmetic development, particularly in the formulation of skin care products. NPN possesses excellent moisturizing properties as well as an anti-aging effect. [49]. The content of NPN resulting from the enzyme treatment was evident through the TCA-soluble nitrogen content, and this result was also observed for 50SFMKUF5 and 50SPMKUF5, which were measured at 0.45 and 0.82 mg N/100 mL, respectively, showing a similar trend to the total nitrogen content.

Table 5. Changes in the trichloroacetic acid (TCA)-soluble nitrogen concentration in various soymilk fractions.

Soymilk Fraction	Total Nitrogen (mg N/100 mL)	TCA-Soluble Nitrogen (mg N/100 mL)	TN/SN (%)
50S	2.82 ± 0.17	0.12 ± 0.03	4.39
50SMKUF5	0.34 ± 0.07	0.12 ± 0.05	36.17
50SMKFUF5	0.60 ± 0.14	0.36 ± 0.03	59.59
50SMKPUF5	0.40 ± 0.09	0.23 ± 0.01	58.02
50SFMKUF5	1.22 ± 0.23	0.45 ± 0.02	36.93
50SPMKUF5	1.60 ± 0.22	0.82 ± 0.06	51.16

Data are presented as mean ± SD of three independent experiments.

3.6. Analysis of Peptides Derived from 50SFMKUF5

3.6.1. Screening of Bioactive Compounds from 50SFMKUF5

Based on these results, 50SFMKUF5, a sample obtained by processing flavourzyme and fractionating soymilk fermented with *L. paracasei* MK1 into UF 5 kDa, was selected as the final fraction. Peptide mapping was performed using LC-MS/MS to search for biologically active compounds present in 50SFMKUF5. The results of the MS analysis are shown in Figure 7, and two major peaks, 726.87 and 755.43 m/z, were identified. The respective molecular weights were confirmed to be 1453.74 and 1510.86 Da, and the amino acid sequences were shown as APEFLKEAFGVN (APE) and QIVTVEGGLSVISPK (QIV), respectively. The APE and QIV were found to be glycinin G2 (NCBI NP_001235810.1) and glycinin G4 (NCBI NP_001238008.1), respectively, as a result of searching using NCBI BLAST, and the full amino acid sequences of glycinin G2 and G4 are shown in Figures S6 and S7, respectively. Soymilk fermented with *L. plantarum* C2 was fractionated at 10, 5, and 3 kDa to confirm the ACE-inhibitory effect, and as a result of LC/MS-MS, a GFAPEFLKEAFGVN sequence of 736.3768 m/z at a retention time of 65.99 min was found [50]. As a result of examining the peptide components present in milk with pea protein added, there have been reports that the measured contents of LQESVIVEISK, FYLAGNQEQEFLK, SQSD-NFEYVSFK, and QIVTVEGGLSVISPK were high [51]. Previous studies have reported that conglycinin and glycinin, the two main protein compounds in soybeans, can yield soy-derived peptides with antioxidant activity that are more effective than those derived from conglycinin [52]. To verify the functionality of the glycinin G2- and G4-derived peptides identified in this study, APE and QIV were synthesized and tested on human dermal fibroblasts, HaCaT keratinocytes, and melanoma cells to assess their ability to stimulate procollagen production, inhibit TNF-α, and reduce melanin production.

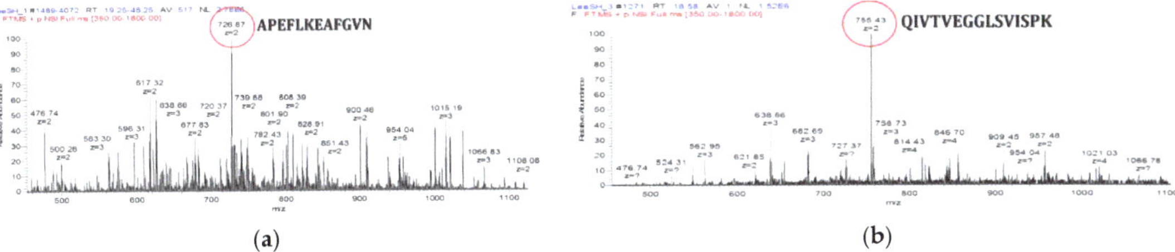

Figure 7. Liquid chromatography–tandem mass spectrometry (LC-MS/MS) chromatograms of 50SFMKUF5. (**a**) APEFLKEAFGVN and (**b**) QIVTVEGGLSVISPK.

3.6.2. Production of Type 1 Procollagen in Fibroblasts after Peptide Treatment

As the molecular weight of peptides is too high (500 Da) to cross the skin barrier, they are not suitable for absorption through or into the skin [53]. Here, the molecular weights of APE and QIV were 1453.74 and 1510.86 Da, respectively, which is much higher than 500 Da. To confirm the anti-wrinkle-like effect of peptides, they must pass through the stratum corneum, which is the skin barrier, and the epidermis to reach the dermis at an optimal concentration [54]. To improve the permeability of peptides, the combination of fatty acids, such as palmitic acids, and peptides holds significant potential as an effective approach for enhancing the permeability and stability of peptides [55]. In this study, UVB-irradiated fibroblasts were treated with 50SFMKUF5, a soymilk fraction, APE, QIV, and Pal-APE. Pal-APE was synthesized by attaching palmitic acid to APE. The objective of this study was to determine the amount of type 1 procollagen production (Figure 8). Cells not treated with anything after UVB irradiation was used as a control, and 50 µM of ascorbic acid was used as a positive control. As the concentrations of 50SFMKUF5, APE, and QIV increased in fibroblasts, the amount of procollagen produced significantly decreased compared to the positive control. Contrary to previous results, when cells were treated with Pal-APE at concentrations of 0.02 and 0.1%, the amount of procollagen produced was significantly increased compared to the positive control. When passing through the skin, palmitoyl-binding polypeptides have been reported to have 100–1000 times improved permeability compared to general polypeptides [56]. In addition, when Lys-Thr-Thr-Lys-Ser (KTTKS), derived from a fragment of type 1 procollagen, was modified with palmitoyl pentapeptide-4 (Pal-KTTKS-OH), its stability and permeability were improved and became better than those of KTTKS [57]. Pal-KTTKS was one of the first polypeptides used in cosmetics, and its application to the skin of 49 women for 4 months significantly improved their skin roughness, wrinkle volume, and wrinkle depth [58]. These results confirmed that the skin permeability of Pal-APE was better than that of 50SFMKUF5, a soymilk-derived fraction, and that of synthesized APE and QIV. This improvement led to enhanced procollagen production.

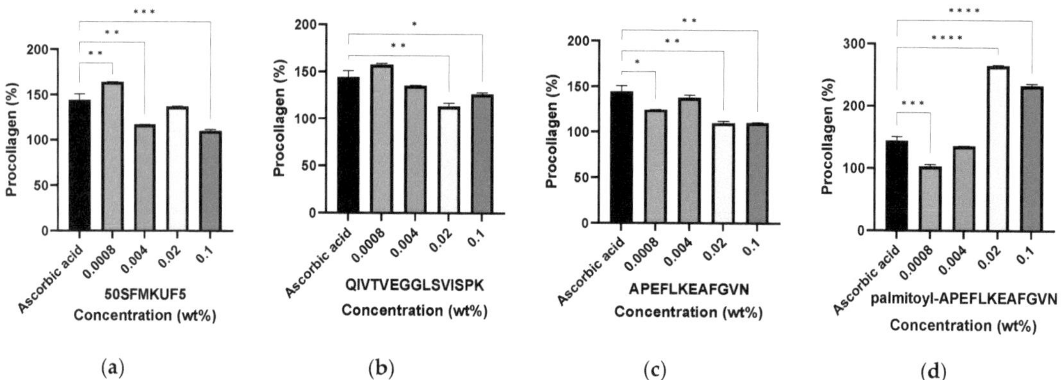

(a) (b) (c) (d)

Figure 8. Production of type 1 procollagen based on the peptide concentration in UVB-irradiated fibroblast cells. Cells were treated with (**a**) 50SFMKUF5, (**b**) QIVTVEGGLSVISPK, (**c**) APEFLKEAFGVN, (**d**) palmitoyl-APEFLKEAFGVN. The mean ± SD values of three independent experiments were analyzed using two-way ANOVA and Dunnett's test (* $p < 0.05$, ** $p < 0.01$, *** $p < 0.001$, and **** $p < 0.0001$).

3.6.3. Inhibition of TNF-α Production in HaCaT Keratinocytes after Peptide Treatment

To assess the degree of TNF-a production inhibition, HaCaT keratinocytes were treated with 50SFMKUF5 and three synthetic peptides at varying concentrations. As a positive control, the anti-inflammatory drug dexamethasone (1 μM) was used for treatment and the inhibition rate of TNF-α was compared (Figure 9). Notably, the production of TNF-α was significantly higher in cells treated with low concentrations (0.0008 and 0.004%) of 50SFMKUF5, QIV, APE, and Pal-APE than in the positive control. When the cells were treated with 0.02 and 0.1% 50SFMKUF5, QIV, and APE, the production of TNF-α was not significantly different from that in the positive control. Furthermore, no significant difference in the amount of TNF-a produced from the positive control was observed at all concentrations when cells were treated with Pal-APE. These results indicated that 50SFMKUF5, QIV, APE, and Pal-APE exhibit anti-inflammatory effects. Specifically, Pal-APE exerted anti-inflammatory effects at all tested concentrations. Among the soy proteins, peptide FLV, derived from β-conglycinin, inhibits the synthesis of pro-inflammatory cytokines, such as TNF-α and IL-6, in 3T3-L1 adipocytes [59]. Additionally, peptide VPY, derived from glycinin, downregulates the level of pro-inflammatory cytokines, such as TNF-α, IL-6, and IL-1β, in Caco-2 intestinal epithelial cells and THP-1 macrophages [60]. Lunasin, a soy-derived peptide consisting of 43–44 amino acids, exerts anti-inflammatory and anticancer effects in RAW264.7 macrophages [61]. However, limited information is available on the anti-inflammatory effects of peptides derived from fermented soymilk fractions and palmitoylated peptides.

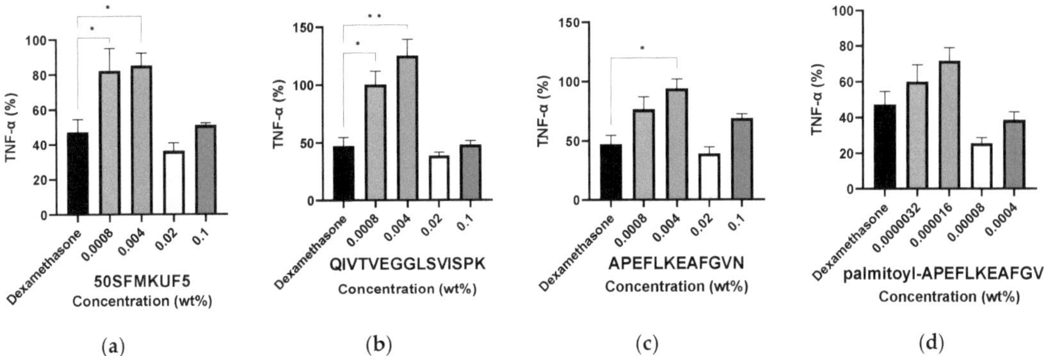

Figure 9. TNF-α inhibition according to the peptide concentration in UVB-irradiated HaCaT keratinocytes. Cells were treated with (**a**) 50SFMKUF5, (**b**) QIVTVEGGLSVISPK, (**c**) APEFLKEAFGVN, and (**d**) palmitoyl-APEFLKEAFGVN. The mean ± SD values of three independent experiments were analyzed using two-way ANOVA and Dunnett's test (* $p < 0.05$ and ** $p < 0.01$).

3.6.4. Inhibition of Melanin Synthesis in B16F1 Melanoma Cells after Peptide Treatment

Hyperpigmentation of the skin occurs when an excessive amount of melanin accumulates in melanocytes within skin cells, primarily due to environmental factors, such as α-melanocyte-stimulating hormone (α-MSH) treatment, UV radiation, cAMP-elevating agents, estrogen, hormonal factors, and genetic factors. Melanin is subsequently transported to the keratinocytes and accumulates in the epithelial layer of the skin [62,63]. Melanin promotes cell death due to the toxicity caused by pigmentation and melanin precursors in terms of beauty. Various substances, such as hydroquinone, arbutin, kojic acid, and vitamin C, prevent melanin deposition, but their actual clinical effects remain unclear [64–66]. In this study, melanin production was induced by treating B16F1 melanoma cells with 5 nM of α-MSH, and 1.6 mM kojic acid was used as a positive control. Treatment with kojic acid, the positive control, resulted in a 37.25% reduction in melanin production (Figure 10). Neither QIV nor APE exhibited any inhibitory effects on melanogenesis at any concentration. However, when 50SFMKUF5 was treated with 0.1%, it was measured as 33.26%, which was lower than the positive control, and melanin formation was significantly reduced, but when Pal-APE was treated with 0.01%, melanin formation was reduced by 47.97%; however, the difference was not significant. In a parallel treatment study involving a mixture of soy- and collagen-derived peptides and *Chrysanthemum morifolium* water-soluble extract, a significant reduction of 46.2% in melanogenesis was observed [67]. Additionally, a separate report revealed that treatment with soybean cell culture extract inhibits melanin synthesis in α-MSH-induced B16F10 melanoma cells in a dose-dependent manner [68]. In this study, 50SFMKUF5 and Pal-APE soy-derived peptides demonstrated the potential to inhibit melanin synthesis, which is consistent with the findings of previous studies.

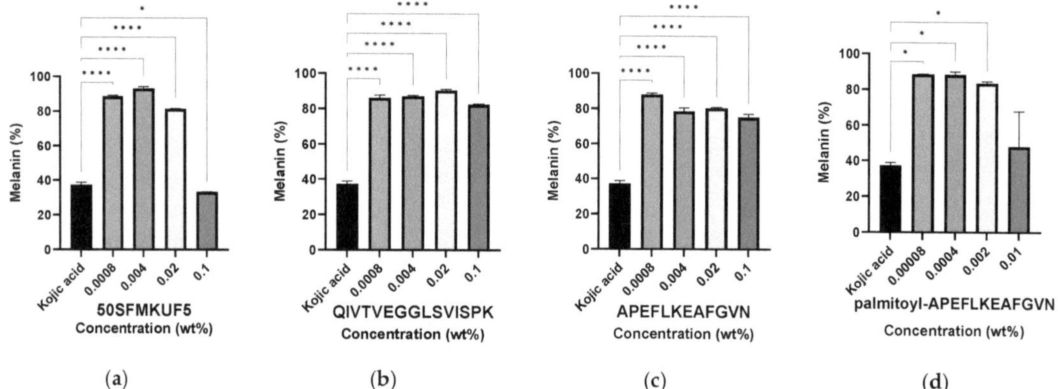

Figure 10. Melanin inhibition according to the peptide concentration in α-melanocyte-stimulating hormone (α-MSH)-treated B16F1 melanoma cells. Cells were treated with (**a**) 50SFMKUF5, (**b**) QIVTVEGGLSVISPK, (**c**) APEFLKEAFGVN, and (**d**) palmitoyl-APEFLKEAFGVN. The mean ± SD values of three independent experiments were analyzed using two-way ANOVA and Dunnett's test (* $p < 0.05$ and **** $p < 0.0001$).

4. Conclusions

In this study, we evaluated the radical scavenging activity and anti-wrinkle effects of the 30 and 5 kDa fractions of soymilk fermented with *L. paracasei* MK1 and subsequently treated with commercial proteases (flavourzyme or protamex). Among the various fractions tested, the fraction obtained by fermenting flavourzyme-treated soymilk with *L. paracasei* MK1 followed by ultrafiltration at 5 kDa (50SFMKUF5) exhibited the best characteristics. This fraction exhibited high radical scavenging capacity, promoted the proliferation and type 1 procollagen production in fibroblasts, and reduced the cytotoxicity and TNF-α production in HaCaT keratinocytes. Moreover, APEFLKEAFGVN, QIVTVEGGLSVISPK, and Pal-APE peptides derived from 50SFMKUF5 exerted anti-inflammatory effects by reducing the TNF-α levels in cells. Both 50SFMKUF5 and Pal-APE inhibited melanin production in melanoma cells, and Pal-APE specifically increased the type 1 procollagen production in cells. Despite its high molecular weight, Pal-APE derived from 50SFMKUF5 exhibited potent anti-inflammatory effects, inhibited melanin production, and induced type 1 procollagen production. Our findings highlight the potential benefits of Pal-APE to human skin.

Supplementary Materials: The following supporting information can be downloaded at: https://www.mdpi.com/article/10.3390/antiox12071392/s1, Table S1: Instrument and operating condition of HPLC for the analysis of free amino acids; Table S2: Instrument and operating condition of LC/MS-MS for the analysis of peptides; Figure S1: DPPH radical scavenging activity of soymilk fractions; Figure S2: Effect of soymilk fractions on the viability of human dermal fibroblasts; Figure S3: Effect of soymilk fractions on the cytotoxicity of HaCaT keratinocytes; Figure S4: Production of type 1 procollagen in UVB-irradiated human dermal fibroblasts treated with soymilk fractions; Figure S5: Inhibition of TNF-α in UVB-irradiated HaCaT keratinocytes treated with soymilk fractions; Figure S6: Amino acid sequence of glycinin G2; Figure S7: Amino acid sequence of glycinin G4; Figure S8: Effect of peptides on the viability of human dermal fibroblasts; Figure S9: Cytotoxic effects of the peptides on HaCaT keratinocytes; Figure S10: Cytotoxic effects of the peptides on B16F1 melanoma cells.

Author Contributions: Conceptualization, S.W.K. and Y.-S.P.; formal analysis, S.L. and S.-P.C.; data curation, S.L., W.K.Y. and H.J.; writing—original draft preparation, S.L.; writing—review and editing, Y.-S.P. and S.-P.C.; visualization, S.L., H.J., W.K.Y. and S.-P.C.; supervision, Y.-S.P.; project administration, Y.-S.P. All authors have read and agreed to the published version of the manuscript.

Funding: This study was funded by the Ministry of Agriculture, Food, and Rural Affairs (MAFRA; grant number: 321035052HD020).

Institutional Review Board Statement: Not applicable.

Informed Consent Statement: Not applicable.

Data Availability Statement: All data generated in this study are presented in this article and the Supplementary Materials file.

Acknowledgments: This work was supported by the Korea Institute of Planning and Evaluation for Technology in Food, Agriculture, and Forestry (IPET) through the High Value-added Food Technology Development Program.

Conflicts of Interest: The authors declare no conflict of interest.

References

1. Qin, P.; Wang, T.; Luo, Y. A review on plant-based proteins from soybean: Health benefits and soy product development. *J. Agric. Food Res.* **2022**, *7*, 100265. [CrossRef]
2. Kumari, M.; Kokkiligadda, A.; Dasriya, V.; Naithani, H. Functional relevance and health benefits of soymilk fermented by lactic acid bacteria. *J. Appl. Microbiol.* **2022**, *133*, 104–119. [CrossRef] [PubMed]
3. Singh, B.P.; Vij, S.; Hati, S. Functional significance of bioactive peptides derived from soybean. *Peptides* **2014**, *54*, 171–179. [CrossRef]
4. Tokudome, Y.; Nakamura, K.; Hashimoto, F. Effects of low molecular weight soybean peptide on mRNA and protein expression levels of differentiation markers in normal human epidermal keratinocytes. *Biosci. Biotechnol. Biochem.* **2014**, *78*, 1018–1021. [CrossRef] [PubMed]
5. Chatterjee, C.; Gleddie, S.; Xiao, C.W. Soybean bioactive peptides and their functional properties. *Nutrients* **2018**, *10*, 1211. [CrossRef]
6. Raveschot, C.; Cudennec, B.; Coutte, F.; Flahaut, C.; Fremont, M.; Drider, D.; Dhulster, P. Production of bioactive peptides by *Lactobacillus* species: From gene to application. *Front. Microbiol.* **2018**, *9*, 2354. [CrossRef]
7. Miri, S.; Hajihosseini, R.; Saedi, H.; Vaseghi, M.; Rasooli, A. Fermented soybean meal extract improves oxidative stress factors in the lung of inflammation/infection animal model. *Ann. Microbiol.* **2019**, *69*, 1507–1515. [CrossRef]
8. Marazza, J.A.; Nazareno, M.A.; de Giori, G.S.; Garro, M.S. Enhancement of the antioxidant capacity of soymilk by fermentation with *Lactobacillus rhamnosus*. *J. Funct. Foods* **2012**, *4*, 594–601. [CrossRef]
9. Wang, Y.C.; Yu, R.C.; Chou, C.C. Antioxidative activities of soymilk fermented with lactic acid bacteria and bifidobacteria. *Food Microbiol.* **2006**, *23*, 128–135. [CrossRef]
10. Masaki, H. Role of antioxidants in the skin: Anti-aging effects. *J. Dermatol. Sci.* **2010**, *58*, 85–90. [CrossRef]
11. Pillai, S.; Oresajo, C.; Hayward, J. Ultraviolet radiation and skin aging: Roles of reactive oxygen species, inflammation and protease activation, and strategies for prevention of inflammation-induced matrix degradation-a review. *Int. J. Cosmet. Sci.* **2005**, *27*, 17–34. [CrossRef] [PubMed]
12. Tobin, D.J. Introduction to skin aging. *J. Tissue Viability* **2017**, *26*, 37–46. [CrossRef] [PubMed]
13. Pandel, R.; Poljšak, B.; Godic, A.; Dahmane, R. Skin photoaging and the role of antioxidants in its prevention. *ISRN Dermatol.* **2013**, *12*, 930164. [CrossRef]
14. De Jager, T.L.; Cockrell, A.E.; Du Plessis, S.S. Ultraviolet light induced generation of reactive oxygen species. *Adv. Exp. Med. Biol.* **2017**, *996*, 15–23. [CrossRef] [PubMed]
15. Ansary, T.M.; Hossain, M.R.; Kamiya, K.; Komine, M.; Ohtsuki, M. Inflammatory molecules associated with ultraviolet radiation-mediated skin aging. *Int. J. Mol. Sci.* **2021**, *22*, 3974. [CrossRef] [PubMed]
16. Han, Y.P.; Tuan, T.L.; Wu, H.; Hughes, M.; Garner, W. TNF-alpha stimulates activation of pro-MMP2 in human skin through NF-(kappa) B mediated induction of MT1-MMP. *J. Cell Sci.* **2001**, *114*, 131–139. [CrossRef]
17. Quan, T.; Little, E.; Quan, H.; Qin, Z.; Voorhees, J.J.; Fisher, G.J. Elevated matrix metalloproteinases and collagen fragmentation in photodamaged human skin: Impact of altered extracellular matrix microenvironment on dermal fibroblast function. *J. Investig. Dermatol.* **2013**, *133*, 1362–1366. [CrossRef] [PubMed]
18. Lee, S.; Jang, D.H.; Choi, H.J.; Park, Y.S. Optimization of soymilk fermentation by the protease-producing *Lactobacillus paracasei*. *Korean J. Food Sci. Technol.* **2013**, *45*, 571–577. [CrossRef]
19. Liu, D.; Shi, J.; Ibarra, A.C.; Kakuda, Y.; Xue, S.J. The scavenging capacity and synergistic effects of lycopene, vitamin E, vitamin C, and β-carotene mixtures on the DPPH free radical. *LWT-Food Sci. Technol.* **2008**, *41*, 1344–1349. [CrossRef]
20. Kumar, P.; Nagarajan, A.; Uchil, P.D. Analysis of cell viability by the MTT assay. *Cold Spring Harb. Protoc.* **2018**, *6*, 469–471. [CrossRef]
21. Tsuboi, T.; Kondoh, H.; Hiratsuka, J.; MIshima, Y. Enhanced melanogenesis induced by tyrosinase gene-transfer increases boron-uptake and killing effect of boron neutron capture therapy for amelanotic melanoma. *Pigment Cell Res.* **1988**, *11*, 275–282. [CrossRef]

22. Murthy, L.; Herreid, E.O. Determination of total nitrogen in stored milk by nesslerization and by the macro-Kjeldahl methods. *J. Dairy Sci.* **1958**, *41*, 314–315. [CrossRef]
23. Rowland, S.J. The determination of the nitrogen distribution in milk. *J. Dairy Res.* **1937**, *9*, 42–46. [CrossRef]
24. Holth, J.G.; Krieg, N.R.; Sneath, P.H.A.; Staley, J.T.; Williams, S.T. *Bergey's Manual of Determinative Bacteriology*, 9th ed.; Hensyl, W.R., Ed.; Lippincott Williams and Wikins: Baltimore, MD, USA, 1994; pp. 243–244.
25. Lee, S.; Kwon, H.K.; Park, H.J.; Park, Y.S. Solid-state fermentation of germinated black bean (*Rhynchosia nulubilis*) using *Lactobacillus pentosus* SC65 and its immunostimulatory effect. *Food Biosci.* **2018**, *26*, 57–64. [CrossRef]
26. Dai, S.; Pan, M.; El-Nezami, H.S.; Wan, J.M.F.; Wang, M.F.; Habimana, O.; Lee, J.C.Y.; Louie, J.C.Y.; Shah, N.P. Effects of lactic acid bacteria-fermented soymilk on isoflavone metabolites and short-chain fatty acids excretion and their modulating effects on gut microbiota. *J. Food Sci.* **2019**, *84*, 1854–1863. [CrossRef] [PubMed]
27. Leksono, B.Y.; Cahyanto, M.N.; Rahayu, E.S.; Yanti, R.; Utami, T. Enhancement of antioxidant activities in black soy milk through isoflavone aglycone production during indigenous lactic acid bacteria fermentation. *Fermentation* **2022**, *8*, 326. [CrossRef]
28. Undhad Trupti, J.; Das, S.; Solanki, D.; Kinariwala, D.; Hati, S. Bioactivities and ACE-inhibitory peptides releasing potential of lactic acid bacteria in fermented soy milk. *Food Prod. Process. Nutr.* **2021**, *3*, 10. [CrossRef]
29. Hati, S.; Patel, N.; Pipaliya, R. Bioactivities and production of antihypertensive peptides during fermentation of soy milk by lactic cultures. *Rev. Res. Med. Microbiol.* **2023**, *34*, 79–88. [CrossRef]
30. Usha Rani, V.; Pradeep, B.V. Antioxidant properties of soy milk fermented with *Lactobacillus paracasei* KUMBB005. *Int. J. Pharm. Sci. Rev. Res.* **2015**, *30*, 39–42.
31. Bhatnagar, M.; Attri, S.; Sharma, K.; Goel, G. *Lactobacillus paracasei* CD4 as potential indigenous lactic cultures with antioxidative and ACE inhibitory activity in soymilk hydrolysate. *J. Food Meas. Charact.* **2018**, *12*, 1005–1010. [CrossRef]
32. Stockert, J.C.; Blázquez-Castro, A.; Cañete, M.; Horobin, R.W.; Villanueva, Á. MTT assay for cell viability: Intracellular localization of the formazan product is in lipid droplets. *Acta Histochem.* **2012**, *114*, 785–796. [CrossRef] [PubMed]
33. Kim, Y.M.; Huh, J.S.; Lim, Y.; Cho, M. Soy isoflavone glycitin (4'-Hydroxy-6-Methoxyisoflavone-7-D-Glucoside) promotes human dermal fibroblast cell proliferation and migration via TGF-β signaling. *Phytother. Res.* **2015**, *29*, 757–769. [CrossRef] [PubMed]
34. Zhang, S.Y.; Hood, M.; Zhang, I.X.; Chen, C.L.; Zhang, L.L.; Du, J. Collagen and soy peptides attenuate contractile loss from UVA damage and enhance the antioxidant capacity of dermal fibroblasts. *J. Cosmet. Dermatol.* **2021**, *20*, 2277–2286. [CrossRef] [PubMed]
35. Kim, E.J.; Han, M.R.; Lee, S.Y.; Kim, A.J. Effect of fermented soybean on the proliferation and growth in HaCaT and fibroblast cell. *J. Korea Acad. -Ind. Coop. Soc.* **2021**, *22*, 326–335. [CrossRef]
36. Cavinato, M.; Jansen-Dürr, P. Molecular mechanisms of UVB-induced senescence of dermal fibroblasts and its relevance for photoaging of the human skin. *Exp. Gerontol.* **2017**, *94*, 78–82. [CrossRef]
37. Hong, G.P.; Min, S.G.; Jo, Y.J. Anti-oxidative and anti-aging activities of porcine by-product collagen hydrolysates produced by commercial proteases: Effect of hydrolysis and ultrafiltration. *Molecules* **2019**, *24*, 1104. [CrossRef]
38. Moure, A.; Domínguez, H.; Parajó, J.C. Antioxidant properties of ultrafiltration-recovered soy protein fractions from industrial effluents and their hydrolysates. *Process Biochem.* **2006**, *41*, 447–456. [CrossRef]
39. Fineschi, S.; Cozzi, F.; Burger, D.; Dayer, J.M.; Meroni, P.L.; Chizzolini, C. Anti-fibroblast antibodies detected by cell-based ELISA in systemic sclerosis enhance the collagenolytic activity and matrix metalloproteinase-1 production in dermal fibroblasts. *Rheumatology* **2007**, *46*, 1779–1785. [CrossRef]
40. Shi, X.; Shang, F.; Zhang, Y.; Wang, R.; Jia, Y.; Li, K. Persimmon oligomeric proanthocyanidins alleviate ultraviolet B-induced skin damage by regulating oxidative stress and inflammatory responses. *Free Radic. Res.* **2020**, *54*, 765–776. [CrossRef]
41. Yang, E.J.; Yun, S.H.; Ko, J.H.; Kang, H.K.; Lee, J.N.; Park, S.M.; Hyun, C.G. Protective effect of *Rhododendron weyrichii* flower extract against UVB-induced proinflammatory cytokine production in human keratinocytes. *J. Appl. Pharm. Sci.* **2019**, *9*, 15–19. [CrossRef]
42. Yang, M.; Kwak, J.S.; Jang, S.; Jia, Y.; Park, I. Fermentation characteristics of soybean yogurt by mixed culture of *Bacillus* sp. and lactic acid bacteria. *Korean J. Food Nutr.* **2013**, *26*, 273–279. [CrossRef]
43. Lee, M.S.; Lee, W.J.; Kim, D.S.; Park, J.H.; Kang, J.H. Production of the bacteriocin from the tofu-residue. *J. Korean Soc. Food Sci. Nutr.* **1999**, *28*, 74–80.
44. Shihata, A.; Shah, N.P. Proteolytic profiles of yogurt and probiotic bacteria. *Int. Dairy J.* **2000**, *10*, 401–408. [CrossRef]
45. Tsai, J.S.; Lin, Y.S.; Pan, B.S.; Chen, T.J. Antihypertensive peptides and γ-aminobutyric acid from prozyme 6 facilitated lactic acid bacteria fermentation of soymilk. *Process Biochem.* **2006**, *41*, 1282–1288. [CrossRef]
46. Kobayashi, M.; Shima, T.; Fukuda, M. Metabolite profile of lactic acid-fermented soymilk. *Food Nutr. Sci.* **2018**, *9*, 1327–1340. [CrossRef]
47. Sun, L.; Tan, K.; Siow, P.; Henry, C. Soya milk exerts different effects on plasma amino acid responses and incretin hormone secretion compared with cows' milk in healthy, young men. *Br. J. Nutr.* **2016**, *116*, 1216–1221. [CrossRef]
48. Boulos, S.; Tännler, A.; Nyström, L. Nitrogen-to-protein conversion factors for edible insects on the Swiss market: *T. molitor*, *A. domesticus*, and *L. migratoria*. *Front. Nutr.* **2020**, *7*, 89. [CrossRef]
49. Draelos, Z.D. The science behind skin care: Moisturizers. *J. Cosmet. Dermatol.* **2018**, *17*, 138–144. [CrossRef]
50. Singh, B.P.; Vij, S. Growth and bioactive peptides production potential of Lactobacillus planatarum strain C2 in soy milk: A LC-MS/MS based revelation for peptides biofunctionality. *LWT-Food Sci. Technol.* **2017**, *86*, 293–301. [CrossRef]

51. Lu, W.; Liu, J.; Gao, B.; Lv, X.; Yu, L. Technical note: Nontargeted detection of adulterated plant proteins in raw milk by UPLC-quadrupole time-of-flight mass spectrometric proteomics combined with chemometrics. *J. Dairy Sci.* **2017**, *100*, 6980–6986. [CrossRef]

52. Vasconcellos, F.C.S.; Woiciechowski, A.L.; Soccol, V.T.; Mantovani, D.; Soccol, C.R. Antimicrobial and antioxidant properties of β-conglycinin and glycinin from soy protein isolate. *Int. J. Curr. Microbiol. Appl. Sci.* **2014**, *3*, 144–157.

53. Mortazavi, S.; Moghimi, H.; Maibach, H. Chemical modification: An important and feasible method for improving peptide and protein dermal and transdermal delivery. In *Percutaneous Absorption: Drugs, Cosmetics, Mechanisms, Methods*, 5th ed.; Dragićević, N., Maibach, H., Eds.; CRC Press: Boca Raton, FL, USA, 2021; pp. 459–468.

54. Abu Samah, N.H.; Heard, C.M. Topically applied KTTKS: A review. *Int. J. Cosmet. Sci.* **2011**, *33*, 483–490. [CrossRef]

55. Benson, H.A.; Namjoshi, S. Proteins and peptides: Strategies for delivery to and across the skin. *J. Pharm. Sci.* **2008**, *97*, 3591–3610. [CrossRef]

56. Dominik, I.; Eileen, J.; Marc, H.; Remo, C.; Eliane, W.; Hugo, Z. Activation of TGF-β: A gateway to skin rejuvenation. *HPC Today* **2015**, *10*, 1–6.

57. Choi, Y.L.; Park, E.J.; Kim, E.; Na, D.H.; Shin, Y.H. Dermal stability and in vitro skin permeation of collagen pentapeptides (KTTKS and palmitoyl-KTTKS). *Biomol. Ther.* **2014**, *22*, 321–327. [CrossRef]

58. Guttman, C. Studies demonstrate value of procollagen fragment Pal-KTTKS. *Dermatol. Times* **2002**, *23*, 68.

59. Kwak, S.J.; Kim, C.S.; Choi, M.S.; Park, T.; Sung, M.K.; Yun, J.W.; Yoo, H.; Mine, Y.; Yu, R. The soy peptide Phe–Leu–Val reduces TNFα-induced inflammatory response and insulin resistance in adipocytes. *J. Med. Food* **2016**, *19*, 678–685. [CrossRef] [PubMed]

60. Kovacs-Nolan, J.; Zhang, H.; Ibuki, M.; Nakamori, T.; Yoshiura, K.; Turner, P.V.; Matsui, T.; Mine, Y. The PepT1-transportable soy tripeptide VPY reduces intestinal inflammation. *Biochim. Biophys. Acta* **2012**, *1820*, 1753–1763. [CrossRef] [PubMed]

61. Shidal, C.; Al-Rayyan, N.; Yaddanapudi, K.; Davis, K.R. Lunasin is a novel therapeutic agent for targeting melanoma cancer stem cells. *Oncotarget* **2016**, *20*, 84128–84141. [CrossRef]

62. Hirobe, T. Role of keratinocyte-derived factors involved in regulating the proliferation and differentiation of mammalian epidermal melanocytes. *Pigment Cell Res.* **2005**, *18*, 2–12. [CrossRef]

63. Riley, P.A. Melanogenesis and melanoma. *Pigment Cell Res.* **2003**, *16*, 548–552. [CrossRef]

64. Benn, E.K.T.; Alexis, A.; Mohamed, N.; Wang, Y.H.; Khan, I.A.; Liu, B. Skin bleaching and dermatologic health of African and Afro-Caribbean populations in the US: New directions for methodologically rigorous, multidisciplinary, and culturally sensitive research. *Dermatol. Ther.* **2016**, *6*, 453–459. [CrossRef]

65. Del Giudice, P.; Yves, P. The widespread use of skin lightening creams in Senegal: A persistent public health problem in West Africa. *Int. J. Dermatol.* **2002**, *41*, 69–72. [CrossRef] [PubMed]

66. Dadzie, O.E.; Petit, A. Skin bleaching: Highlighting the misuse of cutaneous depigmenting agents. *J. Eur. Acad. Dermatol. Venereol.* **2009**, *23*, 741–750. [CrossRef]

67. Gui, M.; Du, J.; Guo, J.; Xiao, B.; Yang, W.; Li, M. Aqueous extract of *Chrysanthemum morifolium* enhances the antimelanogenic and antioxidative activities of the mixture of soy peptide and collagen peptide. *J. Tradit. Complement. Med.* **2014**, *4*, 171–176. [CrossRef] [PubMed]

68. Bodurlar, Y.; Caliskan, M. Inhibitory activity of soybean (*Glycine max* L. Merr.) Cell Culture Extract on tyrosinase activity and melanin formation in alpha-melanocyte stimulating Hormone-Induced B16-F10 melanoma cells. *Mol. Biol. Rep.* **2022**, *49*, 7827–7836. [CrossRef] [PubMed]

Review

Recent Perspective of *Lactobacillus* in Reducing Oxidative Stress to Prevent Disease

Tingting Zhao [1,2,†], Haoran Wang [1,2,†], Zhenjiang Liu [3], Yang Liu [2], DeJi [2], Bin Li [2,*] and Xiaodan Huang [1,2,*]

[1] School of Public Health, Lanzhou University, Lanzhou 730033, China
[2] Institute of Animal Husbandry and Veterinary, Tibet Academy of Agricultural and Animal Husbandry Sciences, Lhasa 850000, China
[3] National Engineering Laboratory for AIDS Vaccine, School of Life Sciences, Jilin University, Changchun 130012, China
* Correspondence: xukesuolibin@163.com (B.L.); huangxiaodan@lzu.edu.cn (X.H.)
† These authors contributed equally to this work.

Abstract: During oxidative stress, an important factor in the development of many diseases, cellular oxidative and antioxidant activities are imbalanced due to various internal and external factors such as inflammation or diet. The administration of probiotic *Lactobacillus* strains has been shown to confer a range of antibacterial, anti-inflammatory, antioxidant, and immunomodulatory effects in the host. This review focuses on the potential role of oxidative stress in inflammatory bowel diseases (IBD), cancer, and liver-related diseases in the context of preventive and therapeutic effects associated with *Lactobacillus*. This article reviews studies in cell lines and animal models as well as some clinical population reports that suggest that *Lactobacillus* could alleviate basic symptoms and related abnormal indicators of IBD, cancers, and liver damage, and covers evidence supporting a role for the Nrf2, NF-κB, and MAPK signaling pathways in the effects of *Lactobacillus* in alleviating inflammation, oxidative stress, aberrant cell proliferation, and apoptosis. This review also discusses the unmet needs and future directions in probiotic *Lactobacillus* research including more extensive mechanistic analyses and more clinical trials for *Lactobacillus*-based treatments.

Keywords: *Lactobacillus*; oxidant stress; IBD; cancer; liver diseases; MAPK; NF-κB; Nrf2

Citation: Zhao, T.; Wang, H.; Liu, Z.; Liu, Y.; DeJi; Li, B.; Huang, X. Recent Perspective of *Lactobacillus* in Reducing Oxidative Stress to Prevent Disease. *Antioxidants* **2023**, *12*, 769. https://doi.org/10.3390/antiox12030769

Academic Editor: Myung-Ji Seo

Received: 16 February 2023
Revised: 19 March 2023
Accepted: 20 March 2023
Published: 21 March 2023

1. Introduction

1.1. Oxidative Stress and Disease

Oxidative stress is associated with the accumulation of excess reactive oxygen and reactive nitrogen species (ROS and RNS, respectively, see Table S1). In cells, ROS/RNS are continuously produced through metabolic processes. Low to moderate ROS levels serve as essential secondary messengers in cell signaling, regulating cell proliferation, differentiation, and migration, and participate in triggering cellular survival mechanisms. However, excessive ROS/RNS can cause oxidative damage to a variety of biomolecules such as unsaturated fatty acids, proteins, and DNA [1]. Endogenous ROS are principally produced in mitochondria, the endoplasmic reticulum, and peroxisomes, with the vast majority of ROS generated by the mitochondrial electron transport chain [2,3]. In addition, numerous enzymes are known to participate in catalyzing the production of endogenous ROS/RNS such as peroxidases, nicotinamide adenine dinucleotide phosphate (NADPH) oxidase (NOX), myeloperoxidase (MPO), nitric oxide synthase (NOS), lipoxygenases (LOXs), and cyclooxygenases (COXs) [4,5]. Beyond these enzymatic pathways for ROS/RNS production, exposure to a wide range of exogenous factors such as trans-fatty acids, iron (Fe), and copper (Cu) in foods, ultraviolet radiation, various drugs and xenobiotics as well as chronic infection and inflammatory diseases can also contribute to inducing ROS accumulation [5–8].

In humans and other mammals, antioxidant defense systems can largely counteract the effects of ROS/RNS. Endogenous antioxidant systems are generally classified as belonging to either the intracellular enzymatic antioxidants, comprising superoxide dismutase (SOD), catalase (CAT), and other such antioxidant enzymes, or the intracellular non-enzymatic antioxidants, which rely on the activity of small molecules in conjunction with enzymes to neutralize ROS/RON. In the latter system, glutathione functions as a significant antioxidant barrier in the gut, along with three related enzymes: glutathione peroxidase (GPX), glutathione reductase (GSR), and glutathione S-transferase (GST) [1,9]. In addition, the melatonin (MEL) and thioredoxin (Trx) system are well-studied, essential, intracellular antioxidants [10,11]. In addition to the above, there are also some extracellular antioxidants such as vitamin A/C or β-carotene obtained from vegetables and fruits, or flavonoids from certain plants [12,13]. Oxidative stress can occur when ROS/RNS generation exceeds the cellular capacity for scavenging activity of the antioxidant systems or as a result of the dysregulation of antioxidant pathways [5], leading to the damage of various physiological systems in the body. Some examples of the pathological damage related to ROS/RNS accumulation include atherosclerosis in the cardiovascular system; neurodegenerative diseases of the nervous system such as Alzheimer's disease and Parkinson's disease, rheumatoid arthritis or systemic lupus erythematosus of the autoimmune system, as well as IBD, stomach cancer, colorectal cancer, and other diseases of the digestive system [14,15].

Alleviating oxidative stress is thus essential for treating many diseases and ensuring systemic health. As our understanding of ROS/RNS and antioxidant mechanisms expands, more precise therapeutic interventions can be developed that focus on disease-related sources and targets of ROS/RNS, some of which are currently entering the clinical trial stages of testing [16]. For instance, free iron is a promising target to control the site and extent of highly aggressive hydroxyl radical generation, which could benefit the treatment of oxidative stress-related diseases [8]. There is also potential for the application of more targeted ROS/RNS inhibitors such as site-specific suppressors of superoxide production (i.e., S1QELs) in quinone-mediated reactions [14,17]. In addition, probiotics such as *Lactobacillus* and *Bifidobacterium* may serve as effective complementary therapies that enhance antioxidant defenses via ROS/RNS removal, the inhibition of pro-oxidative enzymes, and the synthesis of antioxidant enzymes.

1.2. Antioxidant Activity of Lactic Acid Bacteria

Lactobacillus, a genus comprising Gram-positive, aerotolerant, rod-shaped, non-sporulating species, is commonly found among microbiota in the gastrointestinal tract and is also frequently isolated from many fermented food products. This genus currently includes 253 published, validated species [18], and accounts for an estimated 6.0% of total bacterial cell numbers in the human duodenum and ~0.3% of all bacteria in the colon [19]. The beneficial and functional properties of *Lactobacillus* in promoting human health have been extensively documented in recent years. Members of *Lactobacillus* have been shown to adhere to the intestinal epithelium and produce antimicrobial metabolites such as ethanol, hydrogen peroxide (H_2O_2), acetic acid, lactic acid, and/or succinic acid, which can antagonize other, potentially pathogenic, bacteria [20]. Some *Lactobacillus* strains can reportedly enhance cellular immune responses via the activation of macrophages, natural killer (NK) cells, and/or antigen-specific cytotoxic T-lymphocytes, or by stimulating the release of various cytokines. Moreover, some *Lactobacillus* species could also improve the immune response in gut mucosa by promoting the recruitment of IgA(+) cells [21]. Some *Lactobacillus* species may also exhibit direct antioxidant activity, and a few species are known to possess oxidative stress resistance genes and proteins pivotal for redox mechanisms such as *katA* expressed by *L. sakei*, or thioredoxin antioxidant system proteins expressed by some *L. plantarum* and *L. casei* strains [18]. Apart from these effects, *Lactobacillus* species also produce functional products such as exopolysaccharides (EPS), which participate in mitigating oxidative damage or activating the expression of host transcription factors involved in modulating cellular oxidative stress [22,23].

In recent years, several studies have investigated the relationship between *Lactobacillus* and oxidative stress in the host. Due to the purported beneficial role of *Lactobacillus* in human health, research has increasingly focused on the positive impacts of these species in ameliorating oxidative stress and related diseases [18]. Moreover, further in-depth exploration of the role of oxidative stress in disease will provide new perspectives that contribute to improving clinical treatments. This review covers the pathological mechanisms of oxidative stress in inflammatory bowel disease, cancer, and liver-related diseases as well as the preventive and therapeutic effects of *Lactobacillus* in patients with these diseases.

2. Inflammatory Bowel Disease

2.1. Inflammatory Bowel Disease and Oxidative Stress

Inflammatory bowel disease (IBD) including Crohn's disease (CD) and ulcerative colitis (UC) is characterized by chronic, non-specific intestinal inflammation. CD can affect the entire alimentary tract from the oral cavity to the anus (mainly the terminal ileum and adjacent colon) and include lesions that appear in a segmental or stochastic (i.e., not continuous) distribution and commonly lead to complications such as abscesses, fistula, and stenosis. In contrast, these lesions are absent in UC, which is characterized by mucosal inflammation and is limited to the colon [24]. Previous studies have shown that individual genetic susceptibility, external environment, gut microbiota, immune response, and oxidative stress are all contributing factors that can be firmly linked to the pathogenesis of IBD [25,26], with lipid peroxidation (LPO) and ROS accumulation, in particular, serving as important causes of oxidative stress in the intestine.

Several points have been proposed regarding the relationship between oxidative stress and IBD. First, researchers have identified several genetic risk loci associated with IBD, and mutations in genes encoding antioxidant/biotransformation enzymes can negatively impact their activity and increase the risk of IBD. For example, case-control studies found that inter-individual polymorphisms related to a C609T conversion in NAD(P)H: quinone oxidoreductase 1 (NQO1), an enzyme involved in inflammation and oxidative stress response, might play a significant role in the development of colon cancer and could influence steroid resistance in UC patients [27,28]. In addition, a polymorphism in the promoter region of the Nuclear factor E2-related factor 2 (Nrf2) gene, which participates in regulating the expression of detoxifying and antioxidant proteins in the intestine, has been associated with UC development in a Japanese population [29]. Second, mucosal immune cells and intestinal epithelial cells (IECs) are involved in oxidative stress in the intestine and have been shown to play a central role in the pathogenesis of IBD. During mucosal inflammation, IECs as well as neutrophils and macrophages, produce superoxide and nitric oxide by activating NOX and inducible nitric oxide synthase (iNOS), respectively, both induced by inflammatory cytokines, eventually leading to the production of more ROS/RNS. The ROS/RNS overload alters the structure and function of the intestinal epithelium, damages cytoskeletal proteins, and accelerates cellular damage through lipid peroxidation, ultimately resulting in barrier destruction [5,30,31]. Third, mutual interactions between gut microbiota and oxidative stress have been identified in IBD. Notably, the composition of gut microbial taxa is altered in CD patients compared to that in healthy controls, characterized by increased relative abundance of *Bacteroidetes* and decreased *Firmicutes*. In addition, *Enterobacteriaceae*, especially *E. coli*, are enriched in CD patients [32]. ROS/RNS are associated with intestinal dysbiosis. During inflammation, gut microbiota can directly generate ROS/RNS, or indirectly induce the production of excessive ROS/RNS by activating neutrophils or gastrointestinal epithelial cells, stimulating an initial inflammatory response via positive feedback, which leads to further ROS/RNS production, aggravating intestinal oxidative stress and damaging tissues [5,33]. Therefore, restoring the normal structure of gut microbiota has positive implications for the alleviation of inflammation and oxidative stress in IBD. Lifestyle-associated factors can also contribute to gastrointestinal oxidative stress. Although the underlying mechanisms remain unclear, many studies have shown that nicotine intake from smoking can play a dual role in CD and UC, and smoking is a risk factor for the occurrence of CD, but has apparently protective effects in UC, according to epidemiological data [34,35]. Similarly, wine consumption may have various effects in IBD patients. Some chemicals such as polyphenols in red wine appear to confer antioxidant properties [36], whereas the alcohol itself can lead to inflammation and oxidative stress in the liver and intestines, and chronic alcohol consumption leads to increased risk of gastric or colon cancer [37].

At present, treatments for IBD are still based on conventional anti-inflammatory agents (e.g., sulfasalazine and mesalazine, etc.) and immune modulators (e.g., thiopurines and cyclosporine, etc.). However, these drugs have severe side effects [38]. The incidence of UC and CD are higher in regions of Asia with high population density and coastal areas of China also have high rates of IBD, indicating that newer, more effective therapeutic interventions for IBD are urgently needed [39]. The beneficial effects of *Lactobacillus* in IBD have stimulated considerable research attention that has led to increased the clinical application of probiotic *Lactobacillus* strains as alternative or complementary therapies [40] (Figure 1).

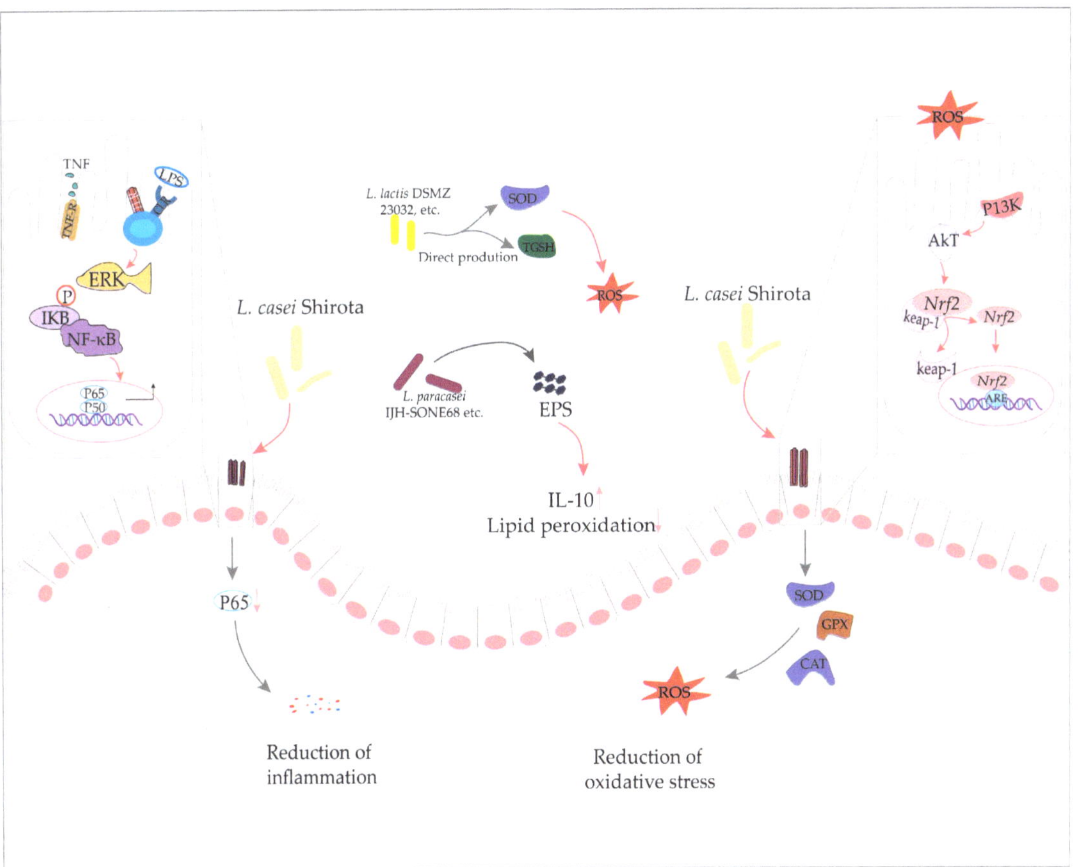

Figure 1. The role of *Lactobacillus* in IBD. *Lactobacillus* can produce EPS or/and related antioxidant enzymes to alleviate intestinal inflammation and oxidative stress, and ameliorate IBD by regulating the NF-κB, Nrf2 signaling pathway.

2.2. The Effects Lactobacillus in Inflammatory Bowel Disease

A recent study exploring the effects of EPS-producing probiotic bacteria found that antioxidant enzyme activities were higher and lipid peroxidation was significantly ameliorated in UC model rats treated with probiotics compared with those in untreated UC model rats. More importantly, improvements to oxidative stress were significantly greater in rats administered with a high EPS producing strain than in those treated with a low EPS strain [41]. Similarly, another study showed that administering a novel EPS produced by *L paracasei* IJH-SONE68 led to reduced macrophage inflammatory protein 2 mRNA levels, and the increased expression of the anti-inflammatory cytokine, interleukin-10 (IL-10) in dextran sulfate sodium (DSS)-induced UC model mice [42] (Figure 1). These results suggest that EPS should be further examined for application in new treatments for IBD in the future, and *Lactobacillus* would be among one of the best choices for EPS production.

In the IBD model mice, administration of *L. plantarum* ZS62 was found to inhibit DSS-induced colonic atrophy. Moreover, mice supplemented with strain ZS62 had lower serum levels of oxidative stress indicators (i.e., Malondialdehyde and MPO) and inflammatory indicators (i.e., IL-1β, IL-6, IL-12, TNF-α, etc.) at both the mRNA and protein levels and elevated protein and mRNA levels of antioxidant enzymes (i.e., CAT, T-SOD) and IL-10 [43] In addition, Chorawala and colleagues reported that exposure to the cell wall components of *L. casei*, *L. acidophilus*, and *L. rhamnosus* could attenuate lipopolysaccharide (LPS)-induced rats colitis by modulating the inflammatory immune response and oxidative stress [44]. Studies have suggested that *Lactobacillus* can improve the hosts' antioxidant level. For example, Erdogan et al. found that feeding fermented Kefir significantly increased the serum total antioxidant status (TAS) levels in Balb/c mice [45]. Similarly, in a healthy volunteer trial (Application No. WO03002131), study subjects who consumed *L. fermentum* ME-3 in fermented goat milk or who received capsules with *L. fermentum* ME-3 displayed significantly higher blood total antioxidative activity and TAS than subjects who did not [46]. Furthermore, a pediatric, randomized, placebo-controlled trial demonstrated that a highly concentrated mixture of probiotic bacterial strains (VSL#3) was both safe and effective in maintaining remission among patients with active UC [47]. However, another trial in which CD patients were administered 1×10^9 CFUs of the probiotic strain *L. johnsonii* LA1 four times daily showed no preventive effects on endoscopic disease recurrence [48].

Assessment of the antioxidant capacity of 34 lactic acid bacteria strains in vitro by Amaretti et al. indicated that strains such as *L. lactis* DSMZ 23032, *L. acidophilus* DSMZ 23033, and *L. brevis* DSMZ 23034 produced relatively high intracellular levels of glutathione and SOD [49], suggesting that LAB might be effective for delivering these antioxidant enzymes to the gut to alleviate oxidative stress in IBD (Figure 1). Furthermore, the co-expression of antioxidant enzyme genes in *Lactobacillus* through genetic engineering could enhance their antioxidant capacity while also increasing their viability in the host by several-fold, supporting their potential development in therapeutics for IBD [50–52].

Finamore et al. showed that pretreatment of the enterocyte-like cell line TC7/human colon carcinoma cell line (Caco-2) with *L. casei* Shirota (LcS) could prevent membrane barrier disruption and intercellular ROS accumulation, significantly increase the gastrointestinal expression of GPX, and reduce p65 phosphorylation. These findings support the involvement of theNfr2 and Nuclear Factor-Kappa B (NF-κB) pathways in the activation of antioxidant cellular defenses [53] (Figure 1). However, the induction of GPX has also been shown to occur in a Nrf2-independent manner via activation of the transcription factor, STAT3 [54]. In short, some *Lactobacillus* strains appear to confer positive effects in maintaining intestinal homeostasis via enhanced antioxidant capacity and inflammatory response, although the specific mechanisms through which *Lactobacillus* can alleviate oxidative stress in IBD remain unclear.

Although *Lactobacillus* is generally regarded as a safe microorganism, several studies have indicated that certain *species* may pose risks such as bacteremia and endocarditis [55,56]. Therefore, the application of *Lactobacillus* for the treatment of IBD raises concerns regarding the presence of virulence genes and antibiotic resistance genes, as their potential transfer to pathogenic microorganisms could constitute a risk [57].

3. Cancer

3.1. ROS and Cancer

Cancer is among the leading causes of death worldwide. According to the Global Cancer Statistics, approximately 19.3 million new cancer cases and nearly 10 million deaths from cancer were reported in 2020 [58]. Tumorigenesis may result from a combination of internal and external factors, some of which influence cancer progression and metastasis through increased ROS production [59]. The generation of ROS in cancer cells is primarily mediated by oxidative phosphorylation in the electron transport chain on the mitochondrial inner membrane. In addition, transition metals such as iron can also generate ROS non-enzymatically via a Fenton reaction, while exposure to various external factors such as air pollutants, radiation, foods, or drugs, etc. can also exogenously induce ROS production [60]. In general, the aberrant and uncontrolled proliferation of cancer cells requires large amounts of ATP as an energy supply, and the resulting metabolic upregulation leads to elevated accumulation of ROS. However, cancer cells have a characteristically high antioxidant capacity due to the Nrf2-mediated activation of various antioxidant response elements (AREs) that can restrict ROS accumulation to levels compatible with cellular biological functions, albeit still higher than that in normal cells [61].

At low to moderate levels, ROS may promote abnormal cell proliferation and differentiation by inducing DNA mutations These mutations mainly include DNA double strand breaks accompanied by abnormalities in their associated repair pathways, or by generating 8-oxo-7-hydrodeoxyguanosine (8-oxodG), which is recognized as an indicator of oxidative damage to DNA [62,63]. Other work has linked Fe-catalyzed ROS accumulation to the development of some cancers. For instance, a multivariate statistical analysis indicated that iron depletion therapy, either through phlebotomy or by maintaining a low-iron diet, could significantly lower the risk of hepatocellular carcinoma compared to that in untreated patients [64,65].

In addition, ROS may also contribute to tumorigenesis through their function as signal molecules (Figure 2). Increased ROS production in cells can induce an enzymatic cascade reaction that results in the activation of extracellular signal-regulated kinase (ERK), c-Jun N-terminal kinase (JNK), and p38 mitogen-activated protein kinases (MAPK), all of which are linked to tumor cell growth and survival [66,67]. Weinberg and colleagues observed that mitochondrial ROS are essential for Kras-mediated tumorigenesis in a murine model of lung carcinoma via ERK-MAPK signaling [59,68]. The phosphatidylinositol-3-kinase (PI3K)/phosphatase and tensin homolog deleted on chromosome 10 (PTEN) signaling pathway is also important for tumorigenesis and cancer metastasis. ROS can stimulate PI3K/protein kinase B (AKT) signaling through the oxidation of a cysteine thiol group on various phosphatases (e.g., PTEN, PTP1B, PP2A) and contribute to dysregulation in a wide range of human cancers (e.g., endometrial, breast, thyroid, and prostate cancers) [69,70]. In addition to the above pathways, NF-κB is increasingly recognized as a key player in many steps of cancer development and progression. Treatment of breast carcinoma cells with IL-1β, TNFα, or sodium arsenite generates H_2O_2 and O^{2-}, which in turn activate NF-κB and enhance uncontrolled cellular growth [71,72]. Constitutive activation of Nrf2 has been observed in various human cancers such as those affecting the lungs, breasts, and skin [73,74]. In addition, deregulation of the Nrf2–Keap1 pathway has been reported in multiple cancer cells, leading to drug resistance, genomic instability, resistance to apoptosis, metastasis, and metabolic [75].

However, high levels of ROS have contrary effects on cancer, promoting cell death and severe cellular damage. Some chemotherapeutic drugs or cytotoxic agents such as anthracyclines, platinum-based drugs (e.g., cisplatin, carboplatin), and other alkylating drugs exert therapeutic effects by stimulating the high intracellular production of ROS [76,77]. In addition, excessive ROS levels are also positively correlated with apoptosis through the

ASK-1/JNK/p38 signaling cascade induced via the H_2O_2-mediated oxidation of cysteine residues on TRX, which leads to the dissociation of ASK-1, the subsequent activation of JNK and p38 signaling, and ultimately apoptosis [59,78] (Figure 2).

Thus, ROS can serve as a "double-edged sword", both causing and suppressing cancer [79], and current studies suggest that modulating the oxidative stress response might be an effective, new potential approach for cancer therapies.

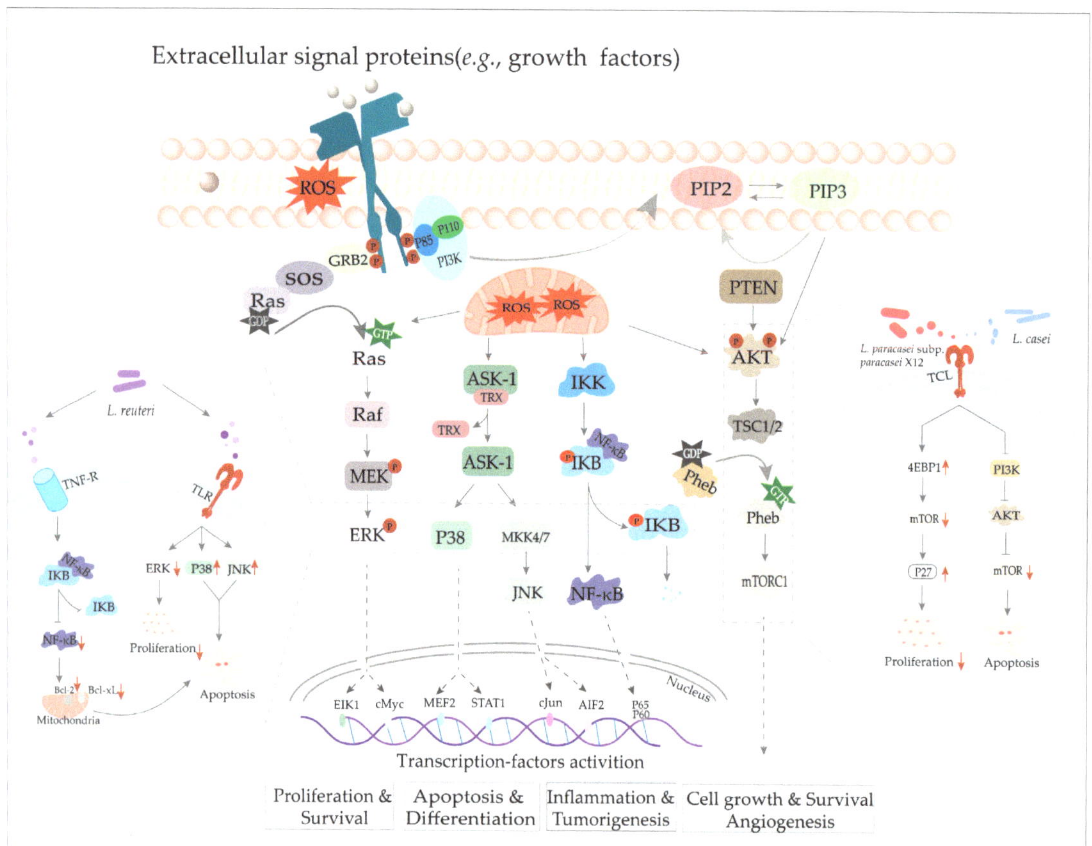

Figure 2. ROS induces the activation of the MAPK, NF-κB, and PI3K/PTEN signaling pathways to promote cancer development, progression, and apoptosis, etc. and aa possible target of LAB [59].

3.2. The Role of Lactobacillus in Cancer

Drug resistance is a serious problem in cancer chemotherapy, and thus effective alternative therapeutic options are desperately needed to improve treatment response in many cancer patients. To address this issue, an increasing number of studies has investigated the potential antitumor activity of *Lactobacillus* and provide evidence supporting its further exploration and development for use in candidate anti-tumor therapies [80,81].

In many cancers, abnormal cell proliferation is tightly linked to the suppression of apoptosis, and many drugs achieve therapeutic efficacy by promoting apoptosis in cancer cells. Notably, *Lactobacillus* has been shown to induce apoptosis [82,83]. For example, oral pretreatment with the probiotic strain, *L. acidophilus* NCFM, was found to stimulate apoptosis in CT-26 cell dorsal flank xenograft tumors or in segmental orthotopic colon cancer in mice [84]. Another study showed that the apoptosis levels increased in the

LS513 colorectal cancer cells exposed to a combination of the *L. acidophilus* and *L. casei* strains in the presence of 5-fluorouracil (5-FU) in a dose-dependent manner. This effect was potentially due to the faster activation of caspase-3 and the downregulation of the p21 protein [85]. Similarly, Altonsy et al. found that the *L. rhamnosus* GG strain (LGG) could induce caspase-9 and caspase-3 activation, promoting apoptosis in the Caco-2 colon cancer cells [86]. Homogenates of strain LGG were also found to induce the apoptosis pathways in HGC-27 human gastric cancer cells via reduced expression of the anti-apoptotic protein, Bcl-2, and elevated expression of the pro-apoptotic protein, Bax [87]. Similarly, ubiquitination of IKB was suppressed in human chronic myeloid leukemia-derived cells (KBM-5 cells) treated with *L. reuteri* ATCC PTA 6475, resulting in the downregulation of the NF-κB-dependent gene products (e.g., Bcl-2, Bcl-xL), while pre-treatment with *L. reuteri* enhanced JNK and p38 phosphorylation, but suppressed ERK1/2 signaling in cells, increasing the apoptosis levels and inhibiting proliferation, respectively [88]. In addition, Hwang and colleagues confirmed that extracts of *L. casei* could induce apoptosis in gastric cancer cells by suppressing PI3K/AKT signaling [89] (Figure 2). Thus, numerous studies have shown that *Lactobacillus* strains can regulate apoptosis in cancer cells via the modulation of endogenous signaling pathways or exogenous cues.

Lactobacillus strains and their metabolites (i.e., components of the cell wall and cytoplasm) have been documented to inhibit cancer cell proliferation. For instance, Orlando and colleagues investigated the anti-proliferative effects of the cell wall and cytoplasmic fractions of strain LGG on the HGC-27 cells and DLD-1 human colonic adenocarcinoma cells. They found that both HGC-27 and DLD-1 cells were resistant to the bacterial cell wall fraction, whereas treatment with the cytoplasmic fraction resulted in apparently reduced proliferation [90]. 4E-Binding protein 1 (4EBP1) is a transcriptional inhibitor in the downstream of the mTOR signaling pathway. Treatment with *L. paracasei* subsp. *paracasei* X12 could induce cell cycle arrest at G1 in the HT-29 cells through the mTOR-4EBP1-p27 signaling pathway, and the mRNA expression of 4EBP1 was increased several folds after treatment by the X12 strain [91]. Hence, blocking hyperactivation of the PI3K/AKT/mTOR signaling pathway has emerged as a plausible target for cancer therapies due to its involvement in cell growth and proliferation (Figure 2).

Polyamines including putrescine, spermidine, and spermine, etc. can serve as nutrient substrates for cancer cell growth and may thus play an important role in cancer cell proliferation. In gastric cancer cells, probiotics such as *Lactobacillus* strains can potentially act as anti-neoplastic agents to inhibit proliferation by modulating polyamine contents or function [92]. *L. salivarius* strains FP25 and FP35, isolated from infant feces, can directly adhere to cancer cells, triggering the biosynthesis of short-chain fatty acids (SCFAs), especially butyric and propionic acids, resulting in the inhibition of colon cancer cell proliferation [93]. In addition, some studies have reported that EPS secreted by *Lactobacillus* might also exert anti-proliferative effects on cancer cells [94–96].

A growing body of evidence supports that *Lactobacillus* strains can also provide antioxidant functions to prevent oxidative stress-related tumor development. A recently identified probiotic strain, *L. salivarius* REN, isolated from fecal samples of centenarians by Zhang and colleagues, was found to inhibit 4NQO (4-nitroquioline 1-oxide)-induced oral cancer in rats by reducing oxidative DNA damage and downregulating COX-2 expression [97]. In addition, *L. reuteri* is capable of regulating intestinal metabolites to promote a growth-repressive effect on cancer and reduce the oxidation level in mice with colorectal cancer [98]. In rats with colon cancer induced by 1,2-dimethylhydrazine (DMH), treatment with *L. plantarum* AS1 led to significant improvements in lipid peroxidation and antioxidant activities in the colon and plasma, and showed promising antioxidant effects in vitro [99]. Choi et al. found that heat-killed (HK) cells and soluble polysaccharide components of *L. acidophilus* 606 exhibited potent antioxidative activity, and further reported that the soluble polysaccharide fraction could induce apoptosis in HT-29 cells in vitro, suggesting that soluble polysaccharides of *Lactobacillus* strains could be explored as possible candidate anticancer agents [100]. Overall, many in vivo and in vitro studies support that *Lactobacillus* can reduce oxidative damage, attenuating DNA mutation and cancer development.

Lactobacillus can also exert immunomodulatory and cytotoxic anticancer effects. For instance, *L. casei* strains 9018 and LcS could activate NK cells by inducing the production of cytokines such as IL-12 and IFN-γ, and highly active NK cells are reportedly associated with a lower risk of cancer development [101–103]. In addition, *L. acidophilus* 36YL shows no apparent cytotoxicity toward normal cells, but displays high toxicity toward cancer cell lines such as HeLa and HT-29 [104].

Although several in vivo and in vitro studies have reported that the administration of various *Lactobacillus* strains leads to the inhibition of cancer development and progression through pro-apoptotic, antioxidant, anti-proliferative, or immunomodulatory pathways, clinical population trials and epidemiological studies are still lacking. Indeed, some issues should be carefully considered in the administration of probiotic *Lactobacillus* strains for the treatment and prevention of cancers. In particular, the potential for undesirable side effects resulting from host–bacteria interactions, individual resistance to particular strains, or variability in the benefits conferred by different *Lactobacillus* strains, all warrant further experimental investigation [80].

4. Oxidative Stress-Related Liver Disease

4.1. Oxidative Stress and Liver Disease

Alcoholic Liver Disease (ALD): Long-term and excessive alcohol consumption is the leading cause of ALD, which includes a broad clinical-histological spectrum ranging from simple steatosis and hepatitis, to liver cirrhosis and hepatocellular carcinoma [105]. Disruption of intestinal microbiota, increased intestinal permeability, inflammatory response, liver oxidative stress, and lipid accumulation are all associated with the progression of ALD [106]. In particular, the role of oxidative stress in the occurrence and development of ALD is currently the focus of increasing research attention. Alcohol is mainly metabolized in the liver, leading to elevated ROS/RNS production, weakening antioxidant defense systems, and finally resulting in oxidative stress in the liver due to excess ROS/RNS. It should be noted that oxidative stress in the liver has been reported to play a key role in hepatocellular carcinogenesis [107].

Non-alcoholic fatty liver disease (NAFLD): NAFLD is mainly due to the accumulation of lipids in the liver, or insulin resistance caused by obesity, type 2 diabetes, and lipid metabolism disorders, among other conditions, resulting in hepatic steatosis. In addition, lipid peroxidation, oxidative stress, and pro-inflammatory cytokines are also involved, culminating in hepatocyte infiltration and necrosis. Although NAFLD has multiple clinical manifestations, most patients present with only simple steatosis [108,109].

4.2. The Preventive Effects of Lactobacillus

Preventive effects on ALD: *Lactobacillus* has been tested in both the treatment and prevention of ALD. Forsyth et al. found that rats fed with alcohol plus *L. rhamnosus* GG (ALC + LGG) had significantly lower severity of alcoholic steatohepatitis (ASH) than rats fed with alcohol alone, while strain LGG was also associated with significantly less alcohol-induced leakiness in the gut, oxidative stress, and inflammation in both the intestinal and liver tissues [110]. Another study comparing treatments of *Lactobacillus* strains with the traditional Chinese medicine Hu-Gan-Pian in a rat model of alcohol-induced enteric dysbiosis found that *L. rhamnosus* CCFM1107 administration could rescue dysbiosis and inhibit the characteristic increase in serum aminotransferase and endotoxin as well as triglyceride (TG) and cholesterol (CHO) levels in the serum and liver. In addition to these findings, the oxidation levels, indicated by the malondialdehyde contents, were lower, while the antioxidant enzyme (GSH/GSH-Px/SOD) levels were elevated in rats treated with *L. rhamnosus* CCFM1107 [111].

In mice with alcoholic subacute liver injury, treatment with *L. plantarum* ZS62, isolated from naturally fermented yogurt, could relieve morphological abnormalities in hepatocytes and reduce markers of liver damage (e.g., AST/ALT/hyaluronidase, etc.). Furthermore, inflammation-related genes were significantly downregulated while lipid- and oxidative-metabolism genes were upregulated, implying that *L. plantarum* ZS62 could potentially function as a beneficial prophylactic supplement for people with frequent alcohol consumption [112]. In a study examining alcoholic liver injury using a zebrafish model, *L. plantarum* was shown to provide some protective effects via the transcriptional activation of the Keap-Nrf2-ARE signal pathway [113]. In work by Li and coworkers investigating the potential mechanism and effects of *Lactobacillus* in alleviating ALD symptoms in mice, treatment with a mixture of *L. plantarum* KLDS1.0344 and *L. acidophilus* KLDS1.0901 resulted in higher SCFA production by the gut microbiota. These SCFAs could potentially inhibit alcohol-induced lipid accumulation, oxidative stress, and inflammation in the liver through hepatic-intestinal circulation, most likely via activation of the AMPK, Nrf2, and TLR4/NF-κB pathways [106] (Figure 3). Collectively, these studies show that *Lactobacillus* strains could provide preventive effects against mild ALD, but long-term experiments in animal models are necessary to determine whether these strains could also alleviate end-stage ALD.

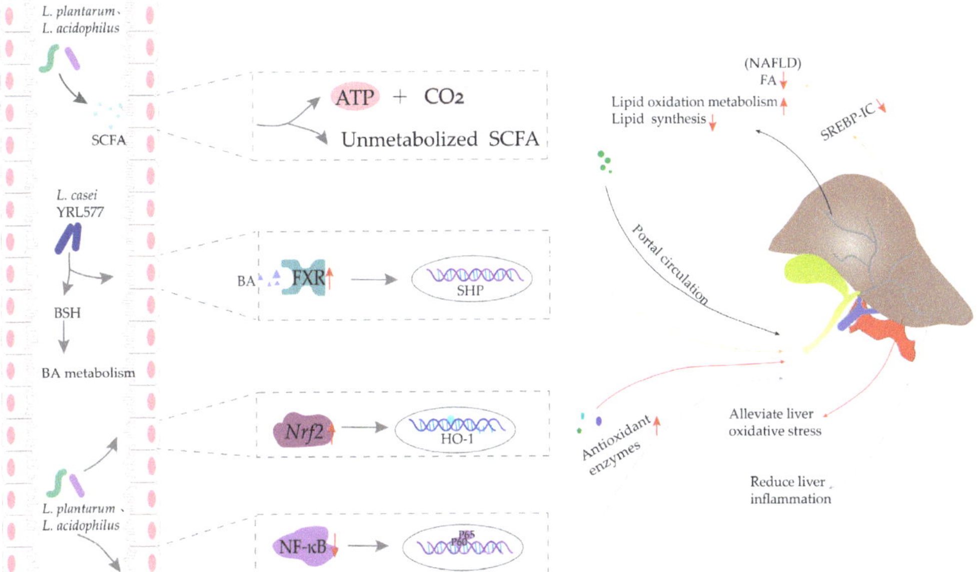

Figure 3. The effects of *Lactobacillus* treatment in oxidative stress-related liver disease. Some *Lactobacillus* regulate hepatic lipid metabolism and synthesis via SCFA production and the FXR/SHP/SREBP1c pathway, and some others reduce liver inflammation and oxidative stress through Nrf2 and NF-κB pathways.

Preventive effects of *Lactobacillus* in NAFLD: One study in rats with induced NAFLD showed that administering a *Lactobacillus*-rich mixture of probiotic strains including *L. acidophilus, L. casei,* and *L. reuteri,* could significantly reverse hepatic and blood triglyceride concentrations and blood glucose levels while suppressing markers of oxidative stress in liver tissue [114]. Zhang et al. isolated *L. casei* YRL577, which exhibited high bile acid hydrolase (BAH) activity that could regulate bile acid metabolism by increasing the transcription of fibroblast growth factor 15 (FGF15), while downregulating the host mRNA levels of Na-dependent bile acid transporter (ASBT). In addition, the YRL577 strain reduced the de novo synthesis of fatty acids (FA) via the FXR/SHP/SREBP1c pathway. These regulatory effects could possibly contribute to alleviating NAFLD in C57BL/6 mice [109] (Figure 3).

Furthermore, a population-based study confirmed that dietary supplementation with probiotic strains *Bifidobacterium longum* and *L. acidophilus* could improve some anthropometric, inflammatory, and oxidative indices in patients with NAFLD [115].

Other studies have also examined the mechanisms by which *Lactobacillus* could affect NAFLD. Nrf2 is a known protective factor against liver damage, and has been implicated in the pathogenesis of chronic conditions such as NAFLD. Notably, a recent study has identified 5-methoxyindoleacetic acid, produced by the human commensal LGG, which can potently activate Nrf2 in both Drosophila liver analog and the murine liver. This activation of Nrf2 has been shown to protect against two models of oxidative liver injury, namely, acetaminophen overdose and acute ethanol toxicity [116]. Chen et al. found that treatment with *L. mali* APS1 led to reduced hepatic lipid accumulation via regulating the SIRT-1/PGC-1α/SREBP-1 pathway, and promoted hepatic antioxidant activity through the Nrf2/HO-1 pathway [117]. Another study reported that *L. plantarum* NA136 could improve NAFLD by regulating fatty acid metabolism and oxidative stress defense pathways via AMPK and Nrf2 signaling, respectively [118]. Similarly, another report suggested that *L. plantarum* NCU116 might ameliorate NAFLD by downregulating lipogenesis while promoting the upregulation of lipolysis and fatty acid oxidation-related gene expression [119]. However, the current body of literature examining the preventive or therapeutic effects of *Lactobacillus* on NAFLD remains limited. Furthermore, most of the available studies have examined mechanisms related to the regulation of lipid metabolism and the alleviation of oxidative stress, although other mechanisms are likely involved in the impacts of *Lactobacillus* on NAFLD.

Lactobacillus strains prevent other types of liver damage: In an animal model of carbon tetrachloride (CCl4)-induced liver fibrosis, *L. plantarum* HFY15 was found to alleviate liver injury through antioxidant, anti-inflammatory, and anti-apoptotic pathways [120]. In addition, selenium-enriched probiotics (SP) might also influence pro- and anti-apoptosis pathways, activate the SIRT1 signaling pathway, or attenuate MAPK signaling, leading to reduced liver fibrosis [121,122]. *Lactobacillus* strains were also shown to alleviate liver injuries caused by other drugs such as D-galactose (D-Gal)/LPS-induced acute liver injury (ALI) in mice or liver toxicity caused by deoxynivalenol (DON), an extremely common environmental pollutant [123,124] (Table 1).

Table 1. Summary of the cited *Lactobacillus* strains and their function in disease.

Lactobacillus Strain	Diseases Involved	Reported Functions	Reference
L. delbrueckii subsp. *bulgaricus* B3, *L. delbrueckii* subsp. *bulgaricus* A13	IBD	Produces EPS and antioxidant	[41]
L. paracasei IJH-SONE68	UC	Produces EPS	[42]
L. plantarum ZS62	IBD	Regulates oxidative stress and immune response	[43]
L. casei, *L. acidophilus*, *L. rhamnosus*	IBD	Modulates inflammatory response and oxidative stress.	[44]
L. fermentum ME-3	/(a healthy volunteer trial)	Antioxidant activity in healthy volunteers	[46]
Probiotic bacterial strains VSL#3	UC	Maintain remission among patients with active UC	[47]
Engineered *L. casei* BL23	CD	Candidate strain for genetic engineering	[50]
L. delbrueckii subsp. *Bulgaricus*, *S. thermophilus* CRL 807	Colitis	Anti-inflammatory effects	[51]

Table 1. *Cont.*

Lactobacillus Strain	Diseases Involved	Reported Functions	Reference
L. casei Shirota	Cellular (caco-2 cells) inflammatory damage and oxidative stress induced by DPPH	Alleviate oxidative stress and inflammation via Nrf2 and NF-κB pathway	[53]
L. acidophilus NCFM	Orthotopic colon cancers	Attenuates tumor growth and pro-apoptotic effects	[84]
L. acidophilus, *L. casei*	Colorectal cancer	Enhances apoptosis	[85]
LGG	Colon cancer	promotes apoptosis	[86]
LGG	HGC-27 human gastric cancer cells	Anti-proliferative effects	[87]
L. reuteri ATCC PTA 6475	Human chronic myeloid leukemia-derived cells	Pro-apoptotic and anti-proliferation effects	[88]
L. casei	Gastric cancer cells	Pro-apoptotic effects	[89]
LGG	Gastric and colonic neoplasms	Anti-proliferative effects	[90]
L. paracasei subsp. *paracasei* X12	HT-29 colon cancer cells	Induce cell cycle arrest at G1 (anti-proliferative)	[91]
L. salivarius FP25, *L. salivarius* FP35	Colon cancer cells	Anti-proliferative effects	[93]
L. gasseri strains	Cervical cancer cells (HeLa)	Inhibits cancer cell growth; Modulates immune response	[94]
L. plantarum NCU116	Colorectal cell line CT26	Regulates cancer cell proliferation and apoptosis by EPS	[95]
L. salivarius REN	Oral cancer	Induces apoptosis and protects against oxidative DNA damage	[97]
L. plantarum AS1	Colorectal cancer	Antioxidant effects	[99]
L. acidophilus 606	HT-29 colon cancer cells	Induces apoptosis; Antioxidant effects	[100]
L. casei Shirota	Normal blood mononuclear cells and splenocytes	Enhances NK cell activity	[101]
L. casei ssp. *casei*	Normal mice	Elicits NK cell activities.	[102]
L. casei Shirota	Cancer	Immunomodulatory effects	[103]
L. acidophilus 36YL	Human cancer cell lines (AGS, HeLa, MCF-7, and HT-29)	Cytoxicity toward cancer cells	[104]
L. plantarum KLDS1.0344; *L. acidophilus* KLDS1.0901	ALD	Inhibits liver lipid accumulation, oxidative stress, and inflammation. Regulates gut microbiota	[106]
L. casei YRL577	NAFLD	Modulates genes in intestinal bile acid pathway	[109]
LGG	ALD	Reduces oxidative stress and inflammation	[110]
L. rhamnosus CCFM1107	ALD	Reduces oxidative stress; Restores the intestinal microbiota	[111]
L. plantarum ZS62	Alcohol-induced subacute hepatic damage	Reduces inflammation; Enhances antioxidative	[112]

Table 1. *Cont.*

Lactobacillus Strain	Diseases Involved	Reported Functions	Reference
L. plantarum	ALD	Antioxidant effects via activation of Keap-Nrf2-ARE pathway	[113]
L. acidophilus, L. casei, L. reuteri	NAFLD	Antioxidant effects	[114]
Bifidobacterium longum; *L. acidophilus*	NAFLD	Mitigates oxidative stress and inflammatory response; Improves lipid profiles	[115]
LGG	NAFLD	Active Nrf2	[116]
L. mali APS1	NAFLD	Modulates lipid metabolism and antioxidant activity	[117]
L. plantarum NA136	NAFLD	Regulates the fatty acid metabolism and defends against oxidative stress via AMPK and Nrf2 pathways	[118]
L. plantarum NCU116	NAFLD	Regulates microbiota and lipid metabolism	[119]
L. plantarum HFY15	Liver fibrosis	Antioxidant; Anti-inflammatory; Anti-apoptotic effects	[120]
Se-enriched probiotics	Liver fibrosis	Attenuate hepatic oxidative stress, ER stress, and inflammation	[121]
Se-enriched probiotics	Liver fibrosis	Attenuate hepatic oxidative stress and inflammation; Induce apoptosis of hepatic stellate cells.	[122]
LGG	Deoxynivalenol exposure induces liver damage	Reduces DON toxicity	[123]
L. plantarum KSFY06	Acute liver injury induced by D-Gal/LPS	Anti-oxidant and anti-inflammatory activities	[124]

5. Conclusions and Prospects

In summary, oxidative damage accompanies inflammation, cancer development, and liver injury, whereas the administration of probiotic *Lactobacillus* strains shows potential for modulating oxidative stress in cells and tissues throughout the human body, especially in the gastrointestinal tract [18]. Treatment with probiotic *Lactobacillus* supplements can also reportedly prevent or ameliorate IBD by downregulating the expression of inflammatory factors while promoting an increase in related antioxidant enzymes, possibly through the activation of various host signaling pathways.

In cancers, it remains largely controversial as to whether targeting ROS, either by antioxidant supplements or chemical/genetic inhibition, is clinically beneficial or detrimental for cancer treatment. Several studies in animal models support the tumor suppressive effects of *Lactobacillus* strains, although it also remains uncertain whether these effects are the result of directly alleviating host ROS or due to stimulating the host antioxidant systems. Further mechanistic insights are therefore necessary to better understand the role of ROS in different types of cancer and to develop more effective, targeted probiotic anti-cancer therapies [60]. Treatment with *Lactobacillus* strains has also been shown to confer prophylactic or therapeutic effects, alleviating inflammation and oxidative damage in ALD, NAFLD, and other types of liver injury in vitro and in vivo.

Since the large majority of the above studies are based on animal and cellular models, multi-level, multi-center population-based trials such as prospective epidemiological studies are necessary for a robust assessment of the effectiveness of probiotic treatments in IBD, cancers, and liver diseases.

In addition, *Lactobacillus* species share conserved microbe-associated molecular patterns (MAMPs) with other Gram-positive bacteria such as cell wall polysaccharides and lipoteichoic acids, which can be recognized by the human immune system and elicit a host inflammatory response. Advances in probiotic development will thus focus on the discovery or engineering of *Lactobacillus* strains with lower MAMP-induced immunogenic effects, but with a high antioxidant capacity to maximize their therapeutic benefits in different diseases.

Supplementary Materials: The following supporting information can be downloaded at: https://www.mdpi.com/article/10.3390/antiox12030769/s1, Table S1: Abbreviations.

Author Contributions: Conceptualizing the content of manuscript, X.H. and B.L.; Writing—all original draft preparation, T.Z. and H.W.; Revising entire manuscript draft, X.H. Supervising, reviewing and editing final version of article, B.L., Z.L., Y.L. and D. All authors have read and agreed to the published version of the manuscript.

Funding: The study was supported by the Key Research and Development Program in Tibet Autonomous Region (XZ202201ZY0004N), the Science and Technology Program of Gansu Province (22JR5RA463), and the Special Item of Regional Collaborative Innovation in Tibet Autonomous Region (QYXTZX-LS2021-01).

Institutional Review Board Statement: Not applicable.

Informed Consent Statement: Not applicable.

Data Availability Statement: Data is contained within the article or Supplementary Material.

Acknowledgments: We gratefully acknowledge Isaac V. Greenhut, ELS for advice with the language and editorial assistance in manuscript preparation.

Conflicts of Interest: The authors declare no conflict of interest.

References

1. Alzoghaibi, M.A. Concepts of oxidative stress and antioxidant defense in Crohn's disease. *World J. Gastroenterol.* **2013**, *19*, 6540–6547. [CrossRef]
2. Li, D.; Ding, Z.; Du, K.; Ye, X.; Cheng, S. Reactive Oxygen Species as a Link between Antioxidant Pathways and Autophagy. *Oxidative Med. Cell Longev.* **2021**, *2021*, 5583215. [CrossRef]
3. Handy, D.E.; Loscalzo, J. Redox regulation of mitochondrial function. *Antioxid. Redox Signal.* **2012**, *16*, 1323–1367. [CrossRef] [PubMed]
4. Swindle, E.J.; Metcalfe, D.D. The role of reactive oxygen species and nitric oxide in mast cell-dependent inflammatory processes. *Immunol. Rev.* **2007**, *217*, 186–205. [CrossRef] [PubMed]
5. Tian, T.; Wang, Z.; Zhang, J. Pathomechanisms of Oxidative Stress in Inflammatory Bowel Disease and Potential Antioxidant Therapies. *Oxid. Med. Cell Longev.* **2017**, *2017*, 4535194. [CrossRef] [PubMed]
6. Oteng, A.B.; Kersten, S. Mechanisms of Action of trans Fatty Acids. *Adv. Nutr.* **2020**, *11*, 697–708. [CrossRef]
7. Bhattacharyya, A.; Chattopadhyay, R.; Mitra, S.; Crowe, S.E. Oxidative stress: An essential factor in the pathogenesis of gastrointestinal mucosal diseases. *Physiol. Rev.* **2014**, *94*, 329–354. [CrossRef]
8. Galaris, D.; Barbouti, A.; Pantopoulos, K. Iron homeostasis and oxidative stress: An intimate relationship. *Biochim. Biophys. Acta Mol. Cell Res.* **2019**, *1866*, 118535. [CrossRef]
9. Iborra, M.; Moret, I.; Rausell, F.; Bastida, G.; Aguas, M.; Cerrillo, E.; Nos, P.; Beltrán, B. Role of oxidative stress and antioxidant enzymes in Crohn's disease. *Biochem. Soc. Trans.* **2011**, *39*, 1102–1106. [CrossRef]
10. Bonnefont-Rousselot, D.; Collin, F. Melatonin: Action as antioxidant and potential applications in human disease and aging. *Toxicology* **2010**, *278*, 55–67. [CrossRef]
11. Lu, J.; Holmgren, A. The thioredoxin antioxidant system. *Free Radic. Biol. Med.* **2014**, *66*, 75–87. [CrossRef] [PubMed]
12. Brown, J.E.; Khodr, H.; Hider, R.C.; Rice-Evans, C.A. Structural dependence of flavonoid interactions with Cu^{2+} ions: Implications for their antioxidant properties. *Biochem. J.* **1998**, *330*, 1173–1178. [CrossRef] [PubMed]
13. Hengstermann, S.; Valentini, L.; Schaper, L.; Buning, C.; Koernicke, T.; Maritschnegg, M.; Buhner, S.; Tillinger, W.; Regano, N.; Guglielmi, F.; et al. Altered status of antioxidant vitamins and fatty acids in patients with inactive inflammatory bowel disease. *Clin. Nutr.* **2008**, *27*, 571–578. [CrossRef] [PubMed]

14. Sies, H.; Jones, D.P. Reactive oxygen species (ROS) as pleiotropic physiological signalling agents. *Nat. Rev. Mol. Cell Biol.* **2020**, *21*, 363–383. [CrossRef]
15. Chiurchiù, V.; Maccarrone, M. Chronic inflammatory disorders and their redox control: From molecular mechanisms to therapeutic opportunities. *Antioxid. Redox Signal.* **2011**, *15*, 2605–2641. [CrossRef]
16. Elbatreek, M.H.; Pachado, M.P.; Cuadrado, A.; Jandeleit-Dahm, K.; Schmidt, H. Reactive Oxygen Comes of Age: Mechanism-Based Therapy of Diabetic End-Organ Damage. *Trends Endocrinol. Metab. TEM* **2019**, *30*, 312–327. [CrossRef]
17. Banba, A.; Tsuji, A.; Kimura, H.; Murai, M.; Miyoshi, H. Defining the mechanism of action of S1QELs, specific suppressors of superoxide production in the quinone-reaction site in mitochondrial complex I. *J. Biol. Chem.* **2019**, *294*, 6550–6561. [CrossRef]
18. Kong, Y.; Olejar, K.J.; On, S.L.W.; Chelikani, V. The Potential of Lactobacillus spp. for Modulating Oxidative Stress in the Gastrointestinal Tract. *Antioxidants* **2020**, *9*, 610. [CrossRef]
19. Heeney, D.D.; Gareau, M.G.; Marco, M.L. Intestinal Lactobacillus in health and disease, a driver or just along for the ride? *Curr. Opin. Biotechnol.* **2018**, *49*, 140–147. [CrossRef]
20. Annuk, H.; Shchepetova, J.; Kullisaar, T.; Songisepp, E.; Zilmer, M.; Mikelsaar, M. Characterization of intestinal lactobacilli as putative probiotic candidates. *J. Appl. Microbiol.* **2003**, *94*, 403–412. [CrossRef]
21. Ashraf, R.; Shah, N.P. Immune system stimulation by probiotic microorganisms. *Crit. Rev. Food Sci. Nutr.* **2014**, *54*, 938–956. [CrossRef] [PubMed]
22. Djavaheri-Mergny, M.; Javelaud, D.; Wietzerbin, J.; Besançon, F. NF-kappaB activation prevents apoptotic oxidative stress via an increase of both thioredoxin and MnSOD levels in TNFalpha-treated Ewing sarcoma cells. *FEBS Lett.* **2004**, *578*, 111–115. [CrossRef] [PubMed]
23. Li, B.; Evivie, S.E.; Lu, J.; Jiao, Y.; Wang, C.; Li, Z.; Liu, F.; Huo, G. Lactobacillus helveticus KLDS1.8701 alleviates d-galactose-induced aging by regulating Nrf-2 and gut microbiota in mice. *Food Funct.* **2018**, *9*, 6586–6598. [CrossRef] [PubMed]
24. Zhang, Y.; Li, Y. Inflammatory bowel disease: Pathogenesis. *World J. Gastroenterol.* **2014**, *20*, 91–99. [CrossRef] [PubMed]
25. Ramos, G.P.; Papadakis, K.A. Mechanisms of Disease: Inflammatory Bowel Diseases. *Mayo Clin. Proc.* **2019**, *94*, 155–165. [CrossRef]
26. Xavier, R.J.; Podolsky, D.K. Unravelling the pathogenesis of inflammatory bowel disease. *Nature* **2007**, *448*, 427–434. [CrossRef] [PubMed]
27. Begleiter, A.; Hewitt, D.; Maksymiuk, A.W.; Ross, D.A.; Bird, R.P. A NAD(P)H:quinone oxidoreductase 1 polymorphism is a risk factor for human colon cancer. *Cancer Epidemiol. Biomark. Prev.* **2006**, *15*, 2422–2426. [CrossRef]
28. Kosaka, T.; Yoshino, J.; Inui, K.; Wakabayashi, T.; Kobayashi, T.; Watanabe, S.; Hayashi, S.; Hirokawa, Y.; Shiraishi, T.; Yamamoto, T.; et al. Involvement of NAD(P)H:quinone oxidoreductase 1 and superoxide dismutase polymorphisms in ulcerative colitis. *DNA Cell Biol.* **2009**, *28*, 625–631. [CrossRef]
29. Arisawa, T.; Tahara, T.; Shibata, T.; Nagasaka, M.; Nakamura, M.; Kamiya, Y.; Fujita, H.; Yoshioka, D.; Okubo, M.; Sakata, M.; et al. Nrf2 gene promoter polymorphism is associated with ulcerative colitis in a Japanese population. *Hepato-Gastroenterol.* **2008**, *55*, 394–397.
30. Banan, A.; Fields, J.Z.; Decker, H.; Zhang, Y.; Keshavarzian, A. Nitric oxide and its metabolites mediate ethanol-induced microtubule disruption and intestinal barrier dysfunction. *J. Pharmacol. Exp. Ther.* **2000**, *294*, 997–1008.
31. Pastorelli, L.; De Salvo, C.; Mercado, J.R.; Vecchi, M.; Pizarro, T.T. Central role of the gut epithelial barrier in the pathogenesis of chronic intestinal inflammation: Lessons learned from animal models and human genetics. *Front. Immunol.* **2013**, *4*, 280. [CrossRef] [PubMed]
32. Wright, E.K.; Kamm, M.A.; Teo, S.M.; Inouye, M.; Wagner, J.; Kirkwood, C.D. Recent advances in characterizing the gastrointestinal microbiome in Crohn's disease: A systematic review. *Inflamm. Bowel Dis.* **2015**, *21*, 1219–1228. [PubMed]
33. Handa, O.; Naito, Y.; Yoshikawa, T. Helicobacter pylori: A ROS-inducing bacterial species in the stomach. *Inflamm. Res.* **2010**, *59*, 997–1003. [CrossRef]
34. Nakajima, A.; Shibuya, T.; Sasaki, T.; Lu, Y.J.; Ishikawa, D.; Haga, K.; Takahashi, M.; Kaga, N.; Osada, T.; Sato, N.; et al. Nicotine Oral Administration Attenuates DSS-Induced Colitis Through Upregulation of Indole in the Distal Colon and Rectum in Mice. *Front. Med.* **2021**, *8*, 789037. [CrossRef] [PubMed]
35. Guo, X.; Ko, J.K.; Mei, Q.B.; Cho, C.H. Aggravating effect of cigarette smoke exposure on experimental colitis is associated with leukotriene B4 and reactive oxygen metabolites. *Digestion* **2001**, *63*, 180–187. [CrossRef]
36. Rodrigo, R.; Miranda, A.; Vergara, L. Modulation of endogenous antioxidant system by wine polyphenols in human disease. *Clin. Chim. Acta* **2011**, *412*, 410–424. [CrossRef] [PubMed]
37. Na, H.K.; Lee, J.Y. Molecular Basis of Alcohol-Related Gastric and Colon Cancer. *Int. J. Mol. Sci.* **2017**, *18*, 1116. [CrossRef]
38. Moon, W.; Loftus, E.V. Review article: Recent advances in pharmacogenetics and pharmacokinetics for safe and effective thiopurine therapy in inflammatory bowel disease. *Aliment. Pharmacol. Ther.* **2016**, *43*, 863–883. [CrossRef]
39. Ng, S.C.; Kaplan, G.G.; Tang, W.; Banerjee, R.; Adigopula, B.; Underwood, F.E.; Tanyingoh, D.; Wei, S.; Lin, W.; Lin, H.; et al. Population Density and Risk of Inflammatory Bowel Disease: A Prospective Population-Based Study in 13 Countries or Regions in Asia-Pacific. *Am. J. Gastroenterol.* **2019**, *114*, 107–115. [CrossRef]
40. Kim, E.S.; Tae, C.H.; Jung, S.A.; Park, D.I.; Im, J.P.; Eun, C.S.; Yoon, H.; Jang, B.I.; Ogata, H.; Fukuhara, K.; et al. Perspectives of East Asian patients and physicians on complementary and alternative medicine use for inflammatory bowel disease: Results of a cross-sectional, multinational study. *Intest. Res.* **2022**, *20*, 192–202. [CrossRef] [PubMed]

41. Sengül, N.; Işık, S.; Aslım, B.; Uçar, G.; Demirbağ, A.E. The effect of exopolysaccharide-producing probiotic strains on gut oxidative damage in experimental colitis. *Dig. Dis. Sci.* **2011**, *56*, 707–714. [CrossRef] [PubMed]
42. Noda, M.; Danshiitsoodol, N.; Kanno, K.; Uchida, T.; Sugiyama, M. The Exopolysaccharide Produced by IJH-SONE68 Prevents and Ameliorates Inflammatory Responses in DSS-Induced Ulcerative Colitis. *Microorganisms* **2021**, *9*, 2243. [CrossRef] [PubMed]
43. Pan, Y.; Ning, Y.; Hu, J.; Wang, Z.; Chen, X.; Zhao, X. The Preventive Effect of ZS62 on DSS-Induced IBD by Regulating Oxidative Stress and the Immune Response. *Oxidative Med. Cell. Longev.* **2021**, *2021*, 9416794. [CrossRef]
44. Chorawala, M.R.; Chauhan, S.; Patel, R.; Shah, G. Cell Wall Contents of Probiotics (Lactobacillus species) Protect Against Lipopolysaccharide (LPS)-Induced Murine Colitis by Limiting Immuno-inflammation and Oxidative Stress. *Probiotics Antimicrob. Proteins* **2021**, *13*, 1005–1017. [CrossRef] [PubMed]
45. Erdogan, F.S.; Ozarslan, S.; Guzel-Seydim, Z.B.; Kök Taş, T. The effect of kefir produced from natural kefir grains on the intestinal microbial populations and antioxidant capacities of Balb/c mice. *Food Res. Int.* **2019**, *115*, 408–413. [CrossRef] [PubMed]
46. Songisepp, E.; Kals, J.; Kullisaar, T.; Mändar, R.; Hütt, P.; Zilmer, M.; Mikelsaar, M. Evaluation of the functional efficacy of an antioxidative probiotic in healthy volunteers. *Nutr. J.* **2005**, *4*, 22. [CrossRef] [PubMed]
47. Miele, E.; Pascarella, F.; Giannetti, E.; Quaglietta, L.; Baldassano, R.N.; Staiano, A. Effect of a probiotic preparation (VSL#3) on induction and maintenance of remission in children with ulcerative colitis. *Am. J. Gastroenterol.* **2009**, *104*, 437–443.
48. Marteau, P.; Lémann, M.; Seksik, P.; Laharie, D.; Colombel, J.F.; Bouhnik, Y.; Cadiot, G.; Soulé, J.C.; Bourreille, A.; Metman, E.; et al. Ineffectiveness of Lactobacillus johnsonii LA1 for prophylaxis of postoperative recurrence in Crohn's disease: A randomised, double blind, placebo controlled GETAID trial. *Gut* **2006**, *55*, 842–847. [CrossRef]
49. Amaretti, A.; di Nunzio, M.; Pompei, A.; Raimondi, S.; Rossi, M.; Bordoni, A. Antioxidant properties of potentially probiotic bacteria: In vitro and in vivo activities. *Appl. Microbiol. Biotechnol.* **2013**, *97*, 809–817. [CrossRef]
50. LeBlanc, J.G.; del Carmen, S.; Miyoshi, A.; Azevedo, V.; Sesma, F.; Langella, P.; Bermudez-Humaran, L.G.; Watterlot, L.; Perdigon, G.; de Moreno de LeBlanc, A. Use of superoxide dismutase and catalase producing lactic acid bacteria in TNBS induced Crohn's disease in mice. *J. Biotechnol.* **2011**, *151*, 287–293. [CrossRef]
51. Del Carmen, S.; de Moreno de LeBlanc, A.; Martin, R.; Chain, F.; Langella, P.; Bermudez-Humaran, L.G.; LeBlanc, J.G. Genetically engineered immunomodulatory Streptococcus thermophilus strains producing antioxidant enzymes exhibit enhanced anti-inflammatory activities. *Appl. Environ. Microbiol.* **2014**, *80*, 869–877. [CrossRef] [PubMed]
52. An, H.; Zhai, Z.; Yin, S.; Luo, Y.; Han, B.; Hao, Y. Coexpression of the superoxide dismutase and the catalase provides remarkable oxidative stress resistance in Lactobacillus rhamnosus. *J. Agric. Food Chem.* **2011**, *59*, 3851–3856. [CrossRef]
53. Finamore, A.; Ambra, R.; Nobili, F.; Garaguso, I.; Raguzzini, A.; Serafini, M. Redox Role of Lactobacillus casei Shirota Against the Cellular Damage Induced by 2,2′-Azobis (2-Amidinopropane) Dihydrochloride-Induced Oxidative and Inflammatory Stress in Enterocytes-Like Epithelial Cells. *Front. Immunol.* **2018**, *9*, 1131. [CrossRef] [PubMed]
54. Hiller, F.; Besselt, K.; Deubel, S.; Brigelius-Flohé, R.; Kipp, A.P. GPx2 Induction Is Mediated Through STAT Transcription Factors During Acute Colitis. *Inflamm. Bowel Dis.* **2015**, *21*, 2078–2089. [CrossRef] [PubMed]
55. Yelin, I.; Flett, K.B.; Merakou, C.; Mehrotra, P.; Stam, J.; Snesrud, E.; Hinkle, M.; Lesho, E.; McGann, P.; McAdam, A.J.; et al. Genomic and epidemiological evidence of bacterial transmission from probiotic capsule to blood in ICU patients. *Nat. Med.* **2019**, *25*, 1728–1732. [CrossRef]
56. Aaron, J.G.; Sobieszczyk, M.E.; Weiner, S.D.; Whittier, S.; Lowy, F.D. Lactobacillus rhamnosus Endocarditis After Upper Endoscopy. *Open Forum Infect. Dis.* **2017**, *4*, ofx085. [CrossRef] [PubMed]
57. Colautti, A.; Arnoldi, M.; Comi, G.; Iacumin, L. Antibiotic resistance and virulence factors in lactobacilli: Something to carefully consider. *Food Microbiol.* **2022**, *103*, 103934. [CrossRef]
58. Sung, H.; Ferlay, J.; Siegel, R.L.; Laversanne, M.; Soerjomataram, I.; Jemal, A.; Bray, F. Global Cancer Statistics 2020: GLOBOCAN Estimates of Incidence and Mortality Worldwide for 36 Cancers in 185 Countries. *CA Cancer J. Clin.* **2021**, *71*, 209–249. [CrossRef]
59. Kirtonia, A.; Sethi, G.; Garg, M. The multifaceted role of reactive oxygen species in tumorigenesis. *Cell Mol. Life Sci.* **2020**, *77*, 4459–4483. [CrossRef]
60. Nakamura, H.; Takada, K. Reactive oxygen species in cancer: Current findings and future directions. *Cancer Sci.* **2021**, *112*, 3945–3952. [CrossRef]
61. Gorrini, C.; Harris, I.S.; Mak, T.W. Modulation of oxidative stress as an anticancer strategy. *Nat. Rev. Drug Discov.* **2013**, *12*, 931–947. [CrossRef] [PubMed]
62. Sallmyr, A.; Fan, J.; Rassool, F.V. Genomic instability in myeloid malignancies: Increased reactive oxygen species (ROS), DNA double strand breaks (DSBs) and error-prone repair. *Cancer Lett.* **2008**, *270*, 1–9. [CrossRef] [PubMed]
63. Shibutani, S.; Takeshita, M.; Grollman, A.P. Insertion of specific bases during DNA synthesis past the oxidation-damaged base 8-oxodG. *Nature* **1991**, *349*, 431–434. [CrossRef]
64. Kato, J.; Miyanishi, K.; Kobune, M.; Nakamura, T.; Takada, K.; Takimoto, R.; Kawano, Y.; Takahashi, S.; Takahashi, M.; Sato, Y.; et al. Long-term phlebotomy with low-iron diet therapy lowers risk of development of hepatocellular carcinoma from chronic hepatitis C. *J. Gastroenterol.* **2007**, *42*, 830–836. [CrossRef] [PubMed]
65. Nakamura, T.; Naguro, I.; Ichijo, H. Iron homeostasis and iron-regulated ROS in cell death, senescence and human diseases. *Biochim. Biophys. Acta Gen. Subj.* **2019**, *1863*, 1398–1409. [CrossRef]
66. Son, Y.; Cheong, Y.K.; Kim, N.H.; Chung, H.T.; Kang, D.G.; Pae, H.O. Mitogen-Activated Protein Kinases and Reactive Oxygen Species: How Can ROS Activate MAPK Pathways? *J. Signal Transduct.* **2011**, *2011*, 792639. [CrossRef] [PubMed]

67. Aggarwal, V.; Tuli, H.S.; Varol, A.; Thakral, F.; Yerer, M.B.; Sak, K.; Varol, M.; Jain, A.; Khan, M.A.; Sethi, G. Role of Reactive Oxygen Species in Cancer Progression: Molecular Mechanisms and Recent Advancements. *Biomolecules* **2019**, *9*, 735. [CrossRef] [PubMed]

68. Weinberg, F.; Hamanaka, R.; Wheaton, W.W.; Weinberg, S.; Joseph, J.; Lopez, M.; Kalyanaraman, B.; Mutlu, G.M.; Budinger, G.R.S.; Chandel, N.S. Mitochondrial metabolism and ROS generation are essential for Kras-mediated tumorigenicity. *Proc. Natl. Acad. Sci. USA* **2010**, *107*, 8788–8793. [CrossRef] [PubMed]

69. Wu, H.; Goel, V.; Haluska, F.G. PTEN signaling pathways in melanoma. *Oncogene* **2003**, *22*, 3113–3122. [CrossRef]

70. Kwon, J.; Lee, S.R.; Yang, K.S.; Ahn, Y.; Kim, Y.J.; Stadtman, E.R.; Rhee, S.G. Reversible oxidation and inactivation of the tumor suppressor PTEN in cells stimulated with peptide growth factors. *Proc. Natl. Acad. Sci. USA* **2004**, *101*, 16419–16424. [CrossRef]

71. Hoesel, B.; Schmid, J.A. The complexity of NF-κB signaling in inflammation and cancer. *Mol. Cancer* **2013**, *12*, 86. [CrossRef] [PubMed]

72. Morgan, M.J.; Liu, Z. Crosstalk of reactive oxygen species and NF-κB signaling. *Cell Res.* **2011**, *21*, 103–115. [CrossRef] [PubMed]

73. Almeida, M.; Soares, M.; Ramalhinho, A.C.; Moutinho, J.F.; Breitenfeld, L.; Pereira, L. The prognostic value of NRF2 in breast cancer patients: A systematic review with meta-analysis. *Breast Cancer Res. Treat.* **2020**, *179*, 523–532. [CrossRef]

74. Rojo de la Vega, M.; Chapman, E.; Zhang, D.D. NRF2 and the Hallmarks of Cancer. *Cancer Cell* **2018**, *34*, 21–43. [CrossRef]

75. Kerins, M.J.; Ooi, A. A catalogue of somatic NRF2 gain-of-function mutations in cancer. *Sci. Rep.* **2018**, *8*, 12846. [CrossRef] [PubMed]

76. Khaitan, D.; Dwarakanath, B.S. Endogenous and induced oxidative stress in multi-cellular tumour spheroids: Implications for improving tumour therapy. *Indian J. Biochem. Biophys.* **2009**, *46*, 16–24. [PubMed]

77. Marullo, R.; Werner, E.; Degtyareva, N.; Moore, B.; Altavilla, G.; Ramalingam, S.S.; Doetsch, P.W. Cisplatin induces a mitochondrial-ROS response that contributes to cytotoxicity depending on mitochondrial redox status and bioenergetic functions. *PLoS ONE* **2013**, *8*, e81162. [CrossRef]

78. Ichijo, H.; Nishida, E.; Irie, K.; ten Dijke, P.; Saitoh, M.; Moriguchi, T.; Takagi, M.; Matsumoto, K.; Miyazono, K.; Gotoh, Y. Induction of apoptosis by ASK1, a mammalian MAPKKK that activates SAPK/JNK and p38 signaling pathways. *Science* **1997**, *275*, 90–94. [CrossRef]

79. Kuo, C.; Ponneri Babuharisankar, A.; Lin, Y.; Lien, H.W.; Lo, Y.K.; Chou, H.; Tangeda, V.; Cheng, L.C.; Cheng, A.N.; Lee, A.Y.-L. Mitochondrial oxidative stress in the tumor microenvironment and cancer immunoescape: Foe or friend? *J. Biomed. Sci.* **2022**, *29*, 74. [CrossRef]

80. Nowak, A.; Paliwoda, A.; Blasiak, J. Anti-proliferative, pro-apoptotic and anti-oxidative activity of Lactobacillus and Bifidobacterium strains: A review of mechanisms and therapeutic perspectives. *Crit. Rev. Food Sci. Nutr.* **2019**, *59*, 3456–3467. [CrossRef]

81. Dasari, S.; Kathera, C.; Janardhan, A.; Praveen Kumar, A.; Viswanath, B. Surfacing role of probiotics in cancer prophylaxis and therapy: A systematic review. *Clin. Nutr.* **2017**, *36*, 1465–1472. [CrossRef] [PubMed]

82. D'Arcy, M.S. Cell death: A review of the major forms of apoptosis, necrosis and autophagy. *Cell Biol. Int.* **2019**, *43*, 582–592. [CrossRef] [PubMed]

83. Daniluk, U.; Alifier, M.; Kaczmarski, M. Probiotic-induced apoptosis and its potential relevance to mucosal inflammation of gastrointestinal tract. *Adv. Med. Sci.* **2012**, *57*, 175–182. [CrossRef] [PubMed]

84. Chen, C.C.; Lin, W.C.; Kong, M.S.; Shi, H.N.; Walker, W.A.; Lin, C.Y.; Huang, C.T.; Lin, Y.C.; Jung, S.M.; Lin, T.Y. Oral inoculation of probiotics Lactobacillus acidophilus NCFM suppresses tumour growth both in segmental orthotopic colon cancer and extra-intestinal tissue. *Br. J. Nutr.* **2012**, *107*, 1623–1634. [CrossRef]

85. Baldwin, C.; Millette, M.; Oth, D.; Ruiz, M.T.; Luquet, F.M.; Lacroix, M. Probiotic Lactobacillus acidophilus and L. casei mix sensitize colorectal tumoral cells to 5-fluorouracil-induced apoptosis. *Nutr. Cancer* **2010**, *62*, 371–378. [CrossRef]

86. Altonsy, M.O.; Andrews, S.C.; Tuohy, K.M. Differential induction of apoptosis in human colonic carcinoma cells (Caco-2) by Atopobium, and commensal, probiotic and enteropathogenic bacteria: Mediation by the mitochondrial pathway. *Int. J. Food Microbiol.* **2010**, *137*, 190–203. [CrossRef]

87. Russo, F.; Orlando, A.; Linsalata, M.; Cavallini, A.; Messa, C. Effects of Lactobacillus rhamnosus GG on the cell growth and polyamine metabolism in HGC-27 human gastric cancer cells. *Nutr. Cancer* **2007**, *59*, 106–114. [CrossRef]

88. Iyer, C.; Kosters, A.; Sethi, G.; Kunnumakkara, A.B.; Aggarwal, B.B.; Versalovic, J. Probiotic Lactobacillus reuteri promotes TNF-induced apoptosis in human myeloid leukemia-derived cells by modulation of NF-kappaB and MAPK signalling. *Cell Microbiol.* **2008**, *10*, 1442–1452. [CrossRef]

89. Hwang, J.W.; Baek, Y.M.; Yang, K.E.; Yoo, H.S.; Cho, C.K.; Lee, Y.W.; Park, J.; Eom, C.Y.; Lee, Z.W.; Choi, J.S.; et al. Lactobacillus casei extract induces apoptosis in gastric cancer by inhibiting NF-κB and mTOR-mediated signaling. *Integr. Cancer Ther.* **2013**, *12*, 165–173. [CrossRef]

90. Orlando, A.; Messa, C.; Linsalata, M.; Cavallini, A.; Russo, F. Effects of Lactobacillus rhamnosus GG on proliferation and polyamine metabolism in HGC-27 human gastric and DLD-1 colonic cancer cell lines. *Immunopharmacol. Immunotoxicol.* **2009**, *31*, 108–116. [CrossRef]

91. Huang, L.; Shan, Y.J.; He, C.X.; Ren, M.H.; Tian, P.J.; Song, W. Effects of *L. paracasei* subp. *paracasei* X12 on cell cycle of colon cancer HT-29 cells and regulation of mTOR signalling pathway. *J. Funct. Foods* **2016**, *21*, 431–439. [CrossRef]

92. Russo, F.; Linsalata, M.; Orlando, A. Probiotics against neoplastic transformation of gastric mucosa: Effects on cell proliferation and polyamine metabolism. *World J. Gastroenterol.* **2014**, *20*, 13258–13272. [CrossRef]
93. Thirabunyanon, M.; Hongwittayakorn, P. Potential probiotic lactic acid bacteria of human origin induce antiproliferation of colon cancer cells via synergic actions in adhesion to cancer cells and short-chain fatty acid bioproduction. *Appl. Biochem. Biotechnol.* **2013**, *169*, 511–525. [CrossRef] [PubMed]
94. Sungur, T.; Aslim, B.; Karaaslan, C.; Aktas, B. Impact of Exopolysaccharides (EPSs) of Lactobacillus gasseri strains isolated from human vagina on cervical tumor cells (HeLa). *Anaerobe* **2017**, *47*, 137–144. [CrossRef] [PubMed]
95. Zhou, X.; Hong, T.; Yu, Q.; Nie, S.; Gong, D.; Xiong, T.; Xie, M. Exopolysaccharides from Lactobacillus plantarum NCU116 induce c-Jun dependent Fas/Fasl-mediated apoptosis via TLR2 in mouse intestinal epithelial cancer cells. *Sci. Rep.* **2017**, *7*, 14247. [CrossRef]
96. Liu, C.F.; Tseng, K.C.; Chiang, S.S.; Lee, B.H.; Hsu, W.H.; Pan, T.M. Immunomodulatory and antioxidant potential of Lactobacillus exopolysaccharides. *J. Sci. Food Agric.* **2011**, *91*, 2284–2291. [CrossRef]
97. Zhang, M.; Wang, F.; Jiang, L.; Liu, R.; Zhang, L.; Lei, X.; Li, J.; Jiang, J.; Guo, H.; Fang, B.; et al. Lactobacillus salivarius REN inhibits rat oral cancer induced by 4-nitroquioline 1-oxide. *Cancer Prev. Res.* **2013**, *6*, 686–694. [CrossRef]
98. Bell, H.N.; Rebernick, R.J.; Goyert, J.; Singhal, R.; Kuljanin, M.; Kerk, S.A.; Huang, W.; Das, N.K.; Andren, A.; Solanki, S.; et al. Reuterin in the healthy gut microbiome suppresses colorectal cancer growth through altering redox balance. *Cancer Cell* **2022**, *40*, 185–200.e6. [CrossRef]
99. Kumar, R.S.; Kanmani, P.; Yuvaraj, N.; Paari, K.A.; Pattukumar, V.; Thirunavukkarasu, C.; Arul, V. Lactobacillus plantarum AS1 isolated from south Indian fermented food Kallappam suppress 1,2-dimethyl hydrazine (DMH)-induced colorectal cancer in male Wistar rats. *Appl. Biochem. Biotechnol.* **2012**, *166*, 620–631. [CrossRef]
100. Choi, S.S.; Kim, Y.; Han, K.S.; You, S.; Oh, S.; Kim, S.H. Effects of Lactobacillus strains on cancer cell proliferation and oxidative stress in vitro. *Lett. Appl. Microbiol.* **2006**, *42*, 452–458. [CrossRef]
101. Hori, T.; Kiyoshima, J.; Yasui, H. Effect of an oral administration of Lactobacillus casei strain Shirota on the natural killer activity of blood mononuclear cells in aged mice. *Biosci. Biotechnol. Biochem.* **2003**, *67*, 420–422. [CrossRef] [PubMed]
102. Ogawa, T.; Asai, Y.; Tamai, R.; Makimura, Y.; Sakamoto, H.; Hashikawa, S.; Yasuda, K. Natural killer cell activities of synbiotic Lactobacillus casei ssp. casei in conjunction with dextran. *Clin. Exp. Immunol.* **2006**, *143*, 103–109. [CrossRef]
103. Shida, K.; Nomoto, K. Probiotics as efficient immunopotentiators: Translational role in cancer prevention. *Indian J. Med. Res.* **2013**, *138*, 808–814. [PubMed]
104. Nami, Y.; Abdullah, N.; Haghshenas, B.; Radiah, D.; Rosli, R.; Khosroushahi, A.Y. Probiotic potential and biotherapeutic effects of newly isolated vaginal Lactobacillus acidophilus 36YL strain on cancer cells. *Anaerobe* **2014**, *28*, 29–36. [CrossRef]
105. Dunn, W.; Shah, V.H. Pathogenesis of Alcoholic Liver Disease. *Clin. Liver Dis.* **2016**, *20*, 445–456. [CrossRef] [PubMed]
106. Li, H.; Shi, J.; Zhao, L.; Guan, J.; Liu, F.; Huo, G.; Li, B. Lactobacillus plantarum KLDS1.0344 and Lactobacillus acidophilus KLDS1.0901 Mixture Prevents Chronic Alcoholic Liver Injury in Mice by Protecting the Intestinal Barrier and Regulating Gut Microbiota and Liver-Related Pathways. *J. Agric. Food Chem.* **2021**, *69*, 183–197. [CrossRef]
107. Ceni, E.; Mello, T.; Galli, A. Pathogenesis of alcoholic liver disease: Role of oxidative metabolism. *World J. Gastroenterol.* **2014**, *20*, 17756–17772. [CrossRef] [PubMed]
108. Buzzetti, E.; Pinzani, M.; Tsochatzis, E.A. The multiple-hit pathogenesis of non-alcoholic fatty liver disease (NAFLD). *Metabolism* **2016**, *65*, 1038–1048. [CrossRef]
109. Zhang, Z.; Zhou, H.; Zhou, X.; Sun, J.; Liang, X.; Lv, Y.; Bai, L.; Zhang, J.; Gong, P.; Liu, T.; et al. Lactobacillus casei YRL577 ameliorates markers of non-alcoholic fatty liver and alters expression of genes within the intestinal bile acid pathway. *Br. J. Nutr.* **2020**, *125*, 521–529. [CrossRef]
110. Forsyth, C.B.; Farhadi, A.; Jakate, S.M.; Tang, Y.; Shaikh, M.; Keshavarzian, A. Lactobacillus GG treatment ameliorates alcohol-induced intestinal oxidative stress, gut leakiness, and liver injury in a rat model of alcoholic steatohepatitis. *Alcohol* **2009**, *43*, 163–172. [CrossRef]
111. Tian, F.; Chi, F.; Wang, G.; Liu, X.; Zhang, Q.; Chen, Y.; Zhang, H.; Chen, W. Lactobacillus rhamnosus CCFM1107 treatment ameliorates alcohol-induced liver injury in a mouse model of chronic alcohol feeding. *J. Microbiol.* **2015**, *53*, 856–863. [CrossRef] [PubMed]
112. Gan, Y.; Chen, X.; Yi, R.; Zhao, X. Antioxidative and Anti-Inflammatory Effects of ZS62 on Alcohol-Induced Subacute Hepatic Damage. *Oxidative Med. Cell Longev.* **2021**, *2021*, 7337988. [CrossRef] [PubMed]
113. Liu, Y.; Liu, X.; Wang, Y.; Yi, C.; Tian, J.; Liu, K.; Chu, J. Protective effect of lactobacillus plantarum on alcoholic liver injury and regulating of keap-Nrf2-ARE signaling pathway in zebrafish larvae. *PLoS ONE* **2019**, *14*, e0222339. [CrossRef]
114. Azarang, A.; Farshad, O.; Ommati, M.M.; Jamshidzadeh, A.; Heydari, R.; Abootalebi, S.N.; Gholami, A. Protective Role of Probiotic Supplements in Hepatic Steatosis: A Rat Model Study. *Biomed. Res. Int.* **2020**, *2020*, 5487659. [CrossRef] [PubMed]
115. Javadi, L.; Khoshbaten, M.; Safaiyan, A.; Ghavami, M.; Abbasi, M.M.; Gargari, B.P. Pro- and prebiotic effects on oxidative stress and inflammatory markers in non-alcoholic fatty liver disease. *Asia Pac. J. Clin. Nutr.* **2018**, *27*, 1031–1039. [PubMed]
116. Saeedi, B.J.; Liu, K.H.; Owens, J.A.; Hunter-Chang, S.; Camacho, M.C.; Eboka, R.U.; Chandrasekharan, B.; Baker, N.F.; Darby, T.M.; Robinson, B.S.; et al. Gut-Resident Lactobacilli Activate Hepatic Nrf2 and Protect Against Oxidative Liver Injury. *Cell Metab.* **2020**, *31*, 956–968.e5. [CrossRef]

117. Chen, Y.T.; Lin, Y.C.; Lin, J.S.; Yang, N.S.; Chen, M.J. Sugary Kefir Strain Lactobacillus mali APS1 Ameliorated Hepatic Steatosis by Regulation of SIRT-1/Nrf-2 and Gut Microbiota in Rats. *Mol. Nutr. Food Res.* **2018**, *62*, e1700903. [CrossRef]
118. Zhao, Z.; Wang, C.; Zhang, L.; Zhao, Y.; Duan, C.; Zhang, X.; Gao, L.; Li, S. Lactobacillus plantarum NA136 improves the non-alcoholic fatty liver disease by modulating the AMPK/Nrf2 pathway. *Appl. Microbiol. Biotechnol.* **2019**, *103*, 5843–5850. [CrossRef]
119. Li, C.; Nie, S.P.; Zhu, K.X.; Ding, Q.; Li, C.; Xiong, T.; Xie, M.Y. Lactobacillus plantarum NCU116 improves liver function, oxidative stress and lipid metabolism in rats with high fat diet induced non-alcoholic fatty liver disease. *Food Funct.* **2014**, *5*, 3216–3223. [CrossRef]
120. Long, X.; Wang, P.; Zhou, Y.; Wang, Q.; Ren, L.; Li, Q.; Zhao, X. Preventive effect of Lactobacillus plantarum HFY15 on carbon tetrachloride (CCl)-induced acute liver injury in mice. *J. Food Sci.* **2022**, *87*, 2626–2639. [CrossRef]
121. Liu, Y.; Liu, Q.; Hesketh, J.; Huang, D.; Gan, F.; Hao, S.; Tang, S.; Guo, Y.; Huang, K. Protective effects of selenium-glutathione-enriched probiotics on CCl-induced liver fibrosis. *J. Nutr. Biochem.* **2018**, *58*, 138–149. [CrossRef]
122. Liu, Y.; Liu, Q.; Ye, G.; Khan, A.; Liu, J.; Gan, F.; Zhang, X.; Kumbhar, S.; Huang, K. Protective effects of Selenium-enriched probiotics on carbon tetrachloride-induced liver fibrosis in rats. *J. Agric. Food Chem.* **2015**, *63*, 242–249. [CrossRef] [PubMed]
123. Bai, Y.; Ma, K.; Li, J.; Li, J.; Bi, C.; Shan, A. Deoxynivalenol exposure induces liver damage in mice: Inflammation and immune responses, oxidative stress, and protective effects of Lactobacillus rhamnosus GG. *Food Chem. Toxicol.* **2021**, *156*, 112514. [CrossRef] [PubMed]
124. Li, C.; Si, J.; Tan, F.; Park, K.Y.; Zhao, X. *Lactobacillus plantarum* KSFY06 Prevents Inflammatory Response and Oxidative Stress in Acute Liver Injury Induced by D-Gal/LPS in Mice. *Drug Des. Dev. Ther.* **2021**, *15*, 37–50. [CrossRef] [PubMed]

 *antioxidants* 

Article

Improving the Aromatic Profiles of Catarratto Wines: Impact of *Metschnikowia pulcherrima* and Glutathione-Rich Inactivated Yeasts

Vincenzo Naselli [1], Rosario Prestianni [1], Natale Badalamenti [2], Michele Matraxia [1], Antonella Maggio [2,*], Antonio Alfonzo [1,*], Raimondo Gaglio [1], Paola Vagnoli [3], Luca Settanni [1], Maurizio Bruno [2], Giancarlo Moschetti [1] and Nicola Francesca [1]

[1] Department of Agricultural, Food and Forest Sciences (SAAF), University of Palermo, Viale delle Scienze Bldg. 5, Ent. C, 90128 Palermo, Italy
[2] Department of Biological, Chemical and Pharmaceutical Sciences and Technologies (STEBICEF), University of Palermo, Viale delle Scienze, Parco d'Orleans II, Bldg. 17, 90123 Palermo, Italy
[3] Lallemand Italia, Via Rossini 14/B, Castel D'Azzano, 37060 Verona, Italy
* Correspondence: antonella.maggio@unipa.it (A.M.); antonio.alfonzo@unipa.it (A.A.)

Abstract: Catarratto is one of the most widely cultivated grape varieties in Sicily. It is an indigenous non-aromatic white grape variety. Despite its widespread use in winemaking, knowledge of the aroma and chemical and microbiological properties of Catarratto wines is quite limited. The influence of *Metschnikowia pulcherrima* combined with *Saccharomyces cerevisiae* on the aromatic expression of Catarratto wines was investigated with and without the addition of glutathione-rich inactivated yeast. The substance is a natural specific inactivated yeast with a guaranteed glutathione level used to limit oxidative processes. The aromatic profiles of the final wines were determined through analysis of the volatile organic compounds using a solid-phase microextraction technique that identified 26 aromatic compounds. The addition of *M. pulcherrima* in combination with the natural antioxidant undoubtedly increased the aromatic complexity of the wines. Dodecanal was exclusively detected in the wines processed with glutathione-rich inactivated yeasts. Furthermore, the presence of this natural antioxidant increased the concentration of six esters above the perception threshold. Sensory analysis was also performed with a panel of trained judges who confirmed the aromatic differences among the wines. These results suggest the suitability of glutathione-rich inactivated yeasts for determining the oxidative stability of Catarratto wines, thus preserving its aromatic compounds and colour.

Keywords: alcoholic fermentation; Catarratto grape variety; glutathione; *Metschnikowia pulcherrima*; VOC's; wine aroma

check for updates

Citation: Naselli, V.; Prestianni, R.; Badalamenti, N.; Matraxia, M.; Maggio, A.; Alfonzo, A.; Gaglio, R.; Vagnoli, P.; Settanni, L.; Bruno, M.; et al. Improving the Aromatic Profiles of Catarratto Wines: Impact of *Metschnikowia pulcherrima* and Glutathione-Rich Inactivated Yeasts. *Antioxidants* **2023**, *12*, 439. https://doi.org/10.3390/antiox12020439

Academic Editor: Myung-Ji Seo

Received: 19 January 2023
Revised: 7 February 2023
Accepted: 8 February 2023
Published: 10 February 2023

1. Introduction

Identifying new strategies for the aromatic enhancement of wine produced from non-aromatic grape varieties is one of the major objectives of oenological microbiology research. During alcoholic fermentation (AF), yeast metabolic activity and winemaking techniques determine the biosynthesis of several products that influence wine aroma [1,2]. The application of non-conventional yeasts isolated during the fermentation of traditional fermented beverages represents an alternative for producing a variety of alcoholic beverages [3], including wine.

Raw materials with a high sugar content if subjected to spontaneous fermentation can provide potential starters with interesting traits. Wild *Saccharomyces* and non-*Saccharomyces* spp. may generate flavour profiles with desirable characteristics to be applied at industrial level [4]. Several authors have isolated strains of *Saccharomyces cerevisiae* from high-sugar-containing matrices such as manna, honey and honey by-products [5–8] that were successfully applied in experimental Catarratto cultivar winemaking [1].

Recently, the controlled inoculation of selected non-*Saccharomyces* and *S. cerevisiae* strains has permitted the production of higher quality wines. The current trend is to exploit non-*Saccharomyces*/*Saccharomyces* sequential inoculation to achieve a positive impact in terms of aroma [9]. The result is closely related to the species of the multi-starter cultures involved in the sequential inoculum [9].

The final aroma of wines can be modulated not only by *Saccharomyces* but also by non-conventional yeasts. The metabolic impact of non-*Saccharomyces* yeasts during the early stages of fermentation is sufficient to trigger significant changes to the wine's volatile profile; they are suitable for inoculation as co-starters with strains of *S. cerevisiae* [10].

Fermentation processes using mixed strains with the sequential addition of non-*Saccharomyces* and *S. cerevisiae* strains tend to reproduce what happens naturally during spontaneous wine fermentation concerning population dynamics [11]. Indeed, the levels of non-*Saccharomyces* yeast populations are reduced over time, leaving space for *S. cerevisiae* to dominate and conclude the AF [12].

The use of non-*Saccharomyces* yeasts has improved the primary aromas of wines, as the production of specific enzymes enables the precursors present in the must to release volatile molecules. Their activity also affects secondary aromas through the production of volatile organic compounds (mainly alcohols and esters) that can influence typical aromatic expressions such as fruity notes [11].

The impact determined by sequential inoculation can influence various aspects of wine characteristics. In recent oenological studies, strains belonging to the species *Lahancea thermotolerans* and *Starmerella bacillaris* in sequential inoculation with *S. cerevisiae* achieved wines with a significant amount of lactic acid and glycerol, respectively, while strains of *Torulospora delbrueckii* and *Hanseniaspora uvarum* in mixed cultures and in sequential inoculation with *S. cerevisiae*, on the other hand, seemed to influence the composition in terms of higher alcohol and ester contents [9].

Among the non-*Saccharomyces* yeasts, *Metschnikowia pulcherrima* is one of the species most abundant in the initial phase of AF of grape musts. In mixed cultures with *S. cerevisiae*, *M. pulcherrima* rapidly declines due to its low resistance to the ethanol produced by *S. cerevisiae* [13]. Some strains of *M. pulcherrima* are known to synthesize fruity esters and can increase the concentrations of terpenes or thiols generally masked by higher alcohols [14]. Non-aromatic grape varieties lack varietal aromatic precursors (terpenes or thiols) and the presence of fruity aromas (pineapple) due to the fact that an increased ethyl octanoate content determines a positive sensorial impact. Some thiol precursors such as 4-methyl-4-sulfanylpentan-2-one, as well as those produced by *S. cerevisiae*, can be synthesized by *M. pulcherrima* at much higher concentrations, thus significantly influencing the characteristics of wine [15]. Recently, *M. pulcherrima* was successfully used in a sequential inoculation with *S. cerevisiae* for a reduction in ethanol content in Merlot wines [16] and to improve the aromatic complexity in Shiraz and Cabernet Sauvignon wines [17].

A solution aimed at limiting the loss of aromaticity in white wines is represented by the addition of a natural antioxidant such as glutathione at the beginning of the vinification process [18]. The application of glutathione during winemaking has positive effects on the colour and aroma stability of white wines [19]. The glutathione content naturally present in musts is relatively low, and its quantities are closely related to the reactions that characterise the fermentation process as well as the metabolic activities of the yeasts [16]. Its use in oenology provides considerable advantages as its antioxidant activity is capable of limiting browning in white grape must as it inhibits polyphenol polymerisation and severely limits the production of compounds such as sotolone that give wine a fenugreek or curry odour [20]. Glutathione's degree of protection also extends to the aromatic molecules in wines, especially the esters, volatile thiols and terpenes produced by yeasts during alcoholic fermentation that are present in greater quantities when glutathione is added to the must [21]. Some sulphite-free wines are produced by exploiting the antioxidant activity of glutathione in place of potassium metabisulphite, meeting the needs of consumers who are more sensitive to the negative health effects of sulphur dioxide [22].

This study focused on the potential of non-*Saccharomyces* and *Saccharomyces* yeasts for the aromatic improvement of wine produced from non-aromatic grape varieties such as Catarratto. We evaluated the sequential inoculation of a commercial non-*Saccharomyces* yeast strain (*M. pulcherrima*) and *S. cerevisiae* SPF52 isolate from honey by-products to simulate what would occur during spontaneous fermentation. Secondly, we assessed the ability of exogenous glutathione addition during fermentation in the form of inactivated yeast to influence the technological and aromatic properties of wine.

The aims of this research were to investigate: (i) the impacts of *M. pulcherrima* associated with *S. cerevisiae*; (ii) the effect of an antioxidant compound on the aroma and sensory profiles of Catarratto wine; and (iii) the volatile organic compound composition of Catarratto white wine.

2. Materials and Methods

2.1. Experimental Drawing and Sampling

The experimental design (Figure 1) was composed of four treatments: T1, sequential inoculum of FLAVIA® MP346/*S. cerevisiae* SPF52; T2, the use of Glutastar™ to the bulk must and sequential inoculum with FLAVIA® MP346/*S. cerevisiae* SPF52; C1, single inoculum of *S. cerevisiae* SPF52; C2, the addition of Glutastar™ and fermentation by *S. cerevisiae* SPF52.

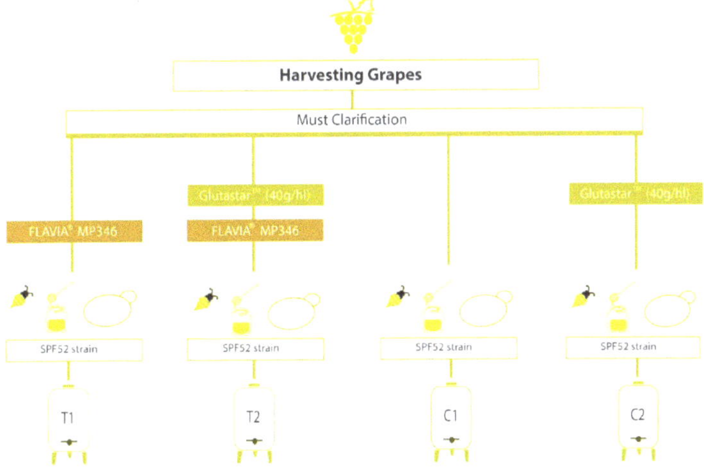

Figure 1. Experimental plan of wines obtained from Catarratto grape must.

FLAVIA® MP346 is a pure culture of *M. pulcherrima* selected by the Universidad de Santiago de Chile (USACH) for its specific capacity to release enzymes with arabinofuranosidase activity [23]. Glutathione-rich inactivated yeast (GIY) is an inactivated yeast mass with a guaranteed glutathione level [24]. GIY and FLAVIA® MP346 were provided by Lallemand Inc. (Castel D'Azzano, Verona, Italy). The *S. cerevisiae* SPF52 strain used in this study belonged to the yeasts collection of the Department of Agricultural, Food and Forestry Sciences (SAAF; University of Palermo, Italy); it was isolated from fermented honey by-products [6] and selected for its high performance in fermenting Catarratto grape must [1].

Samples were collected from clarified bulk must just after the inoculum of *M. pulcherrima* MP346, after the inoculation of *S. cerevisiae* SPF52, during AF (day 3, 6, 12 and 18), during ageing in a steel tank (1, 3 and 5 months) and at bottling. All samples were transported at 4 °C into a portable fridge and subjected to analysis within 24 h after collection.

2.2. Winemaking

After hand harvesting, grapes were stemmer-crushed and treated with 2 g/hL of potassium metabisulphite (Chimica Noto s.r.l., Partinico, Italy). Clarification of the must was carried out at 4 °C for one day by using pectolytic enzymes [Lallzyme® C-Max (Lallemand Inc. Italia, Castel D'Azzano, Verona, Italy); dosage: 4 g/hL].

T1 and T2 were inoculated with FLAVIA® MP346 at 25 g/hL when the clarified must had reached a temperature of 16 °C. The strain *S. cerevisiae* SPF52 was used in a liquid concentrated form [about 7.00×10^{10} colony-forming units (CFU)/g].

After 24 h, T1 and T2 were inoculated with *S. cerevisiae* SPF52 (20 g/hL), while the controls, C1 and C2, were inoculated immediately with the SPF52 strain at the same dose. Before the inoculum of the starter yeast, GIY (40 g/hL) was added to treatments T2 and C2. The organic nutrient Stimula Chardonnay™ (SC; Lallemand Inc. Italia, Castel D'Azzano, Verona, Italy) was added to all tanks (40 g/hL) prior to *S. cerevisiae* yeast inoculation. The use of Stimula Chardonnay™ with *S. cerevisiae* SPF52 was chosen because of the results obtained by previous vinifications on Catarratto wines [1]. The fermentation was carried out at 18 °C in 12 steel tanks with a volume of 2.5 hL each. At the end of AF, the wines were cold-settled, their yeast lees were racked off and they were transferred into stainless-steel tanks at 15° C and topped with nitrogen to avoid oxidation until bottling. During ageing, malolactic fermentation was prevented by keeping the free SO_2 values above 35 mg/L until bottling. Tartaric stability was ensured through the addition of 8 g/hL of metatartaric acid (Chimica Noto s.r.l., Partinico, Italy). Each treatment was performed in triplicate.

2.3. Monitoring Yeast Populations

During the AF, all must samples were microbiologically analysed to determine the total yeast concentration (TY) using the protocol described by Pallmann et al. [25]. *Saccharomyces* and non-*Saccharomyces* yeasts colonies were distinguished as reported by Valera [3]. The analyses were conducted in triplicate.

2.4. Yeast Collection and Genotypic Characterization

Yeasts were isolated from WL medium, purified on the same medium and then subjected to morphological analysis, as reported by Pallmann et al. [25], and genotypic characterisation.

Genomic DNA for PCR assays was prepared from yeast isolates after growth in YPD broth media at 25 °C for 48 h. Cells were harvested, and DNA was extracted using the InstaGene Matrix kit (Bio-Rad Laboratories, Hercules, CA, USA) according to the manufacturer's instructions. According to Sinacori et al. [8], yeasts were discriminated by RFLP of the region spanning the internal transcribed spacers (ITS1 and ITS2) and the 5.8S rRNA gene. Species-level identification of each group was confirmed by sequencing the D1/D2 region of the 26S rRNA gene following the procedure described by Guarcello et al. [7]. DNA sequencing reactions were performed at AGRIVET (University of Palermo, Italy). Sequences were manually corrected using Chromas 2.6.2. (Technelysium Pty Ltd., Brisbane, Australia). Nucleotide sequences were compared to GenBank sequences through BLASTn searches.

2.5. Dominance of S. cerevisiae and M. pulcherrima Isolates

The dominance of the inoculated *S. cerevisiae* and *M. pulcherrima* was verified as reported by Legras et al. [26] and Barbosa et al. [27]. Fingerprinting profiles were analysed as reported by Alfonzo et al. [28].

2.6. Must and Wine Analysis

2.6.1. Chemical Properties

Chemical properties such as sugars (glucose and fructose, g/L) and residual sugars (g/L), yeast-assimilable nitrogen (ammoniacal nitrogen and alpha-amino nitrogen, g/L), organic acids (malic acid, lactic acid and acetic acid, g/L), glycerol (g/L) and ethanol (% *v/v*) were quantified during and at the end of the AF using the methods described by Prestianni et al. [29].

The pH values were measured using a pH 70 Vio FOOD pH meter (XS Instruments, Carpi, Italy), and total acidity (g/L of tartaric acid) was detected through the procedure proposed by OIV-MA-AS313-01 [30]. Free and total sulphur dioxide was determined in accordance with Alfonzo et al. [1].

The analysis of the chemical composition of wines analysed included ash alkalinity, buffering power, total extract, total phenols, flavans reactive to 4-(dimethylamino)cinnamaldehyd, oxidation tests, total phenols and extracts were performed as reported by Alfonzo et al. [1].

2.6.2. Volatile Organic Compounds

All reagents were of analytical grade. Ethyl benzoate was purchased from Sigma-Aldrich (82024 Taufkirchen, Germany). n-Alkane standards (C8 to C40) were purchased from Aldrich Chemical Co. (St. Louis, MO, USA).

An automatic SPME holder (Supelco®, Bellefonte, PA, USA) was used for evaluation of VOC profiles. A fiber 50/30 μm divinylbenzene (DVB)/carbowax (CAR)/polydimethylsiloxane (PDMS) of 1 cm length was used for fractionation of volatile compounds from the headspace (HS) of the conditioned wines. Prior to its use, the fiber was conditioned for 1.5 h at 250 °C in the inlet of the gas chromatograph according to Supelco® Co. Analysis of wine aroma was performed following a slightly modified method proposed by Sagratini et al. [31]. For extraction, each aliquot (10 mL) of the wine samples and 2.2 g of NaCl were placed into a 20 mL vial (75.5 × 22.5 mm) (Supelco, Bellefonte, PA, USA). The samples were equilibrated at 35 °C for 15 min, stirring at 600 rpm. The SPME fiber was exposed to the wine samples for 30 min in the headspace of the sample kept at 35 °C. The flavour compounds were desorbed for 5 min from the fiber to the column through a splitless injector at 250 °C. The SPME fibres were cleaned to prevent cross-contamination by inserting the fibre into the auxiliary injection port at 250 °C for 30 min and were then re-used. All samples were prepared and analysed in triplicates in standard 20 mL volume headspace vials.

Semi-quantification of volatile compounds was performed using an Agilent 7000C GC system fitted with a fused silica apolar DB-5MS capillary column (30 m × 0.25 mm i.d.; 0.25 μm film thickness) (Santa Clara, CA, USA) coupled to an Agilent triple quadrupole Mass Selective Detector MSD 5973. The ionization voltage was 70 eV, the electron multiplier energy was 2000 V and the transfer line temperature was 270 °C. The solvent delay was 0 min. Helium was the carrier gas (1 mL/min). The temperature programme was from 35 °C (0 min) to 270 °C at 8 °C min^{-1}, from 270 °C (2 min) to 300 °C at 15 °C min^{-1} and then 300 °C for 5 min. Volatile compounds were injected at 250 °C automatically in the splitless mode. Linear retention indices were calculated using *n*-alkanes as reference compounds. For the analysis of alkane solutions (C_8-C_{40}) (Sigma-Aldrich, USA), the injector mode was set in the 10:1 split mode. The individual peaks were analysed using the GC-MSolution package, version 2.72. Identification of compounds was carried out using the Adams, NIST 08, Wiley 9 and FFNSC 2 mass spectral databases.

For each volatile organic compound identified, the odour activity value (OAV) as described by Butkhup et al. [32] was calculated in order to assess which VOCs contributed significantly to the odour series characterising each wine.

2.7. Sensory Analysis

A total of 15 judges (7 women and 8 men, ranging from 25 to 46 years old) with previous experience in wine tasting participated in the evaluation of the sensory profile of the wines carried out as described by Jackson [33]. The judges were subjected to preliminary tests to determine their sensory performances in terms of their basic taste and the aromas associated with the wines. The sensory profiles of the wines obtained from Catarratto grapes were constructed using two selected panels each of ten judges trained over several sessions. The fifteen panellists compared the four experimental wines during different sessions. They consensually generated 36 sensory descriptive attributes for appearance, odour, flavour, taste, overall quality and finish in several sessions. The set of attributes were: appearance (green reflexes and yellow colour); odour (banana, citrus, fatty, floral, fruity, grape, green almond, intensity, pear, persistence, pineapple and sweet fruit); taste (bitter, salty, sour and sweet); mouthfeel (body or balance); flavour (banana-like, cherry pit citrus, fruity, intensity, mandarin orange, persistence, pineapple, sweet apple and sweet fruit), overall quality (flavour, mouth-feel, odour and taste) and finish (after-smell and after-taste). The different descriptors were quantified using a 9-point intensity scale as reported by Alfonzo et al. [28].

The sensory test was carried out following the procedures described by Alfonzo et al. [1].

2.8. Statistical Analysis

In order to determine statistically significant differences between the properties monitoring during the AF (chemical and technological data) and in the final wines (sensory analysis and VOCs composition), the ANOVA test was applied. Tukey's test was used for multiple mean comparisons (statistical significance: $p < 0.05$).

Multiple factor analysis (MFA) was carried out in order to distinguish the different treatments from the data acquired during the sensory analysis following the methodology reported by Alfonzo et al. [1]. Agglomerative hierarchical clustering (AHC) was performed to group the trials according to their dissimilarity, as measured by Euclidean distances and Ward's method.

In order to assess the existing correlation between the aromas detected during the sensory analysis and the VOCs with an odour activity value > 1, a principal component analysis (PCA) was performed using the XLstat software version 2019.2.2 (Addinsoft, New York, NY, USA) for Excel.

3. Results and Discussion

3.1. Microbial Growth Dynamic

The concentrations of yeasts (presumptive *Saccharomyces* (PS), non-*Saccharomyces* (NS) and presumptive *Metschnikowia* (PM)) during the alcoholic fermentation (AF) are shown in Figure 2. The PS and NS levels in the Catarratto must were around four logarithmic cycles (Figure 2a,b), while no isolates attributable to the genus *Metschnikowia* were detected (Figure 2c). Catarratto musts are usually poor for the presence of indigenous *Metschnikowia* spp., although in musts from Sicilian Catarratto grapes, *M. pulcherrima* has been isolated at percentages ranging from 0.2 to 1.1% [34].

The *M. pulcherrima* MP346 inoculum concentration in T1 and T2 was close to 6.5 Log CFU/mL. The concentration of PS after the SPF52 inoculum ranged from 7.3 (T1) to 7.6 (C1) Log CFU/mL in all treatments. On day 3 of AF, PS showed an increase to 7.4–8.0 Log CFU/mL for all trials. The NS populations were lower and in the range of 2.3–3.2 Log CFU/mL. The reduction in the NS yeast populations during AF is a known phenomenon attributable to several causes such as metabolite production by *S. cerevisiae*, nutrient limitation and low resistance to ethanol [13]. The PM levels were 3.0 Log CFU/mL for T1 and 4.6 Log CFU/mL for T2 after 3 days of AF and were lower than the limit of detection in the C1 and C2 samples. Indeed, the lower microbial load of the PM populations observed at 3 days of AF in T1 and T2 compared with C1 and C2 could be due to the lower ethanol concentration detected in T1 and T2 (Table S1). At day 6 of AF, when the ethanol reached concentrations above 6% *v/v*, the PS values reached levels in the range of 7.0–8.0 Log CFU/mL, whereas both NS and PM were undetectable in any trials. The absence of *M. pulcherrima* in trials inoculated with the commercial preparation FLAVIA® MP346 (T1 and T2) could be due to the above-mentioned factors. Some authors have recorded a significant decrease in the concentration *of M. pulcherrima* after 9 days of AF when sequential inoculation with *S. cerevisiae* occurred [15].

From the 12th day until the end of AF (18 d), the PS populations decreased slightly from 7.3–8.0 to 6.7–7.0 Log CFU/mL in all treatments. The microbiological count values for *S. cerevisiae* were found by Scacco et al. [35] on Sicilian Catarratto musts inoculated with selected starter strains of the same species.

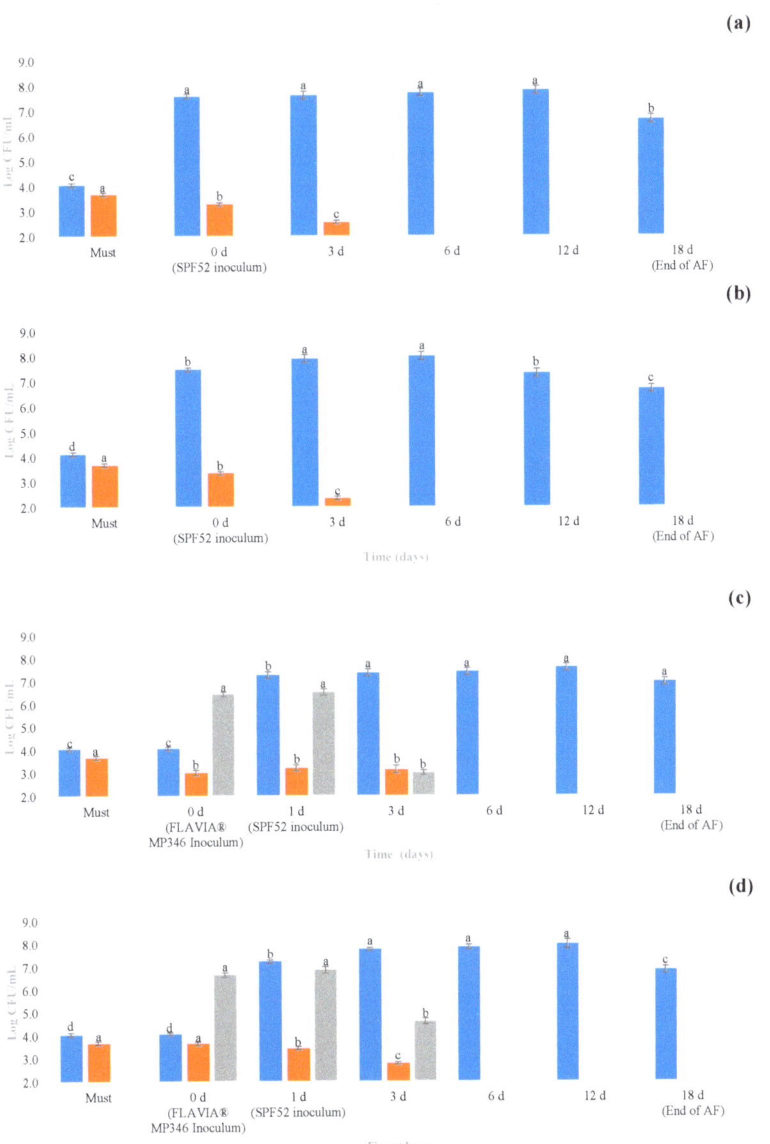

Figure 2. Microbiological concentration (Log CFU/mL) of samples during alcoholic fermentation: (**a**) C1, single inoculum with *S. cerevisiae* SPF52; (**b**) C2, glutathione-rich inactivated yeasts and single inoculum with *S. cerevisiae* SPF52; (**c**) T1, sequential inoculum with *M. pulcherrima* MP346/*S. cerevisiae* SPF52; (**d**) T2, glutathione-rich inactivated yeasts and sequential inoculum with *M. pulcherrima* MP346/*S. cerevisiae* SPF52. For each microbiological group, different letters indicate statistically significant values determined by using Tukey's test ($p \leq 0.05$). Legends: ■, Presumptive *Saccharomyces*; ■, non-*Saccharomyces* (except *Metschnikowia* spp.); ■, Presumptive *Metschnikowia* spp.

3.2. Molecular Analysis

In relation to the macro- and microscopic characteristics, 949 colonies were analysed; from these, 592 isolates showed the typical characteristics of yeasts belonging to the *Saccharomyces* genus. The amplicon size of the 5.8S-ITS region was around 850 bp and confirmed the presumptive species identity of *S. cerevisiae* for all isolates. The other isolates (n = 357) were assigned to the NS yeast group.

A total of 233 isolates were morphologically identified as *Metschnikowia* spp. and showed an ITS amplicon between 380 and 400 bp. The ITS amplicon sizes were equivalent to those reported in the literature for *M. pulcherrima* [36]. The isolates of the PS group (n = 592) and PM (n = 233) were characterised by RFLP analysis of the 5.8S-ITS region.

The PS RFLP profiles were similar to those indicated by Granchi et al. [37]. Consequently, the PS group represented putative *S. cerevisiae*. The sizes of the RFLP profiles of the PM were equivalent to those described in the literature for the species *M. pulcherrima* [37].

The different profiles may have been caused by the presence of native *S. cerevisiae*, although less representative, being present among the isolates obtained. Indeed, the PS count values detected before SPF52 inoculation (4.1 Log CF/mL) clearly explain the presence of eight additional interdelta profiles. The interdelta profile of *S. cerevisiae* SPF52 was the most frequently (>93%) isolated. The strain typing of *M. pulcherrima* was carried out by RAPD-PCR. The results from these analyses showed that all the 233 isolates represented a unique strain.

The genotypic identification of the yeasts was completed by pairwise alignment of the D1/D2 sequence with the type of strain of each species (*S. cerevisiae* CBS 1171^T and *M. pulcherrima* CBS 5833^T). A comparison of the sequences of the D1/D2 region of the two reference strains showed a 100% similarity to the sequences of the type strains of each species, confirming the identification obtained by the RFLP analysis.

3.3. Kinetics of the Main Oenological Properties

The fermentations carried out in the presence of *S. cerevisiae* SPF52 as the only inoculated strain (C1 and C2) and the corresponding trials with *M. pulcherrima* (T1 and T2) were able to conclude the AF as determined by the complete consumption of sugars.

The trends of the principal oenological data during AF are shown in Table S1. The fermentation was concluded in 18 days on average.

After 3 d of AF, differences in pH, TA and the concentrations of sugars, ethanol, ammonia nitrogen and alpha amine nitrogen were observed among the trials. The highest differences in the sugar, glycerol and ethanol contents were registered at day 6 of AF. Specifically, C2 showed the lowest values in residual sugars (58.79 g/L); glucose was 25.80 g/L and fructose was 32.99 g/L, and consequently, it showed the highest values of ethanol (8.44\% *v/v*). The glycerol contents observed in T1, T2 and C2 were similar (5.19–5.28 g/L), whereas the lowest values were found in C1 (5.06 g/L). This trend was observed until the 12th day of AF.

At the end of AF, the glucose concentrations ranged from 1.10 (T1 and T2) to 1.62 g/L (C1), whereas the fructose concentration was slightly higher and within the range of 1.39–2.60 g/L. No differences were observed for TA, whereas VA's values ranged from 0.27 (T1) to 0.31 (C1 and C2) g/L acetic acid. The pH values varied between treatments, where T1 and T2 had slightly lower values (3.41 and 3.43, respectively) when compared to both the control trials C1 (3.47) and C2 (3.51). The ethanol concentrations ranged between 11.35 and 11.43% (*v/v*); the comparison between the T1 and T2 and the C1 and C2 treatments showed no significant differences. In contrast, Contreras et al. [38] reported that some strains of *M. pulcherrima* are able to decrease the amount of ethanol by as much as 1% (*v/v*) during fermentation. An analysis of the ethanol production during AF a revealed lower ethanol production in the T1 and T2 trials after 3 d of AF. After AF, differences in the ethanol concentration between the different trials were not statistically significant. This phenomenon could be attributable to the presence of *M. pulcherrima* up to the 3rd day of AF (3.0–4.6 Log CFU/mL).

The malic acid levels decreased in all the treatments from an initial concentration of 1.90 g/L in the must to 1.28–1.50 g/L at the end of AF. Contrary to the reports of Ruiz et al. [15], no decreases were recorded in T1 and T2 compared to C1 and C2, although these authors showed that in fermentations conducted with *M. pulcherrima*/*S. cerevisiae*, a decrease in the malic acid content of 0.2 g/L occurred in the wines. Lactic acid was absent in all the trials. The highest concentration of glycerol was found in C2 (6.57 g/L), and lower values (>5 g/L) were detected in the other wines. In this case, the sequential inoculum with *M. pulcherrima*/*S. cerevisiae* did not produce an increase in the glycerol concentration, in contrast to what has been observed in white wines made with the Verdejo variety [15].

During the five months of ageing in stainless steel tanks, there were no substantial changes in the monitored chemical properties (Table S2). There was a decrease in residual sugars, glucose and fructose, and all the other properties remained constant or showed minimal variations.

3.4. Oenological Data Analysis

The values of the physico-chemical properties of the wines are reported in Table 1.

Table 1. Oenological properties of four Catarratto wines.

Sample	SO_2 Free (mg/L)	SO_2 Total (mg/L)	Total Extract (g/L)	Total Phenols (mg/L Catechins)	p-DACA Flavans (mg/L Catechins)	Absorbance (420 nm)	Oxidation Test (%)	Buffer Power (meq/L)	Ash Alkalinity (meq/L)
T1	29.00 ± 1.00 [a]	128.00 ± 0.00 [a]	18.80 ± 0.09 [b]	92.74 ± 0.84 [a]	19.80 ± 0.08 [a]	0.079 ± 0.000 [c]	5.74 ± 0.09 [a]	31.25 ± 0.08 [b]	12.28 ± 0.05 [c]
T2	22.00 ± 0.00 [b]	115.00 ± 2.00 [b]	19.10 ± 0.07 [a]	93.47 ± 0.39 [a]	10.23 ± 0.11 [b]	0.080 ± 0.003 [c]	1.12 ± 0.03 [b]	31.25 ± 0.17 [b]	12.58 ± 0.07 [b]
C1	16.00 ± 2.00 [c]	109.00 ± 1.00 [c]	18.50 ± 0.08 [c]	84.38 ± 1.13 [c]	1.63 ± 0.08 [d]	0.101 ± 0.001 [a]	0.00 ± 0.00 [c]	31.32 ± 0.12 [b]	12.11 ± 0.07 [d]
C2	29.00 ± 1.00 [a]	105.00 ± 1.00 [d]	19.00 ± 0.11 [ab]	83.36 ± 0.67 [b]	3.12 ± 0.05 [c]	0.093 ± 0.002 [b]	0.00 ± 0.00 [c]	32.34 ± 0.13 [a]	13.43 ± 0.02 [a]
S.S.	***	***	***	***	***	***	***	***	***

Result indicate mean value ± standard deviation of three determinations. Abbreviations: S.S., statistical significance; T1, sequential inoculum with *M. pulcherrima* MP346/*S. cerevisiae* SPF52; T2, glutathione-rich inactivated yeasts and sequential inoculum with *M. pulcherrima* MP346/*S. cerevisiae* SPF52; C1, single inoculum with *S. cerevisiae* SPF52; C2, glutathione-rich inactivated yeasts and single inoculum with *S. cerevisiae* SPF52. Data in the same column followed by the same letter are not significantly different according to Tukey's test. p value: ***, $p < 0.001$.

The free and total SO_2 values were variable in the different wines. In particular, the highest free SO_2 values were observed in T1 and C2 (29 mg/L), while the highest total SO_2 value was observed in T1 (128 g/L).

The total extract was higher than the minimum legal values, which for white wines are fixed at >14 g/L [39]. In this study, all the wines exceeded this threshold; the values were in the range of 18.50–19.10 g/L for C1 and T2, respectively, which were comparable to the results described in Scacco et al. [35] on Sicilian Catarratto wines.

The T1 and T2 trials retained a greater susceptibility to undergo oxidation than the C1 and C2 controls, which was independent of the use or non-use of GIY with oxidation test values of 5.74 and 1.12% (T1 and T2) and 0% (C1 and C2). The presence of *M. pulcherrima* therefore appeared to exert a bio-protective action by predicting oxidations at the pre-inoculation of *S. cerevisiae*. The decrease in polyphenols was not due to the synthesis of polysaccharides by *M. pulcherrima* but to its bioprotective and inhibiting action against grape tyrosinases. In fact, in the pre-fermentative stage in the C1 and C2 controls, the absence of *M. pulcherrima* favored a significant increase in the optical density at 420 nm. At the same time, in the same controls there would have been a significant decrease in the total polyphenols resulting from the decrease in the phenolic class of the ortho-diphenols detected by means of the p-DACA reagent. The total polyphenol content was independent of the presence/absence of GIY. The null POM test values observed in the controls C1 and C2 may be due to a series of oxidation reactions of polyphenolic compounds that not even the addition of GIY in T2 was able to limit. The colonisation of the must by *M. pulcherrima* in the pre-fermentation phase probably led to a reduction in oxidative activities [40].

Regarding buffering power, there were negligible variations, and only the wine C2 reached statistically significant values compared to the other trials. The highest buffering power value was in C2 (32.34 meq/L), which was comparable to those reported in the literature in Sicilian Catarratto wines [40]. This was similar for ash alkalinity, where C2 had the highest value (13.43 meq/L); the wine values were within the range of 11–17 meq/L, which were similar to those reported in the literature for white wines [41].

3.5. Volatile Organic Compound Composition

The samples showed differences mainly at the quantitative level. Twenty-six compounds were detected, and they were grouped into several classes: alcohols, ethers, aldehydes, ethyl esters of fatty acids (EEFAs), higher alcohol acetates (HAAs), ethyl esters of branched acids (EEBAs), miscellaneous esters (MEs) and other compounds. For clarity, the classification of esters was reported as described by Alfonzo et al. [1]. The most-concentrated compounds in all the samples were EEFAs (2318.98–1401.74 ppb) followed by MEs (233.83–98.84 ppb) and alcohols (36.48–18.84 ppb).

The must inoculated with *M. pulcherrima* MP346 produced less alcohols than the controls. 3-methyl-1-butanol and phenylethyl alcohol were the compounds detected in the highest quantity in C2. A similar condition was observed in Riesling wines fermented by sequential inoculation with *M. pulcherrima*/*S. cerevisiae* [14].

The compound most commonly detected in the aldehyde class was dodecanal. In the wines produced in the absence of GYI, it reached a maximum concentration in C1 (11.06 ppb). Aldehydes, particularly decanal and dodecanal if they are present in high concentrations, can result in the appearance of an unpleasant "green" odour in wines [42].

Esters directly and indirectly influence wine aroma by means of highly varied interactions. The fermentation process applied significantly influences the quality and quantity of esters [43,44]. The wine samples inoculated with *M. pulcherrima* MP346 showed a higher content of esters (2318.98 ppb in T1 and 2056.15 ppb in T2) than the controls (1401.74 ppb in C1 and 1848.45 ppb in C2). Among the esters, the most representative was ethyl decanoate, which was produced in amounts over 1000 ppb in the wines inoculated with *M. pulcherrima* MP346. Indeed, in Riesling musts inoculated with the same commercial strain of *M. pulcherrima*, the quantities detected were half of those present in the Catarratto musts [14,45].

The ethyl decanoate content reported by Benito et al. [14] and Mislata et al. [45] does not appear to have been impacted by the presence of *M. pulcherrima* MP346. However, in the Catarratto wines in this study, the levels of ethyl decanoate were significantly higher in the fermented wines with sequential inoculum.

A different situation was observed for ethyl octanoate, where the second EEFA was detected in greater quantities. Higher levels of ethyl octanoate were found in the experimental wines C1 and T1 without the addition of GIY. The effect of the glutathione-enriched inactivated yeast on ethyl octanoate was unclear, although these highly volatile hydrophobic esters exhibit significant variations in wines containing yeast-derivative products [46]. Among the 2-phenylethyl esters, two opposite situations were found for 2-phenylethyl hexanoate, which was detected only in C1 and C2, while 2-phenylethyl acetate was present exclusively in T1 and T2.

The determination of VOCs in the different wines is reported in Table 2.

Table 2. Volatile organic compounds detected in the four Catarratto wines (all values in ppb).

t_R (min.s)	LRI [1]	Compounds [2]	Aroma Description [47–50]	Odour Threshold [3]	C1 [4] (OAV)	C2 [4] (OAV)	T1 [4] (OAV)	T2 [4] (OAV)	S.s. [5]
		Σ Alcohols			33.00 ± 1.32 [b]	36.48 ± 1.46 [a]	18.84 ± 0.75 [c]	19.19 ± 0.76 [c]	***
10.55	758	3-methyl-1-butanol	Fusel	40,000 [51]	26.65 ± 1.07 [b] (<1)	29.59 ± 1.18 [a] (<1)	12.82 ± 0.51 [d] (<1)	15.59 ± 0.62 [c] (<1)	***
36.49	1110	Phenylethyl alcohol	Floral, rose	125,000 [52]	6.35 ± 0.25 [b] (<1)	6.89 ± 0.28 [a] (<1)	6.02 ± 0.24 [b] (<1)	3.60 ± 0.14 [c] (<1)	***
		Σ Ethers			4.75 ± 0.19 [a]	4.16 ± 0.17 [b]	4.24 ± 0.17 [b]	2.53 ± 0.10 [c]	***
32.14	1042	Ethyl benzyl ether	Tropical fruit, pineapple	unknown	4.75 ± 0.19 [a] (n.d. [6])	4.16 ± 0.17 [b] (n.d. [6])	4.24 ± 0.17 [b] (n.d. [6])	2.53 ± 0.10 [c] (n.d. [6])	***
		Σ Aldehydes			17.37 ± 0.69 [a]	4.91 ± 0.20 [c]	11.85 ± 0.47 [b]	2.73 ± 0.11 [d]	***
24.89	958	Benzaldehyde	Bitter almond, cherry	1500 [53]	6.31 ± 0.25 [a] (<1)	4.91 ± 0.20 [b] (<1)	3.60 ± 0.14 [c] (<1)	2.73 ± 0.11 [d] (<1)	***
37.08	1203	Decanal	Floral, orange peel citrus	0.1 [54]	tr (n.d. [6])	tr (n.d. [6])	tr (n.d. [6])	tr (n.d. [6])	n.d. [6]
56.38	1411	Dodecanal	Citrus, floral	2 [55]	11.06 ± 0.44 [a] (5.53)	0.00 ± 0.00 [c] (<1)	8.25 ± 0.33 [b] (4.13)	0.00 ± 0.00 [c] (<1)	***
		Σ EEFAs			1401.74 ± 56.08 [d]	1848.45 ± 73.94 [c]	2318.98 ± 92.76 [a]	2056.15 ± 82.25 [b]	***
27.64	989	Ethyl hexanoate	Sweet fruity, pineapple, green apple	5 [55]	33.79 ± 1.35 [b] (6.76)	48.86 ± 1.95 [a] (9.77)	27.85 ± 1.11 [c] (5.57)	32.14 ± 1.29 [b] (6.42)	***
37.44	1208	Ethyl octanoate	Fruity, pear	2 [55]	901.19 ± 36.05 [a] (450.60)	730.52 ± 29.22 [b] (365.26)	837.67 ± 33.51 [a] (418.84)	596.78 ± 23.87 [c] (298.39)	***
51.00	1379	Ethyl decanoate	Fruity, grape	200 [55]	273.88 ± 10.96 [c] (1.37)	928.14 ± 37.13 [b] (4.64)	1253.71 ± 50.15 [a] (6.27)	1236.22 ± 49.45 [a] (6.18)	***
54.98	1391	Ethyl 9-decenoate	Fruity, fatty	100 [56]	184.44 ± 7.38 [ab] (1.84)	137.73 ± 5.51 [c] (1.38)	199.75 ± 7.99 [a] (2.00)	178.82 ± 7.15 [b] (1.79)	***
67.44	1599	Ethyl dodecanoate	Sweet, waxy, floral	2000 [55]	8.44 ± 0.34 [b] (<1)	3.20 ± 0.13 [c] (<1)	0.00 ± 0.00 [d] (<1)	12.19 ± 0.49 [a] (<1)	***
		Σ HAAs			15.10 ± 0.60 [b]	19.09 ± 0.76 [a]	6.25 ± 0.25 [d]	9.71 ± 0.39 [c]	***
18.59	882	3-methyl-1-butanol acetate	Sweet fruity, banana	0.75 [52]	15.10 ± 0.60 [b] (20.13)	19.09 ± 0.76 [a] (25.45)	6.25 ± 0.25 [d] (8.33)	9.71 ± 0.39 [c] (12.95)	***
		Σ EEBAs			12.94 ± 0.52 [b]	8.02 ± 0.32 [c]	0.00 ± 0.00 [d]	14.56 ± 0.58 [a]	***
58.69	1447	Isopentyl octanoate	Fruity, pineapple, coconut	125 [57]	12.94 ± 0.52 [b] (<1)	8.02 ± 0.32 [c] (<1)	0.00 ± 0.00 [d] (<1)	14.56 ± 0.58 [a] (<1)	***
		Σ MEs			233.83 ± 9.35 [a]	106.27 ± 4.26 [bc]	118.12 ± 4.74 [b]	98.84 ± 3.96 [c]	***
6.80	611	Ethyl acetate	Ethereal, fruity	7500 [55]	65.36 ± 2.61 [a] (<1)	9.10 ± 0.36 [d] (<1)	33.72 ± 1.35 [c] (<1)	38.29 ± 1.53 [b] (<1)	***
34.79	1089	Methyl benzoate	Green almond	10 [56]	36.94 ± 1.48 [a] (3.69)	25.00 ± 1.00 [b] (2.50)	24.22 ± 0.97 [b] (2.42)	14.93 ± 0.60 [c] (1.49)	***
46.19	1268	2-phenylethyl hexanoate	Sweet, honey, floral	94 [58]	10.28 ± 0.41 [a] (<1)	5.03 ± 0.20 [b] (<1)	0.00 ± 0.00 [c] (<1)	0.00 ± 0.00 [c] (<1)	***
46.24	1542	2-phenylethyl acetate	Rose	250 [55]	0.00 ± 0.00 [c] (<1)	0.00 ± 0.00 [c] (<1)	3.69 ± 0.15 [b] (<1)	5.45 ± 0.22 [a] (<1)	***
		Σ Others			121.25 ± 4.85 [a]	67.62 ± 2.70 [b]	56.49 ± 2.27 [c]	40.17 ± 1.61 [d]	***
7.50	634	Tetrahydrofuran	Butter, caramel	unknown	40.89 ± 1.64 [a] (n.d. [6])	35.68 ± 1.43 [b] (n.d. [6])	26.44 ± 1.06 [c] (n.d. [6])	23.34 ± 0.93 [c] (n.d. [6])	***
18.14	876	1,3-dimethylbenzene	Plastic odour	unknown	12.08 ± 0.48 [a] (n.d. [6])	8.03 ± 0.32 [b] (n.d. [6])	4.14 ± 0.17 [c] (n.d. [6])	2.89 ± 0.12 [d] (n.d. [6])	***
29.59	1023	*o*-cymene	Herb	unknown	15.37 ± 0.61 [a] (n.d. [6])	9.97 ± 0.40 [b] (n.d. [6])	5.41 ± 0.22 [c] (n.d. [6])	3.67 ± 0.15 [d] (n.d. [6])	***
34.04	1097	1-butenyl benzene	unknown	unknown	2.81 ± 0.11 [a] (n.d. [6])	2.05 ± 0.08 [b] (n.d. [6])	1.40 ± 0.06 [c] (n.d. [6])	0.76 ± 0.03 [d] (n.d. [6])	***
44.34	1232	Benzothiazole	Sulfury, rubbery, vegetable	unknown	16.45 ± 0.66 [a] (n.d. [6])	0.00 ± 0.00 [b] (n.d. [6])	0.00 ± 0.00 [b] (n.d. [6])	0.00 ± 0.00 [b] (n.d. [6])	***
50.79	1302	6-ethyltetralin (isomer)	unknown	unknown	6.85 ± 0.27 (n.d. [6])	3.10 ± 0.12 (n.d. [6])	3.44 ± 0.14 (n.d. [6])	tr (n.d. [6])	n.d. [6]
51.29	1311	6-ethyltetralin (isomer)	unknown	unknown	7.66 ± 0.31 (n.d. [6])	0.00 ± 0.00 (n.d. [6])	2.97 ± 0.12 (n.d. [6])	tr (n.d. [6])	n.d. [6]
54.53	1368	2-ethenyl-naphtalene	unknown	unknown	11.50 ± 0.46 [a] (n.d. [6])	6.36 ± 0.25 [c] (n.d. [6])	10.83 ± 0.43 [a] (n.d. [6])	9.51 ± 0.38 [b] (n.d. [6])	***
59.64	1485	2,6-di-tert-butylquinone	unknown	unknown	7.64 ± 0.31 (n.d. [6])	2.43 ± 0.10 (n.d. [6])	1.86 ± 0.07 (n.d. [6])	tr (n.d. [6])	n.d. [6]

[1] Linear retention index obtained through the modulated chromatogram reported for DB-5 MS apolar column; [2] compounds are classified in order of retention time; [3] odor threshold reported in the literature; [4] Relative amounts expressed as ppb with respect to calibration curve of ethyl benzoate; [5] statistical significance; [6] not determinable. Abbreviations: EEFAs: ethyl esters of fatty acids; HAAs: higher alcohol acetates; EEBAs: ethyl esters of branched acids; MEs: miscellaneous esters; OAV, odour activity value; tr: trace amount < 0.05%; T1, sequential inoculum with *M. pulcherrima* MP346/*S. cerevisiae* SPF52; T2, glutathione-rich inactivated yeasts and sequential inoculum with *M. pulcherrima* MP346/*S. cerevisiae* SPF52; C1, single inoculum with *S. cerevisiae* SPF52; C2, glutathione-rich inactivated yeasts and single inoculum with *S. cerevisiae* SPF52. Data in the same line followed by the same letter are not significantly different according to Tukey's test. *p* value: ***, *p* < 0.001.

However, the 2-phenylethyl acetate concentrations were lower than those determined for Riesling wines produced using *M. pulcherrima* MP346. Most likely, the strain of *S. cerevisiae* used as the starter for AF significantly influenced the levels of this ester [59].

Among the twenty-six VOCs, only seven compounds showed an OAV greater than 1 (Table 2), i.e., one aldehyde (dodecanal) and six esters (ethyl exanoate, ethyl octanoate, ethyl decanoate, ethyl 9-decenoate, 3-methyl-1-butanol acetate and methyl benzoate). Esters represent a group of compounds of considerable importance that are formed during AF through yeast metabolism and have a strong impact on the aromatic profile of wine [60].

3.6. Sensory Analysis

The data from the sensory evaluation are shown in Table 3. The trials revealed some differences correlated with the presence/absence of *M. pulcherrima* MP346 and GIY.

The wines showed variability in terms of the attributes that defined appearance. The yellow colour values were in the range of 7.15–7.29, whereas, the green reflexes ranged between 3.63–4.04. The yellow colour values observed were higher than those shown by Scacco et al. [35], while the ratings associated with the green reflections attribute were similar.

The T2 sample displayed a high score for 13 descriptors. The *M. pulcherrima* MP346 and GIY wine (T2) had the highest overall quality score (8.80). With regards to the odour attributes, the T1 and T2 wines showed the highest values for intensity and persistence, respectively. In addition, the T1 wine showed high scores for grape, fruity and fatty odours, the C1 wine showed high scores for citrus, floral, green almond and pineapple odours and the C2 wine was characterised by the presence of odours associated with banana, pear, pineapple and sweet fruit. The T2 wine was characterised by odour attributes with intermediate scores. In wines to which GIY was added (C2 and T2), citrus and floral odours were not perceived. Nevertheless, banana, citrus, floral, fruity and pear aromas were present in the Catarratto wines reported by Scacco et al. [35] but at lower levels.

The descriptors associated with taste enabled discrimination of the wines. T1 and T2 showed high scores for sour flavours, whereas salty flavours showed high values in T2. In terms of mouthfeel, the T2 wine achieved high values for the body and balance attributes. No unpleasant odours or flavours were revealed for all the wines. The GIY increased the flavour intensity and persistence, confirming the results described by Alfonzo et al. [1]. Indeed, the treatment with GIY in combination with *M. pulcherrima* MP346 significantly improved the aromatic complexity of the T2 wine.

The T2 wine showed high intensity and persistence scores for flavours. The sensory descriptors with high flavour values were pineapple (C1), sweet fruit (C2) and fruity (T1 and T2). The T2 wine also excelled compared to the other wines for after-smell (8.50) and after-taste (8.71).

Correlations of the sensory analyses were examined by MFA. The number of sensory attributes (thirty-six variables) for the four wines made it possible to define two factors with an Eigen > 1 that represented a total variance of 89.64%. The correlation between the variables and the MFA factor was expressed by the value of the contribution and $\cos^2$. The incidence of the factors F1 (56.41%) and F2 (33.23%) on the total variance discriminated the different wines. Examining the loading plot (Figure 3), eight variables were located in both quadrants I and IV, ten were located in quadrant II and eleven were located in quadrant III.

Figure 4 reveals that the wines were clustered into three groups. In Figure 4a (MFA) and Figure 4b (AHCA), it is possible to observe how T1 and T2 represented a unique cluster. Interestingly, trial C1 did not cluster with trial C2. Indeed, the C1 and C2 trials represented a different cluster.

Table 3. Sensory score for experimental Catarratto wines.

Attributes	Trial				SEM	Statistical Significance	
	C1	C2	T1	T2		Judges	Wine
Appearance							
Yellow colour	7.28 [a]	7.15 [a]	7.21 [a]	7.29 [a]	0.01	n.s.	n.s.
Green reflexes	4.04 [a]	3.63 [b]	3.74 [b]	3.68 [b]	0.02	***	***
Odour							
Banana	3.63 [b]	3.94 [a]	2.79 [d]	3.15 [c]	0.07	***	***
Citrus	2.40 [a]	1.00 [c]	1.74 [b]	1.00 [c]	0.09	***	***
Fatty	1.35 [b]	1.22 [c]	1.62 [a]	1.32 [b]	0.02	***	***
Floral	2.53 [a]	1.00 [c]	1.97 [b]	1.00 [c]	0.10	***	***
Fruity	8.54 [c]	8.02 [d]	8.88 [a]	8.68 [b]	0.05	***	***
Grape	2.97 [c]	2.99 [c]	4.17 [a]	3.43 [b]	0.07	***	***
Green almond	7.67 [a]	6.84 [b]	6.77 [b]	5.71 [c]	0.11	***	***
Intensity	6.68 [c]	7.19 [b]	8.26 [a]	7.40 [b]	0.09	***	***
Pear	5.14 [b]	5.44 [a]	4.76 [d]	4.91 [c]	0.04	***	***
Persistence	7.11 [d]	8.64 [b]	8.12 [c]	8.97 [a]	0.10	***	***
Pineapple	3.62 [a]	3.63 [a]	2.96 [c]	3.44 [b]	0.04	***	***
Sweet fruit	7.25 [b]	7.57 [a]	5.75 [d]	6.59 [c]	0.10	***	***
Taste							
Sweet	3.48 [a]	3.59 [a]	2.78 [b]	2.68 [b]	0.06	***	***
Sour	5.38 [b]	5.37 [b]	8.11 [a]	8.24 [a]	0.21	***	***
Salty	5.70 [c]	5.85 [c]	7.99 [b]	8.39 [a]	0.18	***	***
Bitter	1.10 [c]	1.25 [b]	1.20 [b]	1.42 [a]	0.02	***	***
Mouthfeel							
Body	7.80 [c]	8.42 [b]	8.55 [b]	8.97 [a]	0.06	***	***
Balance	6.50 [d]	7.49 [c]	8.10 [b]	8.65 [a]	0.12	***	***
Flavour							
Banana-like	2.47 [b]	2.75 [a]	1.93 [d]	2.22 [c]	0.07	***	***
Cherry pit	3.67 [a]	3.84 [a]	3.77 [a]	2.70 [b]	0.07	***	***
Citrus	3.92 [a]	1.00 [b]	3.58 [a]	1.00 [b]	0.21	***	***
Fruity	6.15 [c]	6.26 [c]	7.79 [a]	6.80 [b]	0.10	***	***
Intensity	7.80 [c]	7.85 [c]	8.12 [b]	8.56 [a]	0.04	***	***
Mandarin orange	1.74 [a]	1.00 [c]	1.40 [b]	1.00 [c]	0.05	***	***
Persistence	7.70 [c]	8.78 [a]	7.97 [b]	8.94 [a]	0.08	***	***
Pineapple	7.11 [a]	6.89 [b]	6.86 [b]	6.14 [c]	0.05	***	***
Sweet apple	2.51 [c]	2.66 [c]	3.89 [a]	3.54 [b]	0.09	***	***
Sweet fruit	7.12 [b]	7.56 [a]	5.75 [d]	6.58 [c]	0.10	***	***
Overall quality	7.50 [d]	8.57 [b]	8.25 [c]	8.80 [a]	0.07	***	***
Flavour	6.98 [c]	8.81 [a]	8.11 [b]	8.91 [a]	0.11	***	***
Mouthfeel	7.20 [c]	8.32 [a]	7.88 [b]	7.97 [b]	0.06	***	***
Odour	7.20 [c]	8.86 [a]	8.01 [b]	8.74 [a]	0.10	***	***
Taste	7.01 [d]	7.54 [c]	7.82 [b]	8.11 [a]	0.06	***	***
Finish							
After-smell	6.80 [c]	8.15 [b]	8.21 [b]	8.50 [a]	0.10	***	***
After-taste	7.10 [c]	7.96 [b]	8.22 [b]	8.71 [a]	0.09	***	***

Results indicate mean value of three replicate sessions. Abbreviation: SEM, standard error of the mean; T1, sequential inoculum with *M. pulcherrima* MP346/*S. cerevisiae* SPF52; T2, glutathione-rich inactivated yeasts and sequential inoculum with *M. pulcherrima* MP346/*S. cerevisiae* SPF52; C1, single inoculum with *S. cerevisiae* SPF52; C2, glutathione-rich inactivated yeasts and single inoculum with *S. cerevisiae* SPF52. Data in the same line followed by the same letter are not significantly different according to Tukey's test. *p* value: ***, *p* < 0.001; n.s., not significant.

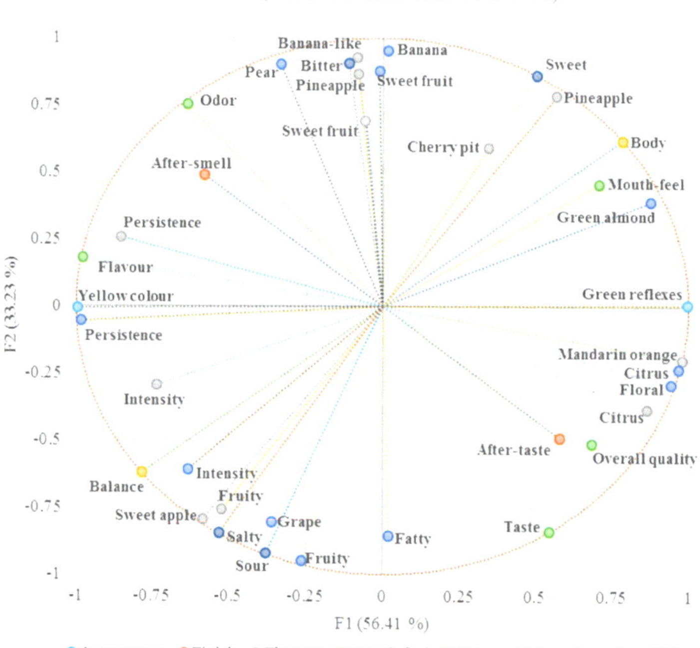

Figure 3. Variable loading plot of MFA.

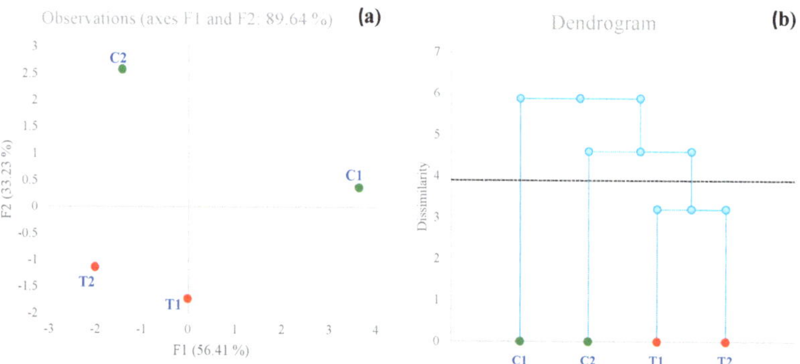

Figure 4. Multiple factor analysis applied to sensory analysis of Catarratto wines: (**a**) sample score; (**b**) agglomerative hierarchical clustering (AHC) dendrogram. Abbreviations: T1, sequential inoc lum with *M. pulcherrima* MP346/*S. cerevisiae* SPF52; T2, glutathione-rich inactivated yeasts and sequential inoculum with *M. pulcherrima* MP346/*S. cerevisiae* SPF52; C1, single inoculum with *S. cerevisiae* SPF52; C2, glutathione-rich inactivated yeasts and single inoculum with *S. cerevisiae* SPF52.

3.7. Sensory Profiles Associated with Volatile Organic Compounds

A PCA was used to evaluate the correlation between VOCs and aroma attributes. According to Figure 5, the F1 factor contributed 66.11% of the total variance, whereas the F2 factor explained 28.60% of the total variance.

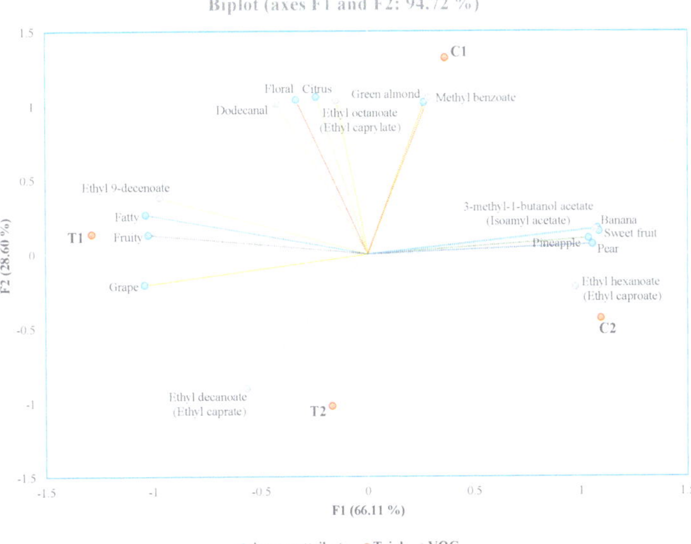

Figure 5. Principal component analysis (PCA) biplot for VOCs and aroma attributes. Abbreviations: T1, sequential inoculum with *M. pulcherrima* MP346/*S. cerevisiae* SPF52; T2, glutathione-rich inactivated yeasts and sequential inoculum with *M. pulcherrima* MP346/*S. cerevisiae* SPF52; C1, single inoculum with *S. cerevisiae* SPF52; C2, glutathione-rich inactivated yeasts and single inoculum with *S. cerevisiae* SPF52.

Each wine, as can be seen from the biplot graph, was separate from the others. The C1 wine was associated with methyl benzoate, which produced green almond aromas [61]. A sensory analysis confirmed this attribute, and the highest scores were achieved in this trial. Ethyl 9-decenoate was the compound closely correlated with the T1 wine. This ester produces fruity and fatty odours [62], which were also detected in the sensory analysis, with the scores of fruity being higher than fatty. The grape aroma emitted by ethyl decanoate [63] represented the T2 wine. The highest sensory analysis attributes detected in the T2 wine were fruity and sweet fruit, and the grape aroma showed modest values. However, fruity and grape aromas are also associated with the presence of ethyl decanoate [64]. Finally, the C2 wine was closely associated with four odour descriptors (pineapple, sweet fruit, banana and pear). Only 3-methyl-1-butanol acetate and ethyl hexanoate were above the odour threshold and were responsible for the odours detected in the C2 wine by sensory analysis [61].

The imperfect correlation between the highest OAV values of VOCs and the sensory analysis might be attributable to the synergistic interaction of odour molecules (high OAVs with low OAVs) with each other. As a result, the odours related to specific compounds were absent or very slightly perceived during the sensory analysis.

4. Conclusions

In this research, four treatments were examined in order to investigate the effect of *M. pulcherrima* and an antioxidant on the aroma and sensory profile of Catarratto wines. The use of *S. cerevisiae* SPF52 from a non-winemaking origin confirmed that yeasts from honey and its derivatives can potentially be used as starter strains in oenology. The combined use of *M. pulcherrima* MP346 and GIY had a positive impact on the taste–olfactory complexity of the wines. These differences were also confirmed by a sensory analysis. The VOC profiles generated by the wines obtained in the presence/absence of *M. pulcherrima* MP346 were correlated to the addition of GIY from the point of view of the quantity–intensity effect.

Dodecanal was only detected in the wines without GIY, whereas six esters had an OAV > 1 and actively contributed to the aroma definition of the different wines. Among the esters, ethyl decanoate was the most abundant in the wines inoculated with *M. pulcherrima* MP346, regardless of the presence/absence of GIY. However, the differences in the VOC profiles enabled the wines produced with the different winemaking protocols to be distinguished.

The modulation of the aromatic profile of each wine was also confirmed by a sensory analysis, which made it possible to differentiate the wines into three groups. The presence of *M. pulcherrima* MP346 and the absence of GIY did not allow the T1 and T2 wines to be discriminated from a sensory profile, while these differences were greater in the C1 and C2 wines, where the only variable was represented by the addition of GIY.

Further studies are needed to evaluate the antioxidant effects of the specific inactive yeast with a guaranteed glutathione content at different times during the pre-fermentation stage (on the crushed-stemmed and drained must during the pressing stage) of Catarratto grapes.

The use of *S. cerevisiae* of a non-oenological origin, *M. pulcherrima* in the pre-fermentation stage and specific inactivated yeast with a high antioxidant power resulted in a better preservation of aromatic the compounds and colour, increasing the positive impact on the oxidative stability of the wines.

Supplementary Materials: The following supporting information can be downloaded at: https://www.mdpi.com/article/10.3390/antiox12020439/s1, Table S1. Chemical parameters determined during alcoholic fermentation time (days): 3, 6, 12 and 18; Table S2. Chemical parameters determined during steel aging time (month): 1, 3 and 5.

Author Contributions: Conceptualization, N.F., G.M., A.M., M.B. and A.A.; methodology, V.N., R.P., N.B., P.V. and M.M.; software, A.A. and A.M.; validation, V.N., R.G., L.S. and P.V.; formal analysis, V.N., R.P., M.M. and N.B.; investigation, V.N., R.P., M.M. and N.B.; resources, N.F.; data curation, V.N., R.G., L.S. and A.A.; writing—original draft preparation, A.A., A.M., L.S., M.B. and G.M.; writing—review and editing, A.A. and A.M.; project administration, N.F.; funding acquisition, N.F. All authors have read and agreed to the published version of the manuscript.

Funding: This work was financially supported by the project for industrial research "Integrated approach to product development innovations in the leading sectors of the Sicilian agri-food sector" Prog. F/050267/03/X32-COR 109494-CUP: B78I17000260008 of the Ministry of the Economic Development, General Management for Business Incentives.

Institutional Review Board Statement: Not applicable.

Informed Consent Statement: Not applicable.

Data Availability Statement: All data included in this study are available upon request by contacting the corresponding author.

Acknowledgments: The authors would like to thank: (i) the wineries Di Bella soc. agr. a.r.l. (San Giuseppe Jato, Italy) and Azienda Agricola Buonivini (Noto, Italy) for their financial contribution and for the harvest and transport of the Catarratto grapes; (ii) LALLEMAND Inc. (Montreal, Canada) for their financial contribution concerning the oenological products; (iii) the experimental winery "G. Dalmasso" (Marsala, Italy) of the Regional Institute of Wine and Oil (IRVO) for their technical assistance during the vinification process; and (iv) Eveline Bartowoski for proofreading the manuscript in English.

Conflicts of Interest: The authors declare no conflict of interest.

References

1. Alfonzo, A.; Prestianni, R.; Gaglio, R.; Matraxia, M.; Maggio, A.; Naselli, V.; Craparo, V.; Badalamenti, N.; Bruno, M.; Vagnoli, P.; et al. Effects of different yeast strains, nutrients and glutathione-rich inactivated yeast addition on the aroma characteristics of Catarratto wines. *Int. J. Food Microbiol.* **2021**, *360*, 109325. [CrossRef]
2. Izquierdo Cañas, P.M.; García-Romero, E.; Heras Manso, J.M.; Fernández-González, M. Influence of sequential inoculation of *Wickerhamomyces anomalus* and *Saccharomyces cerevisiae* in the quality of red wines. *Eur. Food Res. Technol.* **2014**, *239*, 279–286. [CrossRef]

3.   Varela, C. The impact of non-*Saccharomyces* yeasts in the production of alcoholic beverages. *Appl. Microbiol. Biotechnol.* **2016**, *100*, 9861–9874. [CrossRef]

4.   Ciani, M.; Comitini, F. Non-*Saccharomyces* wine yeasts have a promising role in biotechnological approaches to winemaking. *Ann. Microbiol.* **2011**, *61*, 25–32. [CrossRef]

5.   Francesca, N.; Gaglio, R.; Matraxia, M.; Naselli, V.; Prestianni, R.; Settanni, L.; Badalamenti, N.; Columba, P.; Bruno, M.; Maggio, A.; et al. Technological screening and application of *Saccharomyces cerevisiae* strains isolated from fermented honey by-products for the sensory improvement of Spiritu re fascitrari, a typical Sicilian distilled beverage. *Food Microbiol.* **2022**, *104*, 103968. [CrossRef]

6.   Gaglio, R.; Alfonzo, A.; Francesca, N.; Corona, O.; Di Gerlando, R.; Columba, P.; Moschetti, G. Production of the Sicilian distillate "Spiritu re fascitrari" from honey by-products: An interesting source of yeast diversity. *Int. J. Food Microbiol.* **2017**, *261*, 62–72. [CrossRef]

7.   Guarcello, R.; Gaglio, R.; Todaro, A.; Alfonzo, A.; Schicchi, R.; Cirlincione, F.; Moschetti, G.; Francesca, N. Insights into the cultivable microbial ecology of "Manna" ash products extracted from *Fraxinus angustifolia* (Oleaceae) trees in Sicily, Italy. *Front. Microbiol.* **2019**, *10*, 984. [CrossRef]

8.   Sinacori, M.; Francesca, N.; Alfonzo, A.; Cruciata, M.; Sannino, C.; Settanni, L.; Moschetti, G. Cultivable microorganisms associated with honeys of different geographical and botanical origin. *Food Microbiol.* **2014**, *38*, 284–294. [CrossRef]

9.   Lappa, I.K.; Kachrimanidou, V.; Pateraki, C.; Koulougliotis, D.; Eriotou, E.; Kopsahelis. Indigenous yeasts: Emerging trends and challenges in winemaking. *Curr. Opin. Food Sci.* **2020**, *32*, 133–143. [CrossRef]

10.   Rossouw, D.; Bauer, F.F. Exploring the phenotypic space of non-*Saccharomyces* wine yeast biodiversity. *Food Microbiol.* **2016**, *55*, 32–46. [CrossRef]

11.   Gonzalez, R.; Morales, P. Truth in wine yeast. *Microb. Biotechnol.* **2022**, *15*, 1339–1356. [CrossRef]

12.   Fleet, G.H. Yeast interactions and wine flavour. *Int. J. Food Microbiol.* **2003**, *86*, 11–22. [CrossRef]

13.   Wang, C.; Mas, A.; Esteve-Zarzoso, B. The interaction between *Saccharomyces cerevisiae* and non-*Saccharomyces* yeast during alcoholic fermentation is species and strain specific. *Front. Microbiol.* **2016**, *7*, 502. [CrossRef] [PubMed]

14.   Benito, S.; Hofmann, T.; Laier, M.; Lochbühler, B.; Schüttler, A.; Ebert, K.; Frtishm, S.; Röcker, J.; Rauhut, D. Effect on quality and composition of Riesling wines fermented by sequential inoculation with non-*Saccharomyces* and *Saccharomyces cerevisiae*. *Eur. Food Res. Technol.* **2015**, *241*, 707–717. [CrossRef]

15.   Ruiz, J.; Belda, I.; Beisert, B.; Navascués, E.; Marquina, D.; Calderón, F.; Rauhut, D.; Santos, A.; Benito, S. Analytical impact of *Metschnikowia pulcherrima* in the volatile profile of Verdejo white wines. *Appl. Microbiol. Biotechnol.* **2018**, *102*, 8501–8509. [CrossRef]

16.   Aplin, J.J.; Paup, V.D.; Ross, C.F.; Edwards, C.G. Chemical and Sensory Profiles of Merlot Wines Produced by Sequential Inoculation of *Metschnikowia pulcherrima* or *Meyerzyma guilliermondii*. *Fermentation* **2021**, *7*, 126. [CrossRef]

17.   Varela, C.; Bartel, C.; Espinase Nandorfy, D.; Bilogrevic, E.; Tran, T.; Heinrich, A.; Balzan, T.; Bindon, K.; Borneman, A. Volatile aroma composition and sensory profile of Shiraz and Cabernet Sauvignon wines produced with novel *Metschnikowia pulcherrima* yeast starter cultures. *Aust. J. Grape Wine Res.* **2021**, *27*, 406–418. [CrossRef]

18.   Kritzinger, E.C.; Bauer, F.F.; Du Toit, W.J. Role of glutathione in winemaking: A review. *J. Agric. Food Chem.* **2013**, *61*, 269–277. [CrossRef]

19.   Badea, G.A.; Antoce, A.O. Glutathione as a possible replacement of sulfur dioxide in winemaking technologies: A review. *Sci. Pap. Ser. B Hortic.* **2015**, *59*, 123–140.

20.   Binati, R.L.; Larini, I.; Salvetti, E.; Torriani, S. Glutathione production by non-*Saccharomyces* yeasts and its impact on winemaking: A review. *Food Res. Int.* **2022**, *156*, 111333. [CrossRef]

21.   Lyu, X.; Del Prado, D.R.; Araujo, L.D.; Quek, S.Y.; Kilmartin, P.A. Effect of glutathione addition at harvest on Sauvignon Blanc wines. *Aust. J. Grape Wine Res.* **2021**, *27*, 431–441. [CrossRef]

22.   Ferrer-Gallego, R.; Puxeu, M.; Nart, E.; Martín, L.; Andorrà, I. Evaluation of Tempranillo and Albariño SO_2-free wines produced by different chemical alternatives and winemaking procedures. *Food Res. Int.* **2017**, *102*, 647–657. [CrossRef]

23.   Ganga, M.A.; Carriles, P.; Raynal, C.; Heras, J.M.; Ortiz-Julien, A.; Dumont, A. Lots of Latin makes for maximum aroma revelation in white wines. *Aust. N. Z. Grapegrow. Winemak.* **2013**, *598*, 73–77. [CrossRef]

24.   Bahut, F.; Romanet, R.; Sieczkowski, N.; Schmitt-Kopplin, P.; Nikolantonaki, M.; Gougeon, R.D. Antioxidant activity from inactivated yeast: Expanding knowledge beyond the glutathione-related oxidative stability of wine. *Food Chem.* **2020**, *325*, 126941. [CrossRef]

25.   Pallmann, C.L.; Brown, J.A.; Olineka, T.L.; Cocolin, L.; Mills, D.A.; Bisson, L.F. Use of WL medium to profile native flora fermentations. *Am. J. Enol. Vitic.* **2001**, *52*, 198–203. [CrossRef]

26.   Legras, J.L.; Karst, F. Optimisation of interdelta analysis for *Saccharomyces cerevisiae* strain characterisation. *FEMS Microbiol. Lett.* **2003**, *221*, 249–255. [CrossRef]

27.   Barbosa, C.; Mendes-Faia, A.; Mendes-Ferreira, A. The nitrogen source impacts major volatile compounds released by *Saccharomyces cerevisiae* during alcoholic fermentation. *Int. J. Food Microbiol.* **2012**, *160*, 87–93. [CrossRef]

28.   Alfonzo, A.; Francesca, N.; Mercurio, V.; Prestianni, R.; Settanni, L.; Spanò, G.; Naselli, V.; Moschetti, G. Use of grape racemes from Grillo cultivar to increase the acidity level of sparkling base wines produced with different *Saccharomyces cerevisiae* strains. *Yeast* **2020**, *37*, 475–486. [CrossRef]

29. Prestianni, R.; Matraxia, M.; Naselli, V.; Pirrone, A.; Badalamenti, N.; Ingrassia, M.; Gaglio, R.; Settanni, L.; Columba, P.; Maggio, A.; et al. Use of sequentially inoculation of *Saccharomyces cerevisiae* and *Hanseniaspora uvarum* strains isolated from honey by-products to improve and stabilize the quality of mead produced in Sicily. *Food Microbiol.* **2022**, *107*, 104064. [CrossRef]

30. OIV. Office International de la Vigne et du Vin Compendium of International Methods of Wine and Must Analysis. Vol. 2. Paris, France. 2020. Available online: http://www.oiv.int (accessed on 7 July 2022).

31. Sagratini, G.; Maggi, F.; Caprioli, G.; Cristalli, G.; Ricciutelli, M.; Torregiani, E.; Vittori, S. Comparative study of aroma profile and phenolic content of Montepulciano monovarietal red wines from the Marches and Abruzzo regions of Italy using HS-SPME–GC–MS and HPLC–MS. *Food Chem.* **2012**, *132*, 1592–1599. [CrossRef]

32. Butkhup, L.; Jeenphakdee, M.; Jorjong, S.; Samappito, S.; Samappito, W.; Chowtivannakul, S. HS-SPME-GC-MS analysis of volatile aromatic compounds in alcohol related beverages made with mulberry fruits. *Food Sci. Biotechnol.* **2011**, *20*, 1021–1032. [CrossRef]

33. Jackson, R.S. *Wine Tasting: A Professional Handbook*, 3rd ed.; Academic Press: London, UK, 2017; pp. 19–291.

34. Romancino, D.P.; Di Maio, S.; Muriella, R.; Oliva, D. Analysis of non-*Saccharomyces* yeast populations isolated from grape musts from Sicily (Italy). *J. Appl. Microbiol.* **2008**, *105*, 2248–2254. [CrossRef] [PubMed]

35. Scacco, A.; Oliva, D.; Di Maio, S.; Polizzotto, G.; Genna, G.; Tripodi, G.; Lanza, C.M.; Verzera, A. Indigenous *Saccharomyces cerevisiae* strains and their influence on the quality of Cataratto, Inzolia and Grillo white wines. *Food Res. Int.* **2012**, *46*, 1–9. [CrossRef]

36. Esteve-Zarzoso, B.; Belloch, C.; Uruburu, F.; Querol, A. Identification of yeasts by RFLP analysis of the 5.8 S rRNA gene and the two ribosomal internal transcribed spacers. *Int. J. Syst. Evol. Microbiol.* **1999**, *49*, 329–337. [CrossRef]

37. Granchi, L.; Bosco, M.; Messini, A.; Vincenzini, M. Rapid detection and quantification of yeast species during spontaneous wine fermentation by PCR–RFLP analysis of the rDNA ITS region. *J. Appl. Microbiol.* **1999**, *87*, 949–956. [CrossRef]

38. Contreras, A.; Curtin, C.; Varela, C. Yeast population dynamics reveal a potential 'collaboration' between *Metschnikowia pulcherrima* and *Saccharomyces uvarum* for the production of reduced alcohol wines during Shiraz fermentation. *Appl. Microbiol. Biotechnol.* **2015**, *99*, 1885–1895. [CrossRef]

39. Ministero delle Politiche Agricole Alimentari e Forestali Decreto 10 Agosto 2017. Available online: https://www.gazzettaufficiale.it/eli/id/2017/08/29/17A06065/sg (accessed on 31 January 2023).

40. Cinquanta, L.; Zarzana, D.; Planeta, D.; Liguori, L.; Albanese, D.; Di Matteo, M.; Corona, O. Use of potassium polyaspartate for the tartaric stabilization of sicilian white wines. *Chem. Eng. Trans.* **2019**, *75*, 277–282. [CrossRef]

41. Corona, O. Wine-making with protection of must against oxidation in a warm, semi-arid terroir. *S. Afr. J. Enol. Vitic.* **2010**, *31*, 58–63. [CrossRef]

42. Liu, P.T.; Lu, L.; Duan, C.Q.; Yan, G.L. The contribution of indigenous non-*Saccharomyces* wine yeast to improved aromatic quality of Cabernet Sauvignon wines by spontaneous fermentation. *LWT–Food Sci. Technol.* **2016**, *71*, 356–363. [CrossRef]

43. Puertas, B.; Jimenez-Hierro, M.J.; Cantos-Villar, E.; Marrufo-Curtido, A.; Carbú, M.; Cuevas, F.J.; Moreno-Rojas, J.M.; González-Rodríguez, V.E.; Cantoral, J.M.; Ruiz-Moreno, M.J. The influence of yeast on chemical composition and sensory properties of dry white wines. *Food Chem.* **2018**, *253*, 227–235. [CrossRef]

44. Vejarano, R.; Gil-Calderón, A. Commercially available non-*Saccharomyces* yeasts for winemaking: Current market, advantages over *Saccharomyces*, biocompatibility, and safety. *Fermentation* **2021**, *7*, 171. [CrossRef]

45. Mislata, A.M.; Puxeu, M.; Andorrà, I.; Espligares, N.; de Lamo, S.; Mestres, M.; Ferrer-Gallego, R. Effect of the addition of non-*Saccharomyces* at first alcoholic fermentation on the enological characteristics of Cava wines. *Fermentation* **2021**, *7*, 64. [CrossRef]

46. Rigou, P.; Mekoue, J.; Sieczkowski, N.; Doco, T.; Vernhet, A. Impact of industrial yeast derivative products on the modification of wine aroma compounds and sensorial profile. A review. *Food Chem.* **2021**, *358*, 129760. [CrossRef]

47. The Good Scents Company Information System. Available online: http://www.thegoodscentscompany.com/ (accessed on 13 July 2022).

48. Flavornet and Human Odor Space. Available online: http://www.flavornet.org/ (accessed on 13 July 2022).

49. LRI & Odour Database. Available online: http://www.odour.org.uk/ (accessed on 13 July 2022).

50. Vilanova, M.; Martínez, C. First study of determination of aromatic compounds of red wine from *Vitis vinifera* cv. Castanal grown in Galicia (NW Spain). *Eur. Food Res. Technol.* **2007**, *224*, 431–436. [CrossRef]

51. American Industrial Hygiene Association. *Odor Thresholds for Chemicals with Established Occupational Health Standards*; Aiha: Fairfax, VA, USA, 1989; pp. 1–95.

52. Darici, M.E.R.V.E.; Cabaroglu, T.; Ferreira, V.; Lopez, R. Chemical and sensory characterisation of the aroma of Çalkarasi rosé wine. *Aust. J. Grape Wine Res.* **2014**, *20*, 340–346. [CrossRef]

53. Ferreira, V.; Culleré, L.; López, R.; Cacho, J. Determination of important odor-active aldehydes of wine through gas chromatography–mass spectrometry of their O-(2, 3, 4, 5, 6-pentafluorobenzyl) oximes formed directly in the solid phase extraction cartridge used for selective isolation. *J. Chromatogr. A* **2004**, *1028*, 339–345. [CrossRef] [PubMed]

54. Duan, L.L.; Shi, Y.; Jiang, R.; Yang, Q.; Wang, Y.Q.; Liu, P.T.; Duan, P.Q.; Yan, G.L. Effects of adding unsaturated fatty acids on fatty acid composition of *Saccharomyces cerevisiae* and major volatile compounds in wine. *S. Afr. J. Enol. Vitic.* **2015**, *36*, 285–295. [CrossRef]

55. Pino, J.A.; Queris, O. Analysis of volatile compounds of pineapple wine using solid-phase microextraction techniques. *Food Chem.* **2010**, *122*, 1241–1246. [CrossRef]
56. Tao, Y.; Zhang, L. Intensity prediction of typical aroma characters of cabernet sauvignon wine in Changli County (China). *LWT—Food Sci. Technol.* **2010**, *43*, 1550–1556. [CrossRef]
57. Zhang, L.; Tao, Y.S.; Wen, Y.; Wang, H. Aroma evaluation of young Chinese Merlot wines with denomination of origin. *South Afr. J. Enol. Vitic.* **2013**, *34*, 46–53. [CrossRef]
58. Zhang, B.; Xu, D.; Duan, C.; Yan, G. Synergistic effect enhances 2-phenylethyl acetate production in the mixed fermentation of *Hanseniaspora vineae* and *Saccharomyces cerevisiae*. *Process Biochem.* **2020**, *90*, 44–49. [CrossRef]
59. Xu, Y.; Zhao, J.; Liu, X.; Zhang, C.; Zhao, Z.; Li, X.; Sun, B. Flavor mystery of Chinese traditional fermented baijiu: The great contribution of ester compounds. *Food Chem.* **2022**, *369*, 130920. [CrossRef] [PubMed]
60. Tempère, S.; Marchal, A.; Barbe, J.C.; Bely, M.; Masneuf-Pomarede, I.; Marullo, P.; Albertin, W. The complexity of wine: Clarifying the role of microorganisms. *Appl. Microbiol. Biotechnol.* **2018**, *102*, 3995–4007. [CrossRef] [PubMed]
61. Burdock, G.A. *Fenaroli's Handbook of Flavor Ingredients*, 6th ed.; CRC Press: Boca Raton, FL, USA, 2017; pp. 1–2137.
62. Ribéreau-Gayon, P.; Glories, Y.; Maujean, A.; Dubourdieu, D. *Handbook of Enology, Volume 2: The Chemistry of Wine—Stabilization and Treatments*, 2nd ed.; Wiley: Hoboken, FL, USA, 2006; pp. 51–64.
63. Vázquez-Pateiro, I.; Arias-González, U.; Mirás-Avalos, J.M.; Falqué, E. Evolution of the aroma of Treixadura wines during bottle aging. *Foods* **2020**, *9*, 1419. [CrossRef] [PubMed]
64. Fang, Y.; Qian, M. Aroma compounds in Oregon Pinot Noir wine determined by aroma extract dilution analysis (AEDA). *Flavour Fragr. J.* **2005**, *20*, 22–29. [CrossRef]

antioxidants

Article

Isolation and Identification of *Limosilactobacillus reuteri* PSC102 and Evaluation of Its Potential Probiotic, Antioxidant, and Antibacterial Properties

Md. Sekendar Ali [1,2,3], Eon-Bee Lee [2], Suk-Kyung Lim [4], Kyoungho Suk [1] and Seung-Chun Park [2,5,*]

1 Department of Biomedical Science and Department of Pharmacology, School of Medicine, Brain Science and Engineering Institute, Kyungpook National University, Daegu 41944, Republic of Korea
2 Laboratory of Veterinary Pharmacokinetics and Pharmacodynamics, College of Veterinary Medicine, Kyungpook National University, Daegu 41566, Republic of Korea
3 Department of Pharmacy, International Islamic University Chittagong, Kumira, Chittagong 4318, Bangladesh
4 Bacterial Disease Division, Animal and Plant Quarantine Agency, 177 Hyeksin 8-ro, Gimcheon-si 39660, Republic of Korea
5 Cardiovascular Research Institute, Kyungpook National University, Daegu 41566, Republic of Korea
* Correspondence: parksch@knu.ac.kr; Tel.: +82-53-950-5964

Abstract: We isolated and characterized *Limosilactobacillus reuteri* PSC102 and evaluated its probiotic, antioxidant, and antibacterial properties. We preliminarily isolated 154 candidates from pig feces and analyzed their Gram nature, morphology, and lactic acid production ability. Based on the results, we selected eight isolates and tested their ability to produce digestive enzymes. Finally, we identified one isolate using 16S rRNA gene sequencing, namely, *L. reuteri* PSC102. We tested its probiotic properties in vitro, including extracellular enzyme activities, low pH and bile salt tolerance, autoaggregation and coaggregation abilities, adhesion to Caco-2 cells, antibiotic susceptibility, and hemolytic and gelatinase activities. Antioxidant activity was determined using 1-diphenyl-2-picrylhydrazyl and 2-azinobis-(3-ethylbenzothiazoline-6-sulfonic acid) diammonium salt radical scavenging and reducing power assays. The antibacterial activity of this strain and its culture supernatant against enterotoxigenic *Escherichia coli* were evaluated using a time-kill assay and disk diffusion method, respectively. *L. reuteri* PSC102 exhibited tolerance toward low pH and bile salt and did not produce harmful enzymes or possess hemolytic and gelatinase activities. Its intact cells and cell-free extract exhibited potential antioxidant activities, and significantly inhibited the growth of enterotoxigenic *E. coli*. Our results demonstrate that *L. reuteri* PSC102 is a potential probiotic candidate for developing functional feed.

Keywords: *Limosilactobacillus reuteri* PSC102; probiotic properties; antioxidant activity; antibacterial activity

Citation: Ali, M.S.; Lee, E.-B.; Lim, S.-K.; Suk, K.; Park, S.-C. Isolation and Identification of *Limosilactobacillus reuteri* PSC102 and Evaluation of Its Potential Probiotic, Antioxidant, and Antibacterial Properties. *Antioxidants* **2023**, *12*, 238. https://doi.org/10.3390/antiox12020238

Academic Editor: Myung-Ji Seo

Received: 28 November 2022
Revised: 17 January 2023
Accepted: 18 January 2023
Published: 20 January 2023

1. Introduction

Recently, the use of probiotics as growth promoters and alternatives to antibiotics for farm animals, including pigs, has increased significantly [1]. In farm animals, probiotics improve feed utilization efficiency, modulate immunity, and prevent gastrointestinal (GI) infections by developing GI health [2]. Among other probiotics, lactic acid bacteria (LAB) are potential sources of feed supplements for swine nutrition, and are effective for producing functional feeds [3–5]. Hence, it is important to evaluate the probiotic features of LAB strains isolated from various sources before utilizing them in functional feed products [6]. Live probiotics must be safe for consumption, capable of surviving in the GI tract, have beneficial characteristics, and be used effectively [7]. Furthermore, they should be able to adhere and multiply in the gut, tolerate bile salt concentrations and gastric acidity, and possess autoaggregation and coaggregation abilities. Moreover, they should not produce any harmful enzymes or exhibit hemolytic and gelatinase activities [8].

LAB possessing antioxidant properties, including free radical scavenging, reducing power, and enzyme inhibition, might ameliorate stress-related disorders. This increases the growth performance of the host by counteracting the reactive oxygen metabolites created during normal cellular processes, such as protein damage, modification of beneficial lipoproteins, mutation of DNA, and oxidation of phospholipids [9–11].

Infectious diseases, such as GI infections, are serious threats to swine production as they cause severe illness and death in swine populations [12]. Noninfectious bacteria, such as probiotics, can help maintain mucosal integrity and thus prevent infections by reducing paracellular permeability, defending against pathogens, and increasing the physical barrier of the mucosal layer [13,14]. Probiotic LAB and their culture supernatants can be used as antibacterial components for treating bacterial infections [15]. Previous studies have demonstrated that the production of different antibacterial components by *Lactobacillus*, such as organic acids (acetic and lactic acids), proteinaceous compounds (bacteriocins), and miscellaneous compounds (reuterin), can act as selective barriers against GI pathogens [16,17].

Factors such as the safety profile of probiotics, their survivability in standard GI tract conditions, and other potential benefits must be considered while screening probiotic bacteria [18]. Therefore, this study aimed to isolate, characterize, and identify a new probiotic strain, *L. reuteri* PSC102, and evaluate its potential probiotic, antioxidant, and antibacterial properties.

2. Materials and Methods

2.1. Bacterial Strains, Culture Conditions, and Media

Different bacterial strains were used in this study. *Escherichia coli* strains (KVCC1423, KVCC0543, and KVCC0306) were provided by the National Veterinary Research and Quarantine Service (Gimcheon, South Korea) [19]. *E. coli* ATCC 35,218, *Bacillus subtilis* KCTC 1021, *Staphylococcus aureus* ATCC 29,213, *Lactobacillus acidophilus* KCTC 3146, and *Limosilactobacillus reuteri* KCTC 3594 were used as quality control strains. All *E. coli* strains were grown in Luria-Bertani (LB) Broth (Difco, Sparks, MD, USA). *L. reuteri* and *L. acidophilus* were cultured in De Man, Rogosa, and Sharpe (MRS) broth (MB Cell, SeoCho-Gu, Seoul, Korea).

2.2. Isolation and Selection of L. reuteri PSC102

We collected fecal samples from 80 commercial weaning piglets (Duroc × Landrace × Yorkshire) from breeding farms in Kyungsan city, Gyeonsangbuk province, South Korea. Healthy weaning piglets (weighing 8.5 ± 0.7 kg, aged 4–5 weeks) were not administered any antibiotics or probiotics, and were fed a normal diet. Fecal samples were collected by rectal palpation using sterile swabs. Samples were collected in individual sterile flasks, stored under refrigeration, transported to the laboratory, and processed within 3 h of collection. *Lactobacilli* strains were isolated from the fecal samples as previously described [20,21]. The experiment was exempted from review by the institutional animal care and use committee because it did not involve direct experimentation on the animals.

Briefly, 1 g fecal sample was mixed and homogenized with 9 mL of diluent that consisted of 4.5 g dipotassium hydrogen phosphate (Sigma-Aldrich, St. Louis, MO, USA), 0.5 g L-cystein (Duksan, Daejon, Korea), 0.5 g Tween 80 (Difco, Sparks, MD, USA), and 1 g Bacto agar (Becton, Dickinson and Company, Sparks, MD, USA) in 1 L of distilled water. Then, 1 mL of this solution was serially diluted 10-fold using 0.1% Bacto agar saline. Each diluted solution was streaked on an MRS agar (MB Cell, SeoCho-Gu, Seoul, Korea) plate, followed by anaerobic incubation for 48 h at 37 °C. Each colony was examined for appearance, Gram staining, microscopic cell morphology, catalase generation by H_2O_2 (KPL, Gaithersburg, MD, USA), acid production on MRS agar using 0.2% calcium carbonate (Sigma-Aldrich, St. Louis, MO, USA), and lactic acid production using a lactic acid detection kit (Accuvin LLC, Napa, CA, USA). LAB were selected based on their ability to produce dietary enzymes, including protease, lipase, amylase, and phytase, in a modified MRS agar medium (pH 7). Subsequently, they were grown on an MRS agar plate containing 0.2%

methyl cellulose and 0.2% corn starch. After 48 h of anaerobic incubation, 0.2% Congo red reagent was added to the modified MRS agar. After 30 min, 1 M NaCl was added to decolorize the medium, and the colonies were identified by observing a surrounding halo zone.

To isolate LAB producing a specific enzyme, such as protease, lipase, amylase, and phytase, the colonies were assayed for halo formation on specific media as previously described [20]. Protease-producing colonies were screened by incubating them for 48 h on 1.5% agar medium containing 1% beef extract, 0.5% polypeptide, 0.5% milk casein and 0.5% NaCl. Lipase-producing colonies were screened by incubating them for 48 h on 1.5% agar medium containing 0.1% tryptone, 0.5% yeast extract, 0.05% NaCl, and 0.1% tricaprylin. Phytase-producing colonies were screened by incubating them for 48 h on 1.5% agar medium containing 1.5% D-glucose, 0.5% calcium phytate, 0.5% NH_4NO_3, 0.05% $MgSO_4.7H_2O$, 0.01% $MnSO_4.7H_2O$, 0.05% KCl, 0.001% $FeSO_4.7H_2O$, and 0.01% $MnSO_4.4H_2O$. Amylase-producing colonies were screened by incubating them for 48 h on 1.5% agar medium containing 0.5% polypeptone, 0.5% beef extract, 0.2% yeast extract, 0.2% NaCl, and 2% starch. After 48 h of incubation under anaerobic conditions, the media were examined for colonies with a halo zone. The LAB that could produce all the tested enzymes were subcultured and stored at $-70\,^{\circ}C$ for subsequent analysis.

2.3. Identification of L. reuteri PSC102

We identified the strain by analyzing the 16S rRNA gene sequence. Briefly, we extracted the genomic DNA from the isolated strain using a genomic DNA extraction kit (Qiagen Inc., Hilden, Germany) and performed polymerase chain reaction (PCR) using forward and reverse primers (5′-AGGTAACGGCTTACCAAGGC-3′ and 5′-CCACCGCTACACATGGAGTT-3′, respectively). A PCR mixture containing 50 pmole primers, 50 ng template DNA, 5 μL of $10\times$ Taq DNA polymerase buffer, and 1 U of Taq DNA polymerase (Takara, Japan) was denatured at $94\,^{\circ}C$ for 5 min and then at $95\,^{\circ}C$ for 30 s. The PCR included 35 cycles with annealing for 30 s at $56\,^{\circ}C$, elongation at $72\,^{\circ}C$ for 30 s, and then final extension for 7 min at $72\,^{\circ}C$. BLAST software (version 2.8.1) was used to compare closely related sequences retrieved from the GenBank database.

2.4. Scanning Electron Microscopy (SEM) Analysis

SEM was used to observe the morphology of *L. reuteri* PSC102 as previously described [22]. Briefly, *L. reuteri* PSC102 cells were fixed in 2.5% glutaraldehyde in phosphate-buffered saline (PBS) for 2 h and then washed thrice with PBS. After washing, the bacterial cells were dehydrated using graded ethanol concentrations of 30, 50, 70, 80, 90, and 100%. The samples were frozen overnight at $-70\,^{\circ}C$ followed by lyophilization in a vacuum freeze dryer (Operon Advantech Co., Ltd., Gyeonggi, South Korea) for 24 h. The prepared samples were sputter-coated with gold-palladium and analyzed using SEM (S-4300; Hitachi, Tokyo, Japan).

2.5. Characterization of L. reuteri PSC102

To establish a culture system, we tested the glycolytic capacity of *L. reuteri* PSC102 using the API 50 CHL kit according to the manufacturer's instructions (API Biomerieux, Durham, NC, USA). *L. reuteri* PSC102 was suspended in API 50 CHL media, dispensed into strips, and incubated for 48 h at $37\,^{\circ}C$. The reading of the strip was determined as negative (−) or positive (+) based on the color change of each tube.

2.6. Extracellular Enzyme Activities

We used the API-ZYM kit (Biomeriux, Marcy-I'Etoile, France) to measure the extracellular enzymatic activities according to the manufacturer's instructions. Briefly, a single colony of *L. reuteri* PSC102 was inoculated in MRS broth and incubated at $37\,^{\circ}C$ overnight. After centrifugation, the collected cell pellets were mixed with the provided API suspension (0.85% saline), and the turbidity was adjusted with the supplied McFarland 0.5 standard.

Next, 65 μL of cell suspension was loaded into each of the 20 cupules of the supplied strip, and then the strip was inserted into the supplied moisture box to prevent drying. After incubation for 4 h at 37 °C, the ZYM-A and ZYM-B reagents were then added dropwise into each cupule. The color changes were observed after 5 min and compared with the manufacturer's standard response chart. The results were graded from 0 (no activity) to 5 (maximum activity) based on the color intensity per the manufacturer's instructions.

2.7. Acid Tolerance Test

The survivability of *L. reuteri* PSC102 in a low pH milieu was accomplished by measuring the survivable colony counts. Briefly, 1 N HCl was added to 5 mL of MRS medium to adjust the pH to 2, 3, 4, 5, and 7 (control), and then 10^5 CFU/mL *L. reuteri* PSC102 and *L. reuteri* KCCM 40,717 were added, and growth was confirmed at 0, 1, 6, and 12 h intervals by incubation at 37 °C. The number of viable colonies, expressed as log CFU/mL, was determined by evaluating the sample at designated time points and incubating on MRS agar plates.

2.8. Bile Tolerance Test

The resistance of *L. reuteri* PSC102 to bile salts was measured by evaluating its survivability in sterile MRS broth (5 mL) supplemented with different bile salt concentrations (0, 0.1, 0.3, and 1% DifcoTM Oxgall, BD, Franklin Lakes, NJ, SUA) as previously described [23]. Subsequently, *L. reuteri* PSC102 and KCCM 40,717 cells were diluted to 10^5 CFU/mL and incubated at 37 °C. The number of viable colonies (expressed as log CFU/mL) was determined by taking the samples at 0, 1, 6, and 12 h, followed by incubation on MRS agar plates.

2.9. Autoaggregation and Coaggregation Assay

We evaluated the autoaggregation ability of *L. reuteri* PSC102 as previously described, with slight modification [24]. Overnight bacterial culture was harvested by centrifugation ($5000\times g$, 4 °C, 10 min) and washed twice with sterile PBS. The pellets were suspended in PBS and the absorbance was maintained at 0.5 ± 0.05 ($OD_{initial}$) at 600 nm. The bacterial cell suspension (4 mL) was incubated at 37 °C for different time periods. The percentage of autoaggregation was determined using the following equation:

$$Autoaggregation\ (\%) = \left(1 - \frac{ODtime}{ODinitial}\right) \times 100$$

where OD_{time} represents the absorbance at 2, 4, 6, 12, or 24 h and $OD_{initial}$ represents the absorbance at 0 h.

For the coaggregation assay, the bacterial cell suspension (*L. reuteri* PSC102/pathogenic bacteria) was prepared similarly as the autoaggregation assay. OD_{600} of the bacterial suspension was maintained at 0.5 ± 0.05. Equal volumes of *L. reuteri* PSC102 and different pathogenic bacterial cell suspension (2 mL each) were mixed by vortexing for 30 s and incubated for different time periods (2 or 24 h) at 37 °C. The percentage of coaggregation was determined using the following equation:

$$Coaggregation\ (\%) = \left(1 - \frac{ODtime}{ODinitial}\right) \times 100$$

where OD_{time} represents the absorbance at 2 or 24 h and $OD_{initial}$ represents that at 0 h.

2.10. Adhesion to Human Colon Carcinoma (Caco-2) Cells

The adhesion capacity of *L. reuteri* PSC102 was evaluated using the Caco-2 cell line as previously described [25]. Caco-2 cells were obtained from the Korean Cell Line Bank (Seoul, Korea) and cultured in Roswell Park Memorial Institute (RPMI) medium supplemented with 10% fetal bovine serum albumin and 1% penicillin–streptomycin. The cells were then seeded in a 24-well plate at 10^5 cells/well and incubated at 37 °C under 5% CO_2 until they had formed a confluent monolayer. The medium was replaced every alternate day. After the Caco-2 cell monolayer was formed, the media was replaced with antibiotic-free RPMI media. Next, the cell monolayer was washed thrice with PBS before the adhesion assay. *L. reuteri* PSC102 (10^8 CFU/mL) was added to each well to a final volume of 1 mL, followed by incubation for at 37 °C for 3 h. The wells were washed thrice with sterile PBS to eliminate the non-adhered bacteria. Subsequently, 1 mL of 1% (*v/v*) Triton X-100 was added into each well and the mixture was agitated for 10 min to detach the *L. reuteri* PSC102 cells from the wells. The cells were serially diluted 10-fold, streaked on an MRS agar plate, and incubated at 37 °C for 24 h to determine the viable cell count. The percentage adhesion rate of *L. reuteri* PSC102 to Caco-2 cells was determined by dividing the number of adhered bacterial cells (*N3 h*) with the initial number of bacterial cells (*No h*) as follows:

$$Adhesion\ rate\ (\%) = (N3\ h/\ No\ h) \times 100$$

2.11. Antibiotic Sensitivity Test

The antibiotic sensitivity of *L. reuteri* PSC102 was determined against 14 antibiotics, including cephalexin, colistin sulfate, enrofloxacin, cefalonium, amoxicillin trihydrate, penicillin G procaine, norfloxacin, spectinomycin, tylosin base, cefuroxime sodium, florfenicol, penicillin G benzathine, gentamicin sulfate, and streptomycin sulfate (Sigma-Aldrich, St. Louis, MO, USA), as previously described [26]. Briefly, 100 µL of cultured *L. reuteri* PSC102 was mixed with 100 µL of diluted antibiotic solution in a 96-well plate. After finally adjusting the concentration to 10^6 CFU/mL, the bacteria were incubated for 24 h at 37 °C. The minimum inhibitory concentration (MIC) and minimum bactericidal concentration (MBC) were determined by measuring the OD value at 600 nm using Gen5 microplate reader version 3.08 (BioTek, Winooski, VT, USA) and by streaking on Mueller–Hinton agar plate, respectively.

2.12. Hemolytic and Gelatinase Activities

The hemolytic activity of *L. reuteri* PSC102 was assayed using a blood agar plate (BBL Microbiology Systems, Franklin Lakes, NJ, USA). A single colony of *L. reuteri* PSC102 and *S. aureus* ATCC 29,213 (positive control) were cultured overnight. The cultured bacterial cells were streaked on blood agar plate and incubated for 24 h at 37 °C. Finally, the nonhemolytic activity was evaluated based on the no inhibition zone around the colony [8]. Gelatinase activity was assayed following a previously reported method [27]. Briefly, 10 µL of fresh *L. reuteri* PSC102 and *B. subtilis* KCTC 1021 (positive control) cultures were spotted on nutrient agar (23 g/L) supplemented with gelatin (8 g/L), followed by incubation for 24 h at 37 °C. After incubation, the plates were examined to check for the formation of any opaque halos around the colonies, which indicates gelatinase production.

2.13. Antioxidant Activity

2.13.1. Sample Preparation

The antioxidant effects of the cultured *L. reuteri* PSC102 intact cells and their cell-free extracts were measured using a previously described method [8]. Briefly, overnight cultured *L. reuteri* PSC102 was centrifuged ($5000 \times g$, 10 min, 4 °C) to obtain the intact cells. The collected pellets were washed twice and resuspended in PBS (pH 7.4), and the OD_{600} value was adjusted to 0.4 ± 0.05. The cell-free extracts were prepared by ultrasonic (37 kHz for 30 min) disruption of cultured intact cells. The cells were separated by centrifugation for 10 min at $5000 \times g$ and 4 °C. The collected supernatants were regarded as cell-free extracts.

2.13.2. 1-Diphenyl-2-Picrylhydrazyl (DPPH) Radical Scavenging Activity

DPPH radical scavenging activity of the intact cells and cell-free extract of *L. reuteri* PSC102 was measured according to a previously simplified method [8]. Briefly, 2 mL of the prepared sample and 2 mL of 0.4 mM DPPH solution were dissolved in methanol and incubated in the dark for 30 min at 37 °C. The reaction mixtures were incubated and centrifuged for 10 min at 5000× g, and the absorbance of the supernatant was measured at 517 nm. The control consisted of PBS with DPPH but without the sample. The following formula was used to measure the DPPH radical scavenging activity:

$$DPPH\ radical\ scavenging\ activity\ (\%) = (ODControl - ODSample/ODControl) \times 100$$

2.13.3. 2-Azinobis-(3-Ethylbenzothiazoline-6-Sulfonic acid) Diammonium Salt (ABTS) Radical Scavenging Activity

The ABTS free radical scavenging activity of intact cells and their cell-free extracts was determined based on a previously reported method [8]. The ABTS reagent solution was prepared by mixing 7 mM ABTS stock solution with 7 mM potassium persulfate and kept in the dark overnight at room temperature. The absorbance of the ABTS$^+$ solution was adjusted to 0.7 at 734 nm by diluting it with distilled water. Subsequently, 150 µL of the prepared sample solution (intact cells/cell-free extract) was added to 1.35 mL of ABTS$^+$ solution and incubated at 37 °C for 10 min. The reaction mixtures were centrifuged at 5000× g for 10 min to remove the cells and the absorbance was measured at 734 nm. PBS with ABTS and without the sample was used as the control. The following formula was used to determine the ABTS radical scavenging activity:

$$ABTS\ radical\ scavenging\ activity\ (\%) = (ODControl - ODSample/ODControl) \times 100$$

2.13.4. Reducing Power Activity

The reducing power activity of the intact cells and their cell-free extracts were measured based on a previously reported method [28]. Briefly, 500 µL of 1% potassium ferricyanide and an equal volume of samples were mixed and incubated for 20 min at 50 °C. Then, 500 µL of 10% trichloroacetic acid solution was added, and the mixture was centrifuged at 3000× g for 5 min. Finally, after mixing 200 µL of 0.1% FeCl$_3$ (Sigma-Aldrich, St. Louis, MO, USA) with 500 µL of the upper liquid layer from the centrifuged mixture, the absorbance was measured at 700 nm.

2.14. Determination of Antibacterial Activities

2.14.1. Antibacterial Activity of the *L. reuteri* PSC102 Culture Supernatant

We cultured 1% (v/v) *L. reuteri* PSC102 in MRS media for 24 h at 37 °C. The supernatant was collected by centrifugation (6000 rpm, 10 min, 4 °C) and filtered using a 0.45-µm-pore-size filter. The filtered supernatant was concentrated using a vacuum evaporator to reduce the volume by 10 times. The antibacterial activities were tested at different concentrations (×10, ×5, and ×1) against three *E. coli* strains (KVCC0306, KVCC0543, and KVCC1423) using the disk diffusion method. Briefly, 100 µL of overnight cultured *E. coli* (10^8 CFU/mL) were moistened on Mueller–Hinton agar (Becton, Dickinson and Company, Sparks, MD, USA) plates. Subsequently, 6-mm paper disks were soaked with 60 µL of the prepared supernatant sample, dried, and placed on the surface of the bacteria-swabbed plate. A standard ampicillin (10 µg) disk was used as the positive control, and a Mueller–Hinton broth (Becton, Dickinson and Company, Sparks, MD, USA)-treated paper disk was used as the negative control. After overnight incubation at 37 °C, the zone of inhibition was measured using slide calipers and expressed in millimeter (mm).

2.14.2. Time-Kill Assay in Cocultures

The enterotoxigenic *E. coli* strain KVCC0306 was cultured in LB media at 37 °C for 24 h and then adjusted to 10^3 or 10^5 CFU/mL. The *L. reuteri* PSC102 culture was adjusted to 10^3, 10^5, 10^7, or 10^9 CFU/mL and cocultured in tubes containing 10 mL of Iso-sentitest–MRS broth (9:1) at 37 °C for 24 h. The sampling was performed at different time points (0, 1, 6, 12, and 24 h) to determine the viable cell count. Sample aliquots (100 µL) were used to prepare 10-fold serial dilutions and poured onto LB agar plates. The bacterial colonies were enumerated and expressed as log CFU/mL after incubating the plate for 24 h at 37 °C.

E. coli KVCC0306 inoculated in Iso-sentitest–MRS broth without any treatment was used as the normal control, while colistin (0.1 mg/mL) was used as the positive control.

2.15. Statistical Analysis

Statistical significance was determined by one-way analysis of variance using the GraphPad Prism software version 8 (GraphPad Software Inc., San Diego, CA, USA). A *p* value of <0.05 was considered statistically significant.

3. Results

3.1. Screening and Isolation of L.reuteri PSC102

A total of 154 (L0001–L0154) candidate strains were initially isolated from pig feces (100 samples). Then, eight *Lactobacilli* strains (L002, L0006, L0010, L0013, L0014, L0017, L0018, and L0102) were selected and phenotypically characterized (Table 1). The strains were identified as gram-positive, rod-shaped, noncatalase forming, and D- and L-lactic acid producers. To identify the potential probiotic LAB candidates, these strains were tested for their ability to produce protease, lipase, amylase, and phytase (Table 2). The strain L0102, i.e., *Limosilactobacillus* (formerly *Lactobacillus*) *reuteri* PSC102 (*L. reuteri* PSC102), was selected as the final probiotic LAB candidate as it produced the maximum amount of protease, lipase, amylase, and phytase. SEM revealed that *L. reuteri* PSC102 had a rod-shaped morphology (Figure 1).

Table 1. Characterization of *Lactobacilli* strains isolated from pig fecal samples based on Gram staining, cell morphology, and acid production.

Isolates	Gram Staining	Cell Morphology	Catalase	Acid Production	Lactic Acid Production
L0002	+	Rod	−	+	D, L
L0006	+	Rod	−	+	D, L
L0010	+	Rod	−	+	D, L
L0013	+	Rod	−	+	D, L
L0014	+	Rod	−	+	D, L
L0017	+	Rod	−	+	D, L
L0018	+	Rod	−	+	D, L
L0102	+	Rod	−	+	D, L

+, Positive; −, Negative.

Table 2. Enzymatic activities of the screened *Lactobacilli* strains.

Isolates	Protease	Lipase	Amylase	Phytase
L0002	++	+	+	+
L0006	+++	+	+	++
L0010	++	−	++	−
L0013	++	−	+	+
L0014	+++	−	+	+
L0017	+++	+	+++	++
L0018	++	+	+++	+
L0102	+++	++	+++	+++

+++, Maximum; ++, Medium; +, Minimum; −, Negative.

3.2. Identification of L. reuteri PSC102

The final selected strain was *L. reuteri* PSC102 (GenBank accession number: MZ127631.1). Figure 2 shows the base sequence analyzed via 16S rRNA gene sequencing (https://www.ncbi.nlm.nih.gov/nuccore/2032707025 (accessed on 12 May 2021). Comparing the GenBank data homology with sequences from the National Center for Biotechnology Information demonstrated that this strain belongs to *L. reuteri* with >99% sequence similarity [22].

Figure 1. Scanning electron microscope (SEM) image showing the rod-shaped *L. reuteri* PSC102.

CTGGCTCAGGATGAACGCCGGCGGTGTGCCTAATACATGCAAGTCGTACGCACTCGCCCAA
CTCATTCATCCTGCTTGCACCTGATTGACCATGGATCACCAGTGAGTGCCCGGACGCGTGACT
AACACGTAGGTAACCTGCCCCGGAGCGGGCGGATAACATTTGGAAACAGATGCTAATACCGC
ATAACAACAAAAGCCGCATGGCTTTTGTTTGAAAGATGGCTTTGGCTATCACTCTGCGATGG
ACCTGCGGTCCATTAGCTAGTTGGTAAGGTAACGCCTTACCAAGGCCATCATGCATACCCCA
GTTGAGAGACTGATCCGCCACAATGGAACTGAGACACGGTCCATACTCCTACGGGAGGCAG
CAGTAGGGAATCTTCCACAATGGGCGCAAGCCTGATGGAGCAACACCGCGTGAGTCAAGAA
GGGTTTCGGCTCGTAAAGCTCTGTTGTTGGAGAAGAACGTGCCGTGACAGTAACTGTTCACGC
AGTGACGGTATCCAACCAGAAACTCACCGCTAACTACGTGCCAGCAGCCGCGGTAATACGT
AGGTGGCAACCGTTATCCGGATTTATTGGGCGTAAAGCGAGCGCAGGCGGTTGCTTAGGTCT
GATGTGAAAGCCCTTCGGCTTAACCGAAGAAGTGCATCGGAAACCGGGCGACTTGAGTGCAG
AAGACGACAGTGGAACTCCATGTGTAGCGGTGGAATGCGTAGATATATGGAAGAACACCAG
TGGCGAAGGCGGCTGTCTGGTCTGCAACTGACGCTGAGGCTCGAAAGCATGGGTAGCGAAC
AGGATTAGATACCCTGGTAGTCCATGCCGTAAACGATGAGTGCTAGGTGTTGGAGGGTTCC
GCCCTTCAGTGCCGGAGCCTAACGCATTAAGCACTCCGCCTGGGGAGTACGACCGCAAGGTTC
AAACTCAAAGGAATTGACGGGGGCCCGCACAAGCGGTGGACCATGTGGTTTAATTGGAAGC
TACGCGAAGAACCTTACCAGGTCTTGACATCTTGCGCTAACCTTAGAGATAAGGCGTTCCCT
TCGGGCACGCAATGACAGGTGGTGCATGGTGGTCGTCAGCTCGTGTCGTGAGATGTTGGGTT
AAGTCCCGCAACGAGCGCAACCCTTGTTACTAGTTGCCAGCATTAAGTTGGGCACTCTAGTG
AGACTGCCGGTGACAAACCGGAGGAAGGTGGGGACGACGTCAGATCATCATGCCCCTTATG
ACCTGGGCTACACACGTGCTACAATGGACGGTACAACGAGTCGCAAACTCGCGAGAGTAAG
CTAATTTCTTAAAGCCGTTCTCAGTTCGGACTGTAGGCTGCAACTCGCCTACACGAAGTCGG
AATCGCTGGTAATCGCGGATCAGCATGCCGCGGTGAATACGTTCCCGGGCCTTGTACACACC
GCCCGTCACACCATGGGAGTTTGTAACGCCCAAAGTCGGTGGCCTAACCTTTATGGAGGA

Figure 2. The base sequence of *L. reuteri* PSC102 analyzed using 16S rRNA gene sequencing.

3.3. Biochemical Characteristics of L. reuteri PSC102

The carbohydrate fermentation profile of *L. reuteri* PSC102 was determined using API 50 CHL medium with API 50 CH strips (Table 3). *L. reuteri* PSC102 exhibited a positive reaction for most carbohydrates, indicating that it could be used as a fermentation starter to produce active metabolites. Therefore, this strain could be efficiently used in industrial media for fermentation based on its glycolytic properties.

Table 3. Carbohydrate fermentation profile of *L. reuteri* PSC102 using API 50 CHL.

Active Ingredient	Result	Active Ingredient	Result	Active Ingredient	Result
Control	−	Inositol	−	Inulin	+
Glycerol	+	Mannitol	+	Melezitose	−
Erythritol	−	Sorbitol	+	Raffinose	+
D-arabinose	−	α-Methyl-D-mannoside	+	Starch	+
L-arabinose	−	β-Methyl-xyloside	−	Glycogen	+
D-ribose	+	α-Methyl-D-glucoside	−	Xylitol	−
D-xylose	+	N-Acetyl-glucosamine	−	Gentiobiose	−
L-xylose	+	Amygdalin	+	D-Turanose	+
D-adonitol	−	Arbutine	+	D-Lyxose	−
Methyl-β-D-xylopyranoside	−	Esculine	+	D-Tagatose	−
D-galactose	+	Salicine	+	D-fucose	+
D-glucose	+	Cellobiose	+	L-fucose	−
D-fructose	+	Maltose	+	D-arabitol	−
D-mannose	+	Lactose	+	L-arabitol	−
L-sorbose	+	Melibiose	+	Gluconate	−
Rhamnose	−	Sucrose	+	2-keto-Gluconate	+
Dulcitol	−	Trehalose	+		

+, Positive; −, Negative.

3.4. Extracellular Enzyme Activities

The extracellular enzymatic profile of *L. reuteri* PSC102 was evaluated using the API-ZYM kit. As shown in Table 4, *L. reuteri* PSC102 showed the best extracellular enzymatic efficacy. The results demonstrated that among the 19 tested enzymes, leucine arylamidase and α-glucosidase were highly produced by the strain, followed by acid phosphatase, naphthol-AS-BI-phosphohydrolase, and α-galactosidase (moderately produced). The production of beneficial enzymes by *L. reuteri* PSC102 was higher than the production of those by the quality control strains *L. reuteri* KCTC 3594 and *L. acidophilus* KCTC 3146.

Table 4. Enzymatic activities of *L. reuteri* PSC102 as determined using the API-ZYM kit.

No.	Enzymes	*Limosilactobacillus reuteri* **PSC102**	*Limosilactobacillus reuteri* **KCTC 3594**	*Lactobacillus acidophilus* **KCTC 3146**
1	Control	0	0	0
2	Alkaline phosphatase	0	0	0
3	Esterase	3	2	2
4	Esterase lipase	1	1	1
5	Lipase	0	0	0
6	Leucine arylamides	5	5	4
7	Valine arylamides	1	3	1
8	Cystine arylamides	1	0	0
9	Trypsin	0	0	0
10	α-Chymotrypsin	0	0	0
11	Acid phosphatase	4	2	2
12	Naphthol-AS-BI-phosphohydrolase	4	2	2
13	α-Galactosidase	4	4	1
14	β-Galactosidase	1	5	4
15	β-Glucuronidase	0	0	0
16	α-Glucosidase	5	2	1
17	β-Glucosidase	0	0	0
18	N-acyl-glucosaminidase	0	0	0
19	α-Mannosidase	0	0	0
20	α-Fructosidase	0	0	0

3.5. Acid Resistance of L. reuteri PSC102

To function as a probiotic, the bacterial strain should survive low pH conditions (<pH 3) in the stomach. Therefore, we determined the survivability of *L. reuteri* PSC102 at low pH. As shown in Table 5, *L. reuteri* PSC102 strain survived for >6 h at strongly acidic pH (pH 2 and 3), indicating significantly higher acid resistance compared with that of the standard strain *L. reuteri* KCCM 40,717 ($p < 0.05$).

Table 5. Tolerance of *L. reuteri* PSC102 to low pH.

Bacteria	pH 2				pH 3				pH 5				pH 7			
	0 h	1 h	6 h	12 h	0 h	1 h	6 h	12 h	0 h	1 h	6 h	12 h	0 h	1 h	6 h	12 h
	Log CFU/mL				Log CFU/mL				Log CFU/mL				Log CFU/mL			
L. reuteri PSC102	5.55 ± 0.08	3.75 ± 0.16 *	3.52 ± 0.09 *	2.00 ± 0.00	5.55 ± 0.08	4.81 ± 0.12	3.74 ± 0.08 *	2.00 ± 0.00	5.55 ± 0.08	4.95 ± 0.05	4.52 ± 0.08 *	7.65 ± 0.14 *	5.55 ± 0.08	5.97 ± 0.06	6.52 ± 0.23	10.75 ± 0.15 *
L. reuteri KCCM 40,717	5.29 ± 0.16	2.00 ± 0.00 #	2.00 ± 0.00 #	2.00 ± 0.00	5.29 ± 0.16	4.35 ± 0.24	2.00 ± 0.00 #	2.00 ± 0.00	5.29 ± 0.16	4.94 ± 0.05	5.13 ± 0.11 #	5.16 ± 0.07 #	5.29 ± 0.16	5.99 ± 0.15	6.51 ± 0.05	9.17 ± 0.07 #

Data are expressed as mean ± standard deviation (n = 3). Different symbols (*, #) above the values denote significant differences ($p < 0.05$).

3.6. Bile Salt Tolerance of L. reuteri PSC102

As resistance to bile salts is another important criterion for bacteria to be considered a potential probiotic, we determined the bile salt tolerance of *L. reuteri* PSC102 at different concentrations and time intervals. As shown in Table 6, *L. reuteri* PSC102 could survive even in 1% of bile salt, indicating that this strain possesses excellent bile tolerance.

Table 6. Bile salt tolerance of *L. reuteri* PSC102.

Bacteria	Bile Salt (0.1%)				Bile Salt (0.3%)				Bile Salt (1%)				Bile Salt (0%)			
	0 h	1 h	6 h	12 h	0 h	1 h	6 h	12 h	0 h	1 h	6 h	12 h	0 h	1 h	6 h	12 h
	Log CFU/mL				Log CFU/mL				Log CFU/mL				Log CFU/mL			
L. reuteri PSC102	5.23 ± 0.04	5.27 ± 0.09	5.41 ± 0.11	7.28 ± 0.09	5.23 ± 0.09	5.26 ± 0.02	5.38 ± 0.01	6.83 ± 0.08 *	5.23 ± 0.04	5.20 ± 0.06	5.32 ± 0.06	6.53 ± 0.09 *	5.23 ± 0.04	5.36 ± 0.11	6.65 ± 0.08	9.19 ± 0.09
L. reuteri KCCM 40,717	5.19 ± 0.10	5.22 ± 0.08	5.53 ± 0.10	7.27 ± 0.11	5.19 ± 0.05	5.25 ± 0.06	5.37 ± 0.04	6.28 ± 0.06 #	5.19 ± 0.10	5.17 ± 0.06	5.28 ± 0.04	5.18 ± 0.09 #	5.19 ± 0.10	5.75 ± 0.06	6.53 ± 0.06	9.09 ± 0.07

Data are expressed as mean ± standard deviation (n = 3). Different symbols (*, #) above the values denote significant differences ($p < 0.05$).

3.7. Autoaggregation Ability

Autoaggregation was measured over a period of 24 h. The ability of *L. reuteri* PSC102 to autoaggregate increased with increased incubation time (Figures 3 and 4). The strongest autoaggregation ability (84.93%) was observed at 24 h, indicating that *L. reuteri* PSC102 might efficiently adhere to mucosal surfaces and epithelial cells.

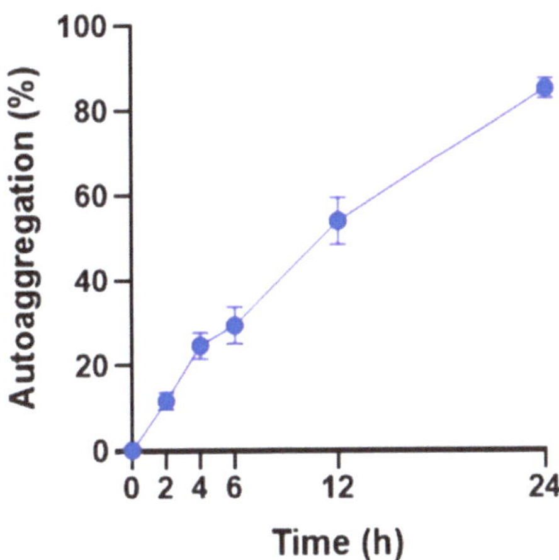

Figure 3. Autoaggregation percentage of *L. reuteri* PSC102 measured at 2, 4, 6, 12, and 24 h of incubation at 37 °C. Data are expressed as mean ± standard deviation (n = 3).

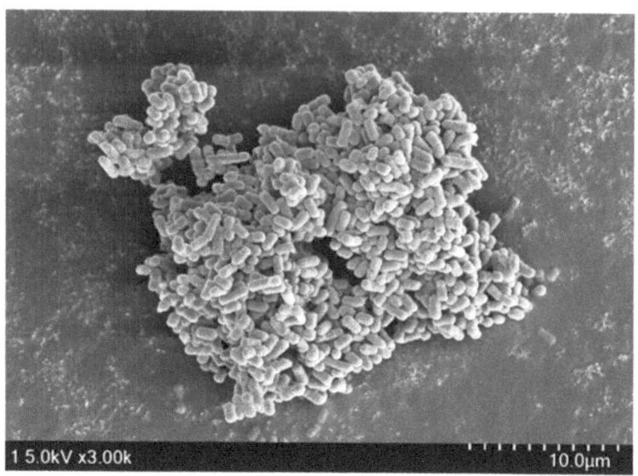

Figure 4. SEM image of autoaggregated *L. reuteri* PSC102.

3.8. Coaggregation Ability

L. reuteri PSC102 was able to coaggregate with pathogens, including enterotoxigenic *E. coli*. After 2 h of incubation, *L. reuteri* PSC102 coaggregated the most with *E. coli* KVCC0306 (9.02%), followed by *S. aureus* ATCC 35,218 (3.35%). After 24 h of incubation, *L. reuteri* PSC102 still coaggregated the most with the same two strains (*E. coli* KVCC0306 [81.13%] and *S. aureus* ATCC 35,218 [72.41%]; Table 7 and Figure 5).

Table 7. Coaggregation percentage of *L. reuteri* PSC102 with different pathogenic bacteria as measured after 2 and 24 h of incubation at 37 °C.

Pathogenic Bacteria	Coaggregation with *L. reuteri* PSC102 (%)	
	2 h	24 h
E. coli ATCC 35,218	6.52 ± 0.75	77.16 ± 1.59
E. coli KVCC0306	9.02 ± 0.91	81.13 ± 0.87
S. aureus ATCC 29,213	3.35 ± 0.51	72.41 ± 0.69

Data are expressed as mean ± standard deviation (n = 3).

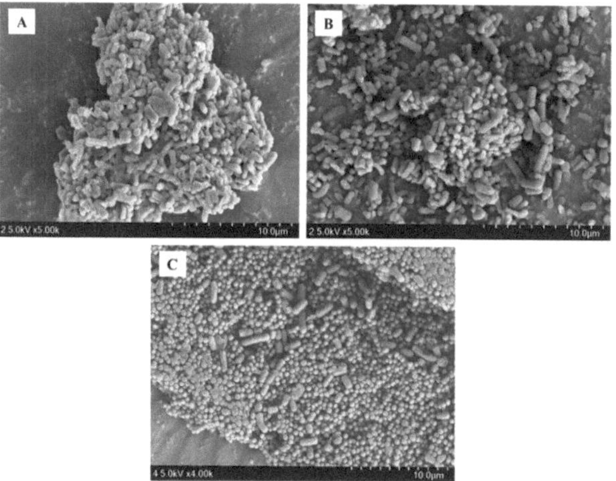

Figure 5. SEM image of *L. reuteri* PSC102 coaggregating with *E. coli* ATCC 35,218 (**A**), *E. coli* KVCC0306 (**B**), and *S. aureus* ATCC 29,213 (**C**).

3.9. Adhesion to Caco-2 Cells

The initial number of *L. reuteri* PSC102 was 1.14 ± 0.08 ($\times 10^8$) CFU/mL. After 3 h, the number of *L. reuteri* PSC102 adhering to Caco-2 cells was 4.6 ± 0.94 ($\times 10^6$) CFU/mL, with a 4.03% (± 0.15) adhesion rate (Table 8). These results are consistent with those of previous studies, which showed that LAB could adhere to Caco-2 cells in the range of 2%–6% [28,29].

Table 8. Adhesion of *L. reuteri* PSC102 to Caco-2 cells.

Initial Concentration (*L. reuteri* PSC102, CFU/mL)	Final Concentration (*L. reuteri* PSC102, CFU/mL)	Adhesion (%)
1.08×10^8	4×10^6	3.70
1.26×10^8	6×10^6	4.76
1.10×10^8	4×10^6	3.63
	Average	
1.14×10^8	4.6×10^6	4.03

3.10. Antibiotic Sensitivity Test

Table 9 shows the sensitivity of *L. reuteri* PSC102 to various antibiotics. *L. reuteri* PSC102 strain was found to be resistant to cephalexin, colistin sulfate, norfloxacin, spectinomycin, gentamicin sulfate, and streptomycin sulfate. Moreover, *L. reuteri* PSC102 exhibited higher MIC and MBC values for most of the tested antibiotics than the control strains (*L. reuteri* KCCM 40,717, *S. aureus* ATCC 25,922, and enterotoxigenic *E. coli* KVCC0306) [30,31].

Table 9. Minimal inhibitory concentration (MIC) and minimum bactericidal concentration (MBC) of *L. reuteri* PSC102 against various antibiotics.

Antibiotics	*Limosilactobacillus reuteri* PSC102		*Limosilactobacillus reuteri* KCCM 40,717		*Staphylococcus aureus* ATCC 25,922		Enterotoxigenic *E. coli* KVCC0306	
	MIC (µg/mL)	MBC (µg/mL)	MIC (µg/mL)	MBC (µg/mL)	MIC (µg/mL)	MBC (µg/mL)	MIC (µg/mL)	MBC (µg/mL)
Cephalexin	>256	>256	>256	>256	2	4	32	64
Colistin sulfate	>64	>128	>64	>128	>64	>128	0.05	1
Enrofloxacin	8	16	4	8	0.05	1	0.025	1
Cefalonium	32	64	64	32	2	4	16	32
Amoxicillin trihydrate	4	2	1	1	0.5	2	1	2
Penicillin G procaine	32	64	16	64	0.05	1	32	64
Norfloxacin	>256	>256	>256	>256	0.5	4	0.5	1
Spectinomycin	64	128	8	128	8	64	2	4
Tylosin base	4	8	2	4	1	8	32	64
Cefuroxime sodium	2	32	1	8	0.05	2	1	2
Florfenicol	2	16	2	16	2	8	4	16
Penicillin G benzathine	4	8	2	4	0.05	2	16	32
Gentamicin sulfate	64	128	32	128	4	16	1	2
Streptomycin sulfate	64	128	32	128	4	16	1	2

3.11. Hemolytic and Gelatinase Activities

For the hemolytic activity assay, *L. reuteri* PSC102 was streaked on blood agar plates and incubated at 37 °C for 24 h. We did not observe clear zones (β-hemolysis) or green-hued zones (α-hemolysis) around the colonies (Figure 6A), indicating that *L. reuteri* PSC102 is not a hemolytic strain. Regarding the gelatinase activity, upon checking the plates after incubation, no opaque halos were observed around the colonies (Figure 6B).

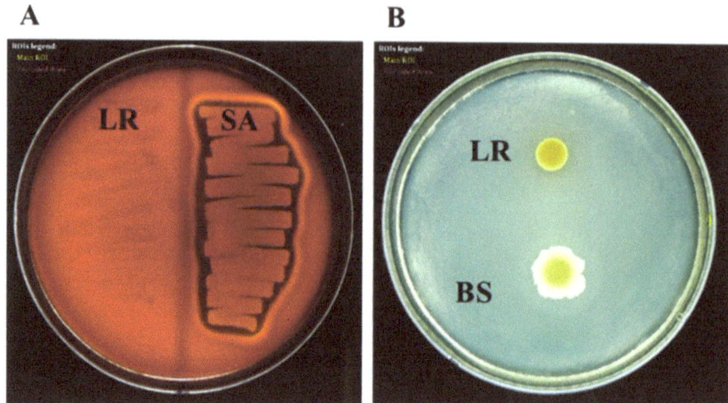

Figure 6. Hemolytic and gelatinase activity test. (**A**) Hemolytic activity: *L. reuteri* PSC102 (LR), positive control; *Staphylococcus aureus* (SA). (**B**) Gelatinase activity: *L. reuteri* PSC102 (LR), positive control; *Bacillus subtilis* (BS).

3.12. Antioxidant Activity

Figure 7 shows the DPPH and ABTS free radical scavenging activities of *L. reuteri* PSC102 intact cells and cell-free extracts. Intact *L. reuteri* PSC102 cells had higher DPPH radical-scavenging ability (34.31%) than the intracellular cell-free extracts (24.04%; Figure 7A). The ABTS radical scavenging effect of intact cells was 17.84%, which was more than double that of the intracellular cell-free extracts (6.86%; Figure 7B). The reducing power of the intact cells was 0.096, whereas that of the cell-free extracts was 0.076 (Figure 7C). The DPPH and ABTS free radical scavenging activities and reducing power of intact cells were higher than those of the cell-free extracts, indicating that more active components were present in the intact cells.

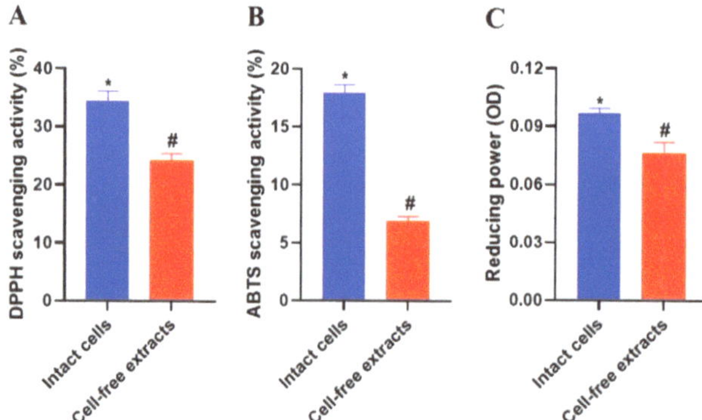

Figure 7. Antioxidant activity of intact cells and cell-free extracts of *L. reuteri* PSC102 measured by (**A**) DPPH radical scavenging, (**B**) ABTS radical scavenging, and (**C**) reducing power assays. Different symbols (*, #) on the bars denote significant differences between groups ($p < 0.05$). Data are expressed as mean ± standard deviation ($n = 3$).

3.13. Antibacterial Activities of L. reuteri PSC102 Culture Supernatant

Evaluation of the potential antibacterial activities of the supernatants of *L. reuteri* PSC102 against three enterotoxigenic *E. coli* strains revealed that the supernatant possessed antibacterial activities against all three *E. coli* strains (Table 10). Among them, *E. coli* KVCC1423 was the most sensitive to ×10 concentrated supernatant (14.72-mm inhibition zone). Interestingly, *E. coli* KVCC0306 was resistant to standard ampicillin but was sensitive to the supernatant.

Table 10. Antibacterial activities of *L. reuteri* PSC102 supernatant against enterotoxigenic pathogens evaluated using the disk diffusion method.

Pathogens	Zone of Inhibition (mm)			
	Concentrated Supernatant of *L. reuteri* PSC102			Positive Control (Ampicillin, 10 µg)
	×10	×5	×1	
E. coli KVCC0306	14.10 ± 0.08	8.96 ± 0.03	-	-
E. coli KVCC0543	14.20 ± 0.06	8.94 ± 0.04	-	20.35 ± 0.35
E. coli KVCC1423	16.17 ± 0.51	9.60 ± 0.41	-	21.47 ± 0.65

Data are expressed as means ± standard deviation ($n = 3$). "-" indicates no inhibition.

3.14. Time-Kill Assay

Figure 8 shows the results of the time-kill assay. After 24 h incubation, the growth of *E. coli* KVCC0306 did not change significantly. However, at 6 and 12 h, the growth of *E. coli* KVCC0306 (10^3 CFU/mL) was significantly inhibited by *L. reuteri* PSC102 (10^3, 10^5, 10^7, or 10^9 CFU/mL) compared with the normal control ($p < 0.05$; Figure 8A). With coculture of 10^5 CFU/mL *E. coli* KVCC0306 and different concentrations of *L. reuteri* PSC102, *E. coli* KVCC0306 was significantly inhibited by *L. reuteri* PSC102 at 6 h ($p < 0.05$; Figure 8B). The CFU/mL of *E. coli* KVCC0306 in the presence or absence of different concentrations of *L. reuteri* PSC102 was adjusted with the inhibitor vs. response model generated using the GraphPad Prism 8 software with the following equation:

$$Y = \text{Bottom} + ((\text{Top} - \text{Bottom})/(1 + 10^{(X - \text{LogIC}_{50})}))$$

where Top and Bottom are *E. coli* KVCC0306 CFU/mL in the absence of *L. reuteri* PSC102 and at maximum growth inhibition in the presence of *L. reuteri* PSC102, and IC_{50} is the minimum concentration of *L. reuteri* PSC102 needed to inhibit the growth of *E. coli* KVCC0306 by 50% in their coculture. For *E. coli* KVCC0306 (10^3 and 10^5 CFU/mL), the IC_{50} values of *L. reuteri* PSC102 were found to be 3.96×10^7 and 1.44×10^5 CFU/mL, respectively (Figure 8C,D).

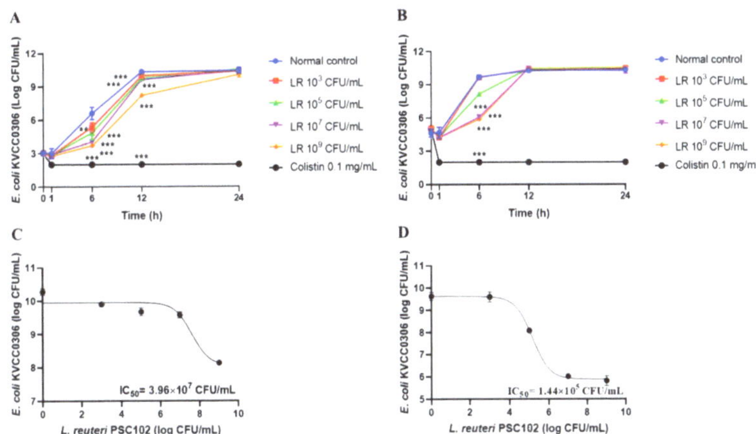

Figure 8. Time-kill assay curves of coculture of *L. reuteri* PSC102 (LR) and enterotoxigenic pathogen *E. coli* KVCC0306. CFU changes of 10^3 CFU/mL (**A**) and 10^5 CFU/mL *E. coli* KVCC0306 (**B**) observed after coculture with four different concentrations of *L. reuteri* PSC102 (10^3, 10^5, 10^7, and 10^9 CFU/mL) for 24 h. The minimum concentration of *L. reuteri* PSC102 needed to inhibit the growth of *E. coli* KVCC0306 by 50% (IC_{50}) when *E. coli* KVCC0306 cocultured at 10^3 CFU/mL (**C**) and 10^5 CFU/mL (**D**). Data are expressed as means ± standard deviation ($n = 3$). ** $p < 0.01$, and *** $p < 0.001$ vs. normal control (*E. coli* KVCC0306 10^3 CFU/mL or 10^5 CFU/mL culture alone).

4. Discussion

Probiotics are obtained after in vitro screening for evaluating several characteristics, such as the ability to produce digestive enzymes, inhibit various pathogens, tolerate bile salts and gastric acids, exhibit antimicrobial sensitivity, and provide safe and beneficial properties for the host [21,32]. In this study, *Lactobacillus* strains were initially isolated, and their phenotypic characteristics were evaluated. Thereafter, their ability to produce digestive enzymes, including lipase, phytase, amylase, and protease, was examined. We isolated 154 potential *Lactobacillus* probiotics from pig fecal samples and cultured them on MRS media. Further, we identified eight candidates based on their phenotypic characteristics and capacity to produce digestive enzymes. Finally, we identified *L. reuteri* PSC102 based on its ability to produce digestive enzymes. Biochemical evaluation of *L. reuteri* PSC102 using API 50 CHL showed that *L. reuteri* PSC102 possesses glycolytic capacity, suggesting that this strain can be successfully used for fermentation.

Probiotic bacteria have several health benefits for humans and other animals [33–37]. These bacteria should not produce enzymes such as β-glucuronidase, α-chymotrypsin, and N-acetyl-glucosaminidase, as these enzymes are potentially harmful [38]. As expected, *L. reuteri* PSC102 did not produce any harmful enzymes. Conversely, acid phosphatase, α-glucosidase, and β-galactosidase released by probiotics have been shown to be beneficial [39]. In particular, acid phosphatase, when used as a supplementary feed additive, increases the absorption of phosphorus in the feed, thereby reducing the use of inorganic phosphorus.

Probiotic strains must be able to survive in conditions similar to those found in the GI tract, by exhibiting tolerance to bile salts and low pH [40]. In addition to resisting low pH and changes in glycolytic flux, probiotic bacteria should maintain intracellular pH [41]. Low pH can increase the ammonium output in the cytoplasm, which is likely liberated during the deamination of amino acid. This reduces the activity of bile salt hydrolase [41]. H^+-ATPase activity is also necessary for survival in an acidic environment [8]. Under acidic conditions, the H^+-ATPase activity of acid-tolerant bacterial strains increases, whereas that of non-acid-tolerant strains decreases [42]. The H^+-ATPase activity in the cells increases rapidly to sustain a stable intracellular pH by releasing H^+ for acid-tolerant bacteria [42]. Tolerance to bile salts is important for *Lactobacillus* to grow, survive, and function during GI tract transit [43]. The most common mechanisms of bile salt tolerance in LAB are active efflux and hydrolysis of bile salts and alterations of the cell membrane and cell wall composition [44]. A previous study showed that different *Lactobacillus* species isolated from goat's milk cheese can hydrolyze bile salts [7]. Interestingly, our tested *L. reuteri* PSC102 strain was resistant to low pH and high bile salt concentration, suggesting that this strain would likely survive the stomach passage and in the small intestine.

Autoaggregation and coaggregation abilities of bacteria are essential for several biological activities [45]. Via aggregation, bacteria may gain sufficient mass to form biofilms or adhere to the host's intestinal mucosal surfaces, allowing them to perform their functions [46]. In our study, the autoaggregation of *L. reuteri* PSC102 increased with an increase in the incubation period. In a previous study, *Lactobacilli* autoaggregation was facilitated by proteins found in the culture supernatant and lipoproteins or proteins on the surface of the cells that were not washed off and resuspended in PBS [47]. Coaggregation enables bacteria to interact intimately with other bacteria [48]. In this study, the percentage of coaggregation of *L. reuteri* PSC102 with *E. coli* KVCC0306 was the highest after 2 and 24 h of incubation. This ability to coaggregate with pathogens might contribute to the probiotic potential of *L. reuteri* PSC102 by allowing it to establish a barrier and prevent the colonization of harmful bacteria [49,50].

The ability of probiotic bacteria to adhere to the intestinal epithelium mucosa and colonize is a significant characteristic as it prevents removal from the intestine by peristalsis [25,51]. In this study, the ability of *L. reuteri* PSC102 to adhere to intestinal mucosa was evaluated using Caco-2 cells, which proved that it could colonize in the intestine and maintain the intestinal microflora homeostasis of the host.

The functional qualities of probiotic strains should be assessed before in vivo administration. Antibiotic resistance is considered detrimental to human and animal health and food safety [28]. However, the antibiotic susceptibility of a probiotic is a critical factor in determining whether it can be coadministered with antibiotics [52]. In this study, LAB strains exhibited antibiotic reactivity, which is consistent with that reported by a previous study [53]. As *L. reuteri* PSC102 was resistant to the tested antibiotics, it can be used in combination with antibiotic-containing feed or feed additives and exert its probiotic effect. Moreover, in the presence of these antibiotics, *L. reuteri* PSC102 might survive and exert useful effects in the host [54].

As most probiotic bacteria are safe for consumption, the chances of infections are rare. However, hemolysis might occur if the ingested bacteria gains access to the blood, resulting in hemolytic symptoms, including fever, anemia, and skin rash [55]. Therefore, the hemolytic activity of probiotics must be assessed to guarantee their safety. Gelatinase activity is considered a risk factor as it indicates the potential to hydrolyze collagen, which may trigger an inflammatory reaction [56]. In this study, *L. reuteri* PSC102 did not show hemolytic and gelatinase activities, thus eliminating the safety issues regarding this strain.

Oxidative stress can harm cells by initiating DNA hydroxylation, lipid peroxidation, and protein denaturation. However, antioxidants can prevent or reduce oxidative damage [57]. LAB exert antioxidant effects by producing various active cell surface components, proteins, and antioxidant enzymes. This prevents or hinders the progress of different oxidative stress-related disorders [58]. LAB have been shown to possess strong DPPH and ABTS free radical scavenging activity, which might enhance the oxidative status of weaned piglets and improve their growth [59]. In our study, *L. reuteri* PSC102 demonstrated a strong free radical scavenging effect and reducing power. Hence, it might be potentially used to alleviate oxidative stress-induced disorders.

The growth-inhibitory activity of probiotics against target pathogens is a desirable property [60]. Our findings demonstrated that *L. reuteri* PSC102 could inhibit all the tested enterotoxigenic *E. coli* pathogenic strains. *L. reuteri* PSC102 cells and culture supernatant exhibited strong antibacterial activity against all three *E. coli* strains in a concentration-dependent manner. Despite *E. coli* KVCC0306 being resistant to standard ampicillin, the supernatant exhibited inhibitory activity against it. Therefore, we selected *E. coli* KVCC0306 for the time-kill assay by coculturing with live *L. reuteri* PSC102. The supernatant might contain different active metabolites with antimicrobial properties, such as organic acids (mainly lactic and acetic acids) and other metabolites, including p-coumaric, 3-phenylpropanoic, 3-phenyl lactic, D-glucuronic, and benzoic acids; cyclic dipeptides; and bacteriocins, all of which are potent antibacterial compounds [8,61]. The antibacterial activity of the supernatant is mediated by the rapid diffusion of the antimicrobial components across the microbial cell membrane. This activity significantly depends on the concentration of antimicrobial components and the number of pathogenic bacteria used for the assay [8]. The most notable metabolite produced by *L. reuteri* is reuterin, a broad-spectrum antimicrobial component that can kill many food-borne pathogens, including *E. coli* and *S. aureus* [62,63]. Rueterin exerts its activity by getting adsorbed by sensitive bacterial cells and disturbing their metabolism by depleting free sulfhydryl groups in bacterial proteins, thereby inducing oxidative stress and ultimately resulting in cell death [64]. Moreover, lactic and acetic acid production can lower the pH inside the cells, which can dissipate membrane function. The acidification of the cytoplasm might restrict bacterial growth by limiting glycolysis [65]. In the time-kill assay, cocultures of *L. reuteri* PSC102 with the enterotoxigenic pathogen *E. coli* inhibited the growth of *E. coli* KVCC0306 for up to 12 h. During coculture, different organic compounds were produced, which might permeabilize the outer membrane of gram-negative bacteria, such as *E. coli*, and enhance the activities of other antibacterial metabolites [66]. Therefore, this *L. reuteri* strain might effectively treat enterotoxigenic *E. coli*-induced diarrhea in pigs.

5. Conclusions

In this study, we isolated and identified *L. reuteri* PSC102, which showed excellent characteristics, including autoaggregation and coaggregation, adhesion to Caco-2 cells, and resistance to in vitro GI tract conditions. The strain did not exhibit any hemolytic and gelatinase activity or produce any undesirable extracellular enzymes. The intact *L. reuteri* PSC102 cells and their cell-free extracts showed promising antioxidant activities. Moreover, both the strain and its supernatant exhibited antibacterial activities against enterotoxigenic *E. coli*. These results suggest that *L. reuteri* PSC102 is an efficient probiotic candidate that can aid in the development of functional feeds. As this strain has been isolated from swine, it could be preferentially used in swine as a growth promoter and an antibiotic alternative.

Author Contributions: Conceptualization, S.-C.P., M.S.A. and E.-B.L.; methodology, M.S.A. and E.-B.L.; software, M.S.A. and S.-K.L.; validation, K.S. and S.-K.L.; investigation, S.-C.P., M.S.A. and E.-B.L.; resources, S.-C.P.; data curation, M.S.A. and E.-B.L.; writing—original draft preparation, M.S.A. and E.-B.L.; writing—review and editing, S.-C.P., K.S. and S.-K.L.; supervision, S.-C.P. and K.S.; and funding acquisition, S.-C.P. All authors have read and agreed to the published version of the manuscript.

Funding: This work was supported by the Korean Institute of Planning and Evaluation for Technology in Food, Agriculture and Forestry (IPET) through Companion Animal Life Cycle Industry Technology Development Program (322098-03) and by a grant (Z-1543081-2020-22-02) from the Animal and Plant Quarantine Agency, Ministry of Agriculture, Food and Rural Affairs, Republic of Korea.

Institutional Review Board Statement: Not applicable.

Informed Consent Statement: Not applicable.

Data Availability Statement: All data generated for this study are contained within the article.

Conflicts of Interest: The authors declare no conflict of interest.

References

1. Mingmongkolchai, S.; Panbangred, W. *Bacillus* Probiotics: An Alternative to Antibiotics for Livestock Production. *J. Appl. Microbiol.* **2018**, *124*, 1334–1346. [CrossRef] [PubMed]
2. Cho, J.H.; Zhao, P.Y.; Kim, I.H. Probiotics as a Dietary Additive for Pigs: A Review. *J. Anim. Vet. Adv.* **2011**, *10*, 2127–2134. [CrossRef]
3. Guerra, N.P.; Bernárdez, P.F.; Méndez, J.; Cachaldora, P.; Castro, L.P. Production of Four Potentially Probiotic Lactic Acid Bacteria and Their Evaluation as Feed Additives for Weaned Piglets. *Anim. Feed Sci. Technol.* **2007**, *134*, 89–107. [CrossRef]
4. Lim, Y.H.; Foo, H.L.; Loh, T.C.; Mohamad, R.; Abdullah, N. Comparative Studies of Versatile Extracellular Proteolytic Activities of Lactic Acid Bacteria and Their Potential for Extracellular Amino Acid Productions as Feed Supplements. *J. Anim. Sci. Biotechnol.* **2019**, *10*, 15. [CrossRef]
5. Hou, C.; Zeng, X.; Yang, F.; Liu, H.; Qiao, S. Study and Use of the Probiotic *Lactobacillus reuteri* in Pigs: A Review. *J. Anim. Sci. Biotechnol.* **2015**, *6*, 14. [CrossRef] [PubMed]
6. Markowiak, P.; Śliżewska, K. The Role of Probiotics, Prebiotics and Synbiotics in Animal Nutrition. *Gut Pathog.* **2018**, *10*, 21. [CrossRef] [PubMed]
7. Lavilla-Lerma, L.; Pérez-Pulido, R.; Martínez-Bueno, M.; Maqueda, M.; Valdivia, E. Characterization of Functional, Safety, and Gut Survival Related Characteristics of *Lactobacillus* Strains Isolated from Farmhouse Goat's Milk Cheeses. *Int. J. Food Microbiol.* **2013**, *163*, 136–145. [CrossRef]
8. Cizeikiene, D.; Jagelaviciute, J. Investigation of Antibacterial Activity and Probiotic Properties of Strains Belonging to *Lactobacillus* and *Bifidobacterium* Genera for Their Potential Application in Functional Food and Feed Products. *Probiotics Antimicrob. Proteins* **2021**, *13*, 1387–1403. [CrossRef]
9. Wang, J.; Ji, H.F.; Wang, S.X.; Zhang, D.Y.; Liu, H.; Shan, D.C.; Wang, Y.M. *Lactobacillus plantarum* ZLP001: In Vitro Assessment of Antioxidant Capacity and Effect on Growth Performance and Antioxidant Status in Weaning Piglets. *Asian Australas. J. Anim. Sci.* **2012**, *25*, 1153. [CrossRef]
10. Firuzi, O.; Miri, R.; Tavakkoli, M.; Saso, L. Antioxidant Therapy: Current Status and Future Prospects. *Curr. Med. Chem.* **2011**, *18*, 3871–3888. [CrossRef]
11. Pisoschi, A.M.; Pop, A. The Role of Antioxidants in the Chemistry of Oxidative Stress: A Review. *Eur. J. Med. Chem.* **2015**, *97*, 55–74. [CrossRef] [PubMed]
12. Valeriano, V.D.V.; Balolong, M.P.; Kang, D. Probiotic Roles of *Lactobacillus* Sp. in Swine: Insights from Gut Microbiota. *J. Appl. Microbiol.* **2017**, *122*, 554–567. [CrossRef] [PubMed]

13. Ohland, C.L.; MacNaughton, W.K. Probiotic Bacteria and Intestinal Epithelial Barrier Function. *Am. J. Physiol. Liver Physiol.* **2010**, *298*, G807–G819. [CrossRef] [PubMed]

14. Pietrangelo, L.; Magnifico, I.; Petronio Petronio, G.; Cutuli, M.A.; Venditti, N.; Nicolosi, D.; Perna, A.; Guerra, G.; Di Marco, R. A Potential "Vitaminic Strategy" against Caries and Halitosis. *Appl. Sci.* **2022**, *12*, 2457. [CrossRef]

15. Yong, C.-C.; Khoo, B.-Y.; Sasidharan, S.; Piyawattanametha, W.; Kim, S.-H.; Khemthongcharoen, N.; Chuah, L.-O.; Ang, M.-Y.; Liong, M.-T. Activity of Crude and Fractionated Extracts by Lactic Acid Bacteria (LAB) Isolated from Local Dairy, Meat, and Fermented Products against *Staphylococcus aureus*. *Ann. Microbiol.* **2015**, *65*, 1037–1047. [CrossRef]

16. Marianelli, C.; Cifani, N.; Pasquali, P. Evaluation of Antimicrobial Activity of Probiotic Bacteria against *Salmonella enterica* subsp. *enterica* Serovar Typhimurium 1344 in a Common Medium under Different Environmental Conditions. *Res. Microbiol.* **2010**, *161*, 673–680. [CrossRef]

17. Servin, A.L. Antagonistic Activities of Lactobacilli and Bifidobacteria against Microbial Pathogens. *FEMS Microbiol. Rev.* **2004**, *28*, 405–440. [CrossRef]

18. Hyronimus, B.; Le Marrec, C.; Hadj Sassi, A.; Deschamps, A. Acid and Bile Tolerance of Spore-Forming Lactic Acid Bacteria. *Int. J. Food Microbiol.* **2000**, *61*, 193–197. [CrossRef]

19. Abbas, M.A.; Lee, E.-B.; Boby, N.; Biruhanu, B.T.; Park, S.-C. A Pharmacodynamic Investigation to Assess the Synergism of Orbifloxacin and Propyl Gallate against *Escherichia coli*. *Front. Pharmacol.* **2022**, *13*, 989395. [CrossRef]

20. Kim, E.-Y.; Kim, Y.-H.; Rhee, M.-H.; Song, J.-C.; Lee, K.-W.; Kim, K.-S.; Lee, S.-P.; Lee, I.-S.; Park, S.-C. Selection of *Lactobacillus* sp. PSC101 That Produces Active Dietary Enzymes Such as Amylase, Lipase, Phytase and Protease in Pigs. *J. Gen. Appl. Microbiol.* **2007**, *53*, 111–117. [CrossRef]

21. Lee, J.-S.; Damte, D.; Lee, S.-J.; Hossain, M.-A.; Belew, S.; Kim, J.-Y.; Rhee, M.-H.; Kim, J.-C.; Park, S.-C. Evaluation and Characterization of a Novel Probiotic *Lactobacillus pentosus* PL11 Isolated from Japanese Eel (*Anguilla japonica*) for Its Use in Aquaculture. *Aquac. Nutr.* **2015**, *21*, 444–456. [CrossRef]

22. Ali, M.S.; Lee, E.-B.; Quah, Y.; Birhanu, B.T.; Suk, K.; Lim, S.-K.; Park, S.-C. Heat-Killed *Limosilactobacillus reuteri* PSC102 Ameliorates Impaired Immunity in Cyclophosphamide-Induced Immunosuppressed Mice. *Front. Microbiol.* **2022**, *13*, 820838. [CrossRef] [PubMed]

23. Kondrotiene, K.; Lauciene, L.; Andruleviciute, V.; Kasetiene, N.; Serniene, L.; Sekmokiene, D.; Malakauskas, M. Safety Assessment and Preliminary in Vitro Evaluation of Probiotic Potential of Lactococcus Lactis Strains Naturally Present in Raw and Fermented Milk. *Curr. Microbiol.* **2020**, *77*, 3013–3023. [CrossRef]

24. Jena, P.K.; Trivedi, D.; Thakore, K.; Chaudhary, H.; Giri, S.S.; Seshadri, S. Isolation and Characterization of Probiotic Properties of Lactobacilli Isolated from Rat Fecal Microbiota. *Microbiol. Immunol.* **2013**, *57*, 407–416. [CrossRef] [PubMed]

25. Barik, A.; Patel, G.D.; Sen, S.K.; Rajhans, G.; Nayak, C.; Raut, S. Probiotic Characterization of Indigenous *Kocuria flava* Y4 Strain Isolated from *Dioscorea villosa* Leaves. *Probiotics Antimicrob. Proteins* **2021**, 1–16. [CrossRef]

26. Hossain, M.A.; Park, H.-C.; Park, S.-W.; Park, S.-C.; Seo, M.-G.; Her, M.; Kang, J. Synergism of the Combination of Traditional Antibiotics and Novel Phenolic Compounds against *Escherichia coli*. *Pathogens* **2020**, *9*, 811. [CrossRef]

27. Perin, L.M.; Miranda, R.O.; Todorov, S.D.; de Melo Franco, B.D.G.; Nero, L.A. Virulence, Antibiotic Resistance and Biogenic Amines of Bacteriocinogenic Lactococci and Enterococci Isolated from Goat Milk. *Int. J. Food Microbiol.* **2014**, *185*, 121–126. [CrossRef]

28. Tian, L.; Liu, R.; Zhou, Z.; Xu, X.; Feng, S.; Kushmaro, A.; Marks, R.S.; Wang, D.; Sun, Q. Probiotic Characteristics of *Lactiplantibacillus Plantarum* N-1 and Its Cholesterol-Lowering Effect in Hypercholesterolemic Rats. *Probiotics Antimicrob. Proteins* **2022**, *14*, 337–348. [CrossRef]

29. Son, S.-H.; Yang, S.-J.; Jeon, H.-L.; Yu, H.-S.; Lee, N.-K.; Park, Y.-S.; Paik, H.-D. Antioxidant and Immunostimulatory Effect of Potential Probiotic *Lactobacillus paraplantarum* SC61 Isolated from Korean Traditional Fermented Food, *jangajji*. *Microb. Pathog.* **2018**, *125*, 486–492. [CrossRef]

30. Clinical and Laboratory Standards Institute. Performance Standards for Antimicrobial Susceptibility Testing. *Clin. Lab Stand. Inst.* **2016**, *35*, 16–38.

31. European Committee on Antimicrobial Susceptibility Testing. Breakpoint Tables for Interpretation of MICs and Zone Diameters. 2019. Available online: https://www.eucast.org/clinical_breakpoints/ (accessed on 27 November 2022).

32. Ghanbari, M.; Rezaei, M.; Jami, M.; Nazari, R.M. Isolation and Characterization of Lactobacillus Species from Intestinal Contents of Beluga (*Huso huso*) and Persian Sturgeon (*Acipenser persicus*). *Iran. J. Vet. Res.* **2009**, *10*, 152–157.

33. Aragón, F.; Carino, S.; Perdigón, G.; de Moreno de LeBlanc, A. Inhibition of Growth and Metastasis of Breast Cancer in Mice by Milk Fermented with *Lactobacillus casei* CRL 431. *J. Immunother.* **2015**, *38*, 185–196. [CrossRef] [PubMed]

34. Park, M.S.; Kwon, B.; Ku, S.; Ji, G.E. The Efficacy of *Bifidobacterium longum* BORI and *Lactobacillus acidophilus* AD031 Probiotic Treatment in Infants with Rotavirus Infection. *Nutrients* **2017**, *9*, 887. [CrossRef] [PubMed]

35. So, S.S.Y.; Wan, M.L.Y.; El-Nezami, H. Probiotics-Mediated Suppression of Cancer. *Curr. Opin. Oncol.* **2017**, *29*, 62–72. [CrossRef]

36. Galdeano, C.M.; Cazorla, S.I.; Dumit, J.M.L.; Vélez, E.; Perdigón, G. Beneficial Effects of Probiotic Consumption on the Immune System. *Ann. Nutr. Metab.* **2019**, *74*, 115–124.

37. Venditti, N.; Vergalito, F.; Magnifico, I.; Cutuli, M.A.; Pietrangelo, L.; Cozzolino, A.; Angiolillo, A.; Succi, M.; Petronio, G.P.; Di Marco, R. The *Lepidoptera Galleria* Mellonella "In Vivo" Model: A Preliminary Pilot Study on Oral Administration of *Lactobacillus plantarum* (Now *Lactiplantibacillus plantarum*). *New Microbiol.* **2021**, *44*, 42–50.

38. Lee, Y.; Choi, Y.; Yoon, Y. Lactic Acid Bacteria in Kimchi Might Be a Cause for Carcinogen Production in Intestine. *Food Control* **2021**, *126*, 108045. [CrossRef]
39. Delgado, S.; O'sullivan, E.; Fitzgerald, G.; Mayo, B. In Vitro Evaluation of the Probiotic Properties of Human Intestinal *Bifidobacterium* Species and Selection of New Probiotic Candidates. *J. Appl. Microbiol.* **2008**, *104*, 1119–1127. [CrossRef]
40. Argyri, A.A.; Zoumpopoulou, G.; Karatzas, K.-A.G.; Tsakalidou, E.; Nychas, G.-J.E.; Panagou, E.Z.; Tassou, C.C. Selection of Potential Probiotic Lactic Acid Bacteria from Fermented Olives by in Vitro Tests. *Food Microbiol.* **2013**, *33*, 282–291. [CrossRef]
41. Sánchez, B.; Champomier-Vergès, M.-C.; Collado, M.D.C.; Anglade, P.; Baraige, F.; Sanz, Y.; de los Reyes-Gavilán, C.G.; Margolles, A.; Zagorec, M. Low-PH Adaptation and the Acid Tolerance Response of *Bifidobacterium longum* Biotype Longum. *Appl. Environ. Microbiol.* **2007**, *73*, 6450–6459. [CrossRef]
42. Matsumoto, M.; Ohishi, H.; Benno, Y. H+-ATPase Activity in *Bifidobacterium* with Special Reference to Acid Tolerance. *Int. J. Food Microbiol.* **2004**, *93*, 109–113. [CrossRef] [PubMed]
43. Bao, Y.; Zhang, Y.; Zhang, Y.; Liu, Y.; Wang, S.; Dong, X.; Wang, Y.; Zhang, H. Screening of Potential Probiotic Properties of *Lactobacillus fermentum* Isolated from Traditional Dairy Products. *Food Control* **2010**, *21*, 695–701. [CrossRef]
44. Ruiz, L.; Margolles, A.; Sánchez, B. Bile Resistance Mechanisms in *Lactobacillus* and *Bifidobacterium*. *Front. Microbiol.* **2013**, *4*, 396. [CrossRef] [PubMed]
45. Cozzolino, A.; Vergalito, F.; Tremonte, P.; Iorizzo, M.; Lombardi, S.J.; Sorrentino, E.; Luongo, D.; Coppola, R.; Di Marco, R.; Succi, M. Preliminary Evaluation of the Safety and Probiotic Potential of *Akkermansia muciniphila* DSM 22959 in Comparison with *Lactobacillus rhamnosus* GG. *Microorganisms* **2020**, *8*, 189. [CrossRef] [PubMed]
46. Grześkowiak, Ł.; Collado, M.C.; Salminen, S. Evaluation of Aggregation Abilities between Commensal Fish Bacteria and Pathogens. *Aquaculture* **2012**, *356*, 412–414. [CrossRef]
47. Kos, B.; Šušković, J.; Vuković, S.; Šimpraga, M.; Frece, J.; Matošić, S. Adhesion and Aggregation Ability of Probiotic Strain *Lactobacillus acidophilus* M92. *J. Appl. Microbiol.* **2003**, *94*, 981–987. [CrossRef]
48. Reuben, R.C.; Roy, P.C.; Sarkar, S.L.; Rubayet Ul Alam, A.S.M.; Jahid, I.K. Characterization and Evaluation of Lactic Acid Bacteria from Indigenous Raw Milk for Potential Probiotic Properties. *J. Dairy Sci.* **2020**, *103*, 1223–1237. [CrossRef]
49. Collado, M.C.; Meriluoto, J.; Salminen, S. Measurement of Aggregation Properties between Probiotics and Pathogens: In Vitro Evaluation of Different Methods. *J. Microbiol. Methods* **2007**, *71*, 71–74. [CrossRef]
50. Schellenberg, J.; Smoragiewicz, W.; Karska-Wysocki, B. A Rapid Method Combining Immunofluorescence and Flow Cytometry for Improved Understanding of Competitive Interactions between Lactic Acid Bacteria (LAB) and Methicillin-Resistant *S. aureus* (MRSA) in Mixed Culture. *J. Microbiol. Methods* **2006**, *65*, 1–9. [CrossRef]
51. Han, S.; Lu, Y.; Xie, J.; Fei, Y.; Zheng, G.; Wang, Z.; Liu, J.; Lv, L.; Ling, Z.; Berglund, B. Probiotic Gastrointestinal Transit and Colonization after Oral Administration: A Long Journey. *Front. Cell. Infect. Microbiol.* **2021**, *11*, 609722. [CrossRef]
52. Śliżewska, K.; Chlebicz-Wójcik, A.; Nowak, A. Probiotic Properties of New *Lactobacillus* Strains Intended to Be Used as Feed Additives for Monogastric Animals. *Probiotics Antimicrob. Proteins* **2021**, *13*, 146–162. [CrossRef] [PubMed]
53. Klare, I.; Konstabel, C.; Werner, G.; Huys, G.; Vankerckhoven, V.; Kahlmeter, G.; Hildebrandt, B.; Müller-Bertling, S.; Witte, W.; Goossens, H. Antimicrobial Susceptibilities of *Lactobacillus*, *Pediococcus* and *Lactococcus* Human Isolates and Cultures Intended for Probiotic or Nutritional Use. *J. Antimicrob. Chemother.* **2007**, *59*, 900–912. [CrossRef] [PubMed]
54. Kumherová, M.; Veselá, K.; Kosová, M.; Mašata, J.; Horáčková, Š.; Šmidrkal, J. Novel Potential Probiotic Lactobacilli for Prevention and Treatment of Vulvovaginal Infections. *Probiotics Antimicrob. Proteins* **2021**, *13*, 163–172. [CrossRef] [PubMed]
55. Kodner, C.; Kudrimoti, A. Diagnosis and Management of Acute Interstitial Nephritis. *Am. Fam. Physician* **2003**, *67*, 2527–2534. [PubMed]
56. Barbosa, J.; Gibbs, P.A.; Teixeira, P. Virulence Factors among Enterococci Isolated from Traditional Fermented Meat Products Produced in the North of Portugal. *Food Control* **2010**, *21*, 651–656. [CrossRef]
57. Wang, Y.; Wu, Y.; Wang, Y.; Xu, H.; Mei, X.; Yu, D.; Wang, Y.; Li, W. Antioxidant Properties of Probiotic Bacteria. *Nutrients* **2017**, *9*, 521. [CrossRef]
58. Nakagawa, H.; Miyazaki, T. Beneficial Effects of Antioxidative Lactic Acid Bacteria. *AIMS Microbiol.* **2017**, *3*, 1–7. [CrossRef]
59. Wang, A.N.; Yi, X.W.; Yu, H.F.; Dong, B.; Qiao, S.Y. Free Radical Scavenging Activity of *Lactobacillus fermentum* in Vitro and Its Antioxidative Effect on Growing–Finishing Pigs. *J. Appl. Microbiol.* **2009**, *107*, 1140–1148. [CrossRef]
60. Halder, D.; Mandal, M.; Chatterjee, S.S.; Pal, N.K.; Mandal, S. Indigenous Probiotic *Lactobacillus* Isolates Presenting Antibiotic like Activity against Human Pathogenic Bacteria. *Biomedicines* **2017**, *5*, 31. [CrossRef]
61. Ma, T.; Suzuki, Y.; Guan, L.L. Dissect the Mode of Action of Probiotics in Affecting Host-Microbial Interactions and Immunity in Food Producing Animals. *Vet. Immunol. Immunopathol.* **2018**, *205*, 35–48. [CrossRef]
62. Langa, S.; Martín-Cabrejas, I.; Montiel, R.; Peirotén, Á.; Arqués, J.L.; Medina, M. Protective Effect of Reuterin-Producing Lactobacillus Reuteri against Listeria Monocytogenes and *Escherichia coli* O157:H7 in Semi-Hard Cheese. *Food Control* **2018**, *84*, 284–289. [CrossRef]
63. Bennett, S.; Ben Said, L.; Lacasse, P.; Malouin, F.; Fliss, I. Susceptibility to Nisin, Bactofencin, Pediocin and Reuterin of Multidrug Resistant *Staphylococcus aureus*, *Streptococcus dysgalactiae* and *Streptococcus uberis* Causing Bovine Mastitis. *Antibiotics* **2021**, *10*, 1418. [CrossRef] [PubMed]

64. Schaefer, L.; Auchtung, T.A.; Hermans, K.E.; Whitehead, D.; Borhan, B.; Britton, R.A. The Antimicrobial Compound Reuterin (3-Hydroxypropionaldehyde) Induces Oxidative Stress via Interaction with Thiol Groups. *Microbiology* **2010**, *156*, 1589. [CrossRef] [PubMed]
65. Šušković, J.; Kos, B.; Beganović, J.; Leboš Pavunc, A.; Habjanič, K.; Matošić, S. Antimicrobial Activity–the Most Important Property of Probiotic and Starter Lactic Acid Bacteria. *Food Technol. Biotechnol.* **2010**, *48*, 296–307.
66. Alakomi, H.-L.; Skytta, E.; Saarela, M.; Mattila-Sandholm, T.; Latva-Kala, K.; Helander, I.M. Lactic Acid Permeabilizes Gram-Negative Bacteria by Disrupting the Outer Membrane. *Appl. Environ. Microbiol.* **2000**, *66*, 2001–2005. [CrossRef]

Article

Development of Green Banana Fruit Wines: Chemical Compositions and In Vitro Antioxidative Activities

Zhichun Li [1,2,*,†] , Cuina Qin [1,2,†] , Xuemei He [1,2] , Bojie Chen [1,3] , Jie Tang [1,2] , Guoming Liu [1,2] , Li Li [1,2] , Ying Yang [1,2] , Dongqing Ye [1,2] , Jiemin Li [1,2] , Dongning Ling [1,2] , Changbao Li [1,2] , Hock Eng Khoo [3] and Jian Sun [1,2,*]

[1] Agro-Food Science and Technology Research Institute, Guangxi Academy of Agricultural Sciences, Nanning 530007, China
[2] Guangxi Key Laboratory of Fruits and Vegetables Storage-Processing Technology, Nanning 530007, China
[3] College of Chemistry and Bioengineering, Guilin University of Technology, Guilin 541006, China
*   Correspondence: lizhichun@gxaas.net (Z.L.); jiansun@gxaas.net (J.S.)
†   These authors contributed equally to this work.

Abstract: This study aimed to develop functional fruit wines using whole fruit, pulp, and peels from green bananas. The boiled banana homogenates were mixed with cane sugar before wine fermentation. Quality parameters, phenolic compounds, flavor components, and antioxidative properties of the green banana peel wine (GBPW), green banana pulp wine (GBMW), and whole banana wine (GBW) were determined. High-performance liquid chromatography was used to determine the phytochemical compounds in three wines, and the flavor components were further analyzed using headspace solid-phase microextraction combined with gas chromatography–mass spectrometry. The flavor components and in vitro antioxidant activities were, respectively, determined using the relative odor activity value and the orthogonal projections on latent structure discrimination analysis (OPLS-DA). In vitro antioxidative capacities for these wines were evaluated using antioxidant chemical assays and cell culture methods. The total phenolic and total tannin content of the GBPW, GBMW, and GBW showed reducing trends with increasing fermentation days, whereas the total flavonoid content of the wine samples exhibited downward trends. The antioxidant capacities of the three wine samples were higher than those of the raw fruit samples, except for the metal chelation rate (%). Additionally, the main flavor component in the wine samples was 3-methyl-1-butanol. Its percentages in the GBPW, GBMW, and GBW were 72.02%, 54.04%, and 76.49%, respectively. The OPLS-DA results indicated that the three wines presented significantly different antioxidant activities. The cell-culture-based antioxidant analysis showed that these wine samples had protective effects against the oxidative stress of the 3T3-L1 preadipocytes induced by hydrogen peroxide. This study provided a theoretical basis for defining the antioxidant characteristics of banana wines and expanding novel channels for using banana peels to develop nutraceuticals.

Keywords: active ingredient; antioxidant activity; aroma components; banana peel; chemical composition

Citation: Li, Z.; Qin, C.; He, X.; Chen, B.; Tang, J.; Liu, G.; Li, L.; Yang, Y.; Ye, D.; Li, J.; et al. Development of Green Banana Fruit Wines: Chemical Compositions and In Vitro Antioxidative Activities. *Antioxidants* **2023**, *12*, 93. https://doi.org/10.3390/antiox12010093

Academic Editor: Stanley Omaye

Received: 16 November 2022
Revised: 27 December 2022
Accepted: 28 December 2022
Published: 30 December 2022

1. Introduction

Bananas are one of the most cultivated and consumed fruits in the world. They are also one of the crucial fruit varieties grown in South China, with an average annual output of 11,130.02 million tons from 2016 to 2020, accounting for 27.23% of the total fruit output. The skin or peel accounts for about 40% of the total fruit weight of a ripe banana. Banana processing often generates many by-products, such as peels, which are rarely eaten directly owing to their bitter taste. Banana peels are by-products or waste, which poses serious environmental hazards [1]. Due to the mass production of banana peels from banana product processing, the fruit peel has been used as compost or produced into more valuable industrial products. Additionally, a high percentage of postharvest green bananas present black spots on the skin. These bananas have a low commercial value and are not marketable. They are commonly discarded as wastes and also pollute the environment.

Bananas have been developed into different functional foods to date. Although banana peels are not commonly used as food, they are typically used as a carbon-based adsorbent for wastewater treatment [2], fuel [3], and animal feed [4]. Previous literature showed that green banana powder contained a high amount of resistant starch, which can be used as a nutritional or functional component [5]. Bananas have high nutritional and medical values due to their colorful pigments and high carbohydrate content [6]. The starch obtained from bananas has high economic values as food and medicine [7]. In banana production and processing, tonnes of residual fruits, tail fruits, and peels are wasted without specific recycling treatment and particular utilization.

Studies on green bananas can be divided into fundamental research and product development, of which the product characteristic determination includes physicochemical and digestive properties [8,9]. The application of bananas is adding banana-based starch as a functional component in developing food products [10,11]. These studies are significant for the processing and use of bananas and their by-products. In addition, the previous reports on banana peels mainly focused on the analyses of physicochemical characteristics and the application of bioactive compounds such as phenolic compounds [12], dietary fibers [13], pectins [14], and tannins [15]. These bioactive compounds have anticancer [16], antibacterial [17,18], antidiabetes [19], and antioxidant capacities [20], and they prevent cardiovascular diseases [21,22]. For example, Durgadevi et al. [23] reported the effect of banana peel samples on the proliferation of human breast cancer cells cultured in vitro.

A large volume of by-products is produced annually during banana production and processing because no special recycling treatment is available. Therefore, fully utilizing these by-products to improve the economic benefits of the banana industry is necessary and urgent. Applying green banana peels in wine making can reduce waste production and increase resource utilization. Furthermore, determining the flavor components and antioxidative capabilities of green banana wines is also helpful for functional product development and recycling banana by-products or alternative uses.

2. Materials and Methods

2.1. Sample Preparation

The unripe green bananas (*Musa acuminate* L.) of the 'Guijiao 9' variety were harvested from the banana germplasm resource nursery of Jinghong city, Xishuangbanna, Yunnan, in April 2021. Green bananas were obtained during the immature stage, packed, and transported to a wine fermentation laboratory. One kilogram of each whole fruit, pulp, and peel was prepared, pulverized, and homogenized with 3.5 L of distilled water. Then, 500 mL of each sample was separately collected from the 3.5 L mixture as a fresh fruit sample. The leftover 3.0 L was boiled for 10 min to denature all the enzymes and destroy the microbes. The hot banana samples were mixed with sugar (136 g/L), homogenized, and cooled to 40 °C before adding 0.5 g/L pectinase and 0.3 g/L cellulase. Then, the mixture was hydrolyzed for 2 h and added with 0.02% dried wine yeast K1-V1116 (*Saccharomyces cerevisiae*). The wine fermentation was conducted at room temperature, and test samples were collected at 0, 3, 6, 9, 15, and 30 days of fermentation. The sterilization was performed by heating the fermentation medium to 65 °C and maintaining it for 15 min.

2.2. Determination of Quality Parameters and Bioactive Components

Banana wine samples were filtrated before further analysis. The selected quality parameters, including the total acid content (TAC), pH value, total soluble solids (TSS) content, and alcohol content, were determined according to the methods described by Li et al. [24]. The total phenolic content (TPC) and total tannin content (TTC) were assayed using the Folin–Ciocalteu method, whereas the total flavonoid content (TFC) was determined using the aluminum nitrate method [25]. The TPC and TTC were expressed as the gallic acid equivalent (mg/g extract). The TFC was expressed as the quercetin equivalent (mg/g extract).

2.3. UHPLC-MS Identification and Quantification of Phenolic Compounds

The phenolic compounds in the banana wine samples were determined using ultrahigh pressure liquid chromatography (UHPLC) coupled with a mass spectrometer (MS) on an Acquity UPLC HSS T3 column (2.1 mm × 100 mm, 1.7 μm, Waters, Milford, MA, USA). The mobile phase A was 0.1% aqueous formic acid, and the mobile phase B was acetonitrile [26]. The flow rate was 0.25 mL/min. The gradient elution procedure was 0–0.8 min, 90% A; 0.8–4.0 min, 90–80% A; 4.0–6.5 min, 80% A; 6.5–7.0 min, 80–55% A; 7.0–13.0 min, 55% A; 13.0–13.5 min, 55–90% A; and 13.5–14.0 min, 90% A. The column temperature was set at 35 °C, and the injection volume was 2 μL. MS scanning was based on electrospray ionization sources (ES+ and ES−), scanning mode multireaction monitoring (MRM), a capillary voltage of 3.0 kV, and an ion source temperature of 350 °C. The solvation gas was nitrogen, with a flow rate of 700 L/h. The flow rate of the cone-hole gas (nitrogen) was 150 L/h, and the impact gas flow rate (Ar) was 0.10 mL/min. The phenolic compounds in the wine samples were analyzed using the 18 phenolic standards. Five different concentrations (0.004–8.0 mg/mL) were selected to plot the standard calibration curves. These standard calibration curves had a very high linearity (>0.99). The screening test was performed using a UV-DAD detector at a wavelength of 280 nm.

2.4. HS-SPME-GC-MS Analysis

The flavor components in the banana wine samples were analyzed using a headspace solid-phase microextraction combined with gas chromatography–mass spectrometry (HS-SPME-GC-MS, Agilent Technologies, Santa Clara, CA, USA) [27]. The GC column was HP-5 MS (30 m × 250 μm × 0.25 μm). The initial column temperature was set to 40 °C and was maintained for 5 min; then, it was increased to 120 °C at 4 °C/min rate and was let to stand for 3 min; then, it was increased to 220 °C at 10 °C/min and was held for 10 min; finally, it was increased to 280 °C at 30 °C/min and was maintained for 15 min. The carrier gas was high-purity helium with a flow rate of 1 mL/min. The inlet temperature was 250 °C. The MS conditions were an EI ion source, electron energy of 70 eV, electron multiplier voltage of 1376 V, scanning range from 15 to 500 amu, ion source temperature of 230 °C, and four-stage rod temperature of 150 °C. The flavor compounds were identified via spectra matching using the GC/MS library. Their ion masses were also compared with the data reported in online databases (PubChem and other databases, accessed on 20 June 2021).

The method for relative odor activity value (ROAV) was used to quantify the flavor components in the banana wine samples. The odor activity value (OAV) was calculated according to Equation (1):

$$OAV = \frac{C}{T} \tag{1}$$

where C is the relative concentration of the flavor substance and T is the sensory threshold of this flavor substance.

The $ROAV_{max}$, with the highest contribution rate, was marked as 100. The ROAV values of the other remaining aroma components were calculated according to Equation (2):

$$ROAV = \frac{C_i}{C_S \tan} \times \frac{T_S \tan}{T_i} \times 100 \tag{2}$$

C_i is the relative content of the aroma components, T_i is the sensory threshold of the aroma components, and $C_S \tan$ and $T_S \tan$ were the relative content and sensory threshold of the aroma components contributing the most, respectively.

The sensory threshold of the flavor components was determined by Shi et al. [28]. According to the calculated result (Equation (2)), a ROAV ≥ 1 denoted that the substance directly impacted the overall flavor components and was a principal compound. If 0.1 ≤ ROAV < 1, it indicated that the flavor substance possessed a modification effect on the total flavor substance and was a modified compound. If the ROAV < 0.1, it showed that the flavor substance had no significant influence on the overall flavor.

2.5. Determination of In Vitro Antioxidant Activity

The antioxidant activities of the banana wine samples were determined using the DPPH radical scavenging rate, hydroxyl radical scavenging ability, superoxide anion radical scavenging ability, and metal chelation rate. The DPPH radical scavenging rate was measured using the DPPH free radical scavenging assay [29], and the metal chelation rate was determined using an iron (II) chloride solution coupled with the ferrozine coloration method [30]. The hydroxyl radical and superoxide anion radical scavenging abilities were also determined with respective test kits according to the manufacturers' instructions. Vitamin C (Vc) was used as a standard for comparison. A metal chelation assay is also known as a FRAP assay.

2.6. Cell-Culture-Based Antioxidant Capacities and Proliferation Inhibition Assays

The antioxidant capacities and inhibition abilities of the banana wine samples were determined using the oxidative stress model of 3T3-L1 preadipocytes induced by 30% hydrogen peroxide. A cell-culture-based oxidative stress model was developed and established by Kim et al. [31], and the antioxidant capacities of the samples were tested using a previously established method [32]. Healthy preadipocytes were treated with 20 µL of the banana wine samples for 48 h with or without adding 20 µL of 30% hydrogen peroxide. The negative control of the cell-culture-based antioxidant capacity was the cells without adding banana wine samples and without being induced by the 30% hydrogen peroxide. The positive control was the cells only induced by hydrogen peroxide. The cell viability was analyzed with an MTT assay.

2.7. Statistical Analysis

All the data are presented as the means of three replicates. The statistically significant differences in the antioxidant activities between the samples were determined using the analysis of variance (ANOVA) coupled with a posthoc LSD test. SPSS version 24 (SPSS Inc., Chicago, IL, USA) was used to analyze the mean differences at a p-value less than 0.05. The antioxidant activities of the three green banana wine samples, determined using four in vitro antioxidant activity assays, were analyzed with OPLS-DA.

3. Results

3.1. Quality Parameters and Bioactive Components

The TSS, alcohol content, pH value, and TAC are important quality indexes of fermented fruit wines. These quality parameters are closely related to the taste and flavor of wines. These quality parameter values for the GBPW, GBMW, and GBW samples are shown in Figure 1. The result showed that the TSS in the fruit wine samples decreased with an increasing fermentation time, with a remarkable reduction observed within six days. The alcohol content increased the most during this fermentation period and then elevated more gradually (Figure 1A). After day 6, the decline in the TSS was less than 1%. The reduction rate remained similar until day 30 of the wine fermentation. The alcohol content in the GBW sample started to reduce on day 9 of fermentation. The average reduction rates of the TSS in the GBPW, GBMW, and GBW samples on day 3 of fermentation were 44.5%, 50.0%, and 46.1%, respectively, while the alcohol contents increased to 3.5%, 4.9%, and 4.5%, respectively.

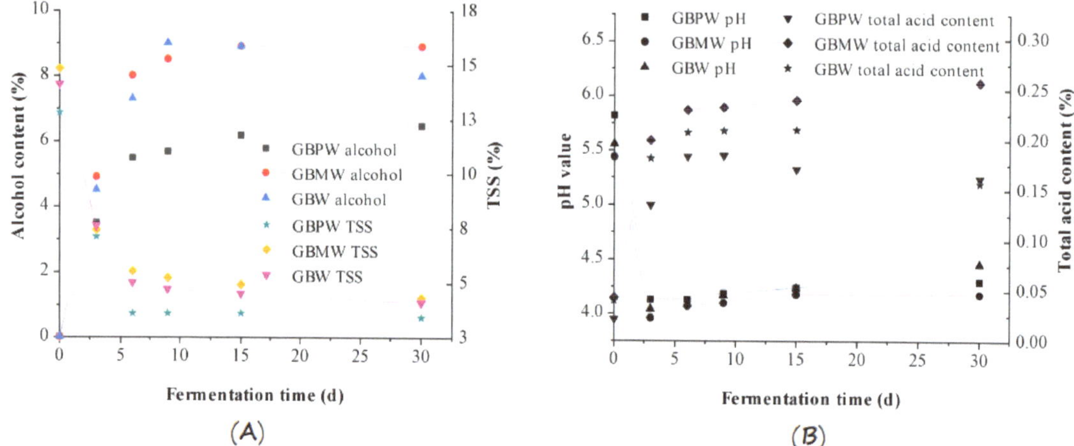

Figure 1. Changes in quality parameters of GBPW, GBMW, and GBW during fermentation of green banana samples. (**A**) Alcohol content (%) and total soluble solids (TSS %); (**B**) pH value and total acid content (%).

After the third day of fermentation, the increment of the alcohol content and the reduction in the TSS became slower, probably because the yeast multiplied rapidly at the early fermentation stage, and the metabolism rate increased. It digested soluble solids in a fermentation solution to provide sufficient nutrients and convert them into alcohol, organic acids, and other flavor substances. At a later fermentation period, soluble solids in the wine samples were lower. The low amount of soluble carbohydrates could not provide enough energy for the growth and survival of the yeast. Therefore, yeast entered the aging period and finally lyzed. The yeast produced increasing amounts of alcohol and organic acids during fermentation. Additionally, the GBMW showed the most notable decreasing TSS level. The TSS in the GBPW was lower than the other wine samples, and the alcohol content in the GBPW increased the least.

The changes in the pH value and TAC of the fruit wine samples are depicted in Figure 1B. The TAC in the wine samples had increasing rates, whereas the pH value showed decreasing values. The increasing and decreasing rates of the TAC in these wine samples varied at different fermentation stages. From day 0 to day 3, the TAC in the GBPW, GBMW, and GBW increased to 13.5%, 19.9%, and 18.1%, while the pH values decreased to 29.0%, 27.2%, and 27.3%, respectively. After the third day of fermentation, the total acid content in these wine samples had a slower increment rate and finally decreased, except for the GBMW. The pH values started to increase on day 3 of fermentation, but the increment rate was slow until day 30.

At the early fermentation stage, the TAC drastically increased because the yeast rapidly multiplied, and a high metabolic rate produced a higher organic acid content. It caused the pH value of the sample to rapidly decrease to a pH of 4. Due to the influences of a low pH value, substrate concentration, and other related factors, yeast metabolism slowed down at the later stage of fermentation. The production rate of the organic acids also decreased; it was easier for the acids to undergo an esterification reaction with alcohols, and the degradation of organic acids occurred and produced ethanol. The reduced concentration of the organic acids led to a lesser increase in the pH value. During wine fermentation, the GBMW had the most notable increase in the TAC, but the pH values of the three fruit wines showed no remarkable differences.

The TPC, TTC, and TFC in the GBPW, GBMW, and GBW during fermentation are shown in Figure 2. The result showed that the TPC in the GBMW and GBW showed overall decreasing values during fermentation (Figure 2A). During 0–3 days of fermentation, the TPC decreased sharply from 15.10 mg/g and 16.71 mg/g to 3.50 mg/g and 9.44 mg/g, respectively. After the third day, the GBW decreased slowly until the end of fermentation, while the GBMW increased with a minor fluctuation. The TPC in the GBPW increased from 20.15 mg/g to 25.28 mg/g from baseline to the sixth day of fermentation and then decreased to 20.93 mg/g on day 9. The TPC finally increased slightly until the end of fermentation. The TPC of the GBPW was significantly higher on day 30 of fermentation than on day 0 ($p < 0.01$). The increased TPC during the early stage of fermentation could be due to the digestion of carbohydrates, thus releasing free phenolic compounds.

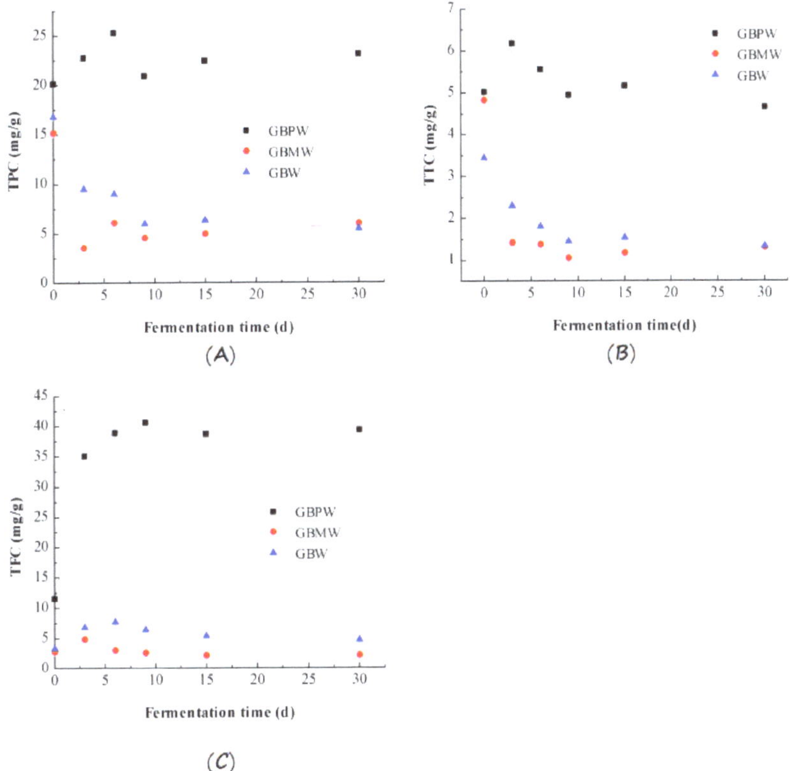

Figure 2. Changes in bioactive components of GBPW, GBMW, and GBW during fermentation of green banana samples. (**A**) Total phenolic content (TPC); (**B**) total tannin content (TTC); (**C**) total flavonoid content (TFC).

The reduction in the TPC of the GBMW and GBW could be due to the increased concentration of the soluble polysaccharides in the extracts. Similarly, the TPC in the GBPW was reduced by about 5 mg/g. This could have been due to the same reason. The further digestion and utilization of banana starch released more free polyphenolic compounds, except for the GBW. The GBW showed a slight decrease in the TPC from day 15 to day 20 of fermentation. It showed that the minor degradation of polyphenols occurred during fermentation at a temperature of 40 °C. The reduced TPC might also be attributed to the binding and adsorption of secondary metabolites produced by microbial fermentation with polyphenolic compounds or the precipitation and oxidation of polyphenols during fermentation [33].

Figure 2B depicts the TTC in the banana wine samples. During the fermentation process, the TTC in the GBMW and GBW showed decreasing rates. A rapid reduction was noted from day 0 to day 3 of fermentation, where the value reduced from 4.81 mg/g and 3.43 mg/g to 1.41 mg/g and 2.30 mg/g, respectively. The decreasing rate was slower after day 3 until the end of fermentation. The TTC in the GBMW and GBW at the end of fermentation was 1.30 mg/g and 1.32 mg/g, respectively. The TTC in the GBPW increased from 5.02 mg/g on day 0 to 6.17 mg/g on day 3 of fermentation and then reduced to 5.56 mg/g on day 6. A slight reduction in the TTC was observed from day 15 until the end of fermentation. The TTC in the GBPW on day 30 of fermentation was significantly higher than on day 0 ($p < 0.01$). The TTC in the GBPW increased on the first three days of fermentation due to the rapid digestion of complex tannin in the banana peel. The reduced TTC could be due to the polymerization reaction of tannins. This result showed that the TTC in the GBPW was significantly higher than in the GBMW and GBW. It could be related to a higher TTC in the green banana peel. Additionally, changes in the TPC and TTC of the fruit wine samples had similar trends.

Figure 2C shows the changes in the TFC in the banana wine samples. The TFC in the GBPW, GBMW, and GBW increased during the first few days of fermentation, then gradually reduced until the end of the fermentation process, except for the GBPW. The TFC in the GBPW, GBMW, and GBW increased from 11.55, 2.69, and 3.20 mg/g on day 0 to 35.10, 4.74, and 6.68 mg/g on day 3, respectively. The TFC in the GBPW rapidly increased until day 9 and then reduced until day 15. It had a minor increase from day 15 until the end of fermentation. The variation in the TFC during fermentation could be because flavonoids are unstable compounds. They were easily affected by changes in the solution pH, reducing agents, light, and other factors. During the fermentation process, the TFC in the GBPW was significantly higher than in the GBMW and GBW. It could be related to the flavonoid content in the banana peel, which was higher than the fresh pulp [34].

3.2. Identification and Quantification of Phenolic Compounds

The mass spectrometric parameters of 18 phenolic standards are shown in Table 1. The concentrations of these phenolic compounds in the GBPW, GBMW, and GBW were determined and are shown in Table 2. Five phenolic acids were identified from three fermented wine samples. They were salicylic, chlorogenic, ferulic, and p-coumaric acids. Among them, salicylic acid accounted for the highest proportion in the GBPW and GBMW, whereas the GBW had the highest concentration of ferulic acid. The ferulic acid concentration in the GBPW and the salicylic acid concentration in the GBMW were significantly higher than in the other two wine samples. Catechin was detected only in the GBPW, and chlorogenic acid was not found in the GBW. Salicylic acid was detected in the banana-peel-based wines because this compound was used for the postharvest treatment of bananas to extend their shelf life [35].

Apigenin, coumarin, and luteolin are flavonoids that were not detectable in these wine samples. Kaempferol was the only flavonoid detected in all the banana wine samples, where the GBPW exhibited the highest concentration. Additionally, the initial screening of the phenolic compounds was performed using the UHPLC method coupled with a UV-DAD detector. The results showed that gallic acid was not detected in the banana wine samples. The differences in the types and phenolic compound concentrations of the banana wine samples might be related to plant origin, microbial species, and several fermentation parameters [36].

Table 1. Mass spectrometric parameters of nine phenolic compounds.

Compounds	Retention Time (min)	$[M + H]^+$ (*m/z*)	$[M - H]^-$ (*m/z*)	MS/MS Fragments	Molecular Weight (g/mol)	Chemical Formula
Gallic acid	1.69	-	168.90	124.91, 78.90	170.12	$C_7H_6O_5$
Gentisic acid	2.83	-	152.90	108.90, 80.92	154.12	$C_7H_6O_4$
Protocatechuic acid	2.83	-	152.90	108.90, 90.88	154.12	$C_7H_6O_4$
Chlorogenic acid	3.70	-	353.03	191.00, 84.88	354.31	$C_{16}H_{18}O_9$
Catechin	3.82	291.07	-	138.96, 123.04	290.27	$C_{15}H_{14}O_6$
Caffeic acid	4.55	-	178.90	134.93, 106.88	180.16	$C_9H_8O_4$
Vanillic acid	4.57	-	166.9	151.93, 107.92	168.14	$C_8H_8O_4$
Epicatechin	4.66	291.13	-	139.09, 123.10	290.27	$C_{15}H_{14}O_6$
Syringic acid	4.75	-	196.97	181.96, 122.89	198.17	$C_9H_{10}O_5$
p-Coumaric acid	5.99	165.11	163.0	119.07, 91.12 (ES+) 119.00, 93.30 (ES−)	164.16	$C_9H_8O_3$
Ferulic acid	6.74	194.99	192.97	145.01, 116.99	194.18	$C_{10}H_{10}O_4$
Salicylic acid	8.33	-	136.90	92.89, 64.91	138.12	$C_7H_6O_3$
p-Hydroxybenzoic acid	8.33	-	136.91	92.89, 64.91	138.13	$C_7H_6O_3$
Luteolin	8.37	287.10	-	153.07, 121.09	286.24	$C_{15}H_{10}O_6$
Coumarin	8.58	147.07	-	91.12, 77.08	146.14	$C_9H_6O_2$
Cinnamic acid	8.97	149.00	-	131.00 103.00	148.16	$C_9H_8O_2$
Apigenin	9.15	271.10	-	153.07, 119.07	270.24	$C_{15}H_{10}O_5$
Kaempferol	9.28	287.09	-	153.07, 121.09	28623	$C_{15}H_{10}O_6$

Table 2. Phenolics content in green banana wine samples.

No	Phenolics	Concentration (µg/mL)			Minimum Detectable Concentration (µg/mL)
		Green Banana Peel Wine	Green Banana Pulp Wine	Green Banana Wine	
1	Gallic acid	4.536	0.46	1.209	4.59
2	Gentisic acid	0.009	0.038	ND	2.53
3	Protocatechuic acid	0.22	ND	ND	1.64
4	Chlorogenic acid	ND	ND	ND	8.38
5	Catechin	0.034	ND	0.007	5.42
6	Caffeic acid	0.782	0.048	0.011	9.27
7	Vanillic acid	ND	ND	ND	5.11
8	Epicatechin	0.109	0.124	0.131	2.03
9	Syringic acid	0.042	0.012	0.021	3.44
10	p-Coumaric acid	0.337	0.013	ND	4.35
11	Ferulic acid	10.493	0.047	0.286	15.49
12	Salicylic acid	0.065	ND	0.040	5.85
13	p-Hydroxybenzoic acid	0.055	ND	0.038	7.17
14	Luteolin	ND	ND	ND	3.19
15	Coumarin	ND	ND	ND	4.06
16	Cinnamic acid	ND	ND	ND	25.87
17	Apigenin	0.0013	ND	ND	0.46
18	Kaempferol	0.007	ND	0.005	2.22

"ND" indicates that the component was not detected.

3.3. HS-SPME-GC-MS Analysis of Flavor Components

The relative contents of the flavor components in the GBPW, GBMW, and GBW are shown in Table 3. A total of 22 flavor substances were identified from the GBPW, GBMW, and GBW. They consisted of four alcohols, thirteen esters, one acid, three alkenes, and one phenol. The relative contents of five different types of aroma substances in three wine samples are shown in Figure 3. The result showed that alcoholic compounds were the main flavor substances in the GBPW, GBMW, and GBW. Among the flavor components, 3-methyl-1-butanol was the highest in the GBPW, GBMW, and GBW, with levels of 72.02%,

54.04%, and 76.49% of the total flavor content, respectively. This alcohol compound has unique grass and a mature fruit aroma [37]. Fatty acid esters in the GBMW were more abundant than in the other two wine samples. Ethyl palmitate, 3-methyl-1-butanol, and 2,4-di-tert-butylphenol were detected in the GBPW, GBMW, and GBW. 2,3-Butanediol was found only in the GBPW and GBMW but not in the wine sample produced from whole bananas. On the contrary, alkenes and acetic acid were not detected in the banana peel wine samples.

Table 3. Flavor components in green banana peel (GBPW), pulp (GBMW), and whole fruit wine (GBW) samples.

No	Compounds	Relative Content (%)			Threshold (mg/L)	Characteristic
		GBPW	GBMW	GBW		
	Fatty Acid Esters					
1	Ethyl palmitate	1.40	3.57	2.03	2.26	Weak, waxy, and creamy aromas
2	Ethyl octanoate	-	0.66	-	0.17	Brandy aroma
3	Ethyl pelargonate	-	0.63	-	0.85×10^{-3}	Cantaloupe flavor
4	Ethyl caprate	-	0.25	-	0.2	Coconut flavor
5	2,6-Diethylphenyl isocyanate	-	1.88	-	-	-
6	Ethyl laurate	-	1.72	1.61	1.5	Floral and fruity aromas
7	Ethyl tridecanoate	-	0.18	-	-	-
8	Ethyl undecanoate	-	0.84	-	1	-
9	Ethyl myristate	-	2.85	2.31	4	Iris aroma
10	13-Methyl-tetradecanoic acid ethyl ester	-	0.10	-	-	-
11	Ethyl pentadecanoate	-	0.67	-	-	-
12	Methyl palmitate	-	0.28	-	4	-
13	Ethyl stearate	-	0.16	-	-	Faint waxy scent
	Acids					
14	Acetic acid	-	0.98	-	-	Pungent smell
	Alcohols					
15	3-Methyl-1-butanol	72.02	54.04	76.49	0.3	Fruity and mellow flavors
16	2,3-Butanediol	4.52	0.70	-	-	-
17	(2R,3R)-(-)-2,3-Butanediol	-	4.79	-	-	-
18	4-Carvomenthenol	-	0.13	-	-	-
	Alkenes					
19	(+)-α-Longipinene	-	0.20	-	-	-
20	Longifolene	-	4.26	2.11	-	-
21	Bicyclo[7.2.0]undec-4-ene,4,11,11-trimethyl-8-methylene-,(1R,4Z,9S)-	-	0.58	-	-	-
	Phenol					
22	2,4-di-tert-butylphenol	19.06	9.16	11.53	-	Phenol odor
	Total Content	**97.00**	**88.63**	**96.08**		

"-" indicates that the compound was not detected.

The ROAV values of the flavor components in the three wine samples are shown in Table 4. The key flavor component in the wine samples was 3-methyl-1-butanol. Ethyl palmitate was another contributor to the aroma of the wines. The flavor of the GBMW was mainly due to 3-methyl-1-butanol and ethyl octanoate. Ethyl palmitate, ethyl caprate, ethyl laurate, ethyl undecanoate, ethyl myristate, and methyl palmitate also enhanced the flavor of the GBMW. Methyl palmitate was another potent flavor compound in the wine sample produced from the banana pulp. The main flavor components in the GBW were ethyl palmitate, ethyl laurate, and ethyl myristate, in addition to 3-methyl-1-butanol. The other flavor components did not contribute to the aroma of these banana wines.

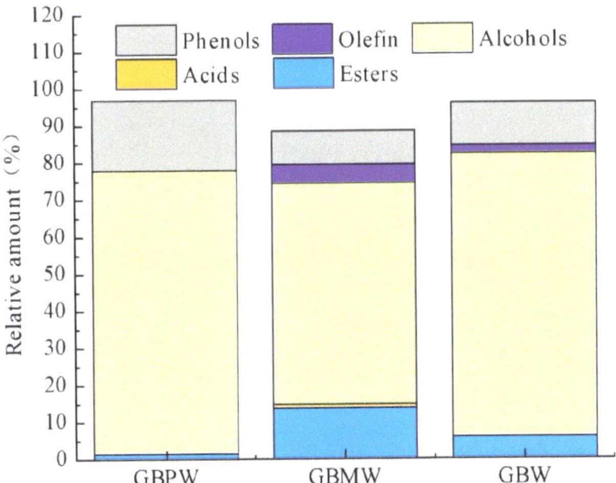

Figure 3. Relative amounts of flavor components in green banana wine samples.

Table 4. ROAV value of flavor substances in green banana peel (GBPW), pulp (GBMW), and whole fruit wine (GBW) samples.

No	Compounds	ROAV		
		GBPW	GBMW	GBW
	Fatty Acid Esters			
1	Ethyl palmitate	0.243	0.620	0.352
2	Ethyl octanoate	-	1.523	-
3	Ethyl pelargonate	-	-	-
4	Ethyl caprate	-	0.490	-
5	2,6-Diethylphenyl isocyanate	-	-	-
6	Ethyl laurate	-	0.450	0.421
7	Ethyl tridecanoate	-	-	-
8	Ethyl undecanoate	-	0.329	-
9	Ethyl myristate	-	0.279	0.227
10	13-Methyl-tetradecanoic acid ethyl ester	-	-	-
11	Ethyl pentadecanoate	-	-	-
12	Methyl palmitate	-	0.027	-
13	Ethyl stearate	-	-	-
	Acids			
14	Acetic acid	-	-	-
	Alcohols			
15	3-Methyl-1-butanol	94.156	70.650	100.000
16	2,3-Butanediol	-	-	-
17	(2R,3R)-(-)-2,3-Butanediol	-	-	-
18	4-Carvomenthenol	-	-	-
	Alkenes			
19	(+)-α-Longipinene	-	-	-
20	Longifolene	-	-	-
21	Bicyclo[7.2.0]undec-4-ene,4,11,11-trimethyl-8-methylene-,(1R,4Z,9S)-	-	-	-
	Phenol			
22	2,4-di-tert-butylphenol	-	-	-

"-" indicates that ROAV of the compound is not determined.

The flavor component is an essential index of fruit wine quality. The sensory quality of a fruit wine is attributed to the volatile and aromatic compounds in the wine. These flavor components indirectly affect the commodity value of the fruit wine. The flavor substances in a fruit wine mainly consist of alcohols, esters, acids, and alkenes, which are related to fruit aroma. Among them, alcohols and acids have relatively high flavor thresholds and contribute little to the flavor of fruit wine. On the other hand, esters have relatively low flavor thresholds and thus contribute more to the aroma of fruit wine.

3.4. In Vitro Antioxidant Capacities

The DPPH radical scavenging activity, hydroxyl radical scavenging ability, superoxide anion radical scavenging ability, and metal chelation rate of the GBPW, GBMW, and GBW are shown in Figure 4. The results showed that different banana parts used in wine fermentation affected DPPH radical scavenging activities (Figure 4A). The scavenging activities were statistically different between the banana fruit and banana wine samples. The DPPH radical scavenging activities of the wine samples were remarkably higher than that of the green banana samples ($p < 0.01$). The green banana peel (GBP) had the significantly lowest DPPH radical scavenging activity ($p < 0.05$), whereas the GBMW had the highest activity. We also found no significant differences between the DPPH radical scavenging activities of green banana pulp (GBM) and whole green banana (GB) ($p > 0.05$). Moreover, no statistical differences were observed between the scavenging activities of Vc and green banana wine samples ($p > 0.05$).

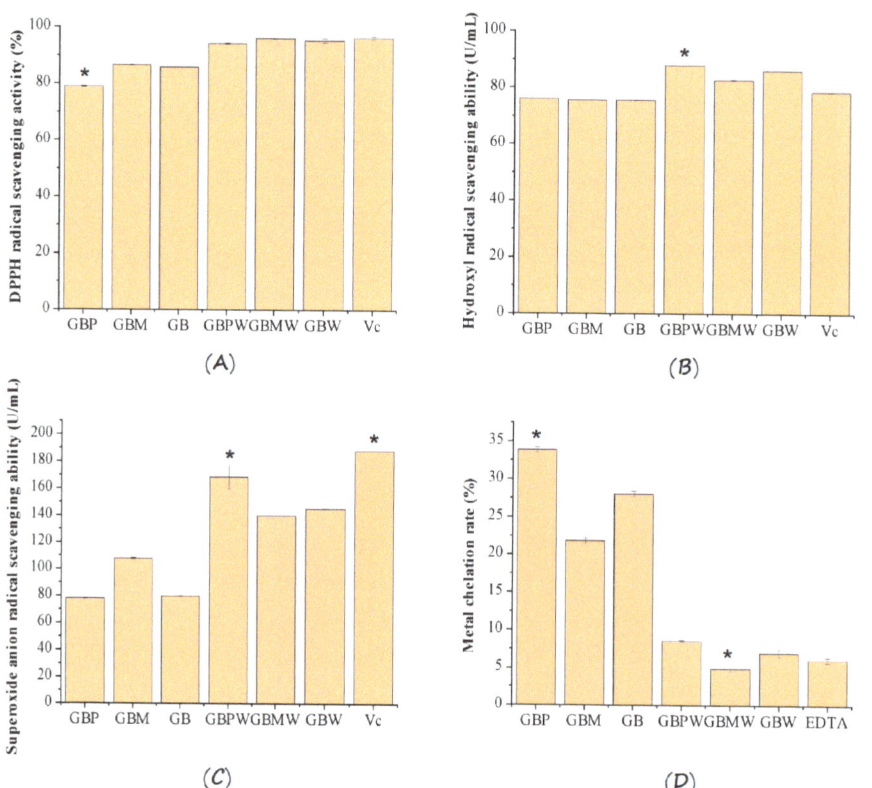

Figure 4. (**A**) DPPH radical scavenging activity, (**B**) hydroxyl radical scavenging ability, (**C**) superoxide anion radical scavenging activity, and (**D**) metal chelation rate of green banana and green banana wine samples. * denotes significant difference from other samples.

The hydroxyl radical scavenging abilities of the banana wine samples are depicted in Figure 4B. The fermentation conditions and different parts of the green banana had a remarkable impact on the hydroxyl radical scavenging ability of the green banana wine samples. The hydroxyl radical scavenging abilities of the banana wine samples were significantly higher than those of the green banana samples ($p < 0.01$). The scavenging ability of the GBPW was also statistically the highest among the banana fruit parts and wine samples ($p < 0.01$). Moreover, hydroxyl radical scavenging ability of Vc was not as high as the green banana wines. Moreover, significant differences were observed between the scavenging abilities of the different green banana wine samples.

Based on the data depicted in Figure 4C, wine fermentation significantly affected the superoxide anion radical scavenging ability. The scavenging abilities of the green banana wines were remarkably higher than those of the green banana fruit samples ($p < 0.01$). The GBPW had the significantly highest scavenging ability ($p < 0.05$). We found no significant differences between the scavenging abilities of the different fruit parts ($p > 0.05$). On the contrary, the scavenging ability of Vc was significantly the highest among all the samples ($p < 0.05$).

The metal chelation rates of the fruit samples were remarkably lower after the wine fermentation (Figure 4D). The metal chelation rates of the green banana wine samples were over three times lower than those of the green banana samples. The GBMW had the significantly lowest metal chelation rate ($p < 0.05$), indicating that the banana pulp had a lower metal chelation rate than the peel. This finding revealed that the fermentation conditions and types of raw materials altered the metal chelation rate of the fermented wine. The metal chelation rate of the green banana fermented wine was significantly lower than that of its raw material ($p < 0.01$), whereas the metal chelation rates of the green banana skin, pulp, and fruit samples were significantly different from each other ($p < 0.01$).

3.5. Comparison of Antioxidant Activities of Wine Samples Based on OPLS-DA

OPLS-DA data are depicted in Figure 5. The total variance of the samples was 99.02%, R2 was 82.3%, and Q2 was 72.2%. Three groups of samples were divided into four quadrants (Figure 5A), with the GBPW located in the first and fourth quadrants, the GBW in the second quadrant, and the GBMW in the third quadrant. A permutation test was subsequently performed to verify the model by randomly rearranging the experiments by changing the ranking order of the categorical variables (Y) and randomly assigning a Q2 of up to 200 times. Figure 5B shows the results of the permutation test. The intersection between the regression line and the vertical axis of point Q2 was lower than zero. It indicated that the discriminant model was not overfitted. Therefore, the initial model outperformed the random permutation model. An antioxidant activity landmark of different wine samples was selected based on the variable influence on projection (VIP), with a VIP value higher than one, namely superoxide anion radical scavenging abilities (Figure 5C). The results showed that the superoxide anion radical scavenging assay was an important antioxidant assay for discriminating the antioxidant activity of these wine samples.

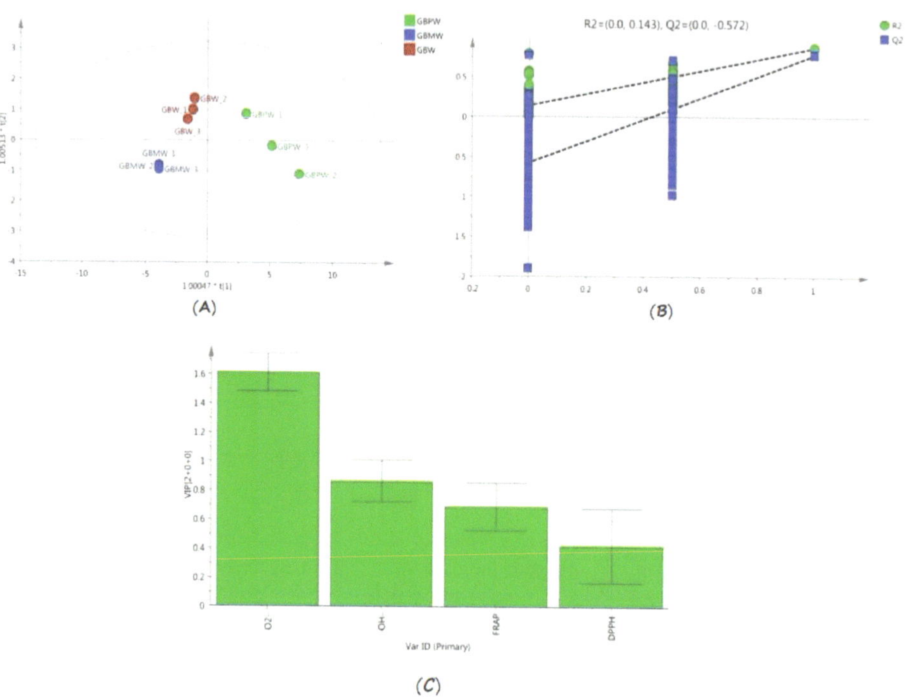

Figure 5. OPLS-DA model of antioxidant activities of green banana wine samples. (**A**) OPLS-DA score plot, (**B**) permutation tests for the OPLS-DA model, and (**C**) column plot of the variable average intensity of the four in vitro antioxidant activity assays.

3.6. Cell-Culture-Based Antioxidant Capacities and Cell Viability of 3T3-L1 Preadipocytes

The antioxidant capacities of the green banana wine samples were determined by an inhibition assay of hydrogen-peroxide-induced oxidative stress on 3T3-L1 preadipocytes. The antioxidant capacities of these wine samples are shown in Figure 6A. The antioxidant capacities of the wine samples were higher with the increase in sample concentrations. At sample concentrations from 0 to 25 mg/mL, the antioxidant capacities of the green banana wine samples were lesser than 50%. Only the GBPW at 25 mg/mL had an antioxidant capacity higher than 50%. At sample concentrations from 50 to 100 mg/mL, all the wine samples had higher antioxidant capacities than 50%, except for the GBW at a concentration of 50 mg/mL. Additionally, the antioxidant capacity of the GBW was significantly the lowest at a sample concentration of 100 mg/mL. The GBMW at lower concentrations had a significantly lower antioxidant capacity than the GBMW and GBW. At higher sample concentrations, the GBMW had an antioxidant capacity comparable to the GBPW. The results showed that the three wine samples could protect 3T3-L1 preadipocytes against oxidative stress induced by hydrogen peroxide. It could be because the GBPW, GBMW, and GBW contained polyphenolic and other compounds with moderate to high antioxidant activities, which could remove free radicals [38].

The protective effects of the three wine samples against hydrogen-peroxide-induced oxidation to 3T3-L1 preadipocytes are shown in Figure 6B. The IC_{50} value of the GBPW was significantly the lowest, followed by the GBW and GBMW. The highest IC_{50} value of the GBMW indicated that it had the least inhibitory effect on the growth of 3T3-L1 preadipocytes. The results showed that the wine produced from the green banana peel had the highest protective effect against cellular oxidative stress compared to the fruit pulp-based wines. Additionally, the in vivo antioxidant capacity of the GBPW was the highest at sample concentrations between 25 and 100 mg/mL. In this study, 3T3-L1 preadipocytes could be specifically induced to differentiate into mature adipocytes [39]. Therefore, it was best to represent a cellular system for an antioxidative assay.

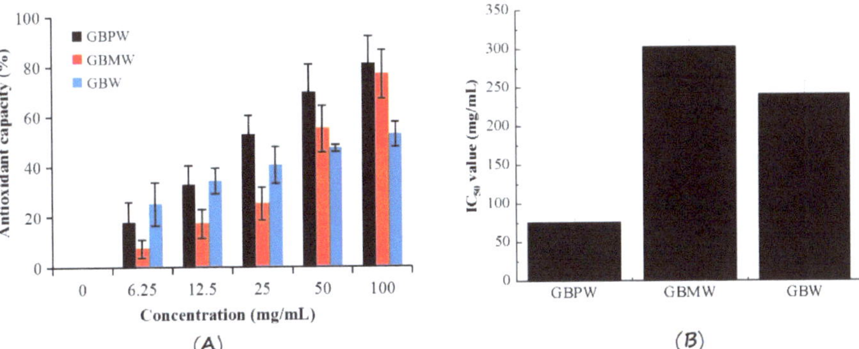

Figure 6. (**A**) Antioxidant capacities and (**B**) IC$_{50}$ values of green banana wine samples on cell viability of 3T3-L1 preadipocytes.

4. Discussion

Green bananas are rich in polyphenolic and other compounds with high antioxidant properties. Green banana waste, such as banana peels, has been used as a raw material to produce functional materials. Banana peels are prebiotics and biosubstrates [22]. They contain biosubstrates, such as pectin and cellulose. The polysaccharides extracted from banana peels can provide energy for microorganisms. Therefore, banana peels have been used to produce alcohol in the past. The use of banana peels in wine production enhanced the antioxidant activity of the fruit wine because the banana peel is high in antioxidants [20]. The ferric-reducing ability of the banana peel was also higher than the fruit pulp. The high antioxidant activities of the banana wine samples could be attributed to their high levels of bioactive antioxidants. Additionally, the findings of this study revealed that the GBPW had remarkably high phenolic compounds, and the antioxidant activities of the wine samples were also higher than the banana fruit samples. Moreover, the DPPH radical scavenging activity of the green banana peel reported in a previous study was remarkably higher than that of a yellow banana peel [40].

The quality of green banana wines was influenced by the other chemical properties in addition to their bioactive antioxidants. The GBPW had a remarkably low alcohol content compared with the other two wine samples. The results showed that wine fermentation using a banana peel had a low alcohol content in the wine sample. The possible reason could be that it had a carbohydrate content lower than the banana pulp. A previous study supported our findings that the acidity and TSS of the GBMW reduced gradually during the 30 days of fermentation [41]. The TSS of the GBPW was notably lower than that of the GBMW. The GBMW also had higher acidity than the GBPW. Therefore, the GBPW had a better quality than the GBMW. Additionally, the green banana wine samples had a lower alcohol content (<10%) compared with the alcohol content of ripe banana pulp wine (15.49%) reported in the literature [42]. It could be due to the ripe banana pulp having a higher sugar content. The alcohol produced during the banana wine fermentation could also be due to the fermentation time, where the banana wine produced from 14-day fermentation had 5% alcohol [42].

Different parts of bananas, such as the peel, pulp, and whole fruit, contain different percentages of phenolic compounds and flavor components. This study compared the fruit wines produced from three parts of bananas because banana pulp wine has a better flavor, but its antioxidant activity could be lower than banana peel wine. The use of the whole banana in wine production somehow improved its aroma and antioxidant activity. The banana peel wine (GBPW) had higher antioxidants than its pulp wine, but the whole banana fermented wine (GBW) had a lower IC_{50} value than the GBMW. The production of fruit wine using bananas with the peel attached maintained a high antioxidant level in addition to improving the wine quality. Therefore, the study on developing banana wines by applying different parts of green bananas warrants the wine quality. This study may also provide a theoretical basis for expanding the channel of banana industrialization, increasing the economic value of bananas, and reducing the waste of resources. The reason for using green bananas was that green banana pulp had higher phenolic content than ripe yellow bananas [43].

The aroma of the GBPW was unique because the main flavor component of the wine produced from banana peels was 3-methyl-1-butanol. The literature reported that isoamyl butyrate was the main volatile in banana peels, but the level of its isoamyl alcohol, such as 3-methyl-1-butanol, was low [44]. The high 3-methyl-1-butanol in the GBPW could be due to the conversion of isoamyl butyrate to isoamyl alcohol during fermentation. Isoamyl butyrate has an intense fruity odor, but the metabolism of this compound to isoamyl butanol during fermentation could change its aroma. This flavor substance has a fusel odor. These wine samples had a banana-like smell, which could be due to the butan-1-ol component. The wine odors of the GBMW and GBW were pleasant because these wine samples had floral and fruity aromas from other volatiles.

We identified 22 flavor substances in the GBMW. Most of these substances were fatty acid esters. This study was preliminary work on developing green banana wines using different parts of green bananas. A few other physicochemical parameters and analyses of the wine samples have not been included in this study. Although we have not performed a sensory evaluation for these wine samples, we noted that each green banana wine sample had a unique color and appearance. The GBPW had a yellowish brown hue, the GBMW had a milky yellowish appearance, and the GBW was a pale golden-yellow-colored liquid. Among the three wine samples, the GBPW had the most pleasant odor. It had a fresh and fruity smell and taste due to its higher percentages of 2,4-di-tert-butylphenol and 3-methyl-1-butanol. Only the banana-pulp-based fermented wines had a pungent smell due to the acetic acid in these wines.

The production of the GBMW is more costly because the banana pulp has a higher commercial value than its peel. The antioxidant quality of the GBMW was also lower than the GBPW, although the GBMW had a better aroma. Since this study did not cover sensory evaluation and some other in vitro bioassays for the wine samples such as the GBPW, they should be included in future studies. The GBPW is one of the potent functional alcoholic beverages with high antioxidants for promoting a healthy drinking lifestyle. The wine sample can be used as an alcoholic tonic drink to prevent metabolic diseases among the population with unhealthy lifestyles because salicylic acid is its key compound.

5. Conclusions

The chemical compositions and bioactive phenolics of the GBPW, GBMW, and GBW had remarkable changes during the first few days of fermentation. The changes in these values were minor during the following days until the end of fermentation. During the fermentation process, the TSS content of the GBPW, GBMW, and GBW showed reducing levels with increasing fermentation time, whereas the alcohol and TAC had increasing rates but with some exceptions. The alcohol and TAC of the GBW had a downward trend. The TPC and TTC of the GBMW and GBW were decreased during the fermentation. However, the changes were not notable after day 10 of fermentation. The GBPW had the highest TPC, TTC, and TFC compared to the GBMW and GBW. This shows that adding a banana peel to the sugar fermentation improved the quality of the fermented wine, such as by further reducing its alcohol and TSS content and increasing its total phenolic components a few times higher than the banana-pulp-based wine.

These wine samples had higher antioxidant activities than the different banana parts, except the wine samples had a lower metal chelation rate than the fruit samples. The OPLS-DA revealed that the antioxidant activities of the green banana wine samples were notably different, and the antioxidant activity assay that discriminated based on the antioxidant activities of these wine samples was the superoxide anion radical scavenging assay. The three fermented wine samples had a protective effect against oxidative stress induced by hydrogen peroxide in 3T3-L1 preadipocytes. The GBPW had a better inhibitory effect than the other two wine samples at the same concentrations. Based on these findings, adding different banana parts with sugar for wine fermentation improved the wine quality. The addition of the whole banana in wine fermentation is recommended because the full utilization of the banana peel reduces waste production, the banana peel is rich in antioxidants, and banana pulp has a more favorable aroma. Wine is good for preventing cardiovascular diseases among heavy drinkers. Future studies can consider performing a large-scale sensory evaluation of these green banana wines and their physicochemical characteristics.

Author Contributions: Conceptualization, Z.L., D.Y. and J.S.; data curation, C.Q., H.E.K. and D.L.; formal analysis, C.Q., G.L. and D.L.; funding acquisition, Z.L., X.H. and J.S.; investigation, B.C., L.L., Y.Y. and G.L.; methodology, J.L. and H.E.K.; project administration, Z.L., X.H. and J.S.; resources, X.H. and C.L.; software, J.T.; supervision, D.Y., J.T. and J.S.; validation, B.C., C.L. and H.E.K.; writing—original draft preparation, C.Q., J.L., Y.Y. and B.C.; writing—review and editing, L.L., H.E.K. and J.S.; All authors have read and agreed to the published version of the manuscript.

Funding: This research was funded by the Earmarked Fund of China Agriculture Research System of MOF and MARA (CARS-31) and the Foundation of Fundamental Research Project from the Guangxi Academy of Agricultural Sciences (Grant Number 2021YT116).

Institutional Review Board Statement: Not applicable.

Informed Consent Statement: Not applicable.

Data Availability Statement: All data are available in the article.

Acknowledgments: We thank the research assistants of the Guangxi Academy of Agricultural Sciences for helping with this study.

Conflicts of Interest: All authors declare no conflict of interest, except for Zhichun Li and Jian Sun, who received the funding.

References

1. Gupta, V.; Dave, S.; Parihar, N. Removal of copper (II) from aqueous solutions using chemically activated banana peels as an adsorbent. *Pollut. Res.* **2020**, *39*, 287–291.
2. Ince, M.; Ince, O.K.; Yonten, V.; Karaaslan, N.M. Nickel, lead, and cadmium removal using a low-cost adsorbent—Banana peel. *At. Spectrosc.* **2016**, *37*, 125–130. [CrossRef]
3. Karimibavani, B.; Sengul, A.B.; Asmatulu, E. Converting briquettes of orange and banana peels into carbonaceous materials for activated sustainable carbon and fuel sources. *Energy Ecol. Environ.* **2020**, *5*, 161–170. [CrossRef]
4. Patel, H.; Patel, A.; Surati, T.; Shah, G. Potential use of banana peels for the production of fermented products. *IJED* **2012**, *9*, 1–7.
5. Mostafa, H.S. Banana plant as a source of valuable antimicrobial compounds and its current applications in the food sector. *J. Food Sci.* **2021**, *86*, 3778–3797. [CrossRef]
6. Mitra, S.K.; Pathak, P.K.; Chakraborty, I. Potential underutilized tropical fruits of India. *Acta Hortic.* **2010**, *2010*, 61–68. [CrossRef]
7. Hernández-Nava, R.G.; De, J.; Berrios, J.; Pan, J.; Osorio-Díaz, P.; Bello-Perez, L.A. Development and characterization of spaghetti with high resistant starch content supplemented with banana starch. *Food Sci. Technol. Int.* **2009**, *15*, 73–78. [CrossRef]
8. Menezes, E.W.; Dan, M.C.T.; Cardenette, G.H.L.; Goñi, I.; Bello-Pérez, L.A.; Lajolo, F.M. In vitro colonic fermentation and glycemic response of different kinds of unripe banana flour. *Plant Foods Hum. Nutr.* **2010**, *65*, 379–385. [CrossRef]
9. Jiang, H.; Zhang, Y.; Hong, Y.; Bi, Y.; Gu, Z.; Cheng, L.; Li, Z.; Li, C. Digestibility and changes to structural characteristics of green banana starch during in vitro digestion. *Food Hydrocoll.* **2015**, *49*, 192–199. [CrossRef]
10. Fasolin, L.H.; de Almeida, G.C.; Castanho, P.S.; Netto-Oliveira, E.R. Cookies produced with banana meal: Chemical, physical and sensorial evaluation. *Food Sci. Technol.* **2007**, *27*, 524–529. [CrossRef]
11. Ramli, S.; Alkarkhi, A.F.M.; Yeoh, S.Y.; Easa, A.M. Utilization of banana peel as a functional ingredient in yellow noodle. *Asian J. Food Agro-Ind.* **2009**, *2*, 321–329.

12. Rebello, L.P.G.; Ramos, A.M.; Pertuzatti, P.B.; Barcia, M.T.; Castillo-Muñoz, N.; Hermosin-Gutierrez, I. Flour of banana (*Musa* AAA) peel as a source of antioxidant phenolic compounds. *Food Res. Int.* **2014**, *55*, 397–403. [CrossRef]
13. Phirom-on, K.; Apiraksakorn, J. Development of cellulose-based prebiotic fiber from banana peel by enzymatic hydrolysis. *Food Biosci.* **2021**, *41*, 101083. [CrossRef]
14. Dmochowska, A.; Czajkowska, J.; Jędrzejewski, R.; Stawiński, W.; Migdał, P.; Fiedot-Toboła, M. Pectin based banana peel extract as a stabilizing agent in zinc oxide nanoparticles synthesis. *Int. J. Biol. Macromol.* **2020**, *165*, 1581–1592. [CrossRef]
15. Wu, T.R.; Wang, H.L.; Jiang, S.W.; Liu, D.D.; Wei, F. Optimization of extraction of tannins from banana peel using response surface methodology. *Appl. Mech. Mater.* **2014**, *678*, 566–571. [CrossRef]
16. Kamal, A.M.; Taha, M.S.; Mousa, A.M. The radioprotective and anticancer effects of banana peels extract on male mice. *J. Food Nutr. Res.* **2019**, *7*, 827–835. [CrossRef]
17. Rita, W.S.; Swantara, I.M.D.; Asih, I.A.R.A.; Puspawati, N.M. Antibacterial activity and antioxidant capacity of selected local banana peel (*Musa* sp.) methanol extracts cultivated in Bali. *Int. J. Agric. Environ. Biotechnol.* **2020**, *5*, 242–251. [CrossRef]
18. Chaudhry, F.; Ahmad, M.L.; Hayat, Z.; Ranjha, M.M.A.N.; Chaudhry, K.; Elboughdiri, N.; Asmari, N.; Uddin, J. Extraction and evaluation of the antimicrobial activity of polyphenols from banana peels employing different extraction techniques. *Separations* **2022**, *9*, 165. [CrossRef]
19. Islam, M.; Haque, A.R.; Kabir, M.; Hasan, M.; Khushe, K.J.; Hasan, S.M. Fruit by-products: The potential natural sources of antioxidants and α-glucosidase inhibitors. *J. Food Sci. Technol.* **2020**, *58*, 1715–1726. [CrossRef]
20. González-Montelongo, R.; Lobo, M.G.; González, M. Antioxidant activity in banana peel extracts: Testing extraction conditions and related bioactive compounds. *Food Chem.* **2010**, *119*, 1030–1039. [CrossRef]
21. Khoozani, A.A.; Birch, J.; Bekhit, A.E.D.A. Production, application and health effects of banana pulp and peel flour in the food industry. *J. Food Sci. Technol.* **2019**, *56*, 548–559. [CrossRef] [PubMed]
22. Vu, H.T.; Scarlett, C.J.; Vuong, Q.V. Phenolic compounds within banana peel and their potential uses: A review. *J. Funct. Foods* **2018**, *40*, 238–248. [CrossRef]
23. Durgadevi, P.K.S.; Saravanan, A.; Uma, S. Antioxidant potential and antitumour activities of Nendran banana peels in breast cancer cell line. *Indian J. Pharm. Sci.* **2019**, *81*, 464–473. [CrossRef]
24. Li, L.; Sun, J.; Gao, H.; Shen, Y.; Li, C.; Yi, P.; He, X.; Ling, D.; Sheng, J.; Li, J.; et al. Effects of polysaccharide-based edible coatings on quality and antioxidant enzyme system of strawberry during cold storage. *Int. J. Polym. Sci.* **2017**, *2017*, 9746174. [CrossRef]
25. Liu, G.; Sun, J.; He, X.; Tang, Y.; Li, J.; Ling, D.; Li, C.; Li, L.; Zheng, F.; Sheng, J.; et al. Fermentation process optimization and chemical constituent analysis on longan (*Dimocarpus longan* Lour.) wine. *Food Chem.* **2018**, *256*, 268–279. [CrossRef]
26. Du, S.B.; Chen, X.W.; Dong, Z.Q.; Xie, H.C. Analysis of chemical components in alcohol extract of *Rosa sertata* × *Rosa rugosa* based on UPLC-Q-TOF-MS. *Nat. Prod. Res. Dev.* **2022**, *34*, 1129–1142. [CrossRef]
27. Chen, G.; Wu, F.; Pei, F.; Cheng, S.; Muinde, B.; Hu, Q.; Zhao, L. Volatile components of white *Hypsizygus marmoreus* detected by electronic nose and HS-SPME-GC-MS: Influence of four drying methods. *Int. J. Food Prop.* **2017**, *20*, 2901–2910. [CrossRef]
28. Shi, J.; Tong, G.; Yang, Q.; Huang, M.; Ye, H.; Liu, Y.; Wu, J.; Zhang, J.; Sun, X.; Zhao, D. Characterization of key aroma compounds in tartary buckwheat (*Fagopyrum tataricum* Gaertn.) by means of sensory-directed flavor analysis. *J. Agric. Food Chem.* **2021**, *69*, 11361–11371. [CrossRef]
29. Sulaiman, S.F.; Yusoff, N.A.M.; Eldeen, I.M.; Seow, E.M.; Sajak, A.A.B.; Ooi, K.L. Correlation between total phenolic and mineral contents with antioxidant activity of eight Malaysian bananas (*Musa* sp.). *J. Food Compos. Anal.* **2011**, *24*, 1–10. [CrossRef]
30. Luna-Vital, D.A.; De Mejía, E.G.; Mendoza, S.; Loarca-Piña, G. Peptides present in the non-digestible fraction of common beans (*Phaseolus vulgaris* L.) inhibit the angiotensin-I converting enzyme by interacting with its catalytic cavity independent of their antioxidant capacity. *Food Funct.* **2015**, *6*, 1470–1479. [CrossRef]
31. Kim, D.; Han, G.D. Ameliorating effects of fermented rice bran extract on oxidative stress induced by high glucose and hydrogen peroxide in 3T3-L1 adipocytes. *Plant Foods Hum. Nutr.* **2011**, *66*, 285–290. [CrossRef]
32. Girard-Lalancette, K.; Pichette, A.; Legault, J. Sensitive cell-based assay using DCFH oxidation for the determination of pro-and antioxidant properties of compounds and mixtures: Analysis of fruit and vegetable juices. *Food Chem.* **2009**, *115*, 720–726. [CrossRef]
33. Escudero-López, B.; Cerrillo, I.; Herrero-Martín, G.; Hornero-Méndez, D.; Gil-Izquierdo, A.; Medina, S.; Ferreres, F.; Berná, G.; Martín, F.; Fernández-Pachón, M.-S. Fermented orange juice: Source of higher carotenoid and flavanone contents. *J. Agric. Food Chem.* **2013**, *61*, 8773–8782. [CrossRef]
34. Fatemeh, S.R.; Saifullah, R.; Abbas, F.M.A.; Azhar, M.E. Total phenolics, flavonoids and antioxidant activity of banana pulp and peel flours: Influence of variety and stage of ripeness. *Int. Food Res. J.* **2012**, *19*, 1041–1046.
35. Debashis, M.; Vanlalruatfeli; Chandra, S.A. Shelf life got extended in salicylic acid treated and waxed Cavendish banana. *Res. Crop.* **2019**, *20*, 712–718. [CrossRef]
36. Hur, S.J.; Lee, S.Y.; Kim, Y.C.; Choi, I.; Kim, G.B. Effect of fermentation on the antioxidant activity in plant-based foods. *Food Chem.* **2014**, *160*, 346–356. [CrossRef]
37. Jung, H.; Lee, S.J.; Lim, J.H.; Kim, B.K.; Park, K.J. Chemical and sensory profiles of *makgeolli*, Korean commercial rice wine, from descriptive, chemical, and volatile compound analyses. *Food Chem.* **2014**, *152*, 624–632. [CrossRef]
38. Khoo, H.E.; Azlan, A.; Ismail, A.; Abas, F.; Hamid, M. Inhibition of oxidative stress and lipid peroxidation by anthocyanins from defatted *Canarium odontophyllum* pericarp and peel using in vitro bioassays. *PLoS ONE* **2014**, *9*, e81447. [CrossRef]

39. Poulos, S.P.; Dodson, M.V.; Hausman, G.J. Cell line models for differentiation: Preadipocytes and adipocytes. *Exp. Biol. Med.* **2010**, *235*, 1185–1193. [CrossRef]
40. Mokbel, M.S.; Hashinaga, F. Antibacterial and antioxidant activities of banana (*Musa*, AAA cv. Cavendish) fruits peel. *Am. J. Biochem. Biotechnol.* **2005**, *1*, 125–131. [CrossRef]
41. Gavimath, C.C.; Kalsekar, D.P.; Raorane, C.J.; Kulkarni, S.M.; Gavade, B.G.; Ravishankar, B.E.; Hooli, V.R. Comparative analysis of wine from different fruits. *Int. J. Adv. Biotechnol. Res.* **2012**, *3*, 810–813.
42. Akubor, P.I.; Obio, S.O.; Nwadomere, K.A.; Obiomah, E. Production and quality evaluation of banana wine. *Plant Foods Hum Nutr.* **2003**, *58*, 1–6. [CrossRef]
43. Mura, K.; Tanimura, W. Change of polyphenol compounds in banana pulp during ripening. *Food Preserv. Sci.* **2003**, *29*, 347–351. [CrossRef]
44. Ji, L.; Srzednicki, G. Extraction of Aromatic Compounds from Banana Peels. In Proceedings of the II Southeast Asia Symposium on Quality Management in Postharvest Systems, Vientiane, Laos, 4–6 December 2013; Acedo, A., Jr., Kanlayanarat, S., Eds.; ISHS Acta Horticulturae: White River, South Africa, 2015; Volume 1088, pp. 541–546. [CrossRef]

antioxidants

MDPI

Review

Integrated Technology for Cereal Bran Valorization: Perspectives for a Sustainable Industrial Approach

Silvia Amalia Nemes [1], Lavinia Florina Călinoiu [2], Francisc Vasile Dulf [3], Anca Corina Fărcas [2] and Dan Cristian Vodnar [2,*]

[1] Institute of Life Sciences, University of Agricultural Sciences and Veterinary Medicine Cluj-Napoca, Manastur 3-5, 400372 Cluj-Napoca, Romania
[2] Faculty of Food Science and Technology, University of Agricultural Sciences and Veterinary Medicine Cluj-Napoca, Manastur 3-5, 400372 Cluj-Napoca, Romania
[3] Faculty of Agriculture, University of Agricultural Sciences and Veterinary Medicine Cluj-Napoca, Manastur 3-5, 400372 Cluj-Napoca, Romania
* Correspondence: dan.vodnar@usamvcluj.ro

Abstract: Current research focuses on improving the bioaccessibility of functional components bound to cereal bran cell walls. The main bioactive components in cereal bran that have major biological activities include phenolic acids, biopeptides, dietary fiber, and novel carbohydrates. Because of the bound form in which these bioactive compounds exist in the bran matrix, their bioaccessibility is limited. This paper aims to comprehensively analyze the functionality of an integrated technology comprising pretreatment techniques applied to bran substrate followed by fermentation bioprocesses to improve the bioaccessibility and bioavailability of the functional components. The integrated technology of specific physical, chemical, and biological pretreatments coupled with fermentation strategies applied to cereal bran previously-pretreated substrate provide a theoretical basis for the high-value utilization of cereal bran and the development of related functional foods and drugs.

Keywords: cereal bran; bioactive compounds; pretreatments; fermentation; health benefits

check for updates

Citation: Nemes, S.A.; Călinoiu, L.F.; Dulf, F.V.; Fărcas, A.C.; Vodnar, D.C. Integrated Technology for Cereal Bran Valorization: Perspectives for a Sustainable Industrial Approach. *Antioxidants* **2022**, *11*, 2159. https://doi.org/10.3390/antiox11112159

Academic Editor: Stanley Omaye

Received: 4 October 2022
Accepted: 29 October 2022
Published: 31 October 2022

Publisher's Note: MDPI stays neutral with regard to jurisdictional claims in published maps and institutional affiliations.

1. Introduction

Cereals are the world's most essential source of nutrients and energy for humans. According to the last report of the FAO (Food and Agriculture Organization of the United Nations), in 2018, the world produced 2.7 billion tonnes of grains, an increase of 18% from 2008. Furthermore, cereal grain output increased faster than the global population. The world's most widely produced agricultural grains are wheat, rice, oat, corn, sorghum, barley, and rye. However, over one-third (1.3 billion tonnes) of the world's food production for human consumption is wasted each year, particularly in the case of cereals, accounting for 30% (286 million tonnes) of total output [1].

At the same time, cereal by-products, such as cereal bran, are frequently devalued and utilized in animal feed, perceived as waste. However, the chemical composition of cereal brans is complex, and their various health benefits could be realized by incorporating them into the human daily diet. In addition to usual nutrients such as proteins, vitamins, minerals, and fats, cereal bran also contains a variety of bioactive compounds such as dietary fiber, phytosterols, biopeptides, novel carbohydrates, and polyphenols, such as phenolic acids and flavonoids [2]. The bioactive compounds have been demonstrated to exhibit a variety of biological functions, including anti-inflammatory, anticancer, antibacterial, and cardiometabolic protective effects [3]. In addition, they are linked to a reduced risk of chronic illnesses, such as chronic gut inflammation [4], obesity [5], cardiovascular disease [6], and cancer [7].

The mechanisms by which cereal bran confers health benefits are still to be fully established, in light of issues related to the digestibility, bioaccessibility, and bioavailability of compounds. Considering these limitations, a priority strategy targets the release of bound phenolic compounds, fiber, and minerals solubilization in situ before consumption, followed by their incorporation in fiber-rich functional food products. The food industry has been attempting to better exploit these by-products. Still, many improvements need to be made to pretreatment techniques to make the accessibility and solubility of the most interesting components valuable, as well as in the fortification into more nutritional and bioactive matrices to ensure sustainability.

Physical, chemical, enzymatic pretreatments, and bioprocesses may free the bound phytochemicals, leading to improved bioavailability and bioaccessibility for the bioactive compounds. The pretreatment method should be selected based on its techno-economic feasibility of integration into an efficient industrial process and its ability to minimize pollution levels in the environment. Therefore, modern and green pretreatments, such as ultrafine grinding, ultrasounds, microwave, steam explosion, and thermal and moisture pretreatments, have gained attention, being economically and environmentally attractive methods [8]. Thus, their optimization for specific matrices or integration for higher efficiency is in the early stage. Particular pretreatments can improve the digestibility of lignocellulosic biomass, with the expected main effects being hemicellulose dissolution and lignin structure modification, resulting in enhanced cellulose accessibility for further microbial enzymatic degradation and, ultimately, the production of new value-added products [9].

Microbial fermentation of bran substrates is an efficient, environmentally friendly bioprocess for increasing the release of bioactive compounds. The microorganisms involved in the bioprocess generate new compounds throughout fermentation. Agro-industrial residues are generally considered the best substrates for the fermentation processes [10,11]. Due to consumers' desire for healthier diets, interest in naturally fortified foods, functional foods, and functional food additives is booming in the marketplace. Cereal bran, which includes wheat bran (WB), oat bran (OB), and rice bran (RB), is one of the most common, cheap, and desired ingredients for enhancing daily nutritional intake through food and supplement consumption [12].

In this respect, the present review aims to focus on the most promising new research progress in the development of an integrated system (Figure 1) consisting of physical, chemical, or biological pretreatments able to enhance the bioaccessibility of cereal bran phytochemicals before the bioprocesses (submerged fermentation and solid-state fermentation), resulting in a complex fortified grain matrix with valuable nutrients and bioactive compounds. The novelty of the proposed article is the approach of complete re-utilization of cereal bran via integrated pretreatments and in situ fortifications.

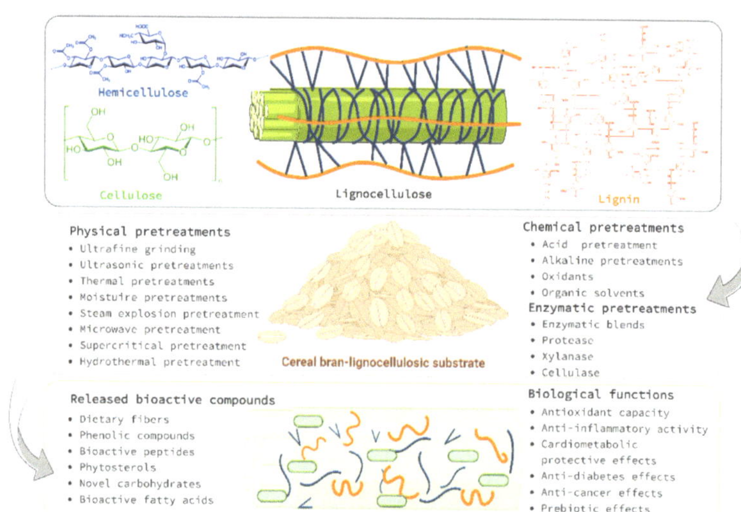

Figure 1. Integrated system to enhance the bioaccessibility of cereal bran phytochemicals.

2. Intracellular Bioactive Phytochemicals of Cereal Bran

The chemical composition (Table 1) of non-wood agro-industrial residues, including cereal bran, which is composed of three layers, pericarp, testa, and aleurone, is dominated by macromolecular components, including cellulose (10.5–23.1%), hemicellulose (26.1–36.2%), and lignin (2.2–12.5%) [13]. Lower quantities of inorganic compounds and extractable components, such as phenolic compounds (0.7–2.7%), bioactive peptides, pectin, and lipids are also present (1.96–22%) [14]. Phytochemicals are bioactive compounds with reduced nutritional value found naturally in fruits, vegetables, and whole grains. Between 5000 and 25,000 distinct phytochemicals exist in these foods [15]. The nutritional compositions of crude WB, RB, and OB per 100 g are presented in Table 1.

Table 1. Cereal bran nutritional composition.

Nutrients	Wheat Bran [16]	Rice Bran [17]	Oat Bran [18]
	Amount (g/100 g)	Amount (g/100 g)	Amount (g/100 g)
Water	9.89	6.13	6.55
Energy	216 kcal	316 kcal	246
Protein	15.6	13.4	17.3
Total lipids	4.25	20.8	7.03
Polyunsaturated fatty acids	2.21	7.46	2.77
Monounsaturated fatty acids	0.63	7.55	2.38
Ash	5.79	9.98	2.89
Carbohydrates	64.5	49.7	66.2
Fibers	42.8	21	15.4
Total sugars	0.41	0.9	1.45

2.1. Dietary Fibers

Non-starch polysaccharides, such as cellulose, hemicellulose, pentosans, gums, and glucans, found in cereal grain cell walls are a significant source of dietary fiber in the human diet. According to Wen Cheng and colleagues, dietary fiber is defined as 'edible parts of plants or analogous carbohydrates resistant to digestion and absorption in the human small intestine with complete or partial fermentation in the large intestine' [19]. Dietary fibers mostly consist of polysaccharides, oligosaccharides, and lignin fractions [19]. Dietary fibers are divided into soluble (primarily pentosans, pectin, gums, and mucilages) and insoluble (cellulose, some hemicellulose, and lignin) classes based on their water solubility [20]. The soluble fibers create a viscous polymer network in the presence of water, while the insoluble fibers remain impenetrable due to their structural rigidity. Cellulose is composed of cellobiose unbranched chains of D-anhydroglucose units β-(1,4)-glycosidic bonded embedded in a matrix of hemicellulose and lignin. Lignin is a component responsible for the rigidity and impermeability of cell walls and the inhibitory effect on cellulase activity. Lignin degradation through specific pretreatments is advantageous because it allows enzymes easier access to compounds of interest [21]. Furthermore, lignin provides a physical barrier to cellulose, and its presence leads to the formation of toxic compounds for fermentative microorganisms in hydrolysis. The cellulose and hemicellulose hydrolysis reaction produces fermentable sugars, which are required for the subsequent fermentation processes. The hard conditions in which cereal brans are pretreated generate sugar breakdown and dehydration of the substrate, resulting in the release of inhibitory substances such as (2-furaldehyde), 5-hydroxy-2-methylfurfural (5-HMF), and organic acids. The presence of β-glucosidases can reduce the inhibitory effect of lignin via the bonds that form between them, boosting the overall efficiency of the bioprocess [21].

Hemicellulose accounts for 20–35% of cereal biomass, is the second most prevalent cereal biomass, and has a significant value in various fields, including polymeric materials, foods, energy, and medicine. Compared to lignin, which is an amorphous polyphenol polymer, hemicellulose is a heterogeneous group of polysaccharides constituted by β-(1→4)-linked backbones of sugars, such as pentoses (xylose and arabinose) and hexoses (glucose, galactose, mannose, rhamnose, and glucuronic and galactouronic acids) [13]. The hemicelluloses in the bran cell wall are closely related to cellulose and lignin since their primary function is to act as a filler between the cellulose and lignin fractions. Furthermore, hemicellulose is more abundant than cellulose in cereal brans, making it the primary insoluble dietary fiber in cereal brans [13]. Arabinoxylans and glucans, particularly β-glucan, are the most studied hemicellulose groups in cereal brans.

Cereal bran pentosans, also known as arabinoxylans, are the predominant hemicellulose in the bran cell wall and influence the water distribution, starch retrogradation, and rheological properties of fermented grain-based food products [22]. Arabinoxylans consists of a linear chain of β-(1–4)-linked D-xylopyranosyl to which α-L-arabinofuranosyl substituents are attached [23]. The fraction of un-, mono- and di-substituted xylose and arabinose residues attached to the chain determines the variety of these molecules. The capacity of the arabinoxylan molecule to interact with other cell wall chemicals is also determined by its complexity, impacting the physicochemical and functional characteristics of the macromolecules [24]. Arabinose residues, which substitute the arabinoxylan molecule, are often composed of phenolic acids such as ferulic, caffeic, and p-coumaric acid. Arabinoxylan is a biomolecule with high antioxidant activity, excellent free radical scavenging capacity, reducing ability, and metal-chelating activity due to its structural complexity. Therefore, it can be well exploited in the food sector as a natural antioxidant [23].

Glucans are widespread D-glucopyranosyl polysaccharides present in cereals bran, including oat bran, wheat bran, and rice bran. The anomeric structure, the position and arrangement of the glycosidic linkages, the molecular size, and the type and degree of branching all differentiate glucans. Except for glucose residues connected by β-1, 3-linkages, known as β-glucans, most glucans have a reduced effect on the human immune system. According to the literature, glucans do not interact with the human immune system. The only interactions that can be observed are due to some glucose residues, found under the name of β-glucans [13]. Whereas β-(1,3) glycosidic connections restrict molecules from packing tightly, they give the β-glucan molecule an irregular shape, flexibility, and partial solubility in water [13,25]. Oat β-glucans provide beneficial health advantages, according to the European Food Safety Authority and the United States Food and Drug Administration, including immunomodulatory effects, the ability to reduce fasting blood cholesterol, and the ability to attenuate post-prandial glycemic reactions [25].

2.2. Phenolic Compounds

Phenolic acids are the primary compounds in cereal bran. There are three types of phenolic compounds and phenolic acids identified mainly in the three layers of bran (pericarp, testa, and aleurone): soluble free, soluble conjugated (glycosides), and insoluble bound states [26]. Most of the phenolic compounds from cereal bran are bonded in insoluble forms (Figure 2), except in the case of rice bran, which contains more free varieties of phenolic acids [2,15,27]. Bound phenolics may only be liberated by acid (H_2SO_4, HNO_3, H_3PO_4, and HCl), alkaline (potassium hydroxide, sodium hydroxide), or enzymatic hydrolysis by applying commercial enzymes (endoxylanase, exoxylanase, or β-glucosidase) or via fermentation bioprocesses [28].

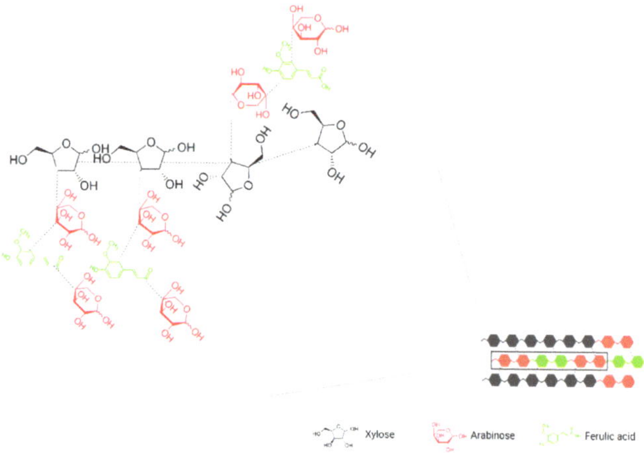

Figure 2. Cell-wall bound ferulic acid in cereal bran.

The two major classes of phenolic acids are hydroxybenzoic acids and hydroxycinnamic acids. Hydroxybenzoic acids are characterized by a C6-C1 structure and include gallic, vanillic, p-hydroxybenzoic, protocatechuic, syringic, and salicylic acids. In contrast, hydroxycinnamic acids, such as caffeic, ferulic, p-coumaric, chlorogenic, and synaptic acids, have a C6-C3 structure [29]. Ferulic acid accounts for most total polyphenols in wheat (90%) and oats (75%). Ferulic acid is commonly found in plants and results from phenylalanine and tyrosine metabolism. It is frequently seen as free ferulic acid, γ-oryzanol, monoesters containing ferulic acid, and several triterpene alcohols [30]. Significant concentrations of vanillic acid, syringic acid, salicylic acid, caffeic acid, and p-coumaric acid have also been detected in WB and OB; dihydroxybenzoic acids being the majority phenolic acids in OB [29]. Bran cell walls contain various complex structures, such as hydrolyzable tannins and lignins, that incorporate hydroxybenzoic and hydroxycinnamic acid derivatives [31]. The cell wall phenolics form ester linkages with structural carbohydrates, such as arabinoxylans, and proteins through their carboxylic group and ether linkages with lignin through their hydroxyl groups [15]. Ferulic acid and p-coumaric acid, the primary polyphenols in wheat bran, are found in the pericarp and aleurone of the grain. These chemicals are linked to the arabinoxylans in the grain layers through ester bonds strengthening the kernel's structural integrity while also providing antibacterial, antifungal, and antioxidant protection. At the same time, some phenolic compounds are caught inside the bran matrix, forming hydrogen or hydrophobic bonds, significantly decreasing bioaccessibility and physiological effects. Moreover, the release and absorption of these compounds, in the form of digestion, encounter many difficulties [32].

Therefore, breaking the bran cell wall is critical for accelerating the release of phenolic compounds and enhancing phenolic bioavailability. The release of insoluble forms of phenolics can be increased by performing specific pretreatments on lignocellulosic cereal bran, such as physical, chemical, and biological/microbial pretreatment.

2.3. Bioactive Peptides

Proteins are well-known for their daily diet significance. Proteins, according to recent studies, are a rich source of biologically active peptides, such as Tyr-Ser-Lys, SSYYPFK, LQAFEPLR, and EFLLAGNNK [33–35]. Bioactive peptides are specific protein fragments formed by short-chain amino acids with health benefits in the human body beyond their essential nutritional value [36]. The positive effects of bioactive peptides include antihypertensive effects [37], antioxidant effects [38], antimicrobial effects [39], hypoglycemic effects [40], and immunomodulatory activity [41]. Compared to protein macromolecules, bioactive peptides have a simplified form and improved stability, while their immunogenicity is low or absent [40]. The presence of bioactive peptides has been noticed in many sources, such as cereals (oat, rice, and wheat), pseudocereals (amaranth, quinoa, and buckwheat), and legumes (soy, pea, and beans) [36]. Oats are grown and consumed worldwide, and their high protein content makes them an excellent source of bioactive peptides (12–24%). Globulins constitute the majority of bioactive protein and have the highest concentration of bioactive peptides, such as SSYYPFK from oat globulin hydrolysates, which can reduce systolic and diastolic blood pressure [34], and Tyr-Ser-Lys from RB powder, which had an antihypertensive effect [33,42]. Feng Wang and colleagues have identified 22 peptides of oat globulin from the tryptic hydrolysates [35]. Oat bioactive peptides provide the nutritional quality and positive impacts mentioned above due to their amino acid composition. Glutamic acid constitutes the majority of non-essential amino acids in oats, accounting for 25% of total non-essential amino acids, followed by aspartic acid, which accounts for 8% [33]. Lunasin is a novel 43-amino-acid biopeptide found originally in soybean cotyledon and later in cereals including rice and wheat bran. Lunasin has chemoprevention, anti-inflammatory, and antioxidant properties, and performs in vivo and in vitro obesity immunomodulation [5]. Tadashi Hatanaka and colleagues also show that bioactive peptides derived from rice and rice bran have anti-diabetic and antioxidant properties. Rice hydrolysates and rice bran were reported to exhibit modest antioxidant activity using the oxygen radical absorbance capacity assay with glutathione as the reference agent [43]. Ile-Pro (1.22 µg/mg), Met-Pro (0.23 µg/mg), Val-Pro (1.59 µg/mg), and Leu-Pro (1.94 µg/mg) dipeptides identified in the rice protein hydrolysate provide inhibitory activity against dipeptidylpeptidase-IV, a necessary preventive measure for type 2 diabetes [43].

2.4. Phytosterols

Phytosterols are bioactive molecules with a structure similar to cholesterol. Phytosterols are naturally found in plants whose central backbone is perhydrocyclopentanophenanthrene. In contrast to cholesterol, phytosterols are only absorbed through the daily diet via intestinal absorption because they cannot be synthesized endogenously in the human body [44]. Phytosterol-rich fractions could be used to increase the consumption of health-promoting compounds from natural sources in the cereal diet. The primary function of phytosterols in the human diet is to minimize the risk of hypercholesterolemia and cardiovascular disease. Clinical research has indicated that consuming 2 g of phytosterols per day is related to an 8–10% reduction in low-density lipoprotein cholesterol levels, lowering the risk of cardiovascular disease [45]. Some phytosterols are present in grains (rice, corn, wheat, and rye) as esters of ferulic acids, such as sterile ferulates. The distribution and composition of bioactive compounds in cereal grain varies within the whole grain, and phytosterols and steryl ferulates are usually concentrated in the bran layers. The most abundant phytosterols found in wheat are sitostanol, campesterol, β-sitosterol, and campestanol [45]. The results obtained in a study conducted by Nurmi Tanja and colleagues confirm that the phytosterols are concentrated in the WB fraction of wheat grain [46]. Moreover, the esterification of phytosterols with ferulic acid in WB can directly impact ferulic acid accumulation in the bran [46]. Rice, particularly RB and RB oil, are also important natural sources of phytosterols. Jiali Zhang and colleagues performed a study to investigate the relationships between the forms of RB and RB oil phytosterols and their bioactivities. Free sterols and sterol conjugates are the two primary structural forms of phytosterols. The structure of sterols is made up of four rings that form the cyclopentane-perhydrophenanthrene cycle that includes a hydroxyl group in the C3 position and a side chain of 8–10 carbon atoms attached to carbon 17 [47]. Free sterols are classified into 4-desmethyl sterols, 4α-monomethyl sterols, and 4, 4-dimethyl sterols, and conjugated sterols are fatty acid steryl esters, hydroxycinnamate sterol esters, steryl glycosides, and acyl steryl glycosides [48]. The conjugated form of sterile esters is characterized by the esterification of the hydroxyl group in the C3 position with a fatty acid. The conjugated form of sterile glycosides also includes the binding of a sugar molecule through a β-glycosidic bond to the C3 hydroxyl group in the sterol structure. Conjugated forms of acylsteryl glycosides include the esterification of the sugar fragment with the hydroxyl group in the C6 position. Moreover, sterol ferulates are a less frequent form of conjugated sterols, which is mostly present in grains [47]. Free sterols are cholesterol-like compounds that have received much interest for their health-promoting properties, such as reducing cholesterol or antioxidant effects. Different phytosterols have various bioactivities; for example, 4-desmethylsteryl ferulates have better antioxidant activity than 4-dimethylsteryl ferulates Field [48]. The structural form of phytosterols directly influences their bioactivity. Therefore, a deep understanding of structural models is required to evaluate phytosterols' functionality.

3. Biological Activities in the Human Body

As stated above, whole grains are rich in polysaccharide compounds, lipids, proteins, and phytochemicals. Regular intake of whole grains or the addition of bran in the human daily diet has been confirmed in many studies to lower the risk of chronic diseases such as cancer, cardiovascular disease, and diabetes. The biological activities (antioxidant capacity, anti-inflammatory activity, cardiometabolic protective activity, anti-diabetes, anticancer, and prebiotic effects) of phytochemicals from WB, RB, and OB are discussed in this section, focusing on chronic illness prevention.

3.1. Antioxidant Capacity

The body's antioxidant defense system can be unbalanced by an excess of free radicals that cause oxidative damage to intracellular biomacromolecules and induce oxidative effects on cells and tissues. Antioxidant substances protect molecules of high biological function, such as DNA, proteins, and lipids, from free radical damage, significantly reducing the risk of chronic diseases [49]. Polyphenols are the most researched compounds in cereal bran with remarkable antioxidant properties. The primary polyphenols having at least one aromatic ring in WB are hydroxycinnamic acid, ferulic acid, synaptic acid, and p-coumaric acid. P-hydroxybenzoic acid, vanillic acid, syringic acid, and gallic acid are the primary derivatives of hydroxybenzoic acid, another polyphenol in WB [50].

The antioxidant activity of phenolic acids in cereal bran can be linked to electron donation and the transfer of hydrogen atoms to free radicals that initiate oxidation processes in the human body responsible for the modification of cellular signaling processes, reduction of cellular destruction or cellular death, and the high incidence risk of chronic diseases [50]. The effectiveness of phenolic compounds in reaching target areas and performing any protective effect is determined by their bioavailability and bioaccessibility. Soluble polyphenols are absorbed into the systemic circulation in the stomach and small intestine, while bound polyphenols reach the colon without significant modification. Under the action of intestinal microbiota and specific enzymes, bound polyphenols can be released in the colon, producing bioavailable phenolic metabolites with potential health benefits, such as m-coumaric, hippuric acids, eriodictyol, 3′,4′,5′-trihydroxyphenylacetic acid, and 3′,5′-dihydroxyphenylacetic acid [29,51]. Chlorogenic acid, for example, was converted into caffeic acid during the first stage of microbial biotransformation (dehydroxylation, dehydrogenation, or ester hydrolysis), and the primary metabolites found included di- and mono-hydroxylated phenylpropionic acids, m-coumaric, and hippuric acid [51]. Recent studies have found that the concentration of phenolic compounds such as phenolic acids, flavonoids, and anthocyanins was positively associated with antioxidant activity [32,52]. Li et al. confirmed that particle size reduction increased the release of phenolic compounds of superfine WB, mainly ferulic acid and that the bioaccessibility of the phenolics increased to 65.51% compared to WB without physical treatment [32]. In addition, following the release of the bound phenolics from extruded rice bran by fungal fermentation, ORAC, and CAA antioxidant activity increased by 1.8 and 4.1, respectively [52]. Moreover, samples with the highest concentrations of phenolic acids, flavonoids, anthocyanins, and total phenolic content extracted with 65% (v/v) ethanol had the highest antioxidant activity based on DPPH radical scavenging activity and ferric reducing antioxidant power (FRAP) assays on two varieties of rice bran [53]. However, the health benefits of phenolic-rich substrates are directly influenced by the bioavailability of the phenolic acids.

The antioxidant potential of cereal fractions is frequently linked to the amount of ferulic acid present. Although bran contains the most ferulic acid, its bioavailability, and intestinal release are limited due to its structural arrangement in the bran matrix, where it is covalently bonded to indigestible cell wall polysaccharides [54]. Yang Li and colleagues investigated the phenolic release from WB during in vitro digestion and the influence of particle size on polyphenols release and antioxidant activity. By strongly decreasing the WB particle size followed by in vitro digestion, the primary phenolic chemicals produced, ferulic acid and p-coumaric acid, increased the capacity of superfine WB to scavenge 2,2-diphenyl-1-picrylhydrazyl radicals [32]. Oxygen radical absorption capacity and cellular antioxidant activity of free and bound phenolics increased significantly after the RB fermentation mediated by Aspergillus oryzae.

Moreover, in vitro gastrointestinal digestion of fungal fermented RB underlines the increasing phenolic content bioaccessibility during the fermentation bioprocess from 347.04 mg GAE/100 g to 567.16 mg GAE/100 g [52]. In white and dark purple RB samples, antioxidant activity was observed by ferric reducing antioxidant power and 2,2-diphenyl-1-picryl-hydrazyl-hydrate (DPPH). In human clinical trials, ferulic acid, vanillic acid, synaptic acid, and 3,4-dimethoxybenzoic acid bioavailability increased 2 to 3 times after consuming 300 g of bioprocessed bran for three days. Phenylpropionic acid and 3-hydroxyphenylpropionic acid were the primary metabolites of the bound phenolic compounds released in the colon [54].

The polysaccharides, bioactive peptides, and phenolic compounds in cereal bran have also been reported to have antioxidant activity. RB biopeptides demonstrated moderate antioxidant activity using the hydrophilic oxygen radical absorbance capacity assay. Inhibiting dipeptidylpeptidase-IV activity is one of the most significant type 2 diabetes prevention methods. Ile-Pro, Met-Pro, Val-Pro, and Leu-Pro are dipeptides with direct inhibitory activity against dipeptidylpeptidase-IV, which metabolizes the insulin-tropic hormone glucagon-like peptide-1, used to control postprandial hyperglycemia in type 2 diabetes, and to prevent and treat type 2 diabetes [43]. Furthermore, biopeptides derived from rice bran intensify the antioxidative responses in gingival epithelial cells, preventing periodontitis, a chronic inflammatory disease of the oral cavity which causes tooth loss in the absence of appropriate treatment [55]. The antioxidant activity of a new wheat bran polysaccharide constituted mostly of arabinose, xylose, mannose, glucose, and galactose fragments was analyzed by Xiao-Lan Shang and colleagues [56]. The investigated polysaccharide provides significant antioxidant activity according to DPPH, 2,2′-and-bis(3-ethylbenzothiazoline-6-sulfonic acid (ABTS), hydroxyl, and superoxide radical scavenging tests [56]. It is expected that introducing foods rich in antioxidant compounds, such as processed cereal bran, into the daily diet will help prevent the free radical-induced diseases that cause oxidative tissue degeneration.

3.2. Anti-Inflammatory Activity

Cytokines modulate the inflammatory process, which impacts the immune response through local and intracellular interactions. The two primary categories of cytokines identified are proinflammatory cytokines and anti-inflammatory cytokines. A clinical trial initiated by Nuria Mateo Anson and colleagues investigated the anti-inflammatory response after consuming processed wheat bran [54]. The proinflammatory cytokines were considerably reduced in blood samples collected from eight male subjects who consumed 300 g of bread containing processed bran for three days compared to the detected anti-inflammatory cytokines [54]. Statistically significant reduction in the inflammatory markers high-sensitivity C-reactive protein (32.66%) and tumor necrosis factor (13.06%) was obtained after ferulic acid diet supplementation for six weeks on hyperlipidemia subjects [57]. Obesity is a serious public health risk worldwide and is linked to low-grade chronic inflammation. Rice fiber components such as phytosterol pherulates and isoprenoids have been found to have anti-inflammatory properties, either directly on immune cells or indirectly through gut microbiota modulation [58]. A valuable clinical trial cross-examined the effects of rice bran on inflammatory factors and 12 weeks of hypocaloric diet in overweight and obese adults. Rice bran consumption combined with an energy-restricted diet has a significant impact on the levels of inflammatory factors due to the ability of fiber to reduce related inflammation factors such as inhibition of hyperglycemia and its effects on lipids, particularly LDL-cholesterol, and also due to the protection of small intestine health and consequently inflammation prevention [59]. Ewelina Kurtys and colleagues critically reviewed the modulation of intestinal microbiota following rice bran consumption, the impact on the central nervous system, and, indirectly, disorders connected to chronic neuroinflammation [58]. Multiple mental and neurodegenerative illnesses, including schizophrenia, depression, and Parkinson's and Alzheimer's diseases, have been linked to chronic neuroinflammation. Through the gut-brain axis, the fiber components in rice bran can lower inflammatory processes in the gastrointestinal tract and the brain [58]. A neuroinflammatory animal model investigated the effect of RB extract administration on memory and cognition after 21 days of treatment [60]. Peroxisome proliferator-activated receptor-gamma modulation was demonstrated to reduce the inflammatory response and improve cognitive dysfunction in neurodegenerative diseases. The study findings show that nuclear brain extracts from mice treated with RB extract increased the peroxisome proliferator-activated receptor-gamma [60]. Blood pressure control necessitates a specified focus on nutritional interventions involving a well-balanced and nutritious diet.

Dietary fiber has been demonstrated to lower blood pressure in hypertensive patients by increasing the relative abundance of SCFA-producing bacteria and beneficial bacteria in the microbiome. Bacteria from the genera *Bifidobacterium*, *Lactobacillus*, *Spirobacter*, and *Eubacteria* ferment the fibers that pass through the gut, creating significant SCFAs. *Trichosporium bacteria*, conversely, have anti-inflammatory properties and control bacterial disorders. The study's findings revealed significant growth in bacteria of the genera *Bifidobacterium* and *Trichosporium* after a 3-month dietary intervention with OB [4]. Supplementing the daily diet with an adequate quantity of cereal bran, such as rice bran, oat bran, or wheat bran, can be used as nutritional prevention therapy for century illnesses. Moreover, cereal bran is a rich source of bioactive compounds used in clinical dietary intervention for patients with chronic disorders, such as non-communicable diseases, including type 2 diabetes, cancer, and cardiovascular disorders.

3.3. Cardiometabolic Protective Activity

Cardiovascular disease (CVD) affects the heart or blood arteries. It is typically related to fatty deposits inside the arteries (atherosclerosis) and an increased risk of coagulated blood. CVD is the leading cause of mortality for people all over the world. Inappropriate nutrition is the primary contributor to the development of CVDs. Oxidative stress, inflammation, hyperlipidemia, and hyperglycemia are major risk factors of inadequate nutrition.

Furthermore, there is a rising interest in consuming foods rich in bioactive phytochemicals that prevent future CVDs [61]. Cereal bran is one of the products with cardio-protective effects. The effects of a high-fat diet supplemented with 1% or 5% rice bran enzymatic extract on mice's daily diet were researched by Cristina Perez-Ternero and colleagues [62]. According to the findings, consuming RB reduces oxidative stress and inflammation associated with atherosclerosis. The ferulic acid moiety of Oryzanol is the primary compound responsible for the observed effects [62]. Long-term dietary supplementation with 5% RB extracts for 23 weeks on wild-type mice diets provided similar findings. The study's main conclusion was that RB extracts decrease cholesterol and avoid the development of atherosclerotic plaques [63].

A comprehensive meta-analysis observed the impact of replacing refined grains with whole grains on CVDs risk factors [6]. There were 25 randomized controlled trials, and 22 provided the criteria to be meta-analyzed. Whole-grain oats improved total cholesterol from SMD (standardized mean difference) = -0.54, 95% CI -0.95 to -0.12 and low-density lipoprotein cholesterol from SMD = -0.57, 95% CI -0.84 to -0.31, whole-grain rice improved triglycerides from SMD = 0.22, 95% CI -0.44 to -0.01, and whole grains, including wheat, oat, and rice, increased hemoglobin A1c from SMD = -0.33, 95% CI -0.61 to -0.04 and C-reactive protein from SMD = -0.22, 95% CI -0.44 to -0.00, according to the meta-analysis [6]. Therefore, when cereal bran is included in a regular and well-balanced diet, it provides considerable nutritional potential in preventing cardiometabolic illnesses.

3.4. Anti-Diabetes Activity

Diabetes is a non-communicable illness defined by a metabolic disorder in which the quantity of blood glucose becomes excessive due to a lack of insulin secretion. Diabetes is caused by various factors, including genetics and improper diet. Diabetes management primarily involves controlling blood glucose levels through proper diet, exercise, and hypoglycemia-prevention drugs [64]. Flour blends of rice bran and oat bran were examined with gluten-free multi-purpose product flour blends with blood glucose-reducing potential. The 70% wheat, 20% soy cake, 5% rice bran, and 5% oat bran mixture exceeded the other examined samples in terms of protein content (23.94 g/100 g), biological value (98.01%), and growth performance in rats [65]. Regarding the anti-diabetic potential of the experimental diets, the mix containing 5% rice bran and 5% oat bran had the strongest blood glucose-reducing action [65]. Timilehin David Oluwajuyitan and colleagues' research produced similar results [66]. A blend with 5% rice bran and 5% oat bran had a low glycaemic index, glycaemic load, and high blood-glucose-lowering potential [66]. A meta-analysis of 15 randomized controlled studies conducted on human subjects from the United States, Canada, and Europe investigated the effect of the oatmeal-enhanced diet on the early stages of diabetes. The study's findings indicate that consuming 3 mg of β-glucan from oats, corresponding to 60 g of oats, for at least eight weeks significantly lowered insulin, blood sugar, and glycosylated hemoglobin [64]. Therefore, personalized nutritional interventions are crucial in preventing and treating diabetes.

3.5. Anti-Cancer Activity

Cancer is one of the world's most severe diseases, and it is one of the significant causes of mortality. Conventional cancer treatment procedures involve cytotoxic treatments such as radiation and chemotherapy, ignoring that both treatments frequently have significant toxic side effects. Natural ingredients are an excellent resource for developing novel and non-invasive anticancer therapies. The OB avenanthramides are a class of polyphenolic alkaloids that have both chemopreventive and chemotherapeutic properties. Avenanthramides' potential therapeutic effect involves modulating many pathways, including stimulating apoptosis and senescence, reducing cell proliferation, and preventing epithelial-mesenchymal transition and metastatization [67]. After analyzing potential anti-cancer bioactivity against hepatocellular carcinoma cells, oat bran mixed with blackcurrant and blueberry powder showed enhanced anticancer benefits [7]. According to Hideyuki Nemoto and colleagues, oral intake of fermented brown rice and fermented rice bran with *Aspergillus oryzae* has a chemopreventive impact in rats, both on the growth of the tumor-associated inflammation and the subsequent development of metastasis [68]. The study found that mice fed with fermented rice bran and fermented brown rice had a reduced frequency of tumor cell development compared to mice fed a different diet. Furthermore, the metastatic capability of mice in the first category was significantly lower than that of mice on a regular diet who generated metastases to distant organs [68]. Additionally, the peptide fraction isolated from rice bran protein hydrolysate has anti-tumor solid activity against colon cancer cells [69].

3.6. Prebiotic Effects

Cereal bran is abundant in dietary fiber. A comprehensive analysis of wheat and rice bran dietary fiber revealed that they could modulate gut microbiota while also boosting the development of SCFA-producing bacteria [70]. The presence of fermentable soluble oligosaccharides, such as galactose-oligosaccharides, fructose-oligosaccharides, and xylooligosaccharides, contributes to bran's prebiotic activity by modulating *Firmicutes*, *Verrucomicrobia*, *Enterobacteriaceae*, *Prevotella*, and *Bacteroides* gut bacteria [70]. An animal-designed study observed a decrease in the *Enterobacteriaceae*, *Streptococcaceae*, and *Enterococcaceae* mainly pathogen bacteria, and an increase in the *Lachnospiraceae* and *Ruminococcaceae* beneficial bacteria after eight weeks of feeding rice bran to the mice [71]. After 14 days of intervention, a randomized-controlled pilot clinical trial reported that 30 g/day of heat-stabilized rice bran resulted in a lower Firmicutes: Bacteroidetes ratio (2.7 down to 1.4) and increased SCFA (propionate and acetate) in feces [72]. Bacterial metabolic by-products, particularly SCFAs, are frequently regarded as essential mediators of gut-brain communication, and altered SCFA synthesis has been linked to allergies, asthma, cancers, autoimmune illnesses, metabolic diseases, and neurological diseases [73]. Arabinoxylan-oligosaccharide extracted from WB provided prebiotic effects in a double-blind, randomized, controlled, crossover trial. After three weeks of intervention, subjects who consumed 4.8 g/day of arabinoxylan as part of ready-to-eat cereal had increased bifidobacteria concentration in their fecal samples [74]. Therefore, cereal bran can be a natural source of prebiotic compounds that can be integrated as ingredients in food products and supplements to improve host gut health.

However, the health benefits of cereal bran phytochemicals are directly related to the number of phytochemicals available for absorption in the upper or lower section of the digestive system, known as bioaccessibility. It is critical to degrade the structural composition and liberate more bioactive compounds to increase bioaccessibility and modulate the functional value of cereal bran. Several pretreatments are widely required to support the release of bioactive from cereal bran and boost their bioavailability. Table 2 shows the main biological activities of cereal bran as well as the pretreatments done prior to the examinations.

Table 2. Biological activities of wheat bran, rice bran, and oat bran.

Source	Study	Pretreatments	Biological Activity	Key Findings	References
Wheat bran	in vitro	Ultrafine grinding: Coarse wheat bran, Medium wheat bran, Fine wheat bran, Superfine wheat bran	Antioxidant activity; Digestive enzymes inhibitory activities	*p*-coumaric acid content in superfine WB was five times higher; Phenolic compounds bioaccessibility was increased by 65.51%; Starch digestibility was reduced.	[32]
	in vivo	Enzyme pretreatment (xylanase, cellulose, β-glucanase, and feruloyl-esterase) Yeast fermentation (Baker's Yeast)	Antioxidant capacity; Anti-inflammatory properties	Phenylpropionic acid and 3-hydroxyphenylpropionic acid were the main colonic metabolites identified; The pro-anti-inflammatory cytokines ratio was significantly lower after the consumption of 300 g of bioprocessed bran-based bread for 3 days.	[54]
	in vitro	Grinding	Antioxidant activity	The extracted xylose, mannose, glucose, and galactose exhibited remarkable antioxidant activities (DPPH, ABTS, hydroxyl, and superoxide radical scavenging tests)	[56]
	in vivo	-	Prebiotic effect	Arabino-xylan-oligosaccharide provided prebiotic properties increasing fecal bifidobacteria and postprandial ferulic acid concentrations.	[74]
	in vitro	Extrusion	Antioxidant activity;	Extractability of the bound phenolics was increased; Increased free bound and total phenolic content by 23.0%, 50.7%, and 36.3%; No effect was observed on CAA antioxidant activity; Bioaccessible phenolics increased by 40.5%	[52]
	in vitro	Fungal fermentation	Antioxidant activity;	Extractability of the bound phenolics was increased; Increased free bound and total phenolic content by 99.4%, 40%, and 71.6%; ORAC antioxidant activity increased 1.8-fold; CAA antioxidant activity increased 4.1-fold; Bioaccessible phenolics increased by 64.5%	[52]
Rice bran	in vitro	Enzymatic pretreatment (protease from *Aspergillus oryzae*)	Anti-diabetic activity; Antioxidant activity;	Dipeptides Ile-Pro, Met-Pro, Val-Pro, and Leu-Pro had shown inhibitory activity against DPP-IV (dipeptidylpeptidase-IV); No inhibitory activity against human maltase–glucoamylase was observed.	[43]
	in vivo (animal study-mouse model)	Maceration in ethanol	Neuroinflammatory responses (memory and cognitive performance)	Spatial working, reference memory, and non-spatial recognition memory were improved	[60]
	in vivo (animal study-mouse model)	Enzymatic hydrolysis	Antioxidant capacity; Anti-inflammatory properties	Oxidative stress was reduced; Ferulic acid and γ-oryzanol reduced pro-inflammatory monocyte phenotype; Atherosclerosis-related oxidative stress and inflammation were reduced.	[62]
	in vivo (animal study-mouse model)	Fermentation with *Aspergillus oryzae*	Anti-inflammatory activity; Anti-mutagenic effects	Fermented brown rice and rice bran provided chemopreventive effectiveness against inflammation-related carcinogenesis	[68]
Oat bran	in vivo (animal study-mouse model)	Grinding	Antidiabetic activity; Carbohydrate hydrolyzing inhibitory activity; Antihyperglycaemia activity.	The samples with 70% wheat, 20% soy cake, 5% rice bran, and 5% oat bran provided low glycaemic index, high carbohydrate hydrolyzing enzyme inhibitory potential, and blood glucose lowering potential.	[65]
	in vivo	Maceration	Antidiabetic activity	The samples with plantain 60%, soy cake 30%, rice bran 5%, and oat bran 5% had the highest blood glucose-reducing activity.	[66]

4. Pretreatments' Effect on Cereal Bran Phytochemicals

The aim of the biomass pretreatments is to fractionate the structure of the lignocellulose (Figure 3) in order to accelerate the chemical and biological processes of recovering the bioactive components connected to the cell wall. Untreated lignocellulosic materials are highly resistant due to cellulose crystallinity, matrix heterogeneity, limited surface accessibility, and lignin protection, as evidenced by their limited convertibility to fermentable sugars by enzymatic hydrolysis. At the same time, the pretreated biomass possesses characteristics that boost the accessibility and yield of cellulolytic enzymes in the conversion processes, such as higher porosity, a higher contact surface, and a low concentration of lignin [21]. Physical pretreatments consume a lot of energy; however, they are the most practical approach for treating biomass for additional bioprocessing.

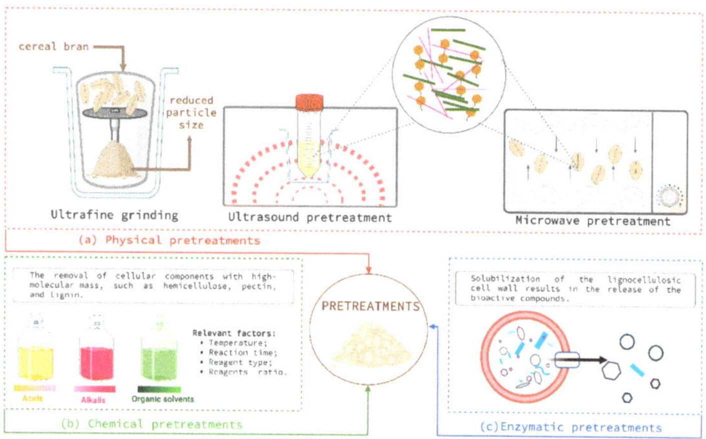

Figure 3. Key pretreatments for cereal bran processing.

Physical pretreatments (Figure 3a) include processes that reduce particle size significantly (e.g., ultrafine grinding) to increase conversion yield in bioprocesses such as fermentations [21,75]; processes that destroy and remove the protection provided by the rigid structures of hemicellulose and lignin (e.g., microwave pretreatments) [76]; processes that facilitate molecular interactions that cause mass transfer through cell membranes (e.g., ultrasound pretreatments) [77]; and forced separation (e.g., explosive steam pretreatments) [78]. Chemical pretreatments (Figure 3b) involve the removal of cellular components with high-molecular-mass, such as hemicellulose, pectin, and lignin, and the formation of a cellulose-based suspension for further enzymatic hydrolysis. Biological pretreatments (Figure 3c), on the other hand, are exceedingly specific but often time-consuming, with the primary objective of preparing the biomass before the application of a second bioprocessing step, such as microbial fermentation [21].

4.1. Physical Pretreatments

4.1.1. Ultrafine Grinding

Cereal milling has been traditionally performed by physical procedures involving mechanical forces. Chemical changes, including changes in enzymatic sensitivity, occur due to particle size reduction operations and material changes. Ball milling is an efficient method of intensively reducing particles and macromolecules like cellulose into micro colloidal particles. The ball grinding method is applied for ultrafine grinding in wet and dry environments. Using balls of varying sizes and crushing the fibers in multiple directions helps increase the fibers' exposed surface and their solubility level [75,79]. The reduction of the bran particle size supposes the growth of the internal surface of the cell wall, and the contact between the enzymes and the existing substrate is enhanced. The increased interactions between enzyme and substrate directly influence the release process of the bounded compounds. The increased access of enzymes to the exposed compounds as a result of the substantial particle size reduction boosts the extraction yield and, consequently, the bioaccessibility of the bioactive compounds. The grinding process is a physical method of modifying the physicochemical characteristics and improving the release of phenolic compounds by destroying the resistant structure of the bran [32]. The number of phenolic compounds released, particularly ferulic acid, increased significantly after in vitro digestion due to decreasing particle size. The content of free p-coumaric acid in superfine wheat bran (19.16 μm particle size) was almost five times higher (373%) than in coarse wheat bran (1110.39 μm particle size), medium wheat bran (235.68 μm particle size), and refined wheat bran (83.73 μm particle size). The aleurone layer of the superfine WB was entirely degraded, substantially increasing polyphenol extraction capacity and bioavailability. Following in vitro digestion of superfine WB, a significant release of ferulic acid was observed during intestinal digestion, and an improvement in the antioxidant capacity and inhibitory activity of the digestive enzyme of starch in the intestine [32].

Similar results were found when RB was exposed to intense milling. Compared to fine and coarsely milled powders, superfine RB powder had a higher content of bound flavonoids (30.98%). However, the concentration of extracted free flavonoids didn't change the field considerably [80]. As a result of the intense decreasing particle pretreatments of cereal bran, the contact surface is enhanced. Therefore, the accessibility of phytochemicals contained in the fibrous matrix can be improved.

4.1.2. Ultrasonic Pretreatments

The ultrasound principle is defined by mechanical vibrations between particles in an elastic medium generated by an ultrasonic wave [77,81]. The ultrasonic frequency is between 20 kHz and 10 MHz, higher than the human hearing range (16 Hz to 20 kHz) [77]. The energy released by the vibrations influences the interaction of the particles, resulting in the generation of thermal, cavitation, and mechanical effects [81]. Ultrasonic waves generate molecular interactions, which break cell wall structure, reduce particle size, and facilitate mass transfer across cell membranes. Cracks caused by cell wall disintegration enhance tissue permeability and increase the release of targeted compounds [77]. Chao Chen and colleagues investigated the influence of ultrasonic application and temperature on bound phenolics and β-glucan from defatted oat bran [82]. The pretreatments were followed by conventional extractions of the bioactive compounds and the determination of antioxidant capacity through the ORAC method and HPLC-DAD to identify and quantify the main phenolic compounds. Ultrasonic treatment enhanced β-glucan production by about 37%, and the phenolic content showed a higher yield following ultrasound application than maceration [82]. Maceration is the primary extraction method used for one to four days at various temperatures (from room temperature to boiling temperature) and solid-to-solvent ratios (from 1:1 to 1:20 g/mL) [83]. The ultrasound pretreatment method has been used to remove lignocellulose material and allow access to the nutrients by microorganisms for the growth and production of enzymes during rice bran solid-state fermentation (SSF) with Rhizopus oligosporus. Different ultrasonic parameters were studied and optimized, including ultrasound timing, cycling, liquid-to-solid ratio, and ultrasound amplitude conditions. Following the ultrasonic treatments, the required compounds for increased fungal growth were released, resulting in the liberation of the targeted bioactive compounds. The phytic acid concentration in the RB was reduced after ultrasonic pretreatment, while phytase production was increased after SSF [84]. Three minutes of high-intensity ultrasound (40 kHz, 11 W/cm^2) pretreatment of fermented whey and oat beverages led to substantial Lactobacillus casei 431 growth, high antioxidative activity, and favorable consumer acceptability [85]. Using ultrasonic pretreatment on RB substrate supplemented with lingonberry pulp, followed by microbial fermentation, was an efficient strategy that enhanced the fermentation process, improved fermentation efficiency, and reduced fermentation time in Ruta Vaitkeviciene and colleagues' study [86]. The quantity of soluble dietary fiber in RB increased by 17.5% after ultrasound pretreatment at 850 kHz and 160 W, for 20 min at 40 °C [86]. Therefore, ultrasonic pretreatments have an increasing potential to destabilize the rigidity of the cell wall to boost the bioavailability of the relevant compounds.

4.1.3. Thermal and Moisture Pretreatments

The release of bioactive compounds from cereal bran and the increase in their bioavailability is also supported by temperature and moisture pretreatments [31]. Before the SSF, black RB was exposed to a physical pretreatment that included moisturizing to 30% and lifting in repose overnight, followed by autoclaving for sterilization. The major anthocyanin of RB, cyanidin-3-glucoside, was 99.58% reduced after physical pretreatment. Heat treatment causes anthocyanin's circular structure to open, converting it to chalcone, a quickly degraded molecule. The reduction of cyanidin-3-glucoside due to increased protocatechuic acid content from 151.10 µg/g to 1055 µg/g by PP, possibly due to anthocyanin degradation during PP. The increasing of moisture to 30% activated some of the inherent enzymes of RB, which support the release of extractable phenolic acids during SSF [31]. Furthermore, the thermal process of 10 min at 80 °C boosted the total phenolic content of oat bran by 25.84% and wheat bran by 22.49%. Both WB (ferulic acid + 39.18%, vanillic acid + 95.68%, apigenin-glucoside + 71.96%, and p-coumaric acid + 71.91) and OB (avenanthramide 2c + 52.17% and dihydroxybenzoic acids + 38.55%) indicated a considerable ratio percentage increase in phenolic content following the thermal process [87].

4.1.4. Steam Explosion Pretreatment

Steam explosion technology is a new high-efficiency modification approach. Steam explosion is a pretreatment procedure that involves exposing bran to saturated steam at high pressure for some time, followed by an immediate pressure reduction. Following steam explosion pretreatments, the crystalline structure of cellulose and hemicellulose is degraded due to cell rupture, mechanical fracture, and the structural rearrangement of biological macromolecules. Moreover, the conversion of insoluble fibers to soluble fibers occurs, while the solubility of soluble fibers is increased [88]. Steam explosion of wheat bran, followed by superfine grinding, increased water solubility index (211.75 mg/g), oil holding capacity (1.68 g/g), bile salts, and cholesterol adsorbing capacity (1.31 and 21.87 mg/g). Furthermore, total phenolic and flavonoid content increased significantly and indicated the highest DPPH radical scavenging activity (87.68%) [88]. The pretreatment efficacy is influenced by the type and physical accessibility of the tested samples. The samples are usually pretreated at a temperature of 160–260 °C, corresponding to a pressure of 0.69–4.83 MPa, for a short duration (30 s and 120 s) before being treated at atmospheric pressure. The principal impacts of steam explosion pretreatment are lignocellulosic structural deterioration, hemicellulose hydrolysis, and lignin fraction depolymerization. According to Liu and colleagues, steam explosion pretreatment (60 s at 185, 205, 215, and 225 °C, with corresponding pressure 1.02, 1.62, 2.05, and 2.45 MPa, and 215 °C for 30, 60, 90, and 120 s) successfully released wheat bran phenolic acids such as p-hydroxybenzoic, vanillic, syringic, p-coumaric, and ferulic acid [89]. Following pretreatment, the production and antioxidant capacity of soluble free and conjugated phenolic acids in wheat bran improved considerably [89]. The main economic benefits of steam explosion pretreatment are low costs, reduced energy consumption, high efficiency, and no pollution [88].

4.1.5. Microwave Pretreatment

Pretreatments that include microwaves (MWs) on solid and liquid matrices induce the dipole to rotate, resulting in molecular friction and heat generation. Heat stress and high pressure in cellular tissues promote cellular structure breakdown and the release of targeted compounds [76]. Furthermore, cell disruption due to MW irradiation enhances the accessibility of polysaccharides in the cell wall and polymeric macromolecules are more fermentable during bacterial fermentation [76]. MW pretreatment before in vitro fecal fermentation enhanced the fermentability of previously extracted and isolated insoluble dietary fibers such as xylan A, xylan T, and arabinan, and the production of the SCFAs acetate, butyrate, and propionate [76]. MW irradiation was applied post-treatment after mixed fermentation of soybean by-products by Lactobacillus bulgaricus and Neurospora crassa, mixed in equal proportions as inoculum and fermented at 30 °C for two days with an inoculum size of 5.5%. The study's findings indicate that SSF followed by MW irradiation at 600 W for 2.5 min significantly improved the soluble dietary fiber amount and composition from 9.55% to 20.92% while dispersing and stabilizing the fermented matrices [90]. Therefore, MW treatments may be used before and after bacterial fermentation, with beneficial results in both circumstances. MW treatments can be an attractive option from an economic perspective since processing time and energy consumption decrease, while also having favorable effects on bioactive compounds.

4.1.6. Supercritical Pretreatment

Supercritical fluids can be used to pretreat lignocellulosic biomass under milder pretreatment conditions, resulting in higher sugar yields, reduced generation of fermentation inhibitors, and increased susceptibility to enzymatic hydrolysis while consuming fewer chemicals such as solvents, reagents, and catalysts [21]. Supercritical treatments use a technique in which supercritical fluids are introduced into the biomass, and bioactive compounds are separated or extracted based on solubility differences. Under regulated conditions, the compression and heating of a gas, such as carbon dioxide, change the gas's physical properties, producing a supercritical fluid. The advantages of the supercritical fluid technique cover the elimination of toxic organic solvents, complete process automation, and sample oxidation prevention. At the same time, the technology's limitations include the equipment's high price, the inability to extract polar compounds, and high operating pressure, which requires a high energy consumption [78]. Carbon dioxide (CO_2), along with ammonia, water, and hydrocarbons such as propane and butane, are among the most often used supercritical fluids [21,91]. CO_2's main advantage is the critical temperature of 31.1 °C, which allows for a wide range of applications in heat-sensitive bioactive compounds. Supercritical CO_2 advantages include non-toxicity, non-flammability, reduced cost, minimal thermal energy consumption, and complete removal of the toxic residual solvent from the final extracts [91]. The supercritical fluid technique is performed as both pretreatment and extraction. The application of supercritical fluids as pretreatments improves the substrate's accessibility to enzymatic hydrolysis by physical or chemical degradation of the lignocellulosic matrix. Furthermore, the fermentable sugar yield is significantly higher after pretreatment with supercritical fluids. In contrast, synthesizing fermentation inhibitory compounds, such as acetic acid, levulinic acid, formic acid, furfural, and 5-(hydroxymethyl) furfural, is minimal [21,91].

4.1.7. Hydrothermal Pretreatment

Hydrothermal pretreatment is cost-effective because of its low energy consumption and the use of water as a solvent, which is inexpensive and does not cause environmental pollution [92,93]. Hydrothermal pretreatments can be subcritical or supercritical based on the critical points of water temperature and pressure, 374 °C and 22.1 MPa, respectively [92]. Changes in the physical characteristics of water around the critical temperature and pressure points allow the exact selection of the chemical reactions that occur. As a result, when water reaches the required temperature and pressure, the dielectric constant reduces, increasing the solubility of organic compounds. Secondly, when the temperature rises from 300 °C to 400 °C at a pressure of 25 MPa, the ionic product of water changes, and therefore the reaction mechanism changes [93]. Wheat bran was pretreated with supercritical water at a temperature of 400 °C, a pressure of 25 MPa, and a reaction rate of 0.19 s. Following pretreatment, high cellulose and hemicellulose hydrolysis efficiency was attained, with minimum degraded compound content. The essential part of preventing glucose deterioration was rigorous reaction time optimization [93].

4.2. Chemical Pretreatments

For many years, chemical pretreatments (CPs) have been used to increase the bioavailability of plant phytochemicals. However, because of high concentrations of ecologically hazardous composites, such as acids, alkalis, or oxidants, CPs have been observed to have unfavorable environmental consequences over time. Another result of applying CPs is using solid acids with a high degree of severity, such as H_2SO_4, HNO_3, H_3PO_4, and HCl. They increase the synthesis of furfural and its derivatives, which are inhibitory compounds [94].

CPs are based on using acids and alkalis to solubilize bioactive compounds from cereal bran cell walls. The main factors in chemical pretreatments are the number of used reagents (acids and alkalis), temperature, reaction time, and the type and concentration of reagents. Appling acid-alkaline treatment (6.0 N HCl for the acid treatment and 6.0 N NaOH for alkaline treatment) on wheat bran, Huma Bader Ul Ain and colleagues brokedown the large polysaccharides into smaller oligosaccharide fractions and obtained increased amounts of soluble dietary fiber [20]. The main conclusions of the study were that the mode of action of consecutive acid-alkaline pretreatments opened the structure of the wheat cell wall and increased the fiber surface porosity, increasing the hydroxyl groups' access to spread inside and perform the hydrolysis reaction during the subsequent alkaline treatment [20]. Lei Wang and colleagues obtained comparable results, which achieved acid treatment (H_2SO_4 and potassium hydroxide) on wheat bran to increase the concentration of soluble dietary fiber [95]. Following acid treatment, the structure of soluble fiber was considerably degraded, and the starch content was reduced. By increasing the concentration of H_2SO_4, the hydrolysis of hemicelluloses and amorphous cellulose was performed, increasing the crystallinity of the soluble fibers. The water retention, swelling, and oil-binding properties of the soluble fibers improved after the H_2SO_4 concentration was lowered below 2%. CPs influence the hypoglycemic properties of rice bran insoluble dietary fiber. Jing Qi and colleagues have found that the hypoglycemic effect of rice bran varies based on the pretreatment applied to rice bran [96]. After H_2SO_4 CPs on RB, the glucose fraction increased from 40.95% to 60.84%, increasing the CPs rigor, resulting in depolymerization of cellulose and noncellulosic components (e.g., starch). Different intensities of H_2SO_4 CPs decreased the xyloglucan fraction from RB hemicelluloses, resulting in a reduction in xylose, rhamnose, and fructose composition. As a result, applying acid in association with alkaline treatment will significantly boost rice bran fiber's ability to absorb glucose while inhibiting α-amylase activity. Acid-alkaline treatments can increase the hypoglycemic properties of insoluble rice bran fibers, allowing them to be used as promising low-calorie functional components for fiber fortification [96].

CP applied to WB at the laboratory level costs about 4$ per kilogram of bran [97]. This cost can be recovered considerably due to the high-value biocomposites, such as dietary fiber and phenolic acids extracted from fermented wheat bran, a future-oriented perspective for cereal processors and growers.

4.3. Enzymatic Pretreatment

Most pretreatment procedures, including physical, chemical, and thermal procedures, require significant energy and volumes of chemical solvents. Furthermore, the rigorous environments they generate have adverse consequences on biomass, such as inhibitory by-products. Therefore, from these perspectives, biological pretreatments are more favorable since they do not require a significant consumption of energy and chemicals and do not generate inhibitory substances. However, industrial enzymes that break down lignocellulosic components, such as ligninolytic enzymes and cellulase, have a high unit price. Additionally, assuring the efficacy of pretreatments necessitates a high quantity of enzymes [94]. Conversely, enzymatic hydrolysis of lignocellulosic biomass is a safe and environmentally friendly pretreatment that may be successfully used in the food industry. Enzymatic hydrolysis involves the solubilization of insoluble components under heterogeneous environments by applying enzymatic blends of enzymes such as α-amylase, glucoamylase, protease, lysing enzyme, β-glucanase, xylanase, cellulase, feruloesterase, or pectinesterase. The hydrolysis of polysaccharide glycosidase linkages generates structural changes in the cereal bran matrices, lowering the rigidity and mechanical strength of cell walls [27]. Xue et al. applied β-endoxylanase and α-arabinofuranosidase enzymatic hydrolysis on WB. WB was first heated at 121 °C for 30 min and then incubated at 60 °C and pH 6.0 with the purified enzyme blend (β-endoxylanase and arabinofuranosidase) [98]. The quantity of xylooligosaccharides and phenolic acids, together with the antioxidant capacity, and water or oil retention capacity of WB rises after enzymatic processing. Furthermore, the results revealed that the degree of WB hydrolysis was enhanced, and a substantial synergetic effect was detected in the bran-containing dough and bread [98].

Due to their expanded application and industrial-scale efficiency, commercial enzyme pretreatments are among the most common treatments for phytochemical release. The fundamental limitation of this pretreatment is the high cost of enzymes. Therefore, further efficient methods for naturally releasing similar enzymes, such as fermentation processes, are required.

5. Cereal Bran Fermentation

Cereal bran bioprocessing is applied to enhance bran's nutritional and technological value. The application of bioprocesses is designed to take advantage of the biological activity of cells and their parts to maximize the value of the bran matrix. Microbial fermentation of bran substrates is an efficient, environmentally friendly bioprocess for increasing the release of bioactive compounds. The microorganisms involved in the bioprocess generate new compounds throughout fermentation, the most frequent of which are microbial enzymes, biomass, and primary and secondary metabolites [99]. Among the microorganisms used in the fermentation of cereal bran can be mentioned molds or fungi, such as *Aspergillus* spp. [30,31,100], *Monascus* spp. [101], and *Rhizopus* spp. [102], bacteria such as *Lactobacillus* spp. [103], *Bifidobacterium* spp. and *Streptococcus* spp., and yeasts such as *Saccharomyces* spp. Microbial fermentation may be accomplished by two primary methods: solid-state fermentation (SSF) and submerged fermentation (SmF) (Figure 4).

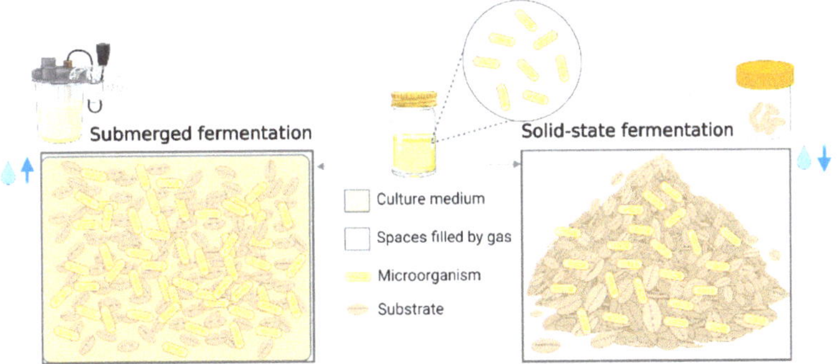

Figure 4. Solid-state fermentation and submerged fermentation.

Solid-state fermentation (SSF) is a food processing method in which microorganisms grow and multiply without free water. It has been used for thousands of years and recently has received increased interest because of the compositional changes it may generate in agro-industrial wastes, particularly their bioactive components. Microorganism-produced enzymes and endogenous enzymes implicated in SSF, and those involved in SSF can liberate bound phenolic compounds and bioconvert the phenolic metabolites, enhancing cereal bran bioavailability and biological activities (Table 3) [100,103]. Cereal bran represents a matrix with low free water content.

To maximize the efficiency of the SSF bioprocess, several parameters must be optimized, including the fermentation microorganisms selection, substrate composition to ensure the carbon source required for microorganism growth and multiplication, potential pretreatments to the matrix, and bioprocess temperature, aeration, pH, and moisture [28]. Călinoiu and colleagues evaluated the potential of SSF to enhance the phenolic content and antioxidant activity of WB and OB using the Saccharomyces cerevisiae yeast [10]. The maximum concentration of total phenolic compounds was reached after the first three days of fermentation in the case of WB, and the fourth day of fermentation in the case of OB. WB recorded a 56.6% increase in ferulic acid, 259.3% vanillinic acid, and 161.2% dihydroxybenzoic acids, whereas OB fermentation increased the ferulic acid by 21.2%, and the avenanthramide by 48.5% [10].

Similarly, after RB SSF with Rhizopus oryzae, the biological activities (antioxidant activity, anti-inflammatory activity, and ferric-reducing power) and main bioactive compounds (phenolics, flavonoids, carotenoids, and anthocyanin) significantly increased [102]. Similarly, following the SSF on WB substrate, the soluble phenolic content of wheat bran was nearly triplicated compared to the raw bran [104]. The significant increase in total phenolic compound concentration is primarily due to acid hydrolysis during fermentation, which accelerates the breakdown of the bonds that link the phenolic compounds to the cell wall polysaccharides [103]. At the initiation of SSF, microorganisms come into contact with the bran cell wall, mainly composed of cellulose, hemicellulose, and lignin. Therefore, during the first fermentation stage, the microorganisms naturally produce a large number of enzymes to break down the cell wall and get accessibility to the lignocellulosic compounds [28]. According to Bei and colleagues, xylanase activity produced by *Monascus anka* throughout SSF of OB was highly related to phenolic content. Xylan is a significant component of the cell wall. The high activity of xylanase during the first days of fermentation resulted in the breakdown of OB cell walls and the consequent release of phenolic compounds [101]. Therefore, in addition to phytochemical compound synthesis and extraction efficiency enhancement, SSF has excellent potential to be applied to bran biomass to produce microbial enzymes and develop ingredients with added value for food, pharmaceutical, and cosmetic applications.

Submerged fermentation is a bioprocess wherein microorganisms grow inside a fluid medium rich in essential nutrients and with an appropriate oxygen concentration for microorganism growth and multiplication [99]. The SmF bioprocess, like the SSF bioprocess, involves the release of bound bioactive compounds from the bran matrix via microorganisms involved in fermentation and the bioconversion and production of additional high-value compounds as secondary metabolites. Ferulic acid was liberated and biotransformed by lactic acid bacteria into phenolic derivatives such as 4-ethylphenol, vanillin, vanillic acid, and vanillyl alcohol after 24 h of rice bran SmF. Ferulic acid esterase and ferulic acid decarboxylase were the enzymes that increased the efficiency of the bioprocess. The study's key finding was the potential of lactic acid bacteria to synthesize in situ biovanillin through Smf of agro-industrial lignocellulosic waste, such as cereal bran [105]. Nowadays, synthetic vanillin (4-hydroxy–3-methoxy benzaldehyde) dominates the vanilla-flavoring industry in the food, beverage, pharmaceutical, cosmetic, and tobacco areas. Therefore, SmF and SSF biotechnological processes using low-cost substrates like agro-industrial by-products are likely to develop new vanilla synthesis pathways via strain metabolism [106]. The primary advantages of SSF over SmF are high yield, enhanced product stability, low protein degradation, and minimal contamination risk. Another advantage is the reduced production cost. According to economic analysis, the production cost of cellulase enzyme via SSF (15.67 USD/kg) is three times lower than obtaining a similar enzyme using SmF (40.36 USD/kg), but also advantageous because the enzyme's marketing price is 90 USD/kg [104]. SmF, on the other hand, provides many advantages that support its use in many applications, such as efficient control of fermentation parameters such as temperature, humidity, pH monitoring, and optimal aeration due to continuous homogenization through the fermentation. Furthermore, the SmF bioprocess can minimize the danger of fungal hyphae drying up when fermentation occurs in molds [99].

Table 3. SSF and SmF applied to cereal bran.

Fermentation Method	Inoculum	Substrate	Fermentation Condition	Results	References
SSF	*Enterococcus faecalis* M2	WB	Inoculation rate: 10%; Moisture content: 60%; Time: 36 h; Temperature: 37 °C.	Soluble dietary fiber ↑ Phenols ↑ Flavonoids ↑ Alkylresorcinols ↑ Free amino acid ↑ Protein ↓ Phytic acid ↓ Antioxidant capacity ↑	[103]
	Bacillus sp. TMF–2		Inoculation: 0.5 mL of bacterial suspension; Solid-to-liquid ratio: 1:1; Time: 11 days; Temperature: 30 °C.	Soluble phenolic content ↑ Antioxidant capacity ↑ Free radical scavenging rate ↑ Activity of hydrolytic enzymes (amylase, cellulase, pectinase, mannanase, protease, and phytase) ↑ Inorganic phosphorus ↑ Phytic acid ↓	[104]
	Aspergillus strains: *A. brasiliensis, A. awamori*, and *A. sojae*.		Moisture: 1:1 (w/w); Fungal spores: 1% (v/w); Time: 8 days for *A. brasiliensis* and 14 days for *A. awamori* and *A. sojae*; Temperature: 25 °C.	Radical scavenging activity ↑ Tyrosinase inhibitory activity ↑ Lactase inhibitory activity ↑ Kojic acid ↑ Free phenolic acids ↑ Total flavonoid content ↑	[100]
	Rhizopus oryzae	RB	Moisture: 50%; Spore concentration: 4×10^6 spores/g bran; Time: 96 h; Temperature: 30 °C.	Total phenolic ↑ Total flavonoid ↑ Total carotenoid ↑ Total anthocyanin ↑ Antioxidant capacity ↑ Ferric reducing power ↑ Anti-inflammatory properties ↑ Anti-diabetic properties ↓ Radical scavenging ability ↓	[102]
	Aspergillus awamori and *Aspergillus oryzae*		Moisture: 30%; Inoculation: 1 mL of the spore suspension; Time: 5 days; Temperature: 30 °C.	Total phenolic content ↑ Protocatechuic acid ↑ Ferulic acid ↑ Radical scavenging activity ↑ Tyrosinase inhibitory activity ↑	[31]

Table 3. *Cont.*

Fermentation Method	Inoculum	Substrate	Fermentation Condition	Results	References
	Saccharomyces cerevisiae	OB	Moisture: 45% (w/w); Inoculation: 5 mL of yeast suspensions (10^7 CFU/mL) per 100 g of dry weight; Time: 6 days; Temperature: 30 °C; Static conditions.	DPPH radical activity ↑ Total phenolic content ↑ Avenanthramides ↑ Ferulic acid ↑ Protocatechuic acid ↑ Caffeic acid ↑ Vanillic acid ↑	[10]
	Monascus anka		Moisture: 60% (w/w); Inoculation: 0.1 mL of spore suspension per 1 g dry oats; Time: 14 days; Temperature: 30 °C	Ferulic acid ↑ Vanillic acid ↑ α-amylase activity ↑ Xylanase activity ↓ Total cellulase activity ↓ β-glucosidase activity ↓	[101]
	3-member consortium of *Bacillus subtilis, Bacillus coagulans, Bacillus cereus*	WB	Wheat bran 2% (w/v); Time: 7 days; Temperature: 30 ± 2 °C; Agitation speed: 140 rpm.	Cellulases activities ↑ Digestibility of solid substrates ↑ Cellulose bioconversion ↑ Lignocellulose degradation ↑ β-glucosidase ↓	[107]
SmF	*Aspergillus phoenicis (Aspergillus saitoi)*		Wheat bran: 1% (w/v); Temperature: 40 °C; pH: 6; Inoculation: 10^5 spores/mL.	Production of thermo-toleran mycelial β-D-fructofuranosidase and raffinose;	[108]
	Pediococcus acidilactici		Inoculation: 3% (v/v); Time: 24 h; Temperature 37 °C; pH: 5.6	Bioconversion of ferulic acid into phenolic derivatives such as 4-ethylphenol, vanillin, vanillic acid, and vanillyl alcohol.	[105]
	Saccharomyces cerevisiae	RB	Inoculation: 25 mL (1×10^8 cells/mL); Constant aeration; Agitation speed: 150 rpm; Time: 24 h; Temperature: 35–45 °C; pH: 3.5–4.5.	Antioxidant properties ↑ Cyanidin-3-glucoside and peonidin-3-glucoside were bioconverted to cyanidin and peonidin Bioactivity ↑	[109]
	Bacillus mojavensis	OB	Inoculation: 0.5%, 1%, and 2% (w/v); pH: 8.0 Time: 120 h; Temperature: 37 °C; Agitation speed: 180 rpm.	Xylanase yield of about 249.308 IU/mL	[110]

6. Conclusions and Future Perspectives

The fortification and re-use of cereal milling industry by-products to develop new functional foods is a research direction of great interest and practicality from the perspective of food–health relations, as well as from the environment protection and waste management perspectives. Due to the high concentration of bioactive compounds in cereal bran that have valuable biological functions, such as antioxidant activity, anti-inflammatory activity, cardiometabolic protective activity, anti-diabetes activity, anti-cancer activity, and prebiotic effects, cereal bran phytochemicals have lately attracted a lot of interest from the food and medical industries. The proportion of phytochemicals in cereal bran that are bioaccessible, or available for absorption in the digestive system, is closely correlated with the health benefits of cereal bran. SSF and SmF bioprocesses significantly increase the production of phytochemicals, particularly antioxidant phytochemicals, such as phenolic compounds, through the secondary metabolic pathway of fermentation microorganisms or following extracellular enzymatic activity after the release from the substrate matrix. The accessibility of microorganisms to bran phytochemical compounds contributes to the efficacy of fermentative bioprocesses. The rigid structure of cell wall polysaccharides such as lignin, hemicellulose, and cellulose protects the high-value bioactive compounds. Specific pretreatments are required to break the lignocellulosic structure and make cereal polysaccharides accessible to hydrolytic enzymes, allowing the fermentable sugars to be freed. Moreover, the porosity and specific surface of the insoluble fiber of the bran can be enhanced by applying chemical, physical, and biological pretreatments, resulting in increased bioavailability of cell walls phytochemicals. Pretreatments that are modern and environmentally friendly, such as steam explosion, ultrafine grinding, and ultrasound and microwave pretreatments, have gained attention because they are both efficient and cost-effective. As a result, more work should be done on optimizing these pretreatments for particular matrices or integrating them in the early stage of the bioprocesses for greater efficiency. However, to scale the pretreatment's applicability, more research is required to evaluate these integrated methods' environmental effects, cost efficiency, and energy balance.

Author Contributions: Data for the paper were collected by S.A.N., L.F.C., F.V.D., A.C.F. and D.C.V., who also wrote, reviewed and edited the manuscript. S.A.N., L.F.C. and D.C.V. significantly improved the discussion of the manuscript's content and edited and revised it. All authors have read and agreed to the published version of the manuscript.

Funding: This research was funded by MCI-UEFISCDI, project number PN-III-P4-PCE-2021-0750, project number PN-III-P4-ID-PCE-2020-2306 and project number PN-III-P1-1.1-TE-2021-1052.

Conflicts of Interest: The authors declare no conflict of interest.

References

1. FAO. *Food Outlook—Biannual Report on Global Food Markets*; FAO: Rome, Italy, 2019.
2. Martín-Diana, A.B.; García-Casas, M.J.; Martínez-Villaluenga, C.; Frías, J.; Peñas, E.; Rico, D. Wheat and Oat Brans as Sources of Polyphenol Compounds for Development of Antioxidant Nutraceutical Ingredients. *Foods* **2021**, *10*, 115. [CrossRef] [PubMed]
3. Călinoiu, L.F.; Vodnar, D.C. Whole Grains and Phenolic Acids: A Review on Bioactivity, Functionality, Health Benefits and Bioavailability. *Nutrients* **2018**, *10*, 1615. [CrossRef] [PubMed]
4. Xue, Y.; Cui, L.; Qi, J.; Ojo, O.; Du, X.; Liu, Y.; Wang, X. The effect of dietary fiber (oat bran) supplement on blood pressure in patients with essential hypertension: A randomized controlled trial. *Nutr. Metab. Cardiovasc. Dis.* **2021**, *31*, 2458–2470. [CrossRef] [PubMed]
5. Hsieh, C.-C.; Wang, Y.-F.; Lin, P.-Y.; Peng, S.-H.; Chou, M.-J. Seed peptide lunasin ameliorates obesity-induced inflammation and regulates immune responses in C57BL/6J mice fed high-fat diet. *Food Chem. Toxicol.* **2021**, *147*, 111908. [CrossRef]
6. Marshall, S.; Petocz, P.; Duve, E.; Abbott, K.; Cassettari, T.; Blumfield, M.; Fayet-Moore, F. The Effect of Replacing Refined Grains with Whole Grains on Cardiovascular Risk Factors: A Systematic Review and Meta-Analysis of Randomized Controlled Trials with GRADE Clinical Recommendation. *J. Acad. Nutr. Diet.* **2020**, *120*, 1859–1883.e31. [CrossRef]
7. Xue, B.; Zhao, B.; Luo, S.; Wu, G.; Hui, X. Inducing apoptosis in human hepatocellular carcinoma cell lines via Nrf2/HO-1 signalling pathway of blueberry and blackcurrant powder manipulated oat bran paste extracts. *J. Funct. Foods* **2022**, *89*, 104967. [CrossRef]

8. Carpentieri, S.; Soltanipour, F.; Ferrari, G.; Pataro, G.; Donsì, F. Emerging Green Techniques for the Extraction of Antioxidants from Agri-Food By-Products as Promising Ingredients for the Food Industry. *Antioxidants* **2021**, *10*, 1417. [CrossRef]

9. Yan, J.K.; Wu, L.X.; Cai, W.D.; Xiao, G.S.; Duan, Y.; Zhang, H. Subcritical water extraction-based methods affect the physicochemical and functional properties of soluble dietary fibers from wheat bran. *Food Chem.* **2019**, *298*, 124987. [CrossRef]

10. Călinoiu, L.F.; Cătoi, A.-F.; Vodnar, D.C. Solid-State Yeast Fermented Wheat and Oat Bran as A Route for Delivery of Antioxidants. *Antioxidants* **2019**, *8*, 372. [CrossRef]

11. Martău, G.-A.; Unger, P.; Schneider, R.; Venus, J.; Vodnar, D.C.; López-Gómez, J.P. Integration of Solid State and Submerged Fermentations for the Valorization of Organic Municipal Solid Waste. *J. Fungi* **2021**, *7*, 766. [CrossRef]

12. Spaggiari, M.; Ricci, A.; Calani, L.; Bresciani, L.; Neviani, E.; Dall'Asta, C.; Lazzi, C.; Galaverna, G. Solid state lactic acid fermentation: A strategy to improve wheat bran functionality. *LWT* **2020**, *118*, 108668. [CrossRef]

13. Arzami, A.N.; Ho, T.M.; Mikkonen, K.S. Valorization of cereal by-product hemicelluloses: Fractionation and purity considerations. *Food Res. Int.* **2022**, *151*, 110818. [CrossRef] [PubMed]

14. Ndolo, V.U.; Beta, T. Comparative Studies on Composition and Distribution of Phenolic Acids in Cereal Grain Botanical Fractions. *Cereal Chem.* **2014**, *91*, 522–530. [CrossRef]

15. Acosta-Estrada, B.A.; Gutiérrez-Uribe, J.A.; Serna-Saldívar, S.O. Bound phenolics in foods, a review. *Food Chem.* **2014**, *152*, 46–55. [CrossRef]

16. U.S. Department of Agriculture. Wheat Bran, Crude. Available online: https://fdc.nal.usda.gov/fdc-app.html#/food-details/169722/nutrients (accessed on 26 October 2022).

17. U.S. Department of Agriculture. Rice Bran, Crude. Available online: https://fdc.nal.usda.gov/fdc-app.html#/food-details/169713/nutrients (accessed on 26 October 2022).

18. U.S. Department of Agriculture. Oat Bran, Raw. Available online: https://fdc.nal.usda.gov/fdc-app.html#/food-details/168872/nutrients (accessed on 26 October 2022).

19. Cheng, W.; Sun, Y.; Fan, M.; Li, Y.; Wang, L.; Qian, H. Wheat bran, as the resource of dietary fiber: A review. *Crit. Rev. Food Sci. Nutr.* **2021**, *62*, 7269–7281. [CrossRef]

20. Bader Ul Ain, H.; Saeed, F.; Khan, M.A.; Niaz, B.; Khan, S.G.; Anjum, F.M.; Tufail, T.; Hussain, S. Comparative study of chemical treatments in combination with extrusion for the partial conversion of wheat and sorghum insoluble fiber into soluble. *Food Sci. Nutr.* **2019**, *7*, 2059–2067. [CrossRef]

21. Escobar, E.L.N.; da Silva, T.A.; Pirich, C.L.; Corazza, M.L.; Pereira Ramos, L. Supercritical Fluids: A Promising Technique for Biomass Pretreatment and Fractionation. *Front. Bioeng. Biotechnol.* **2020**, *8*, 252. [CrossRef]

22. Teleky, B.-E.; Martău, G.A.; Vodnar, D.C. Physicochemical Effects of *Lactobacillus plantarum* and *Lactobacillus casei* Cocultures on Soy–Wheat Flour Dough Fermentation. *Foods* **2020**, *9*, 1894. [CrossRef]

23. Li, S.; Chen, H.; Cheng, W.; Yang, K.; Cai, L.; He, L.; Du, L.; Liu, Y.; Liu, A.; Zeng, Z.; et al. Impact of arabinoxylan on characteristics, stability and lipid oxidation of oil-in-water emulsions: Arabinoxylan from wheat bran, corn bran, rice bran, and rye bran. *Food Chem.* **2021**, *358*, 129813. [CrossRef]

24. Revanappa, S.B.; Nandini, C.D.; Salimath, P.V. Structural variations of arabinoxylans extracted from different wheat (*Triticum aestivum*) cultivars in relation to *chapati*-quality. *Food Hydrocoll.* **2015**, *43*, 736–742. [CrossRef]

25. Ballance, S.; Lu, Y.; Zobel, H.; Rieder, A.; Knutsen, S.H.; Dinu, V.T.; Christensen, B.E.; Ulset, A.-S.; Schmid, M.; Maina, N.; et al. Inter-laboratory analysis of cereal beta-glucan extracts of nutritional importance: An evaluation of different methods for determining weight-average molecular weight and molecular weight distribution. *Food Hydrocoll.* **2022**, *127*, 107510. [CrossRef]

26. Plamada, D.; Vodnar, D.C. Polyphenols-Gut Microbiota Interrelationship: A Transition to a New Generation of Prebiotics. *Nutrients* **2021**, *14*, 137. [CrossRef] [PubMed]

27. Bannikova, A.; Zyainitdinov, D.; Evteev, A.; Drevko, Y.; Evdokimov, I. Microencapsulation of polyphenols and xylooligosaccharides from oat bran in whey protein-maltodextrin complex coacervates: In-vitro evaluation and controlled release. *Bioact. Carbohydr. Diet. Fibre* **2020**, *23*, 100236. [CrossRef]

28. Cano y Postigo, L.O.; Jacobo-Velázquez, D.A.; Guajardo-Flores, D.; Garcia Amezquita, L.E.; García-Cayuela, T. Solid-state fermentation for enhancing the nutraceutical content of agrifood by-products: Recent advances and its industrial feasibility. *Food Biosci.* **2021**, *41*, 100926. [CrossRef]

29. Roasa, J.; De Villa, R.; Mine, Y.; Tsao, R. Phenolics of cereal, pulse and oilseed processing by-products and potential effects of solid-state fermentation on their bioaccessibility, bioavailability and health benefits: A review. *Trends Food Sci. Technol.* **2021**, *116*, 954–974. [CrossRef]

30. Yin, Z.N.; Wu, W.J.; Sun, C.Z.; Liu, H.F.; Chen, W.B.; Zhan, Q.P.; Lei, Z.G.; Xin, X.; Ma, J.J.; Yao, K.; et al. Antioxidant and Anti-inflammatory Capacity of Ferulic Acid Released from Wheat Bran by Solid-state Fermentation of *Aspergillus niger*. *Biomed. Environ. Sci.* **2019**, *32*, 11–21. [CrossRef]

31. Shin, H.-Y.; Kim, S.-M.; Lee, J.H.; Lim, S.-T. Solid-state fermentation of black rice bran with *Aspergillus awamori* and *Aspergillus oryzae*: Effects on phenolic acid composition and antioxidant activity of bran extracts. *Food Chem.* **2019**, *272*, 235–241. [CrossRef]

32. Li, Y.; Li, M.; Wang, L.; Li, Z. Effect of particle size on the release behavior and functional properties of wheat bran phenolic compounds during in vitro gastrointestinal digestion. *Food Chem.* **2022**, *367*, 130751. [CrossRef]

33. Sánchez-Velázquez, O.A.; Cuevas-Rodríguez, E.O.; Mondor, M.; Ribéreau, S.; Arcand, Y.; Mackie, A.; Hernández-Álvarez, A.J. Impact of in vitro gastrointestinal digestion on peptide profile and bioactivity of cooked and non-cooked oat protein concentrates. *Curr. Res. Food Sci.* **2021**, *4*, 93–104. [CrossRef]

34. Zheng, Y.; Wang, X.; Zhuang, Y.; Li, Y.; Shi, P.; Tian, H.; Li, X.; Chen, X. Isolation of novel ACE-inhibitory peptide from naked oat globulin hydrolysates in silico approach: Molecular docking, in vivo antihypertension and effects on renin and intracellular endothelin-1. *J. Food Sci.* **2020**, *85*, 1328–1337. [CrossRef]

35. Wang, F.; Zhang, Y.; Yu, T.; He, J.; Cui, J.; Wang, J.; Cheng, X.; Fan, J. Oat globulin peptides regulate antidiabetic drug targets and glucose transporters in Caco-2 cells. *J. Funct. Foods* **2018**, *42*, 12–20. [CrossRef]

36. Jahandideh, F.; Bourque, S.L.; Wu, J. A comprehensive review on the glucoregulatory properties of food-derived bioactive peptides. *Food Chem. X* **2022**, *13*, 100222. [CrossRef] [PubMed]

37. Shobako, N.; Ogawa, Y.; Ishikado, A.; Harada, K.; Kobayashi, E.; Suido, H.; Kusakari, T.; Maeda, M.; Suwa, M.; Matsumoto, M.; et al. A Novel Antihypertensive Peptide Identified in Thermolysin-Digested Rice Bran. *Mol. Nutr. Food Res.* **2018**, *62*, 1700732. [CrossRef]

38. Martín-del-Campo, S.T.; Martínez-Basilio, P.C.; Sepúlveda-Álvarez, J.C.; Gutiérrez-Melchor, S.E.; Galindo-Peña, K.D.; Lara-Domínguez, A.K.; Cardador-Martínez, A. Production of Antioxidant and ACEI Peptides from Cheese Whey Discarded from Mexican White Cheese Production. *Antioxidants* **2019**, *8*, 158. [CrossRef] [PubMed]

39. Elbarbary, H.A.; Ejima, A.; Sato, K. Generation of antibacterial peptides from crude cheese whey using pepsin and rennet enzymes at various pH conditions. *J. Sci. Food Agric.* **2019**, *99*, 555–563. [CrossRef]

40. Wang, J.-b.; Liu, X.-r.; Liu, S.-q.; Mao, R.-x.; Hou, C.; Zhu, N.; Liu, R.; Ma, H.-j.; Li, Y. Hypoglycemic Effects of Oat Oligopeptides in High-Calorie Diet/STZ-Induced Diabetic Rats. *Molecules* **2019**, *24*, 558. [CrossRef]

41. Wen, L.; Jiang, Y.; Zhou, X.; Bi, H.; Yang, B. Structure identification of soybean peptides and their immunomodulatory activity. *Food Chem.* **2021**, *359*, 129970. [CrossRef]

42. Wang, X.M.; Chen, H.X.; Fu, X.G.; Li, S.Q.; Wei, J. A novel antioxidant and ACE inhibitory peptide from rice bran protein: Biochemical characterization and molecular docking study. *LWT-Food Sci. Technol.* **2017**, *75*, 93–99. [CrossRef]

43. Hatanaka, T.; Uraji, M.; Fujita, A.; Kawakami, K. Anti-oxidation Activities of Rice-Derived Peptides and Their Inhibitory Effects on Dipeptidylpeptidase-IV. *Int. J. Pept. Res. Ther.* **2015**, *21*, 479–485. [CrossRef]

44. Feng, S.; Wang, L.; Shao, P.; Lu, B.; Chen, Y.; Sun, P. Simultaneous Analysis of Free Phytosterols and Phytosterol Glycosides in Rice Bran by SPE/GC–MS. *Food Chem.* **2022**, *387*, 132742. [CrossRef]

45. Cabral, C.E.; Klein, M. Phytosterols in the Treatment of Hypercholesterolemia and Prevention of Cardiovascular Diseases. *Arq. Bras. Cardiol.* **2017**, *109*, 475–482. [CrossRef] [PubMed]

46. Nurmi, T.; Lampi, A.-M.; Nyström, L.; Hemery, Y.; Rouau, X.; Piironen, V. Distribution and composition of phytosterols and steryl ferulates in wheat grain and bran fractions. *J. Cereal Sci.* **2012**, *56*, 379–388. [CrossRef]

47. Ferrer, A.; Altabella, T.; Arró, M.; Boronat, A. Emerging roles for conjugated sterols in plants. *Prog. Lipid Res.* **2017**, *67*, 27–37. [CrossRef]

48. Zhang, J.; Zhang, T.; Tao, G.; Liu, R.; Chang, M.; Jin, Q.; Wang, X. Characterization and determination of free phytosterols and phytosterol conjugates: The potential phytochemicals to classify different rice bran oil and rice bran. *Food Chem.* **2021**, *344*, 128624. [CrossRef]

49. Deroover, L.; Tie, Y.; Verspreet, J.; Courtin, C.M.; Verbeke, K. Modifying wheat bran to improve its health benefits. *Crit. Rev. Food Sci. Nutr.* **2020**, *60*, 1104–1122. [CrossRef] [PubMed]

50. Si, D.; Shang, T.; Liu, X.; Zheng, Z.; Hu, Q.; Hu, C.; Zhang, R. Production and characterization of functional wheat bran hydrolysate rich in reducing sugars, xylooligosaccharides and phenolic acids. *Biotechnol. Rep.* **2020**, *27*, e00511. [CrossRef]

51. Mosele, J.I.; Macià, A.; Motilva, M.-J. Metabolic and Microbial Modulation of the Large Intestine Ecosystem by Non-Absorbed Diet Phenolic Compounds: A Review. *Molecules* **2015**, *20*, 17429–17468. [CrossRef]

52. Chen, Y.; Ma, Y.; Dong, L.; Jia, X.; Liu, L.; Huang, F.; Chi, J.; Xiao, J.; Zhang, M.; Zhang, R. Extrusion and fungal fermentation change the profile and antioxidant activity of free and bound phenolics in rice bran together with the phenolic bioaccessibility. *LWT* **2019**, *115*, 108461. [CrossRef]

53. Peanparkdee, M.; Patrawart, J.; Iwamoto, S. Effect of extraction conditions on phenolic content, anthocyanin content and antioxidant activity of bran extracts from Thai rice cultivars. *J. Cereal Sci.* **2019**, *86*, 86–91. [CrossRef]

54. Mateo Anson, N.; Aura, A.M.; Selinheimo, E.; Mattila, I.; Poutanen, K.; van den Berg, R.; Havenaar, R.; Bast, A.; Haenen, G.R. Bioprocessing of wheat bran in whole wheat bread increases the bioavailability of phenolic acids in men and exerts antiinflammatory effects ex vivo. *J. Nutr.* **2011**, *141*, 137–143. [CrossRef]

55. Mineo, S.; Takahashi, N.; Yamada-Hara, M.; Tsuzuno, T.; Aoki-Nonaka, Y.; Tabeta, K. Rice bran-derived protein fractions enhance sulforaphane-induced anti-oxidative activity in gingival epithelial cells. *Arch. Oral Biol.* **2021**, *129*, 105215. [CrossRef] [PubMed]

56. Shang, X.-L.; Liu, C.-Y.; Dong, H.-Y.; Peng, H.-H.; Zhu, Z.-Y. Extraction, purification, structural characterization, and antioxidant activity of polysaccharides from Wheat Bran. *J. Mol. Struct.* **2021**, *1233*, 130096. [CrossRef]

57. Bumrungpert, A.; Lilitchan, S.; Tuntipopipat, S.; Tirawanchai, N.; Komindr, S. Ferulic Acid Supplementation Improves Lipid Profiles, Oxidative Stress, and Inflammatory Status in Hyperlipidemic Subjects: A Randomized, Double-Blind, Placebo-Controlled Clinical Trial. *Nutrients* **2018**, *10*, 713. [CrossRef] [PubMed]

58. Kurtys, E.; Eisel, U.L.M.; Hageman, R.J.J.; Verkuyl, J.M.; Broersen, L.M.; Dierckx, R.; de Vries, E.F.J. Anti-inflammatory effects of rice bran components. *Nutr. Rev.* **2018**, *76*, 372–379. [CrossRef] [PubMed]

59. Edrisi, F.; Salehi, M.; Ahmadi, A.; Fararoei, M.; Rusta, F.; Mahmoodianfard, S. Effects of supplementation with rice husk powder and rice bran on inflammatory factors in overweight and obese adults following an energy-restricted diet: A randomized controlled trial. *Eur. J. Nutr.* **2018**, *57*, 833–843. [CrossRef]

60. Mostafa, A.O.; Abdel-Kader, R.M.; Heikal, O.A. Enhancement of cognitive functions by rice bran extract in a neuroinflammatory mouse model via regulation of PPARγ. *J. Funct. Foods* **2018**, *48*, 314–321. [CrossRef]

61. Perez-Ternero, C.; Alvarez de Sotomayor, M.; Herrera, M.D. Contribution of ferulic acid, γ-oryzanol and tocotrienols to the cardiometabolic protective effects of rice bran. *J. Funct. Foods* **2017**, *32*, 58–71. [CrossRef]

62. Perez-Ternero, C.; Bermudez Pulgarin, B.; Alvarez de Sotomayor, M.; Herrera, M.D. Atherosclerosis-related inflammation and oxidative stress are improved by rice bran enzymatic extract. *J. Funct. Foods* **2016**, *26*, 610–621. [CrossRef]

63. Perez-Ternero, C.; Herrera, M.D.; Laufs, U.; Alvarez de Sotomayor, M.; Werner, C. Food supplementation with rice bran enzymatic extract prevents vascular apoptosis and atherogenesis in ApoE−/− mice. *Eur. J. Nutr.* **2017**, *56*, 225–236. [CrossRef] [PubMed]

64. Bao, L.; Cai, X.; Xu, M.; Li, Y. Effect of oat intake on glycaemic control and insulin sensitivity: A meta-analysis of randomised controlled trials. *Br. J. Nutr.* **2014**, *112*, 457–466. [CrossRef]

65. Ijarotimi, O.S.; Fakayejo, D.A.; Oluwajuyitan, T.D. Nutritional Characteristics, Glycaemic Index and Blood Glucose Lowering Property of Gluten-Free Composite Flour from Wheat (*Triticum aestivum*), Soybean (*Glycine max*), Oat-Bran (*Avena sativa*) and Rice-Bran (*Oryza sativa*). *Appl. Food Res.* **2021**, *1*, 100022. [CrossRef]

66. Oluwajuyitan, T.D.; Ijarotimi, O.S.; Fagbemi, T.N.; Oboh, G. Blood glucose lowering, glycaemic index, carbohydrate-hydrolysing enzyme inhibitory activities of potential functional food from plantain, soy-cake, rice-bran and oat-bran flour blends. *J. Food Meas. Charact.* **2021**, *15*, 3761–3769. [CrossRef]

67. Turrini, E.; Maffei, F.; Milelli, A.; Calcabrini, C.; Fimognari, C. Overview of the Anticancer Profile of Avenanthramides from Oat. *Int. J. Mol. Sci.* **2019**, *20*, 4536. [CrossRef]

68. Nemoto, H.; Otake, M.; Matsumoto, T.; Izutsu, R.; Jehung, J.P.; Goto, K.; Osaki, M.; Mayama, M.; Shikanai, M.; Kobayashi, H.; et al. Prevention of tumor progression in inflammation-related carcinogenesis by anti-inflammatory and anti-mutagenic effects brought about by ingesting fermented brown rice and rice bran with *Aspergillus oryzae* (FBRA). *J. Funct. Foods* **2022**, *88*, 104907. [CrossRef]

69. Kannan, A.; Hettiarachchy, N.; Narayan, S. Colon and breast anti-cancer effects of peptide hydrolysates derived from rice bran. *Open Bioact. Compd. J.* **2009**, *2*, 17–20. [CrossRef]

70. Yao, W.; Gong, Y.; Li, L.; Hu, X.; You, L. The effects of dietary fibers from rice bran and wheat bran on gut microbiota: An overview. *Food Chem. X* **2022**, *13*, 100252. [CrossRef]

71. Zhang, X.; Dong, L.; Jia, X.; Liu, L.; Chi, J.; Huang, F.; Ma, Q.; Zhang, M.; Zhang, R. Bound Phenolics Ensure the Antihyperglycemic Effect of Rice Bran Dietary Fiber in db/db Mice via Activating the Insulin Signaling Pathway in Skeletal Muscle and Altering Gut Microbiota. *J. Agric. Food Chem.* **2020**, *68*, 4387–4398. [CrossRef]

72. Sheflin, A.M.; Borresen, E.C.; Kirkwood, J.S.; Boot, C.M.; Whitney, A.K.; Lu, S.; Brown, R.J.; Broeckling, C.D.; Ryan, E.P.; Weir, T.L. Dietary supplementation with rice bran or navy bean alters gut bacterial metabolism in colorectal cancer survivors. *Mol. Nutr. Food Res.* **2017**, *61*, 1500905. [CrossRef] [PubMed]

73. Tan, J.; McKenzie, C.; Potamitis, M.; Thorburn, A.N.; Mackay, C.R.; Macia, L. Chapter Three—The Role of Short-Chain Fatty Acids in Health and Disease. In *Advances in Immunology*; Alt, F.W., Ed.; Academic Press: Cambridge, MA, USA, 2014; Volume 121, pp. 91–119.

74. Maki, K.C.; Gibson, G.R.; Dickmann, R.S.; Kendall, C.W.C.; Chen, C.Y.O.; Costabile, A.; Comelli, E.M.; McKay, D.L.; Almeida, N.G.; Jenkins, D.; et al. Digestive and physiologic effects of a wheat bran extract, arabino-xylan-oligosaccharide, in breakfast cereal. *Nutrition* **2012**, *28*, 1115–1121. [CrossRef]

75. Tian, X.; Wang, Z.; Wang, X.; Ma, S.; Sun, B.; Wang, F. Mechanochemical effects on the structural properties of wheat starch during vibration ball milling of wheat endosperm. *Int. J. Biol. Macromol.* **2022**, *206*, 306–312. [CrossRef]

76. Cantu-Jungles, T.M.; Zhang, X.; Kazem, A.E.; Iacomini, M.; Hamaker, B.R.; Cordeiro, L.M.C. Microwave treatment enhances human gut microbiota fermentability of isolated insoluble dietary fibers. *Food Res. Int.* **2021**, *143*, 110293. [CrossRef] [PubMed]

77. Khadhraoui, B.; Ummat, V.; Tiwari, B.K.; Fabiano-Tixier, A.S.; Chemat, F. Review of ultrasound combinations with hybrid and innovative techniques for extraction and processing of food and natural products. *Ultrason. Sonochem.* **2021**, *76*, 105625. [CrossRef]

78. Jha, A.K.; Sit, N. Extraction of bioactive compounds from plant materials using combination of various novel methods: A review. *Trends Food Sci. Technol.* **2022**, *119*, 579–591. [CrossRef]

79. Song, L.-w.; Qi, J.-r.; Liao, J.-s.; Yang, X.-q. Enzymatic and enzyme-physical modification of citrus fiber by xylanase and planetary ball milling treatment. *Food Hydrocoll.* **2021**, *121*, 107015. [CrossRef]

80. Zhao, G.; Zhang, R.; Dong, L.; Huang, F.; Tang, X.; Wei, Z.; Zhang, M. Particle size of insoluble dietary fiber from rice bran affects its phenolic profile, bioaccessibility and functional properties. *LWT* **2018**, *87*, 450–456. [CrossRef]

81. Wang, T.; Chen, X.; Wang, W.; Wang, L.; Jiang, L.; Yu, D.; Xie, F. Effect of ultrasound on the properties of rice bran protein and its chlorogenic acid complex. *Ultrason. Sonochem.* **2021**, *79*, 105758. [CrossRef] [PubMed]

82. Chen, C.; Wang, L.; Wang, R.; Luo, X.; Li, Y.; Li, J.; Li, Y.; Chen, Z. Ultrasound-assisted extraction from defatted oat (*Avena sativa* L.) bran to simultaneously enhance phenolic compounds and β-glucan contents: Compositional and kinetic studies. *J. Food Eng.* **2018**, *222*, 1–10. [CrossRef]

83. Milić, P.S.; Rajković, K.M.; Stamenković, O.S.; Veljković, V.B. Kinetic modeling and optimization of maceration and ultrasound-extraction of resinoid from the aerial parts of white lady's bedstraw (*Galium mollugo* L.). *Ultrason. Sonochem.* **2013**, *20*, 525–534. [CrossRef]

84. Suresh, S.; Sivaramakrishnan, R.; Radha, K.V.; Incharoensakdi, A.; Pugazhendhi, A. Ultrasound pretreated rice bran for *Rhizopus* sp. phytase production as a feed. *Food Biosci.* **2021**, *43*, 101281. [CrossRef]

85. Herrera-Ponce, A.L.; Salmeron-Ochoa, I.; Rodriguez-Figueroa, J.C.; Santellano-Estrada, E.; Garcia-Galicia, I.A.; Alarcon-Rojo, A.D. High-intensity ultrasound as pre-treatment in the development of fermented whey and oat beverages: Effect on the fermentation, antioxidant activity and consumer acceptance. *J. Food Sci. Technol.* **2022**, *59*, 796–804. [CrossRef]

86. Vaitkeviciene, R.; Zadeike, D.; Gaizauskaite, Z.; Valentaviciute, K.; Marksa, M.; Mazdzieriene, R.; Bartkiene, E.; Lele, V.; Juodeikiene, G.; Jakstas, V. Functionalisation of rice bran assisted by ultrasonication and fermentation for the production of rice bran–lingonberry pulp-based probiotic nutraceutical. *Int. J. Food Sci. Technol.* **2022**, *57*, 1462–1472. [CrossRef]

87. Călinoiu, L.F.; Vodnar, D.C. Thermal Processing for the Release of Phenolic Compounds from Wheat and Oat Bran. *Biomolecules* **2020**, *10*, 21. [CrossRef] [PubMed]

88. Kong, F.; Wang, L.; Gao, H.; Chen, H. Process of steam explosion assisted superfine grinding on particle size, chemical composition and physico-chemical properties of wheat bran powder. *Powder Technol.* **2020**, *371*, 154–160. [CrossRef]

89. Liu, L.; Zhao, M.; Liu, X.; Zhong, K.; Tong, L.; Zhou, X.; Zhou, S. Effect of steam explosion-assisted extraction on phenolic acid profiles and antioxidant properties of wheat bran. *J. Sci. Food Agric.* **2016**, *96*, 3484–3491. [CrossRef] [PubMed]

90. Lin, D.; Long, X.; Huang, Y.; Yang, Y.; Wu, Z.; Chen, H.; Zhang, Q.; Wu, D.; Qin, W.; Tu, Z. Effects of microbial fermentation and microwave treatment on the composition, structural characteristics, and functional properties of modified okara dietary fiber. *LWT* **2020**, *123*, 109059. [CrossRef]

91. Alonso, E. The role of supercritical fluids in the fractionation pretreatments of a wheat bran-based biorefinery. *J. Supercrit. Fluids* **2018**, *133*, 603–614. [CrossRef]

92. Scapini, T.; dos Santos, M.S.N.; Bonatto, C.; Wancura, J.H.C.; Mulinari, J.; Camargo, A.F.; Klanovicz, N.; Zabot, G.L.; Tres, M.V.; Fongaro, G.; et al. Hydrothermal pretreatment of lignocellulosic biomass for hemicellulose recovery. *Bioresour. Technol.* **2021**, *342*, 126033. [CrossRef]

93. Cantero, D.A.; Martínez, C.; Bermejo, M.; Cocero, M. Simultaneous and selective recovery of cellulose and hemicellulose fractions from wheat bran by supercritical water hydrolysis. *Green Chem.* **2015**, *17*, 610–618. [CrossRef]

94. Paudel, S.R.; Banjara, S.P.; Choi, O.K.; Park, K.Y.; Kim, Y.M.; Lee, J.W. Pretreatment of agricultural biomass for anaerobic digestion: Current state and challenges. *Bioresour. Technol.* **2017**, *245*, 1194–1205. [CrossRef]

95. Wang, L.; Tian, Y.; Chen, Y.; Chen, J. Effects of acid treatment on the physicochemical and functional properties of wheat bran insoluble dietary fiber. *Cereal Chem.* **2021**, *99*, 343–354. [CrossRef]

96. Qi, J.; Li, Y.; Masamba, K.G.; Shoemaker, C.F.; Zhong, F.; Majeed, H.; Ma, J. The effect of chemical treatment on the In vitro hypoglycemic properties of rice bran insoluble dietary fiber. *Food Hydrocoll.* **2016**, *52*, 699–706. [CrossRef]

97. Rahman, A.; Fehrenbach, J.; Ulven, C.; Simsek, S.; Hossain, K. Utilization of wheat-bran cellulosic fibers as reinforcement in bio-based polypropylene composite. *Ind. Crops Prod.* **2021**, *172*, 114028. [CrossRef]

98. Xue, Y.; Cui, X.; Zhang, Z.; Zhou, T.; Gao, R.; Li, Y.; Ding, X. Effect of β-endoxylanase and α-arabinofuranosidase enzymatic hydrolysis on nutritional and technological properties of wheat brans. *Food Chem.* **2020**, *302*, 125332. [CrossRef]

99. Garrido-Galand, S.; Asensio-Grau, A.; Calvo-Lerma, J.; Heredia, A.; Andrés, A. The potential of fermentation on nutritional and technological improvement of cereal and legume flours: A review. *Food Res. Int.* **2021**, *145*, 110398. [CrossRef]

100. Ritthibut, N.; Oh, S.-J.; Lim, S.-T. Enhancement of bioactivity of rice bran by solid-state fermentation with *Aspergillus* strains. *LWT* **2021**, *135*, 110273. [CrossRef]

101. Bei, Q.; Chen, G.; Lu, F.; Wu, S.; Wu, Z. Enzymatic action mechanism of phenolic mobilization in oats (*Avena sativa* L.) during solid-state fermentation with Monascus anka. *Food Chem.* **2018**, *245*, 297–304. [CrossRef] [PubMed]

102. Janarny, G.; Gunathilake, K.D.P.P. Changes in rice bran bioactives, their bioactivity, bioaccessibility and bioavailability with solid-state fermentation by Rhizopus oryzae. *Biocatal. Agric. Biotechnol.* **2020**, *23*, 101510. [CrossRef]

103. Mao, M.; Wang, P.; Shi, K.; Lu, Z.; Bie, X.; Zhao, H.; Zhang, C.; Lv, F. Effect of solid state fermentation by Enterococcus faecalis M2 on antioxidant and nutritional properties of wheat bran. *J. Cereal Sci.* **2020**, *94*, 102997. [CrossRef]

104. Tanasković, S.J.; Šekuljica, N.; Jovanović, J.; Gazikalović, I.; Grbavčić, S.; Đorđević, N.; Sekulić, M.V.; Hao, J.; Luković, N.; Knežević-Jugović, Z. Upgrading of valuable food component contents and anti-nutritional factors depletion by solid-state fermentation: A way to valorize wheat bran for nutrition. *J. Cereal Sci.* **2021**, *99*, 103159. [CrossRef]

105. Kaur, B.; Chakraborty, D.; Kaur, G.; Kaur, G. Biotransformation of Rice Bran to Ferulic Acid by Pediococcal Isolates. *Appl. Biochem. Biotechnol.* **2013**, *170*, 854–867. [CrossRef]

106. Martău, G.A.; Călinoiu, L.-F.; Vodnar, D.C. Bio-vanillin: Towards a sustainable industrial production. *Trends Food Sci. Technol.* **2021**, *109*, 579–592. [CrossRef]

107. Vu, V.; Farkas, C.; Riyad, O.; Bujna, E.; Kilin, A.; Sipiczki, G.; Sharma, M.; Usmani, Z.; Gupta, V.K.; Nguyen, Q.D. Enhancement of the enzymatic hydrolysis efficiency of wheat bran using the *Bacillus strains* and their consortium. *Bioresour. Technol.* **2022**, *343*, 126092. [CrossRef] [PubMed]

108. Rustiguel, C.B.; Jorge, J.A.; Guimarães, L.H.S. Characterization of a thermo-tolerant mycelial β-fructofuranosidase from *Aspergillus phoenicis* under submerged fermentation using wheat bran as carbon source. *Biocatal. Agric. Biotechnol.* **2015**, *4*, 362–369. [CrossRef]

109. Chaiyasut, C.; Pengkumsri, N.; Sirilun, S.; Peerajan, S.; Khongtan, S.; Sivamaruthi, B.S. Assessment of changes in the content of anthocyanins, phenolic acids, and antioxidant property of Saccharomyces cerevisiae mediated fermented black rice bran. *AMB Express* **2017**, *7*, 114. [CrossRef] [PubMed]
110. Akhavan Sepahy, A.; Ghazi, S.; Akhavan Sepahy, M. Cost-Effective Production and Optimization of Alkaline Xylanase by Indigenous *Bacillus mojavensis* AG13 Fermented on Agricultural Waste. *Enzym. Res.* **2011**, *2011*, 593624. [CrossRef]

Review

Antioxidant Potential and Capacity of Microorganism-Sourced C$_{30}$ Carotenoids—A Review

Inonge Noni Siziya [1,2], Chi Young Hwang [3] and Myung-Ji Seo [1,2,3,*]

1   Division of Bioengineering, Incheon National University, Incheon 22012, Korea
2   Research Center for Bio Material & Process Development, Incheon National University, Incheon 22012, Korea
3   Department of Bioengineering and Nano-Bioengineering, Incheon National University, Incheon 22012, Korea
*   Correspondence: mjseo@inu.ac.kr; Tel.: +82-32-835-8267

Abstract: Carotenoids are lipophilic tetraterpenoid pigments produced by plants, algae, arthropods, and certain bacteria and fungi. These biologically active compounds are used in the food, feed, and nutraceutical industries for their coloring and the physiological benefits imparted by their antioxidant properties. The current global carotenoid market is dominated by synthetic carotenoids; however, the rising consumer demand for natural products has led to increasing research and development in the mass production of carotenoids from alternative natural sources, including microbial synthesis and plant extraction, which holds a significant market share. To date, microbial research has focused on C$_{40}$ carotenoids, but studies have shown that C$_{30}$ carotenoids contain similar—and in some microbial strains, greater—antioxidant activity in both the physical and chemical quenching of reactive oxygen species. The discovery of carotenoid biosynthetic pathways in different microorganisms and advances in metabolic engineering are driving the discovery of novel C$_{30}$ carotenoid compounds. This review highlights the C$_{30}$ carotenoids from microbial sources, showcasing their antioxidant properties and the technologies emerging for their enhanced production. Industrial applications and tactics, as well as biotechnological strategies for their optimized synthesis, are also discussed.

Keywords: C$_{30}$ carotenoid; antioxidant; microorganism; production

Citation: Siziya, I.N.; Hwang, C.Y.; Seo, M.-J. Antioxidant Potential and Capacity of Microorganism-Sourced C$_{30}$ Carotenoids—A Review. *Antioxidants* **2022**, *11*, 1963. https://doi.org/10.3390/antiox11101963

Academic Editor: Stanley Omaye

Received: 7 September 2022
Accepted: 27 September 2022
Published: 30 September 2022

Publisher's Note: MDPI stays neutral with regard to jurisdictional claims in published maps and institutional affiliations.

1. Introduction

Carotenoids are a broad class of lipophilic pigments that are naturally synthesized by photosynthetic and nonphotosynthetic organisms, including plants and certain species of algae, bacteria, fungi, and yeasts [1]. De novo carotenoid synthesis has also been observed within certain species of arthropods such as aphids, cicadas, and stink bugs [2–4]. Although noncarotenogenic organisms such as humans and crustaceans cannot synthesize these compounds, they can obtain carotenoids from their diet and accumulate them in their tissues [1,5]. Carotenoids have antioxidant, anti-inflammatory, and anticancer properties attributed to their biochemical characteristics. These properties have led to a rise in the demand for carotenoids in the food, feed, cosmetic, and pharmacological industries, where they are employed as colorants, biofortification compounds, immunoregulators, and preventative supplements for conditions associated with metabolic syndrome [6,7].

Commercially available carotenoids are mainly obtained from plant extracts or chemical synthesis, both of which are not sustainable and can be detrimental to the environment. As such, there is an inclination toward safer, more environmentally friendly options for consumer products. Carotenoids from plant extracts are limited by seasonal and geographic conditions, whereas those obtained from chemical synthesis yield undesired byproducts and hazardous waste.

Conversely, microbial production is a practical alternative from a commercial and ecological perspective. Bacterial sources of carotenoids include nonphototrophic and phototrophic bacteria. Anoxygenic phototrophs are commonly found in aquatic environments and require solar energy for growth and metabolism, performing photosynthesis without evolving oxygen, whereas oxygenic phototrophs, namely Cyanobacteria, produce oxygen during the photosynthetic process [8]. Carotenoid production in microorganisms can occur to mitigate oxidative damage as part of the microbial preventative and responsive defense mechanisms against reactive oxygen species (ROS). Various bacterial strains contain distinct pathways for carotenoid production. The final products of these pathways have different structures depending on the pathways, enzymes, and environmental conditions.

Carotenoids generally possess a linear hydrocarbon backbone composed of C_5 isoprenoid units present as six, eight, nine, and ten units, which are classified as C_{30}, C_{40}, C_{45}, and C_{50}, respectively. C_{40} carotenoids are the most abundant in nature, and thus the most researched [9]. Carotenoids contain double bonds in their structures, which allow for numerous stereochemical arrangements, and account for their strong antioxidant potential [10]. The extent of the double bonds largely defines the carotenoid absorption of light and spectral properties, which typically fall between 400 and 550 nm, from violet to green light [11]. Within this range, carotenoid colors are detected from yellow to red to violet. However, unpigmented precursors are also present within carotenoid production pathways [12,13]. Within the carotenoid groups, C_{30} carotenoids are distributed among Euryarchaeota, Firmicutes, Cyanobacteria, Alphaproteobacteria, and Gammaproteobacteria [14]. Currently, few natural sources of C_{30} carotenoids are known, and they are more commonly found in Gram-positive bacteria such as *Lactiplantibacillus plantarum* [15], *Cytobacillus firmus* (originally named *Bacillus firmus*) [16], and *Staphylococcus aureus* [17].

The majority of carotenoid research is centered on the more naturally abundant C_{40} carotenoids, resulting in a gap in the knowledge of other carotenoid classes. Disseminating information on the production of C_{30} carotenoids with health benefits comparable to C_{40} carotenoids is advantageous in expanding the current market share of carotenoids. The antioxidant properties of C_{30} carotenoids, where similar or greater than their C_{40} counterparts, provide alternative sources of natural carotenoids with varying applications in the food, feed, and health sectors. This article discusses the current knowledge on microbial production of C_{30} carotenoids and their commercial applications considering their antioxidant content and potential in the food and pharmaceutical industries.

2. Microbial C_{30} Carotenoids and Their Derivatives

Like most carotenoids, the isoprene units that comprise C_{30} carotenoids are constructed from the isopentyl pyrophosphate (IPP) and dimethylallyl pyrophosphate (DMAPP) precursors from the mevalonate (MVA) pathway in plants and microalgae (Figure 1), and the methylerythritol phosphate (MEP) pathway in bacteria and fungi (Figure 2) [18]. The biosynthesis of C_{30} carotenoids occurs via either the condensation of two C_{15} farnesyl diphosphate (FPP) molecules, or the condensation of C_{10} geranyl diphosphate (GPP) and C_{20} geranylgeranyl diphosphate (GGPP). The former results in symmetric 4,4'-diapocarotenoids, whereas the latter produces asymmetric apo-8'-carotenoids [10,19,20].

The C_{30} 4,4'-diapocarotenoids include 4,4'-diapocarotene-4-oic acid and fatty acid esters di-(β,D-glucosyl)-4,4'-diapocarotene-4,4'-dioate from *Methylobacterium rhodinum* (formerly *Pseudomonas rhodos*) [21]; 4,4'-diaponeurosporene and OH-diaponeurosporene from Heliobacteria [22,23]; and staphyloxanthin, 4,4'-diapophytoene, 4,4'-diapophytofluene, 4-4'-diapo-zeta-carotene, 4,4'-diaponeurosporene, and several of its derivatives from *S. aureus* [24,25] (Figure 3). Biosynthetic pathways of known carotenoids can be extended or altered through the inclusion of genes that further alter the forms of the carotenoids. Additionally, novel derivatives of these compounds can be produced by modifying reactions such as methylation, cyclization, and oxygenation.

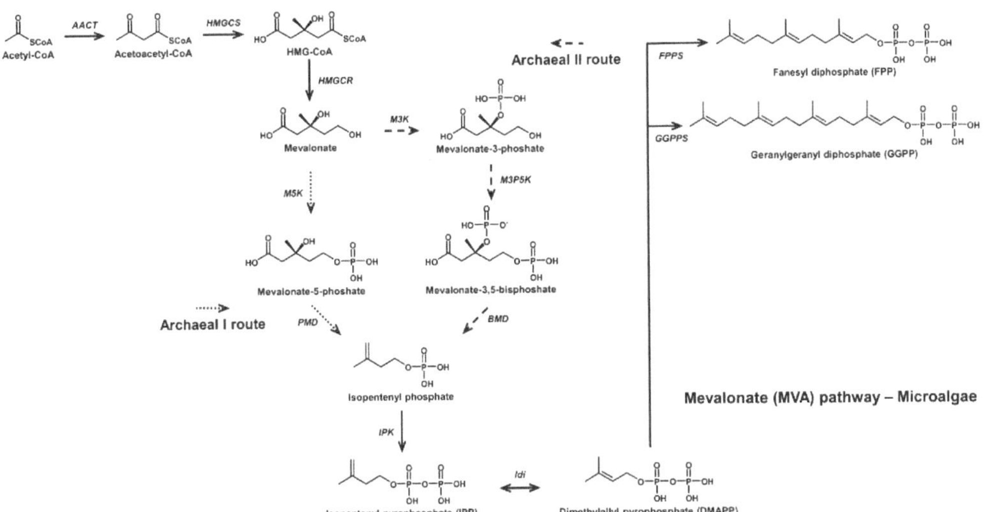

Figure 1. Mevalonate (MVA) pathway as taken by microalgae. AACT: Acetoacetyl-CoA thiolase. HMGCS: HMG-CoA synthase. HMGCR: HMG-CoA reductase. M3K: Mevalonate-3-kinase. M5K: Mevalonate-5-kinase. M3P5K: Mevalonate-3-phosphate-5-kinase. PMD: Phosphomevalonate decarboxylase. BMD: Bisphosphomevalonate decarboxylase. IPK: Isopentenyl phosphate kinase. Idi: Isopentenyl pyrophosphate isomerase. FPPS: Farnesyl pyrophosphate synthase. GGPPS: Geranylgeranyl pyrophosphate synthase.

Figure 2. Methyl erythritol 4-phosphate (MEP) pathway as taken by bacteria and fungi. DXS: Deoxyxylulose 5-phosphate synthase. DXR: Deoxyxylulose 5-phosphate reductoisomerase. MCT: 2-C-methyl-d-erythritol 4-phosphate cytidylyltransferase. CMK: 4-diphosphocytidyl-2-C-methyl-D-erythritol kinase. MDS: 2-C-methyl-d-erythritol 2,4-cyclodiphosphate synthase. HDS: 4-hydroxy-3-methylbut-2-en-1-yl diphosphate synthase. HDR: (E)-4-hydroxy-3-methylbut-2-enyl diphosphate reductase. IPPI: Isopentenyl diphosphate isomerase. GPPS: Geranyl diphosphate synthase. GGPPS: Geranylgeranyl diphosphate synthase.

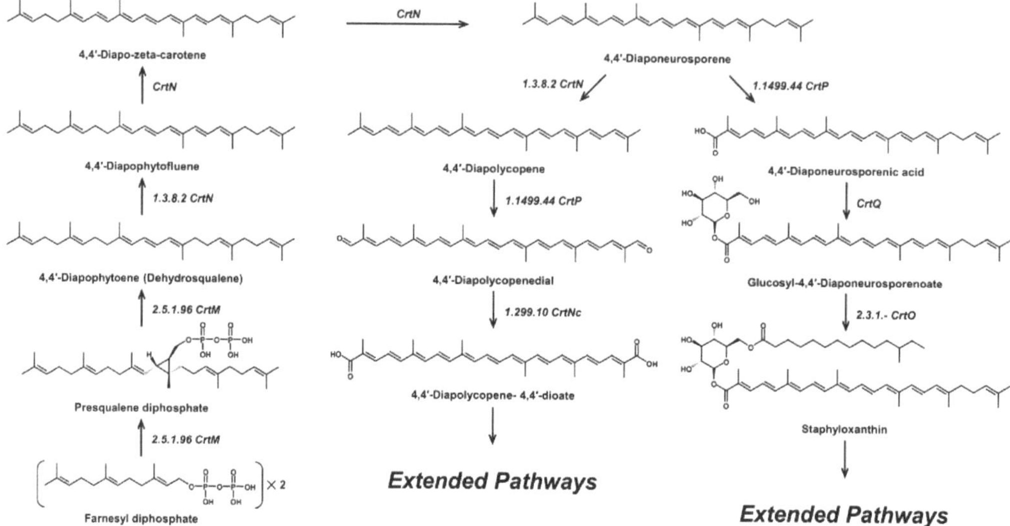

Figure 3. Biosynthetic pathway of 4,4′-diapocarotenoids. CrtM: 4,4′-Diapophytoene synthase. CrtN: 4,4′-Diapophytoene desaturase. CrtP: 4,4′-Diaponeurosporene oxidase. CrtNc: 4,4′-diapolycopene aldehyde oxidase. CrtQ: 4,4′-diaponeurosporenoic acid glycosyltransferase. CrtO: Glycosyl-4,4′-diaponeurosporenoate acyltransferase.

The asymmetrical C_{30} apo-8′-carotenoids have three conjugated double bonds not found in the center of their skeletal structure (Figure 4), and include methyl glucosyl-3,4-dehydro-apo-8′-lycopenoate from *Planococcus* [26], and hydroxy-3,4-dehydro-apo-8′-lycopene and methyl hydroxy-3,4-dehydro-apo-8′-lycopenoate from *Halobacillus* [27]. Those of *Planococcus* were isolated products of natural biosynthesis beyond the known pathways. This suggests the possibility of gene clusters with unconfirmed genes that allow for their synthesis or other reactions within the bacteria. However, hydroxy-3,4-dehydro-apo-8′-lycopene and methyl hydroxy-3,4-dehydro-apo-8′-lycopenoate were produced after chemical mutagenesis of the microorganisms.

Different microbial C_{30} carotenoids originate from a limited number of microorganisms. Their properties are dependent on their structures, which can be altered further by changes or extensions in the biosynthetic pathways. Several C_{30} carotenoids are presented in Table 1 with their structures, bacterial sources, and antioxidant activities compared to those of C_{40} carotenoids and other standard compounds for the determination of antioxidant capacity.

4,4′-Diaponeurosporene ($C_{30}H_{42}$; also known as 7,8-dihydro-4,4′-diapo-ψ,ψ-carotene and *all-trans*-4,4′-diaponeurosporene) has been isolated from *S. aureus*, Heliobacteria, and *L. plantarum* subsp. *plantarum*, among others [17,22,28–30]. Research on the deep-yellow pigment revealed its potential against *Salmonella typhimurium* in a mouse model. The compound enhanced the immune system response and T-cell stimulation. It is also highly resistant to external stresses, consistent with its antioxidant potency [31]. These properties highlight its potential for medical or nutraceutical applications in immunocompromised individuals [30,32].

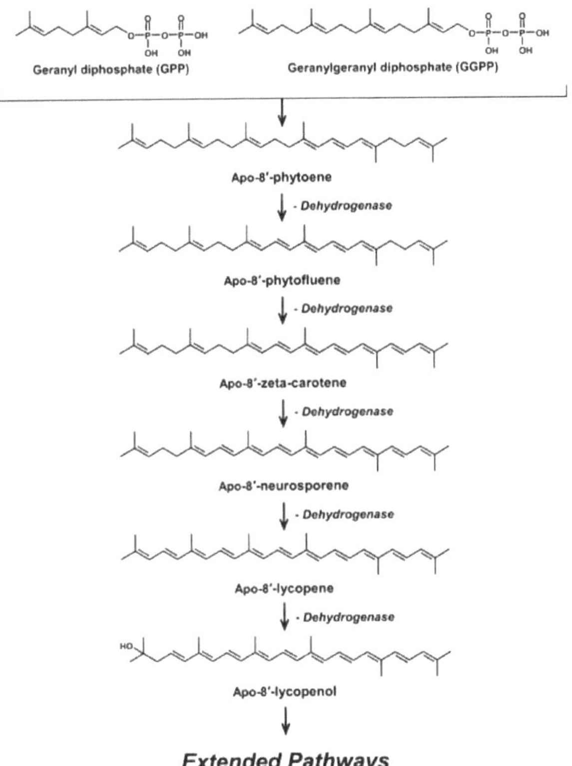

Figure 4. Biosynthetic pathway of Apo-8′-carotenoids.

Colorless 4,4′-diapophytoene ($C_{30}H_{48}$; also called *all-trans*-4,4′-diapophytoene and dehydrosqualene) was the first C_{30} intermediate in the staphyloxanthin synthesis pathway [24,25]. Perez-Fons et al. [19] synthesized and isolated the C_{30} carotenoid apophytoene ($C_{30}H_{48}$) from spore-forming Bacillus species. Additionally, the C_{30} apocarotenoid derivatives of 1-glycosyl-3-4-dehydro-8′-apolycopene ester and methyl 1-glycosyl-3,4-dehydro-8′-apolycopenate ester were obtained in pigmented *Bacillus* species. The former was more abundant in vegetative cells, whereas the latter was dominant during spore formation. The carotenoids were considered apolycopene derivatives, which are different from those of diaponeurosporene (Table 1).

The bacteria *Planococcus maritimus* produces the acyclic C_{30} carotenoid methyl 5-glucosyl-5,6-dihydro-4,4′-diapolycopenoate, as well as C_{30} intermediates 5-hydroxy-5,6-dihydro-4,4′-diaponeurosporene, 5-glucosyl-5,6-dihydro-4,4′-diapolycopene, and 5-hydroxy-5,6-dihydro-4,4′-diapolycopene [33].

Staphyloxanthin ($C_{51}H_{78}O_8$, also called 8′-apo-ψ,ψ-carotenoic acid) is a golden yellow pigment used to initiate immune system responses for the survival of infected host cells. The compound is more commonly associated with *S. aureus*, although several other species have been found to contain it, including *S. gallinarum* [34] and *S. carnosus*, from which it was identified as β-D-glucopyranosyl 1-O-(4,4′-diaponeurosporen-4-oate)-6-O-(12-methyltetradecanoate) [35].

Table 1. Antioxidant contents and capacities of C_{30} carotenoids.

C_{30} Carotenoid	Quenching Type	Antioxidant Activity		Conjugated Double Bond	Group Contained	H Bond Donor /Acceptor	Microbial Sources [Ref.]
		Carotenoid Content	Antioxidant Control				
4,4′-Diapolycopenedial (4,4′-diapolycopen-4,4′-dial)/$C_{30}H_{36}O_2$	DPPH scavenging	7.5 μM (IC_{50})	DL-α-tocopherol, 33.2 μM (IC_{50})	13	Aldehyde (2)	0/2	*Staphylococcus* spp. [36], *Methylomonas* spp. [37]
4,4′-Diapocaroten-4′-al-4-oic acid/$C_{30}H_{36}O_3$	-	-	-	13	Carboxylic acid, aldehyde	1/3	*Pseudomonas rhodos* [21]
4,4′-Diapolycopene-4,4′-dioic acid/$C_{30}H_{36}O_4$	Singlet oxygen quenching	5.8 μM	Astaxanthin, 9.3 μM	13	Carboxylic acid (2)	2/4	*Cytobacillus firmus* [16]
4,4′-Diapolycopen-4-al (4, 4′-Diapolycopenal)/$C_{30}H_{38}O$	-	-	-	12	Aldehyde	0/1	*Staphylococcus aureus* [24]
4,4′-Diapocarotenoic acid/$C_{30}H_{38}O_2$	-	-	-	12	Carboxylic acid	1/2	*Pseudomonas rhodos* [21]
4,4′-Diapolycopene/$C_{30}H_{40}$	DPPH scavenging	8.7 μM (IC_{50})	DL-α-tocopherol, 33.2 μM (IC_{50})	11	Hydrocarbon	0/0	*Heliobacteria* spp. [36]

Table 1. *Cont.*

C_{30} Carotenoid	Quenching Type	Antioxidant Activity		Conjugated Double Bond	Group Contained	H Bond Donor/Acceptor	Microbial Sources [Ref.]
		Carotenoid Content	Antioxidant Control				
Methyl hydroxy-3,4-dehydro-apo-8′-lycopenoate /$C_{31}H_{42}O_3$	Singlet oxygen quenching	44 μM (IC$_{50}$)		11	Ester, hydroxyl	1/3	*Halobacillus halophilus* [27]
Methyl glucosyl-3,4-dehydro-apo-8′-lycopenoate /$C_{37}H_{52}O_8$	Singlet oxygen quenching	5.1 μM (IC$_{50}$)	Astaxanthin, 8.9 μM (IC$_{50}$); β-carotene, >100 μM (IC$_{50}$)	11	Ester, glucosyl	4/8	*Planococcus maritimus* [26], *Halobacillus halophilus* [27]
Hydroxy-3,4-dehydro-apo-8′-lycopene /$C_{30}H_{42}O$	Singlet oxygen quenching	79 μM (IC$_{50}$)		10	Hydroxyl	1/1	*Halobacillus Halophilus* [27]
4,4′-Diaponeurosporen-4-al (Diaponeurosporenal)/$C_{30}H_{40}O$	DPPH scavenging	10.2 μM (IC$_{50}$)	DL-α-tocopherol, 33.2 μM (IC$_{50}$)	10	Aldehyde	0/1	*Staphylococcus aureus* [36]
Hydroxy-diaponeurosporenal/$C_{30}H_{40}O_2$	-	-	-	10	Aldehyde, hydroxyl	1/2	*Staphylococcus aureus* [38]
4,4′-Diaponeurosporen-4-oic acid/$C_{30}H_{40}O_2$	DPPH scavenging	9.7 μM (IC$_{50}$)	DL-α-tocopherol, 33.2 μM (IC$_{50}$)	10	Carboxylic acid	1/2	*Staphylococcus aureus* [36]

Table 1. *Cont.*

C_{30} Carotenoid	Quenching Type	Antioxidant Activity		Conjugated Double Bond	Group Contained	H Bond Donor /Acceptor	Microbial Sources [Ref.]
		Carotenoid Content	Antioxidant Control				
Staphyloxanthin/$C_{51}H_{78}O_8$	DPPH scavenging	54.22 µg/mL (IC_{50})	Ascorbic acid, 35.54 µg/mL (IC_{50})	10	Ketone	3/8	*Staphylococcus* spp. [34]
4,4′-Diaponeurosporene/$C_{30}H_{42}$	DPPH scavenging, singlet oxygen quenching	27.1% 11.6 µM (IC_{50}) 45 µM	BHT, 50 µg/mL DL-α-tocopherol, 33.2 µM (IC_{50}) astaxanthin, 3.7 µM	9	Hydrocarbon	0/0	*Staphylococcus aureus* [17], *Heliobacteria* spp. [22], *Lactiplantibacillus plantarum* [29,36]
4-Hydroxy-4,4′-diaponeurosporene/$C_{30}H_{42}O$	-	-	-	9	Hydroxyl	1/1	*Streptococcus faecium* [39]
4-(D-Glucopyranosyloxy)-4,4′-diaponeurosporene/$C_{36}H_{52}O_6$	-	-	-	9	Ester, glycosyl	4/6	*Streptococcus faecium* [39]
OH-Diaponeurosporene glucoside ester /$C_{40}H_{60}O_7$	-	-	-	9	Ester	3/7	*Heliobacteria* spp. [23]
4,4′-Diapo-7,8,11,12-tetrahydrolycopene /$C_{30}H_{44}$	-	-	-	7	Hydrocarbon	0/0	*Staphylococcus aureus* [24]

Table 1. *Cont.*

C$_{30}$ Carotenoid	Quenching Type	Antioxidant Activity		Conjugated Double Bond	Group Contained	H Bond Donor/Acceptor	Microbial Sources [Ref.]
		Carotenoid Content	Antioxidant Control				
4,4′-Diapo-ζ-carotene (4,4′-Diapo-zeta-carotene) /C$_{30}$H$_{44}$	-	-	-	7	Hydrocarbon	0/0	*Staphylococcus aureus* [24]
4,4′-Diapophytofluene/C$_{30}$H$_{46}$	-	-	-	5	Hydrocarbon	0/0	*Staphylococcus aureus* [24]
4,4′-Diapophytoene (Dehydrosqualene) /C$_{30}$H$_{48}$	-	-	-	3	Hydrocarbon	0/0	*Staphylococcus aureus* [24], *Bacillus* spp. [19]

Dashes (-) are designated as not yet studied or determined.

The orange pigment 4,4′-diapolycopene ($C_{30}H_{40}$; also known as 4,4′-diapo-ψ,ψ-caroten *all-trans*-4,4′-diapolycopene) is an acyclic carotenoid produced in trace amounts by several species of anoxygenic photosynthetic Heliobacteria and by recombinant *B. subtilis* and *E. coli* [22,36,40].

4,4′-Diapolycopen-4-al ($C_{30}H_{38}O$; also known as *all-trans*-4,4′-diapolycopen-4-al, 4,4′-Diapo-ψ,ψ-caroten-4-al (International Union of Pure and Applied Chemistry, IUPAC), 4,4′-diapolycopenal and 4,4′-diapocaroten-4-al) is an apo carotenoid whose parent molecule is 4,4′-diapolycopene. 4,4′-Diapolycopen-4-al is found in trace amounts in *S. aureus* mutants [20,24,38,41]. 4,4′-Diapolycopenedial ($C_{30}H_{36}O_2$; also known as 4,4′-diapo-ψ,ψ-carotenedial, 4,4′-diapolycopen-4,4′-dial, 4,4′-diapolycopene dialdehyde and *all-trans*-4,4′-diapolycopene-4,4′-dial) is a dialdehyde apocarotenoid also formed from 4,4′-diapolycopene. It has been isolated from *Staphylococcus* and methylotrophic bacteria [37,38].

Although a few studies have confirmed the isolation of C_{30} carotenoids, the antioxidant potential and microbial yields of these carotenoids can be underestimated owing to the difficulties faced during analysis. The identification of specific microbial-sourced carotenoids is hindered by their interactions with the materials used in standard protocols or by the limitations of separating intermediates and novel compounds that have variable characteristics from the more common carotenoid standards [42]. Low recovery and high carotenoid losses tend to occur due to their instability in the presence of light, heat, acids, alkali, and oxygen, as well as during chromatography. Therefore, all carotenoids necessitate special care in their handling. Additionally, stereochemical modifications influence their purification and analysis, as the *cis/trans* configurations affect their biochemistry and stereoisomerism affects their solubility and biofunctionality. *All-trans* isomers are predominant in naturally occurring carotenoids but are easily isomerized to their *cis* forms in the presence of light and heat. *Trans* forms have a propensity for crystallization due to their rigidity, and when converted to their *cis* isomeric forms, there is a slight reduction in the color saturation of the pigments [43–45].

Some carotenoids exhibit inconsistent retention in silica and octadecylsilyl (ODS) chromatography. To counter this problem, Osawa et al. [46] purified microbial carotenoid extracts using polystyrenic synthetic adsorbents and a CHP20/C10 packed column to separate the pigments. Their method was validated since the hydrophobic interactions between 4,4′-diapolycopene-4,4′-dioic acid and the polystyrene resin allowed for identification of the carotenoid using high-resolution electrospray ionization mass spectrometry (HRESI-MS) and nuclear magnetic resonance (NMR) analyses.

3. Antioxidant Properties of C_{30} Carotenoids

The antioxidant activity of a compound is the constant reaction rate between the compound and reactive species that include both radicals and nonradicals [13]. The physiological health benefits of carotenoids are attributed to the length of their conjugated backbones, which makes them highly reactive molecules capable of scavenging free radicals and quenching singlet oxygen. Carotenoids respond to ROS by donating electrons and, through oxidation, creating carotenoid–radical cations, thus lowering the oxidation state of the oxidizer [47]. Carotenoids also scavenge reactive oxygen species through the hydrogen atom transfer (HAT) mechanism in which a chemical transformation occurs, whereby there is movement of a proton and an electron between two substrates [48].

Currently, there are no universal measures for the determination of antioxidant activity. The antioxidant activity of C_{30} carotenoids is more commonly determined using the 1,1-diphenyl-2-picrylhydrazyl (DPPH) free-radical scavenging activity assay and the singlet oxygen quenching assay, which is performed by measuring the methylene blue-sensitized photooxidation of linoleic acid. While the DPPH assay is commonly used for antioxidant capacity determination with absorbance typically read at 515 nm, some carotenoids such as β-carotene and lycopene produce dark brown colors when mixed with DPPH. The color has been observed to interfere with the absorbance readings at 515 nm, but a change to 540 nm allowed for readings as there was still significant absorption at that wavelength [49,50]. The carotenoid antioxidant activity and reactivity in quenching ROS stem from the high electron density of the compounds, especially C=C bonds, and occur via physical or chemical ROS quenching. Singlet oxygen ($^1O_2$) originates from triplet molecular oxygen ($^3O_2$) in the presence of light, or superoxide and hydrogen peroxide in the absence of light. During physical quenching, carotenoids convert singlet oxygen to its triplet ground state, which is less reactive. The energy taken on by the carotenoid leaves it in an excited state, which dissipates as it releases the energy as heat into the environment and returns to its ground state. During chemical quenching (or scavenging), there is a chemical reaction between the carotenoid and the ROS, and barring enzymatic recycling of the carotenoid, it is oxidized and consumed completely. Carotenoid antiradical capabilities have also been studied by analyzing the carotenoid–hydrogen bond dissociation energy via the HAT antiradical mechanism [51]. Nakanishi et al. [52] developed a method to distinguish between hydrogen and electron transfer by using a galvinoxyl radical ($G^\bullet$) and magnesium ions. If electron transfer is the rate-determining step, the mechanism is accelerated by the magnesium via stabilization of the reduced radical anion. A subsequent proton transfer occurs from the radical cation, resulting in a neutral antioxidant radical and GH. However, information on this as applied to C_{30} carotenoids is lacking.

The structure of a carotenoid has significant bearing on its antioxidative efficiency. Depending on their structure, carotenoids can be classified as carotenes, which only have hydrocarbon backbones; and xanthophylls, which possess functional groups such as hydroxy, keto, and methoxy groups [53]. In certain microorganisms, the arrangement and yield of carotenoids are affected by their growth conditions, with different parameters eliciting different effects [54].

Stereoisomerism also plays a role in carotenoid biofunctionality and affects both solubility and absorbance. The rigidity of the *trans* forms increases their propensity for aggregation and crystallization. Additionally, certain *trans* forms have greater antioxidant activity than their *cis* counterparts, although exceptions such as *cis*-lycopene have greater bioavailability than *trans*-lycopene [55].

Carotenoids with higher numbers of double bonds tend to have higher scavenging activities, and acyclic compounds have greater scavenging activities than cyclic carotenoids [36]. Additionally, carotenoid functional groups influence antioxidant activity and solubility. Aldehyde groups increase carotenoid solubility, which in turn affects bioavailability and stability, two critical factors in carotenoid end-product manufacture.

Functional groups have a clear effect on C_{30} carotenoid reactivity. Scavenging activity is greater in carotenoids with aldehyde or carboxylic groups. Carboxyl groups have slightly higher antioxidant capacities than aldehyde groups, but aldehyde–protein crosslinking increases the timely effects of cosmetic applications, which improves the viability of commercial production [37]. Carbonyl and hydroxyl groups have significant roles in the potency of the quenching capacity of a carotenoid, which is attributed to the affinity of the carotenoid to the ROS and the hydrophobicity of the experimental solvent. As such, the antioxidant activities of carotenoids can be enhanced by increasing the number of conjugated double bonds (C=C and C=O) or by lowering the hydrophobicity of the compound through the addition of more hydroxyl groups [56].

According to Kim et al. [36], the activity of 4,4′-diapolycopene-4,4′-dial, which has 13 double bonds and two aldehyde groups, is greater than that of 4,4′-diaponeurosporene which only has 11 double bonds. Compared to DL-α-tocopherol, which had a DPPH half-maximum inhibitory concentration (IC_{50}) value of 33.2 μM, the tested C_{30} carotenoids and their corresponding DPPH IC_{50} values were: diapolycopen-dial, 7.5 μM; diapolycopene, 8.7 μM; diaponeurosporenoic acid, 9.7 μM; diaponeurosporen-al, 10.2 μM; diaponeurosporene, 11.6 μM; diapotorulene, 70.3 μM; and diapo-β-carotene, 77.8 μM. Their double bonds were 13, 11, 10, 10, 9, 8, and 5, respectively. The cyclic compounds diapotorulene and diapo-β-carotene had one and two additional double bonds, respectively, within their β-end groups [36].

In another study, 4,4′-diaponeurosporene was isolated from *Lactiplantibacillus plantarum* subsp. *plantarum* KCCP11226^T, a strain that showed high tolerance to oxidative stress, a high survival rate after H_2O_2 treatment, and a DPPH free-radical scavenging ability of 27.41%, which was greater than that of butylated hydroxytoluene (BHT, 50 μg/mL) [29]. Staphyloxanthin from *S. gallinarum* KX912244 has a DPPH free-radical scavenging activity of IC$_{50}$ 54.22 μg/mL, a vitamin C equivalent antioxidant capacity (VEAC) while ascorbic acid had an IC$_{50}$ value of 35.54 μg/mL [34].

Strong $^1O_2$-quenching activity was observed in methyl 5-glucosyl-5,6-dihydro-4,4′-diapolycopenoate (5.1 μM), while the following intermediate carotenoids exhibited moderate $^1O_2$-quenching activity compared to astaxanthin (3.7 μM): 5-glucosyl-5,6-dihydro-4,4′-diapolycopene (30 μM), 5-hydroxy-5,6-dihydro-4,4′-diapolycopene (30 μM), 4,4′-diaponeurosporene (45 μM), 5-hydroxy-5,6-dihydro-4,4′-diaponeurosporene (56 μM), and 15-*cis*-4,4′-diapophytoene (>100 μM) [33].

4,4′-Diapolycopene-4,4′-dioic acid and its methyl esters (monomethyl ester and dimethyl ester) were tested for their antioxidant activity according to their quenching capabilities of singlet oxygen ($^1O_2$). Their IC$_{50}$ values were 5.8 μM, 6.0 μM, and 6.2 μM, respectively, compared to that of astaxanthin (9.3 μM) [46].

In addition to natural C_{30} carotenoids, several studies have reported the synthesis of novel C_{30} carotenoids, such as glycol-C_{30} carotenoids, which have different antioxidative properties. Mijts et al. [38] synthesized aldehyde and carboxylic acid C_{30} carotenoid derivatives along with 4,4′-diapolycopen-4,4′-dial using a carotenoid oxygenase (CrtOx, diapocarotenal synthase) from *S. aureus* to produce oxygenated linear C_{30} carotenoids. Methyl glucosyl-3,4-dehydro-apo-8′-lycopenoate ($C_{37}H_{52}O_8$), a C_{30} carotenoid from *Halobacillus halophilus*, underwent chemical mutagenesis to isolate its intermediate compounds through a novel biosynthetic pathway. The intermediates, hydroxy-3,4-dehydro-apo-8′-lycopene and methyl hydroxy-3,4-dehydro-apo-8′-lycopenoate, were identified as 8′-apo derivatives that differed from 4,4′-diapo derivatives due to the nonsymmetric placement of the methyl groups rather than through FPP synthesis. Diapocarotenoid end groups are usually susceptible to oxidation by free peroxyl radicals formed in lipid membranes, possibly enhancing their reactivity [57,58]. The antioxidant activity of the three carotenoids was measured according to singlet oxygen quenching, with an IC$_{50}$ value of 5.1 μM for methyl glucosyl-3,4-dehydro-apo-8′-lycopenoate, and 44 μM and 79 μM for its intermediates methyl hydroxy-3,4-dehydro-apo-8′-lycopenoate and hydroxy-3,4-dehydro-apo-8′-lycopene, respectively. The disparity between these three compounds was attributed to the lower hydrophobicity of the intermediate and to a greater number of conjugated double bonds ascribed to greater singlet oxygen quenching [27].

Shindo et al. [26] isolated methyl glucosyl-3,4-dehydro-apo-8′-lycopenoate, a C_{30} glycocarotenoic acid ester, from the marine bacterium *P. maritimus*. The singlet oxygen-suppressing activity of this carotenoid yielded an IC$_{50}$ of 5.1 μM which was more potent than that of astaxanthin (8.9 μM) and β-carotene (>100 μM). Another marine bacterium, *Rubritalea squalenifaciens*, produces novel acyl glycocarotenoic acid called diapolycopene-dioic acid xylosyl ester, which exhibits similar singlet oxygen suppression with an IC$_{50}$ value of 5.1 μM. The esters isolated from this study were the first C_{30} carotenoids shown to have D-xylose groups, although others have reported acyl glucose within the compound structures. The potency of the antioxidant activity was attributed to the presence of carboxylic acids at either end of the aglycone structure [59].

The synthesis and isolation of rare and novel carotenoids from different types of bacteria, including lactic acid and marine bacteria, are steadily increasing the number of characterized carotenoids. Two novel C_{30} carotenoids, 4-[2-O-11Z-octadecenoyl-β-glucopyranosyl]-4,4′-diapolycopene-4,4′-dioic acid (IC$_{50}$ 3.6 μM) and 4-[2-O-9Z-hexadecenoyl-β-glucopyranosyl]-4,4′-diapolycopene-4,4′-dioic acid (IC$_{50}$ 3.2 μM), isolated from Methylobacterium strains, had antioxidative activities comparable to that of astaxanthin (IC$_{50}$ 4.1 μM) and greater than that of β-carotene (IC$_{50}$ >100 μM) [60]. Additionally, the antioxidant activity of the novel compound methyl 5-glucosyl-5,6-dihydro-apo-4,4′-lycopenoate yielded an IC_{50} of 5.1 µM from $^1O_2$ suppression activity, while the IC_{50} values of astaxanthin and β-carotene were 8.9 µM and >100 µM, respectively [61].

4. Microbial Fermentation for Commercial Production of C_{30} Carotenoids

Only a few of the known carotenoids are commercially available, with most being C_{40} carotenoids such as astaxanthin and lycopene [11]. Chemical synthesis is the most common method used for industrial carotenoid production; however, it is associated with risks such as the generation of hazardous waste, and possible adverse effects include allergic reactions in humans [62]. Carotenoids obtained from plant extractions have environmental and geographic limitations and are relatively costly to purchase [63]. Microbial synthesis of carotenoids is increasingly preferable because of their sustainable nature, viable options for low substrate costs, simplicity of production, and monitoring of practices and applications [10,64].

A recent report on the global market for carotenoids stated that it was expected to reach USD 2.7 billion in 2027, increasing from USD 2.0 billion in 2022. This is consistent with their previous report on predicted growth. However, the majority of carotenoids are still supplied through chemical synthesis and are predominantly composed of C_{40} carotenoids such as β-carotene ($C_{40}H_{56}$), lutein ($C_{40}H_{56}O_2$), astaxanthin($C_{40}H_{52}O_4$), and lycopene ($C_{40}H_{56}$). The C_{30} apocarotenoid, β-apo-8-carotenal ($C_{30}H_{40}O$), does have a part in the market value of carotenoids and can be sourced from the cyanobacterium *Arthrospira platensis* (or Spirulina), a microalgae species that can also synthesize various C_{40} carotenoids [65].

Carotenoid prices are dependent on origin and purity. Astaxanthin prices range from 2500 to 7000 USD/kg from microalgae *Haematococcus pluvialis*, but synthetic astaxanthin prices are typically less than 2000 USD/kg, with a production cost of only 1000 USD/kg, making them much more competitive and leaving the natural supply for more conscientous consumers who desire natural products [66,67]. β-carotene is priced between 300 and 3000 USD/kg when sourced from microalgae *Dunaliella salina* and has biomass production costs of about 17 USD/kg dry weight calculated from a process model [68–70]. Microalgae carotenoid production is highly costly, hindered by high capital, labor, and processing costs. Astaxanthin from *Haematococcus* has an estimated production cost of 718 USD/kg based on a conceptually designed facility, compared to costs of over 3000 USD/kg incurred by established firms producing the carotenoid from the microalgae [71]. Adaptive and recombinant microbial strains can reduce production costs in theory, but there is a lack of detailed economic assessment studies on carotenoid production by source that cover production, capital, labor, and processing costs. In addition, the studies that do tackle financial feasibility and requirements also focus on C_{40} carotenoids rather than other types.

Employing microorganisms as natural sources of carotenoids is beneficial to meet the growing demand for natural products. The short life cycles of microorganisms and lower land requirements for their production make them advantageous over plants for carotenoid extraction. In addition, many microorganisms can grow in stressful environments and use agricultural waste as nutrient sources, showcasing their environmental benefits. Microbes can be exploited to produce multiple desired products. The ease of genetic manipulation for enhanced microbial biosynthesis further highlights the potential of microbes for carotenoid production over chemical synthesis and plant extraction.

Carotenoids are lipid-soluble compounds typically extracted using organic solvents, with nonpolar solvents utilized for nonpolar compounds and polar solvents for polar carotenoids. Bacterial carotenoids can also be extracted by relatively costly enzymatic action. This process includes cell membrane hydrolysis in a greener, more sustainable process, although yield remains a concern. Another extraction method is microwave-assisted extraction, which uses microwave radiation to destroy cellular structures and release the desired product. Supercritical fluid extraction (SFE) is an improvement of traditional extraction as it uses water and carbon dioxide, thereby being more environmentally friendly.

It also exhibits higher permeability and diffusivity, faster processing, and higher carotenoid extraction yields [44].

Utilizing microorganisms as commercial carotenoid sources is relatively cost-effective. However, the cost of substrates and running costs for large-scale production vary by country and region. Production costs are relatively lower in countries where substrate-manufacturing firms are present than in those where imports are required. In addition, the versatility of microorganisms in substrate usage has led to an increase in the use of low-cost agricultural waste, including husks, straw, and fruit pulp, as substrates that could readily curtail production costs [72]. Although viability studies considering the economic practicality of microbial carotenoid production include factors that affect running costs (such as electricity and water tariffs incurred from production), research tends to focus on optimizable factors that affect microbial growth and product yield, such as media composition and temperature.

4.1. Growth Factor Effects on C_{30} Carotenoid Production

The main regulatory factors for carotenoid production in nonphotosynthetic bacteria are temperature, aeration, agitation, and culture media composition. Conversely, carotenoid synthesis by phototrophic bacteria is modulated by oxygen and light, with the bacteria producing less oxidized forms of carotenoids in oxygen-deprived environments and synthesizing ketocarotenoids in oxygenated environments [73–75]. Modifications in the quantity and quality of these factors have various effects on different species of microorganisms when initiating carotenoid production. *Rhodotorula glutinis* DFR-PDY, a yeast strain, produces higher carotenoid yields when using fructose as a carbon source, but exhibits greater cell growth in galactose-supplemented media [76]. Although the lactic acid bacteria (LAB) *L. plantarum* subsp. *plantarum* KCCP11226^T showed maximum cell growth in maltose, it produced the greatest C_{30} carotenoid 4,4′-diaponeurosporene yield in lactose-supplemented media. Nitrogen sources also play a role in microbial growth and carotenoid yield, with beef extract being optimal for both cell growth and 4,4′-diaponeurosporene production [31,77,78].

In lieu of conventional supplementation, the use of waste products for fermentation in carotenoid production is cost-effective and sustainable. Fermentation can be carried out in a solid state (SSF), or via submerged fermentation (SmF). IN SSF, the substrate consumption rate is slower, and fermentation requires a longer time and less moisture (such as in wheat bran and fruit pulp). In SmF, substrate consumption occurs rapidly, and the resupply of substrates, which include soluble sugars, molasses, and fruit juice, must be constant [79].

4.2. Abiotic Stresses for the Optimization of C_{30} Carotenoid Production

Various methods have been used to enhance carotenoid yield, including the application of stress factors, progressive nutrient deprivation, and recombinant techniques for the production of high-yield strains (Table 2) [44].

The processes employed in biotechnological applications depend on the stability and synthesis requirements of the desired products. Carotenoid production and cell growth are affected by the nutrient composition of the culture medium and culture conditions. For industrial production, the costs associated with substrates and the running costs of large-scale fermenters must be minimized while optimizing final product yields. Optimization studies that use response surface methodology often analyze the main factors for scaled-up production and maximize desired parameters, such as carotenoid production, while mitigating undesirables, such as high cost. An example of this is the optimization of nitrogen and carbon sources, particularly when standard media components are substituted by natural compounds or agricultural waste. This optimization reduces costs while also increasing sustainability, yielding environmentally friendly production methods [72].

Table 2. Effects of applied factors on C_{30} carotenoid production.

Controlled Factor	Action	Effect	Ref.
Oxidative stress	Diphenylamine and duroquinone application	Complete inhibition of C_{30} carotenoids	[80]
Oxidative stress	Diphenylamine application	Complete inhibition of C_{30} carotenoids	[16]
Oxidative stress	Hydrogen peroxide exposure	5.9-fold increase in carotenoid production	[29]
Temperature	Low temperature	Increased 4,4′-diaponeurosporene antioxidant activity– DPPH: 1.7-fold vs BHT; ABTS: 7.5-fold vs BHT; FRAP: 8-fold vs BHT	[31]
Recombinant technique	Constitutive lac promoter	Increased 4,4′-diapolycopene production	[81]
Recombinant technique	Optimizing engineered pathways in heterologous host	Increased 4,4′-diapolycopene and 4,4′-diaponeurosporene production	[82]
Recombinant technique	Gene coexpression	20-fold increase in 4,4′-diapolycopene dialdehyde production	[37]
Recombinant technique	Farnesyl diphosphate (FDP) synthase overexpression	3- to 5-fold increase in diapotorulene-to-diaponeurosporene ratio	[58]

Microbial growth or carotenoid synthesis can be maximized within a specific range of conditions. Environmental conditions optimal for cell growth are not necessarily optimal for carotenogenesis. In certain bacterial species, high temperatures are ideal, whereas in others, lower temperatures provide better carotenoid synthesis. This is also true for pH, salinity, and carbon and nitrogen sources. However, with biotechnological advancements, the effects of these modulating parameters on carotenoid production can be adjusted. Metabolically engineered strains of *E. coli* have been compared to determine the effect of recombinant hosts and carotenoid structure on carotenoid yield and ascertain whether heterologous carotenoid formation was strain-dependent.

The initial pH of the culture media influences bacterial cell growth and carotenoid production, but as the bacteria grow, the acidity/alkalinity levels of the media change due to by-products of bacterial metabolism [83], as is the case with LAB. In yeast cells, the initial decrease in pH is not permanent and the pH rises as the yeast proliferates and intensively produces carotenes, followed by stabilization of the pH values [84]. Acid or alkaline tolerance is highly strain-dependent, with certain microorganisms accumulating more carotenoids at neutral pH, while others exhibit optimum productivity in slightly acidic or alkaline conditions [85,86].

Temperature is an important factor in bacterial growth and carotenoid synthesis and affects biosynthetic pathways, stability, reactivity of substrates, enzymes, and the products involved in these pathways [87]. Temperature also plays a role in enzyme concentrations within biosynthetic pathways and in internal temperature regulation [88,89].

Aeration has a known effect on aerobic microorganisms, with agitation increasing the movement and availability of oxygen and nutrients within the culture. For aerobic microorganisms, culture aeration is important for the efficient supply of oxygen to the growing biomass, which in turn promotes carotenoid synthesis. Intensive aeration and agitation stimulate carotenoid production as long as the threshold at which the mechanical shear forces could cause cell damage is not crossed [90,91].

The substrate source and composition in the culture media also influence carotenoid production and yield. The most common carbon sources for carotenoid biosynthesis are sucrose and glucose. Cell growth maintenance requires available carbon, and there is a relationship between carotenoid production and carbon:nitrogen ratios in certain bacteria [92].

White-light irradiation has a positive effect on carotenoid synthesis in algae, fungi, and bacteria [93]. The carotenoid yield depends on the microorganism strain and light intensity. White light acts as a stimulatory inducer for cell growth, which then follows carotenoid accumulation, owing to increased enzyme activity [94]. White light increases carotenoid production by regulating biosynthetic genes in the pathway. High light intensity can restrict cell development and growth while inducing carotenoid synthesis [85].

Stress can be utilized to improve either cell growth or carotenoid production, and previous studies have shown that the two are not necessarily related or proportional. Different species respond differently to stress factors, either increasing or decreasing carotenoid concentrations [72]. Carotenoid accumulation can be enhanced at lower incubation temperatures [95], although carotenogenesis is not necessarily related to cell growth or biomass yield [72,96]. In microalgae, stress can stimulate carotenoid yields but production is limited by slow growth rates, required culture volumes, and bacterial contamination risks [97].

Salinity, oxidative stress, and pH have been used to increase carotenoid concentrations [10]. This increase is a result of microbial defense mechanisms triggered via gene regulation to modulate and alleviate the effects of stress factors [98]. However, while stress can trigger and stimulate carotenoid accumulation, stressful environments past the tolerance point will more likely trigger cell death than carotenoid accumulation.

5. Biosynthetic Pathway Engineering for Novel C_{30} Carotenoid Generation

Both natural and engineered extended pathways have been identified and developed for the synthesis of C_{30} carotenoids [99]. Metabolic engineering and directed evolution allow the development of new pathways for carotenoid production in noncarotenogenic microorganisms through the amalgamation of carotenoid genes from various sources. The enzymes expressed from these genes or gene clusters work in tandem to produce the desired compound. Additionally, the promiscuity of these carotenoid enzymes is based on end-group recognition rather than on the whole carotenoid structure (Figure 5) [100].

The genes involved in the synthesis of the C_{30} carotenoid skeleton are dehydrosqualene synthase (*crtM*) and dehydrosqualene desaturase (*crtN*) [14]. However, several novel C_{30} carotenoids have been created through combinatorial biosynthesis using enzymes from different microorganisms. For instance, C_{40} carotenoids were modified to produce cyclic C_{30} diapocarotenoids using lycopene cyclase, CrtY, which catalyzes the reaction on the ψ-end groups [101].

Various carotenoid structures can be obtained by manipulating biosynthetic pathways. Kim et al. [36] synthesized the acyclic, monocyclic, and bicyclic forms of C_{30} carotenoids from *E. coli* that had undergone directed evolution and a combination of biosynthetic processes. The extension of the diapolycopene pathway led to the formation of diaponeurosporene and the novel cyclic carotenoid, diapotorulene [58,99]. Following comparisons with other lycopene cyclase producers, Kim et al. [36] selected C_{40} CrtY from *Brevibacterium linens* (CrtYBL) to catalyze the cyclization of a saturated end of 4,4′-diaponeurosporene and produce the structurally novel monocyclic 4,4′-diapotorulene. CrtN evolution was used to optimize the 4,4′-diaponeurosporene or 4,4′-diapo-ζ-carotene biosynthesis pathways for cyclization to produce cyclic C_{30} carotenoids not found in nature. Modifications to the enzymes and pathways also led to the formation of other structurally novel compounds. These compounds include bicyclic C_{30} 4,4′-diapo-β-carotene; monocyclic C_{30} carotenoids harboring modified β-ionone rings, 7-hydroxy-4,4′-diapotorulene and 8-keto-4,4′-diapotorulene; monocyclic C_{30} carotenoids with modified acyclic ends, 4,4′-diapotorulen-4′-al and 4,4′-diapotorulen-4′-oic acid; and bicyclic compounds, 4,4′-diapo-β-cryptoxanthin, 4,4′-diapozeaxanthin,4,4′-diapoechinenone, and 4,4′-diapo-β-cryptoxanthin glucoside [36].

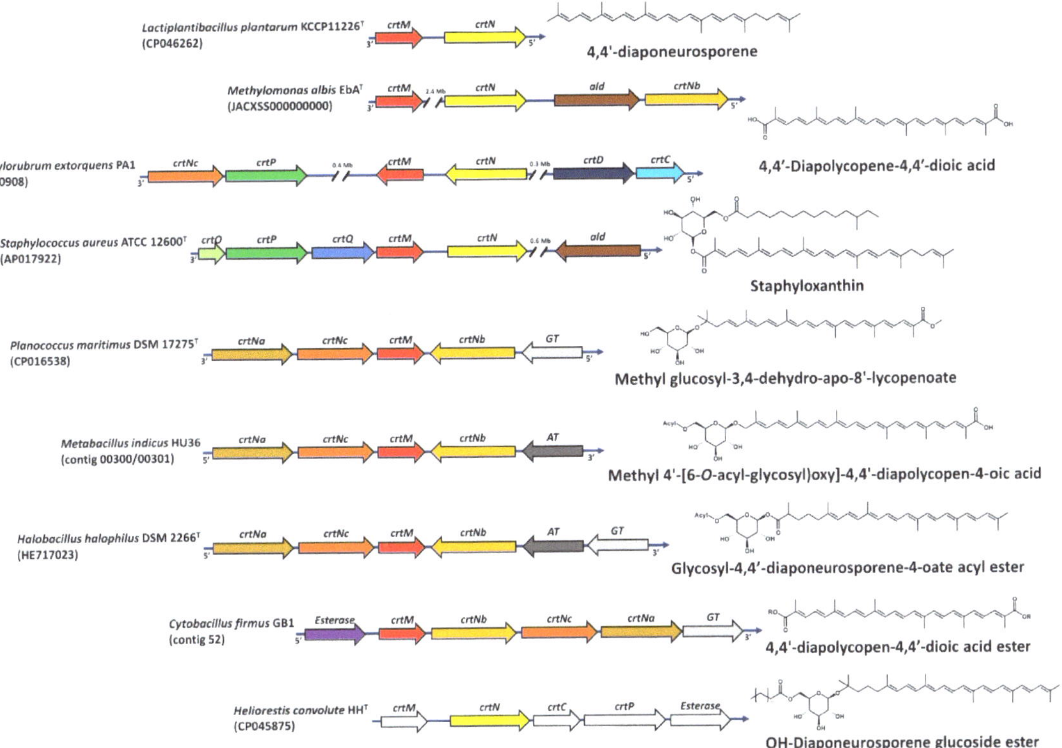

Figure 5. Gene clusters of 4,4′-diapocarotenoids and apo-8′-carotenoids from various microbial sources. White arrows represent putative genes in the biosynthetic pathway, including new and novel carotenoids.

Tao et al. [37] identified a novel gene (*crtNb*) from *Methylomonas*, and its homologue in Staphylococcus, which converts 4,4′-diapolycopene to 4,4′-diapolycopene aldehyde, Additionally, they identified an aldehyde dehydrogenase gene (*ald*) involved in the oxidation of 4,4′-diapolycopene aldehyde to 4,4′-diapolycopene acid.

New acyclic carotenoids, diapolycopenedioc acid xylosylesters A–C, and the C_{30} aglycone, methyl 5-glucosyl-5,6-dihydro-apo-4,4′-lycopenoate, were produced by *Rubritalea squalenifaciens* and *P. maritimus* strains. The diapolycopenedioc acids xylosylesters A, B, and C had 2-acyl-D-xylose in their structures and were derivatives of the aglycone diapolycopenedioc acid [61]. In another study on pathway manipulation, diaponeurosporene was formed when CrtN introduced four double bonds into dehydrosqualene to produce a completely conjugated diapolycopene in recombinant *E. coli*, substituting the three-step desaturation [17]. Lee et al. [58] extended the acyclic C_{30} diaponeurosporene pathway using the C_{40} carotenoid enzymes spheroidene monooxygenase and lycopene cyclase to produce new oxygenated acyclic products and the novel cyclic C_{30} compound, diapotorulene.

Host strains also play a role in the biosynthetic pathway and final carotenoid structure, although the reason for this has not yet been determined. The acyclic C_{30} carotenoids diaponeurosporene and diapolycopene were produced by seven *E. coli* host strains (Top10, MG1655, MDS42, JM109, SURE, DH5a, and XL1-Blue). However, monocyclic diapotorulene was heavily strain-dependent and favored the SURE *E. coli* strain in a manner similar to that of heterologous lycopene [57].

With regard to bioengineered pathways, engineered microbial hosts designed for high carotenoid yields require optimization of their isoprenoid precursor pool. Moreover, microbes must be able to contain lipophilic carotenoids and the expression of carotenogenic genes need to be modulated for the efficient transformation of precursors into target compounds. Optimization of the precursor pool requires increasing the isoprenoid flux via overexpression of enzymes in the nonmevalonate isoprenoid pathway. Balancing isoprenoid expression and carotenoid-synthesizing enzymes enhances carotenoid production by alleviating the effects of growth inhibition caused by enzyme overexpression.

6. Future Predictions and Expectations

The limitations of microbial carotenoid production are mainly insufficient yields and poor consumer acceptance, especially with a bias against genetically modified organisms. Industry regulations in the food, feed, and nutraceutical fields may also hinder their commercial availability. Additionally, production inconsistency and optimization requirements may hamper their ability to compete with plant extractions and chemical synthetic sources [44].

Biotechnological advances will likely lead to the development of carotenoids with enhanced antioxidant capacities, capable of tackling metabolic syndrome-associated conditions and debilitating diseases. As research progresses toward the industrialized and commercial microbial production of carotenoids, it is imperative that consumer acceptance and industry regulations be considered.

With the discovery of carotenoid biosynthetic pathways in different microorganisms and the advancement of metabolic engineering, a vast number of C_{30} carotenoids will likely be added to the current pool of known carotenoid compounds. Gene editing and biosynthetic pathway expansion have led to novel C_{30} carotenoids and intermediates with different characteristics from their counterparts and derivatives. The formation of carotenoids with superior antioxidant activity could propel further developments in microbial carotenoid production and consumer acceptance.

7. Conclusions

The review highlights the antioxidant potential of C_{30} carotenoids from various microbial sources and the synthesis of novel C_{30} carotenoid compounds through metabolic engineering. It validates the prospect of commercialization of these compounds as alternatives for the more common C_{40} carotenoids that take up a majority of the global market. The synthesis of C_{30} carotenoids that possess similar or greater antioxidant activities than C_{40} carotenoids would be a worthwhile path for novel marketable antioxidant products in the sectors of food, feed, cosmetics, and pharmaceuticals. Microbial synthesis provides faster and safer production conditions with nontoxic waste by-products as opposed to chemical synthesis. Additionally, the simplicity of genetic manipulation and process optimization can aid in cutting costs that are detrimental to large-scale production. The isolation of 4,4′-diaponeurosporene from *Lactiplantibacillus plantarum* subsp. *plantarum* is particularly noteworthy because of the "generally regarded as safe" (GRAS) status of the microorganism and its role as a probiotic. The supplementation of foods with probiotics that can synthesize carotenoids with high antioxidant potential can offer great health benefits by promoting digestive health in addition to preventing cellular damage through the reduction of oxidative stress in the human body. More research is required to improve and optimize the development and applications of microbial carotenoid sources, by obtaining more thorough knowledge on even the less common carotenoid types.

Author Contributions: Conceptualization, I.N.S. and M.-J.S.; investigation and writing—original draft preparation, I.N.S.; writing—review and editing, C.Y.H.; supervision and funding acquisition, M.-J.S. All authors have read and agreed to the published version of the manuscript.

Funding: This work was supported by the National Research Foundation of Korea (NRF) grants funded by the Korea government (MIST) (Grant No. 2019R1A2C1006038 and 2022R1F1A1062699). This work was also carried out with the support of "Cooperative Research Program for Agriculture

Science & Technology Development (Project No. PJ01708502)" Rural Development Administration, Republic of Korea.

Institutional Review Board Statement: Not applicable.

Informed Consent Statement: Not applicable.

Data Availability Statement: Not applicable.

Conflicts of Interest: The authors declare no conflict of interest.

References

1. Alcaíno, J.; Baeza, M.; Cifuentes, V. Carotenoid distribution in nature. *Subcell. Biochem.* **2016**, *79*, 3–33.
2. Maoka, T.; Kawase, N.; Hironaka, M.; Nishida, R. Carotenoids of Hemipteran insects, from the perspective of chemo-systematic and chemical ecological studies. *Biochem. Syst. Ecol.* **2021**, *95*, 104241. [CrossRef]
3. Nováková, E.; Moran, N.A. Diversification of genes for carotenoid biosynthesis in aphids following an ancient transfer from a fungus. *Mol. Biol. Evol.* **2012**, *29*, 313–323. [CrossRef]
4. Rodriguez-Concepcion, M.; Avalos, J.; Bonet, M.L.; Boronat, A.; Gomez-Gomez, L.; Hornero-Mendez, D.; Limon, M.C.; Meléndez-Martínez, A.J.; Olmedilla-Alonso, B.; Palou, A. A global perspective on carotenoids: Metabolism, biotechnology, and benefits for nutrition and health. *Prog. Lipid Res.* **2018**, *70*, 62–93. [CrossRef]
5. Britton, G.; Liaaen-Jensen, S.; Pfander, H.; Mercadante, A.; Egeland, E. *Carotenoids-Handbook*; Birkhäuser Verlag: Basel, Switzerland, 2004; pp. 1–625.
6. Maoka, T. Carotenoids as natural functional pigments. *J. Nat. Med.* **2020**, *74*, 1–16. [CrossRef]
7. Mezzomo, N.; Ferreira, S.R. Carotenoids functionality, sources, and processing by supercritical technology: A review. *J. Chem.* **2016**, *2016*, 3164312. [CrossRef]
8. Schweiggert, R.M.; Carle, R. Carotenoid production by bacteria, microalgae, and fungi. In *Carotenoids: Nutrition, Analysis and Technology*; John Wiley & Sons, Ltd.: Chichester, UK, 2016; pp. 217–240.
9. Armstrong, G.A.; Hearst, J.E. Genetics and molecular biology of carotenoid pigment biosynthesis. *FASEB J.* **1996**, *10*, 228–237. [CrossRef]
10. Foong, L.C.; Loh, C.W.L.; Ng, H.S.; Lan, J.C.-W. Recent development in the production strategies of microbial carotenoids. *World J. Microbiol. Biotechnol.* **2021**, *37*, 12. [CrossRef]
11. Cardoso, L.A.; Karp, S.G.; Vendruscolo, F.; Kanno, K.Y.; Zoz, L.I.; Carvalho, J.C. Biotechnological production of carotenoids and their applications in food and pharmaceutical products. *Carotenoids* **2017**, *8*, 125–141.
12. Fernandes, A.S.; do Nascimento, T.C.; Jacob-Lopes, E.; De Rosso, V.V.; Zepka, L.Q. Carotenoids: A brief overview on its structure, biosynthesis, synthesis, and applications. *Prog. Carotenoid Res.* **2018**, *1*, 1–17.
13. Pérez-Gálvez, A.; Viera, I.; Roca, M. Carotenoids and chlorophylls as antioxidants. *Antioxidants* **2020**, *9*, 505. [CrossRef]
14. Yabuzaki, J. Carotenoids Database: Structures, chemical fingerprints and distribution among organisms. *Database* **2017**, *2017*, bax004. [CrossRef]
15. Garrido-Fernández, J.; Maldonado-Barragán, A.; Caballero-Guerrero, B.; Hornero-Méndez, D.; Ruiz-Barba, J.L. Carotenoid production in *Lactobacillus plantarum*. *Int. J. Food Microbiol.* **2010**, *140*, 34–39. [CrossRef]
16. Steiger, S.; Perez-Fons, L.; Fraser, P.; Sandmann, G. Biosynthesis of a novel C_{30} carotenoid in *Bacillus firmus* isolates. *J. Appl. Microbiol.* **2012**, *113*, 888–895. [CrossRef]
17. Wieland, B.; Feil, C.; Gloria-Maercker, E.; Thumm, G.; Lechner, M.; Bravo, J.-M.; Poralla, K.; Götz, F. Genetic and biochemical analyses of the biosynthesis of the yellow carotenoid 4,4′-diaponeurosporene of *Staphylococcus aureus*. *J. Bacteriol.* **1994**, *176*, 7719–7726. [CrossRef]
18. Carruthers, D.N.; Lee, T.S. Diversifying Isoprenoid Platforms via Atypical Carbon Substrates and Non-model Microorganisms. *Front. Microbiol.* **2021**, *12*, 3725. [CrossRef]
19. Perez-Fons, L.; Steiger, S.; Khaneja, R.; Bramley, P.M.; Cutting, S.M.; Sandmann, G.; Fraser, P.D. Identification and the developmental formation of carotenoid pigments in the yellow/orange *Bacillus* spore-formers. *Biochim. Biophys. Acta* **2011**, *1811*, 177–185. [CrossRef]
20. Steiger, S.; Perez-Fons, L.; Cutting, S.M.; Fraser, P.D.; Sandmann, G. Annotation and functional assignment of the genes for the C_{30} carotenoid pathways from the genomes of two bacteria: *Bacillus indicus* and *Bacillus firmus*. *Microbiology* **2015**, *161*, 194–202. [CrossRef]
21. Kleinig, H.; Schmitt, R.; Meister, W.; Englert, G.; Thommen, H. New C_{30}-carotenoic acid glucosyl esters from *Pseudomonas rhodos*. *Z. Naturforsch.* **1979**, *34*, 181–185. [CrossRef]
22. Takaichi, S.; Inoue, K.; Akaike, M.; Kobayashi, M.; Oh-oka, H.; Madigan, M.T. The major carotenoid in all known species of heliobacteria is the C_{30} carotenoid 4,4′-diaponeurosporene, not neurosporene. *Arch. Microbiol.* **1997**, *168*, 277–281. [CrossRef]
23. Takaichi, S.; Oh-Oka, H.; Maoka, T.; Jung, D.O.; Madigan, M.T. Novel carotenoid glucoside esters from alkaliphilic heliobacteria. *Arch. Microbiol.* **2003**, *179*, 95–100. [CrossRef] [PubMed]
24. Marshall, J.H.; Wilmoth, G.J. Pigments of *Staphylococcus aureus*, a series of triterpenoid carotenoids. *J. Bacteriol.* **1981**, *147*, 900–913. [CrossRef] [PubMed]

25. Marshall, J.H.; Wilmoth, G.J. Proposed pathway of triterpenoid carotenoid biosynthesis in *Staphylococcus aureus*: Evidence from a study of mutants. *J. Bacteriol.* **1981**, *147*, 914–919. [CrossRef] [PubMed]

26. Shindo, K.; Endo, M.; Miyake, Y.; Wakasugi, K.; Morritt, D.; Bramley, P.M.; Fraser, P.D.; Kasai, H.; Misawa, N. Methyl glucosyl-3,4-dehydro-apo-8′-lycopenoate, a novel antioxidative glyco-C_{30}-carotenoic acid produced by a marine bacterium *Planococcus maritimus*. *J. Antibiot.* **2008**, *61*, 729–735. [CrossRef] [PubMed]

27. Osawa, A.; Ishii, Y.; Sasamura, N.; Morita, M.; Koecher, S.; Mueller, V.; Sandmann, G.; Shindo, K. Hydroxy-3,4-dehydro-apo-8′-lycopene and methyl hydroxy-3,4-dehydro-apo-8′-lycopenoate, novel C_{30} carotenoids produced by a mutant of marine bacterium *Halobacillus halophilus*. *J. Antibiot.* **2010**, *63*, 291–295. [CrossRef] [PubMed]

28. Jing, Y.; Liu, H.; Xu, W.; Yang, Q. 4,4′-Diaponeurosporene-producing *Bacillus subtilis* promotes the development of the mucosal immune system of the piglet gut. *Anat. Rec.* **2019**, *302*, 1800–1807. [CrossRef]

29. Kim, M.; Seo, D.-H.; Park, Y.-S.; Cha, I.-T.; Seo, M.-J. Isolation of *Lactobacillus plantarum* subsp. *plantarum* producing C30 carotenoid 4,4′-diaponeurosporene and the assessment of its antioxidant activity. *J. Microbiol. Biotechnol.* **2019**, *29*, 1925–1930.

30. Liu, H.; Xu, W.; Yu, Q.; Yang, Q. 4,4′-Diaponeurosporene-producing *Bacillus subtilis* increased mouse resistance against *Salmonella typhimurium* infection in a CD36-dependent manner. *Front. Immunol.* **2017**, *8*, 483. [CrossRef]

31. Kim, M.; Jung, D.-H.; Seo, D.-H.; Park, Y.-S.; Seo, M.-J. 4,4′-Diaponeurosporene from *Lactobacillus plantarum* subsp. *plantarum* KCCP11226: Low Temperature Stress-Induced Production Enhancement and in Vitro Antioxidant Activity. *J. Microbiol. Biotechnol.* **2021**, *31*, 63–69.

32. Liu, H.; Xu, W.; Chang, X.; Qin, T.; Yin, Y.; Yang, Q. 4,4′-diaponeurosporene, a C_{30} carotenoid, effectively activates dendritic cells via CD36 and NF-κB signaling in a ROS independent manner. *Oncotarget* **2016**, *7*, 40978–40991. [CrossRef]

33. Takemura, M.; Takagi, C.; Aikawa, M.; Araki, K.; Choi, S.-K.; Itaya, M.; Shindo, K.; Misawa, N. Heterologous production of novel and rare C_{30}-carotenoids using *Planococcus* carotenoid biosynthesis genes. *Microb. Cell Fact.* **2021**, *20*, 194. [CrossRef] [PubMed]

34. Barretto, D.A.; Vootla, S.K. In vitro anticancer activity of staphyloxanthin pigment extracted from *Staphylococcus gallinarum* KX912244, a gut microbe of *Bombyx mori*. *Indian J. Microbiol.* **2018**, *58*, 146–158. [CrossRef] [PubMed]

35. Pelz, A.; Wieland, K.-P.; Putzbach, K.; Hentschel, P.; Albert, K.; Götz, F. Structure and biosynthesis of staphyloxanthin from *Staphylococcus aureus*. *J. Biol. Chem.* **2005**, *280*, 32493–32498. [CrossRef]

36. Kim, S.H.; Kim, M.S.; Lee, B.Y.; Lee, P.C. Generation of structurally novel short carotenoids and study of their biological activity. *Sci. Rep.* **2016**, *6*, 21987. [CrossRef] [PubMed]

37. Tao, L.; Schenzle, A.; Odom, J.M.; Cheng, Q. Novel carotenoid oxidase involved in biosynthesis of 4,4′-diapolycopene dialdehyde. *Appl. Environ. Microbiol.* **2005**, *71*, 3294–3301. [CrossRef] [PubMed]

38. Mijts, B.N.; Lee, P.C.; Schmidt-Dannert, C. Identification of a carotenoid oxygenase synthesizing acyclic xanthophylls: Combinatorial biosynthesis and directed evolution. *Chem. Biol.* **2005**, *12*, 453–460. [CrossRef] [PubMed]

39. Taylor, R.F.; Davies, B.H. Triterpenoid carotenoids and related lipids. Triterpenoid carotenoid aldehydes from *Streptococcus faecium* UNH 564P. *Biochem. J.* **1976**, *153*, 233–239. [CrossRef]

40. Maeda, I. Genetic modification in *Bacillus subtilis* for production of C_{30} carotenoids. *Methods Mol. Biol.* **2012**, *892*, 197–205.

41. Kleinig, H.; Schmitt, R. On the biosynthesis of C_{30} carotenoic acid glucosyl esters in *Pseudomonas rhodos*. Analysis of *car*-mutants. *Z. Naturforsch.* **1982**, *37*, 758–760. [CrossRef]

42. Butnariu, M. Methods of analysis (extraction, separation, identification and quantification) of carotenoids from natural products. *J. Ecosyst. Ecograph.* **2016**, *6*, 193. [CrossRef]

43. Ngamwonglumlert, L.; Devahastin, S.; Chiewchan, N. Natural colorants: Pigment stability and extraction yield enhancement via utilization of appropriate pretreatment and extraction methods. *Crit. Rev. Food Sci. Nutr.* **2017**, *57*, 3243–3259. [CrossRef] [PubMed]

44. Ram, S.; Mitra, M.; Shah, F.; Tirkey, S.R.; Mishra, S. Bacteria as an alternate biofactory for carotenoid production: A review of its applications, opportunities and challenges. *J. Func. Foods* **2020**, *67*, 103867. [CrossRef]

45. Valla, A.R.; Cartier, D.L.; Labia, R. Chemistry of natural retinoids and carotenoids: Challenges for the future. *Curr. Org. Syn.* **2004**, *1*, 167–209. [CrossRef]

46. Osawa, A.; Iki, K.; Sandmann, G.; Shindo, K. Isolation and Identification of 4,4′-diapolycopene-4,4′-dioic acid produced by *Bacillus firmus* GB1 and its singlet oxygen quenching activity. *J. Oleo Sci.* **2013**, *62*, 955–960. [CrossRef] [PubMed]

47. Gao, Y.; Focsan, A.L.; Kispert, L.D. Antioxidant activity in supramolecular carotenoid complexes favored by nonpolar environment and disfavored by hydrogen bonding. *Antioxidants* **2020**, *9*, 625. [CrossRef]

48. Han, R.-M.; Zhang, J.-P.; Skibsted, L.H. Reaction dynamics of flavonoids and carotenoids as antioxidants. *Molecules* **2012**, *17*, 2140–2160. [CrossRef]

49. Müller, L.; Fröhlich, K.; Böhm, V. Comparative antioxidant activities of carotenoids measured by ferric reducing antioxidant power (FRAP), ABTS bleaching assay (αTEAC), DPPH assay and peroxyl radical scavenging assay. *Food Chem.* **2011**, *129*, 139–148. [CrossRef]

50. Liu, D.; Shi, J.; Ibarra, A.C.; Kakuda, Y.; Xue, S.J. The scavenging capacity and synergistic effects of lycopene, vitamin E, vitamin C, and β-carotene mixtures on the DPPH free radical. *LWT-Food Sci. Technol.* **2008**, *41*, 1344–1349. [CrossRef]

51. Martínez, A.; Barbosa, A. Antiradical power of carotenoids and vitamin E: Testing the hydrogen atom transfer mechanism. *J. Phys. Chem. B* **2008**, *112*, 16945–16951. [CrossRef]

52. Nakanishi, I.; Miyazaki, K.; Shimada, T.; Ohkubo, K.; Urano, S.; Ikota, N.; Ozawa, T.; Fukuzumi, S.; Fukuhara, K. Effects of metal ions distinguishing between one-step hydrogen-and electron-transfer mechanisms for the radical-scavenging reaction of (+)-catechin. *J. Phys. Chem. A* **2002**, *106*, 11123–11126. [CrossRef]
53. Perera, C.O.; Yen, G.M. Functional properties of carotenoids in human health. *Int. J. Food Prop.* **2007**, *10*, 201–230. [CrossRef]
54. D'Alessandro, E.B.; Antoniosi Filho, N.R. Concepts and studies on lipid and pigments of microalgae: A review. *Renew. Sustain. Energy Rev.* **2016**, *58*, 832–841. [CrossRef]
55. Rodriguez-Amaya, D.B.; Kimura, M. *HarvestPlus Handbook for Carotenoid Analysis*; International Food Policy Research Institute (IFPRI): Washington, DC, USA; International Center for Tropical Agriculture (CIAT): Cali, CO, USA, 2004.
56. Shimidzu, N.; Goto, M.; Miki, W. Carotenoids as singlet oxygen quenchers in marine organisms. *Fish. Sci.* **1996**, *62*, 134–137. [CrossRef]
57. Chae, H.S.; Kim, K.-H.; Kim, S.C.; Lee, P.C. Strain-dependent carotenoid productions in metabolically engineered *Escherichia coli*. *Appl. Biochem. Biotechnol.* **2010**, *162*, 2333–2344. [CrossRef]
58. Lee, P.C.; Momen, A.Z.R.; Mijts, B.N.; Schmidt-Dannert, C. Biosynthesis of structurally novel carotenoids in *Escherichia coli*. *Chem. Biol.* **2003**, *10*, 453–462. [CrossRef]
59. Shindo, K.; Asagi, E.; Sano, A.; Hotta, E.; Minemura, N.; Mikami, K.; Tamesada, E.; Misawa, N.; Maoka, T. Diapolycopenedioic acid xylosyl esters A, B, and C, novel antioxidative glyco-C_{30}-carotenoic acids produced by a new marine bacterium *Rubritalea squalenifaciens*. *J. Antibiot.* **2008**, *61*, 185–191. [CrossRef]
60. Osawa, A.; Kaseya, Y.; Koue, N.; Schrader, J.; Knief, C.; Vorholt, J.A.; Sandmann, G.; Shindo, K. 4-[2-O-11Z-Octadecenoyl-β-glucopyranosyl]-4,4′-diapolycopene-4,4′-dioic acid and 4-[2-O-9Z-hexadecenoyl-β-glucopyranosyl]-4,4′-diapolycopene-4,4′-dioic acid: New C_{30}-carotenoids produced by *Methylobacterium*. *Tetrahedron Lett.* **2015**, *56*, 2791–2794. [CrossRef]
61. Shindo, K.; Misawa, N. New and rare carotenoids isolated from marine bacteria and their antioxidant activities. *Mar. Drugs* **2014**, *12*, 1690–1698. [CrossRef]
62. Gultekin, F.; Doguc, D.K. Allergic and immunologic reactions to food additives. *Clin. Rev. Allergy Immunol.* **2013**, *45*, 6–29. [CrossRef]
63. Brauch, J. Underutilized fruits and vegetables as potential novel pigment sources. In *Handbook on Natural Pigments in Food and Beverages: Industrial Applications for Improving Food Color*; Carle, R., Schweiggert, R., Eds.; Elsevier: Amsterdam, The Netherlands, 2016; pp. 305–335.
64. Córdova, P.; Baeza, M.; Cifuentes, V.; Alcaíno, J. Microbiological synthesis of carotenoids: Pathways and regulation. In *Progress in Carotenoid Research Pigments*; Queiroz Zepka, L., Jacob-Lopes, E., Rosso, V.V., Eds.; IntechOpen: London, UK, 2018; pp. 63–83.
65. McWilliams, A. The Global Market for Carotenoids | BCC Research. Available online: https://www.bccresearch.com/marketresearch/food-and-beverage/the-global-market-for-carotenoids.html (accessed on 16 February 2022).
66. Guerin, M.; Huntley, M.E.; Olaizola, M. *Haematococcus* astaxanthin: Applications for human health and nutrition. *Trends Biotechnol.* **2003**, *21*, 210–216. [CrossRef]
67. Stachowiak, B.; Szulc, P. Astaxanthin for the food industry. *Molecules* **2021**, *26*, 2666. [CrossRef] [PubMed]
68. Koller, M.; Muhr, A.; Braunegg, G. Microalgae as versatile cellular factories for valued products. *Algal Res.* **2014**, *6*, 52–63. [CrossRef]
69. Spolaore, P.; Joannis-Cassan, C.; Duran, E.; Isambert, A. Commercial applications of microalgae. *J. Biosci. Bioeng.* **2006**, *101*, 87–96. [CrossRef] [PubMed]
70. Pirwitz, K.; Flassig, R.J.; Rihko-Struckmann, L.K.; Sundmacher, K. Energy and operating cost assessment of competing harvesting methods for *D. salina* in a β-carotene production process. *Algal Res.* **2015**, *12*, 161–169. [CrossRef]
71. Li, J.; Zhu, D.; Niu, J.; Shen, S.; Wang, G. An economic assessment of astaxanthin production by large scale cultivation of *Haematococcus pluvialis*. *Biotechnol. Adv.* **2011**, *29*, 568–574. [CrossRef]
72. Igreja, W.S.; Maia, F.d.A.; Lopes, A.S.; Chisté, R.C. Biotechnological production of carotenoids using low cost-substrates is influenced by cultivation parameters: A review. *Int. J. Mol. Sci.* **2021**, *22*, 8819. [CrossRef]
73. da Costa Cardoso, L.A.; Kanno, K.Y.F.; Karp, S.G. Microbial production of carotenoids a review. *Afr. J. Biotechnol.* **2017**, *16*, 139–146.
74. Frigaard, N.-U. Biotechnology of anoxygenic phototrophic bacteria. *Adv. Biochem. Eng. Biotechnol.* **2016**, *156*, 139–154.
75. Seckbach, J.; Oren, A. Oxygenic photosynthetic microorganisms in extreme environments. In *Algae and Cyanobacteria in Extreme Environments*; Springer: Berlin/Heidelberg, Germany, 2007; pp. 3–25.
76. Latha, B.; Jeevaratnam, K.; Murali, H.; Manja, K. Influence of growth factors on carotenoid pigmentation of *Rhodotorula glutinis* DFR-PDY from natural source. *Indian J. Biotechnol.* **2005**, *4*, 353–357.
77. Kim, M.; Jung, D.-H.; Seo, D.-H.; Chung, W.-H.; Seo, M.-J. Genome analysis of *Lactobacillus plantarum* subsp. *plantarum* KCCP11226 reveals a well-conserved C30 carotenoid biosynthetic pathway. *3 Biotech* **2020**, *10*, 150.
78. Li, S.; Zhao, Y.; Zhang, L.; Zhang, X.; Huang, L.; Li, D.; Niu, C.; Yang, Z.; Wang, Q. Antioxidant activity of *Lactobacillus plantarum* strains isolated from traditional Chinese fermented foods. *Food Chem.* **2012**, *135*, 1914–1919. [CrossRef] [PubMed]
79. Panesar, R.; Kaur, S.; Panesar, P.S. Production of microbial pigments utilizing agro-industrial waste: A review. *Curr. Opin. Food Sci.* **2015**, *1*, 70–76. [CrossRef]
80. Köcher, S.; Breitenbach, J.; Müller, V.; Sandmann, G. Structure, function and biosynthesis of carotenoids in the moderately halophilic bacterium *Halobacillus halophilus*. *Arch. Microbiol.* **2009**, *191*, 95–104. [CrossRef] [PubMed]

81. Umeno, D.; Tobias, A.V.; Arnold, F.H. Evolution of the C_{30} carotenoid synthase CrtM for function in a C_{40} pathway. *J. Bacteriol.* **2002**, *184*, 6690–6699. [CrossRef] [PubMed]

82. Kim, J.W.; Choi, B.H.; Kim, J.H.; Kang, H.-J.; Ryu, H.; Lee, P.C. Complete genome sequence of *Planococcus faecalis* AJ003T, the type species of the genus *Planococcus* and a microbial C_{30} carotenoid producer. *J. Biotechnol.* **2018**, *266*, 72–76. [CrossRef]

83. Mata-Gómez, L.C.; Montañez, J.C.; Méndez-Zavala, A.; Aguilar, C.N. Biotechnological production of carotenoids by yeasts: An overview. *Microb. Cell Fact.* **2014**, *13*, 12. [CrossRef]

84. Frengova, G.; Simova, E.; Pavlova, K.; Beshkova, D.; Grigorova, D. Formation of carotenoids by *Rhodotorula glutinis* in whey ultrafiltrate. *Biotechnol. Bioeng.* **1994**, *44*, 888–894. [CrossRef]

85. Gallego-Cartagena, E.; Castillo-Ramírez, M.; Martínez-Burgos, W. Effect of stressful conditions on the carotenogenic activity of a Colombian strain of *Dunaliella salina*. *Saudi J. Biol. Sci.* **2019**, *26*, 1325–1330. [CrossRef]

86. Sowmya, R.; Sachindra, N. Carotenoid production by *Formosa* sp. KMW, a marine bacteria of Flavobacteriaceae family: Influence of culture conditions and nutrient composition. *Biocatal. Agric. Biotechnol.* **2015**, *4*, 559–567. [CrossRef]

87. Hayman, E.P.; Yokoyama, H.; Chichester, C.O.; Simpson, K.L. Carotenoid biosynthesis in *Rhodotorula glutinis*. *J. Bacteriol.* **1974**, *120*, 1339–1343. [CrossRef]

88. Mihalcea, A.; Ungureanu, C.; Ferdes, M.; Chirvase, A.A.; Tanase, C. The influence of operating conditions on the growth of the yeast *Rhodotorula rubra* ICCF 209 and on torularhodin formation. *Rev. Chim.* **2011**, *11*, 12.

89. Takaichi, S.; Ishidsu, J.-I. Influence of Growth Temperature on Compositions of Carotenoids and Fatty Acids from Carotenoid Glucoside Ester and from Cellular Lipids in *Rhodococcus rhodochrous* RNMSI. *Biosci. Biotechnol. Biochem.* **1993**, *57*, 1886–1889. [CrossRef]

90. Simova, E.D.; Frengova, G.I.; Beshkova, D.M. Effect of aeration on the production of carotenoid pigments by *Rhodotorula rubra-lactobacillus casei* subsp. *casei* co-cultures in whey ultrafiltrate. *Z. Naturforsch. C* **2003**, *58*, 225–229. [CrossRef]

91. Tinoi, J.; Rakariyatham, N.; Deming, R. Simplex optimization of carotenoid production by *Rhodotorula glutinis* using hydrolyzed mung bean waste flour as substrate. *Process Biochem.* **2005**, *40*, 2551–2557. [CrossRef]

92. Marova, I.; Carnecka, M.; Halienova, A.; Certik, M.; Dvorakova, T.; Haronikova, A. Use of several waste substrates for carotenoid-rich yeast biomass production. *J. Environ. Manag.* **2012**, *95*, S338–S342. [CrossRef] [PubMed]

93. Valduga, E.; Tatsch, P.; Vanzo, L.T.; Rauber, F.; Di Luccio, M.; Treichel, H. Assessment of hydrolysis of cheese whey and use of hydrolysate for bioproduction of carotenoids by *Sporidiobolus salmonicolor* CBS 2636. *J. Sci. Food Agric.* **2009**, *89*, 1060–1065. [CrossRef]

94. Bhosale, P. Environmental and cultural stimulants in the production of carotenoids from microorganisms. *Appl. Microbiol. Biotechnol.* **2004**, *63*, 351–361. [CrossRef]

95. Shi, F.; Zhan, W.; Li, Y.; Wang, X. Temperature influences β-carotene production in recombinant *Saccharomyces cerevisiae* expressing carotenogenic genes from *Phaffia rhodozyma*. *World J. Microbiol. Biotechnol.* **2014**, *30*, 125–133. [CrossRef]

96. El-Banna, A.A.; El-Razek, A.M.A.; El-Mahdy, A.R. Some factors affecting the production of carotenoids by *Rhodotorula glutinis* var. *glutinis*. *Food Nutr. Sci.* **2012**, *2012*, 17082.

97. Guedes, A.C.; Amaro, H.M.; Malcata, F.X. Microalgae as sources of carotenoids. *Mar. Drugs* **2011**, *9*, 625–644. [CrossRef]

98. Ram, S.; Paliwal, C.; Mishra, S. Growth medium and nitrogen stress sparked biochemical and carotenogenic alterations in *Scenedesmus* sp. CCNM 1028. *Bioresour. Technol. Rep.* **2019**, *7*, 100194. [CrossRef]

99. Wang, F.; Jiang, J.G.; Chen, Q. Progress on molecular breeding and metabolic engineering of biosynthesis pathways of C_{30}, C_{35}, C_{40}, C_{45}, C_{50} carotenoids. *Biotechnol. Adv.* **2007**, *25*, 211–222. [CrossRef] [PubMed]

100. Lee, P.; Schmidt-Dannert, C. Metabolic engineering towards biotechnological production of carotenoids in microorganisms. *Appl. Microbiol. Biotechnol.* **2002**, *60*, 1–11. [PubMed]

101. Takaichi, S.; Sandmann, G.; Schnurr, G.; Satomi, Y.; Suzuki, A.; Misawa, N. The carotenoid 7,8-dihydro-Ψ end group can be cyclized by the lycopene cyclases from the bacterium *Erwinia uredovora* and the higher plant *Capsicum annuum*. *Eur. J. Biochem.* **1996**, *241*, 291–296. [CrossRef] [PubMed]

MDPI AG
Grosspeteranlage 5
4052 Basel
Switzerland
Tel.: +41 61 683 77 34

Antioxidants Editorial Office
E-mail: antioxidants@mdpi.com
www.mdpi.com/journal/antioxidants